RTL HARDWARE DESIGN USING VHDL

RTL HARDWARE DESIGN USING VHDL
Coding for Efficiency, Portability, and Scalability

PONG P. CHU
Cleveland State University

WILEY-
INTERSCIENCE

A JOHN WILEY & SONS, INC., PUBLICATION

Library of Congress Cataloging-in-Publication Data:

Chu, Pong P., 1959–
    RTL hardware design using VHDL / by Pong P. Chu.
        p. cm.
    Includes bibliographical references and index.
    "A Wiley-Interscience publication."
    ISBN-13: 978-0-471-72092-8 (alk. paper)
    ISBN-10: 0-471-72092-5 (alk. paper)
    1. Digital electronics—Data processing.   2. VHDL (Computer hardware description
    language). I. Title.

    TK7868.D5C46 2006
    621.39'2—dc22   2005054234

Printed in the United States of America.

10 9 8 7 6 5 4

To my parents Chia-Chi and Chi-Te, my wife Lee, and my daughter Patricia

CONTENTS

Preface   xix

Acknowledgments   xxiii

1 Introduction to Digital System Design   **1**

   1.1   Introduction   1
   1.2   Device technologies   2
         1.2.1   Fabrication of an IC   2
         1.2.2   Classification of device technologies   2
         1.2.3   Comparison of technologies   5
   1.3   System representation   8
   1.4   Levels of Abstraction   9
         1.4.1   Transistor-level abstraction   10
         1.4.2   Gate-level abstraction   10
         1.4.3   Register-transfer-level (RT-level) abstraction   11
         1.4.4   Processor-level abstraction   12
   1.5   Development tasks and EDA software   12
         1.5.1   Synthesis   13
         1.5.2   Physical design   14
         1.5.3   Verification   14
         1.5.4   Testing   16
         1.5.5   EDA software and its limitations   16

1.6	Development flow	17
	1.6.1 Flow of a medium-sized design targeting FPGA	17
	1.6.2 Flow of a large design targeting FPGA	19
	1.6.3 Flow of a large design targeting ASIC	19
1.7	Overview of the book	20
	1.7.1 Scope	20
	1.7.2 Goal	20
1.8	Bibliographic notes	21
	Problems	22

2   Overview of Hardware Description Languages   **23**

2.1	Hardware description languages	23
	2.1.1 Limitations of traditional programming languages	23
	2.1.2 Use of an HDL program	24
	2.1.3 Design of a modern HDL	25
	2.1.4 VHDL	25
2.2	Basic VHDL concept via an example	26
	2.2.1 General description	27
	2.2.2 Structural description	30
	2.2.3 Abstract behavioral description	33
	2.2.4 Testbench	35
	2.2.5 Configuration	37
2.3	VHDL in development flow	38
	2.3.1 Scope of VHDL	38
	2.3.2 Coding for synthesis	40
2.4	Bibliographic notes	40
	Problems	41

3   Basic Language Constructs of VHDL   **43**

3.1	Introduction	43
3.2	Skeleton of a basic VHDL program	44
	3.2.1 Example of a VHDL program	44
	3.2.2 Entity declaration	44
	3.2.3 Architecture body	46
	3.2.4 Design unit and library	46
	3.2.5 Processing of VHDL code	47
3.3	Lexical elements and program format	47
	3.3.1 Lexical elements	47
	3.3.2 VHDL program format	49
3.4	Objects	51
3.5	Data types and operators	53

3.5.1	Predefined data types in VHDL	53
3.5.2	Data types in the IEEE std_logic_1164 package	56
3.5.3	Operators over an array data type	58
3.5.4	Data types in the IEEE numeric_std package	60
3.5.5	The std_logic_arith and related packages	64
3.6	Synthesis guidelines	65
3.6.1	Guidelines for general VHDL	65
3.6.2	Guidelines for VHDL formatting	66
3.7	Bibliographic notes	66
	Problems	66

4 Concurrent Signal Assignment Statements of VHDL   **69**

4.1	Combinational versus sequential circuits	69
4.2	Simple signal assignment statement	70
4.2.1	Syntax and examples	70
4.2.2	Conceptual implementation	70
4.2.3	Signal assignment statement with a closed feedback loop	71
4.3	Conditional signal assignment statement	72
4.3.1	Syntax and examples	72
4.3.2	Conceptual implementation	76
4.3.3	Detailed implementation examples	78
4.4	Selected signal assignment statement	85
4.4.1	Syntax and examples	85
4.4.2	Conceptual implementation	88
4.4.3	Detailed implementation examples	90
4.5	Conditional signal assignment statement versus selected signal assignment statement	93
4.5.1	Conversion between conditional signal assignment and selected signal assignment statements	93
4.5.2	Comparison between conditional signal assignment and selected signal assignment statements	94
4.6	Synthesis guidelines	95
4.7	Bibliographic notes	95
	Problems	95

5 Sequential Statements of VHDL   **97**

5.1	VHDL process	97
5.1.1	Introduction	97
5.1.2	Process with a sensitivity list	98
5.1.3	Process with a wait statement	99
5.2	Sequential signal assignment statement	100

5.3   Variable assignment statement   101
5.4   If statement   103
    5.4.1   Syntax and examples   103
    5.4.2   Comparison to a conditional signal assignment statement   105
    5.4.3   Incomplete branch and incomplete signal assignment   107
    5.4.4   Conceptual implementation   109
    5.4.5   Cascading single-branched if statements   110
5.5   Case statement   112
    5.5.1   Syntax and examples   112
    5.5.2   Comparison to a selected signal assignment statement   114
    5.5.3   Incomplete signal assignment   115
    5.5.4   Conceptual implementation   116
5.6   Simple for loop statement   118
    5.6.1   Syntax   118
    5.6.2   Examples   118
    5.6.3   Conceptual implementation   119
5.7   Synthesis of sequential statements   120
5.8   Synthesis guidelines   120
    5.8.1   Guidelines for using sequential statements   120
    5.8.2   Guidelines for combinational circuits   121
5.9   Bibliographic notes   121
    Problems   121

6   Synthesis Of VHDL Code   **125**

6.1   Fundamental limitations of EDA software   125
    6.1.1   Computability   126
    6.1.2   Computation complexity   126
    6.1.3   Limitations of EDA software   128
6.2   Realization of VHDL operators   129
    6.2.1   Realization of logical operators   129
    6.2.2   Realization of relational operators   129
    6.2.3   Realization of addition operators   130
    6.2.4   Synthesis support for other operators   130
    6.2.5   Realization of an operator with constant operands   130
    6.2.6   An example implementation   131
6.3   Realization of VHDL data types   133
    6.3.1   Use of the `std_logic` data type   133
    6.3.2   Use and realization of the 'Z' value   133
    6.3.3   Use of the '-' value   137
6.4   VHDL synthesis flow   139
    6.4.1   RT-level synthesis   139
    6.4.2   Module generator   141

	6.4.3 Logic synthesis	142
	6.4.4 Technology mapping	143
	6.4.5 Effective use of synthesis software	148
6.5	Timing considerations	149
	6.5.1 Propagation delay	150
	6.5.2 Synthesis with timing constraints	154
	6.5.3 Timing hazards	156
	6.5.4 Delay-sensitive design and its dangers	158
6.6	Synthesis guidelines	160
6.7	Bibliographic notes	160
	Problems	160

7 Combinational Circuit Design: Practice   **163**

7.1	Derivation of efficient HDL description	163
7.2	Operator sharing	164
	7.2.1 Sharing example 1	165
	7.2.2 Sharing example 2	166
	7.2.3 Sharing example 3	168
	7.2.4 Sharing example 4	169
	7.2.5 Summary	170
7.3	Functionality sharing	170
	7.3.1 Addition–subtraction circuit	171
	7.3.2 Signed–unsigned dual-mode comparator	173
	7.3.3 Difference circuit	175
	7.3.4 Full comparator	177
	7.3.5 Three-function barrel shifter	178
7.4	Layout-related circuits	180
	7.4.1 Reduced-xor circuit	181
	7.4.2 Reduced-xor-vector circuit	183
	7.4.3 Tree priority encoder	187
	7.4.4 Barrel shifter revisited	192
7.5	General circuits	196
	7.5.1 Gray code incrementor	196
	7.5.2 Programmable priority encoder	199
	7.5.3 Signed addition with status	201
	7.5.4 Combinational adder-based multiplier	203
	7.5.5 Hamming distance circuit	206
7.6	Synthesis guidelines	208
7.7	Bibliographic notes	208
	Problems	208

8 Sequential Circuit Design: Principle   **213**

8.1	Overview of sequential circuits		213
	8.1.1	Sequential versus combinational circuits	213
	8.1.2	Basic memory elements	214
	8.1.3	Synchronous versus asynchronous circuits	216
8.2	Synchronous circuits		217
	8.2.1	Basic model of a synchronous circuit	217
	8.2.2	Synchronous circuits and design automation	218
	8.2.3	Types of synchronous circuits	219
8.3	Danger of synthesis that uses primitive gates		219
8.4	Inference of basic memory elements		221
	8.4.1	D latch	221
	8.4.2	D FF	222
	8.4.3	Register	225
	8.4.4	RAM	225
8.5	Simple design examples		226
	8.5.1	Other types of FFs	226
	8.5.2	Shift register	229
	8.5.3	Arbitrary-sequence counter	232
	8.5.4	Binary counter	233
	8.5.5	Decade counter	236
	8.5.6	Programmable mod-m counter	237
8.6	Timing analysis of a synchronous sequential circuit		239
	8.6.1	Synchronized versus unsynchronized input	239
	8.6.2	Setup time violation and maximal clock rate	240
	8.6.3	Hold time violation	243
	8.6.4	Output-related timing considerations	243
	8.6.5	Input-related timing considerations	244
8.7	Alternative one-segment coding style		245
	8.7.1	Examples of one-segment code	245
	8.7.2	Summary	250
8.8	Use of variables in sequential circuit description		250
8.9	Synthesis of sequential circuits		253
8.10	Synthesis guidelines		253
8.11	Bibliographic notes		253
	Problems		254

9 Sequential Circuit Design: Practice — **257**

9.1	Poor design practices and their remedies		257
	9.1.1	Misuse of asynchronous signals	258
	9.1.2	Misuse of gated clocks	260
	9.1.3	Misuse of derived clocks	262
9.2	Counters		265

9.2.1	Gray counter	265
9.2.2	Ring counter	266
9.2.3	LFSR (linear feedback shift register)	269
9.2.4	Decimal counter	272
9.2.5	Pulse width modulation circuit	275
9.3	Registers as temporary storage	276
9.3.1	Register file	276
9.3.2	Register-based synchronous FIFO buffer	279
9.3.3	Register-based content addressable memory	287
9.4	Pipelined design	293
9.4.1	Delay versus throughput	294
9.4.2	Overview on pipelined design	294
9.4.3	Adding pipeline to a combinational circuit	297
9.4.4	Synthesis of pipelined circuits and retiming	307
9.5	Synthesis guidelines	308
9.6	Bibliographic notes	309
	Problems	309

10 Finite State Machine: Principle and Practice — **313**

10.1	Overview of FSMs	313
10.2	FSM representation	314
10.2.1	State diagram	315
10.2.2	ASM chart	317
10.3	Timing and performance of an FSM	321
10.3.1	Operation of a synchronous FSM	321
10.3.2	Performance of an FSM	324
10.3.3	Representative timing diagram	325
10.4	Moore machine versus Mealy machine	325
10.4.1	Edge detection circuit	326
10.4.2	Comparison of Moore output and Mealy output	328
10.5	VHDL description of an FSM	329
10.5.1	Multi-segment coding style	330
10.5.2	Two-segment coding style	333
10.5.3	Synchronous FSM initialization	335
10.5.4	One-segment coding style and its problem	336
10.5.5	Synthesis and optimization of FSM	337
10.6	State assignment	338
10.6.1	Overview of state assignment	338
10.6.2	State assignment in VHDL	339
10.6.3	Handling the unused states	341
10.7	Moore output buffering	342
10.7.1	Buffering by clever state assignment	342

10.7.2 Look-ahead output circuit for Moore output	344
10.8 FSM design examples	348
10.8.1 Edge detection circuit	348
10.8.2 Arbiter	353
10.8.3 DRAM strobe generation circuit	358
10.8.4 Manchester encoding circuit	363
10.8.5 FSM-based binary counter	367
10.9 Bibliographic notes	369
Problems	369
11 Register Transfer Methodology: Principle	**373**
11.1 Introduction	373
11.1.1 Algorithm	373
11.1.2 Structural data flow implementation	374
11.1.3 Register transfer methodology	375
11.2 Overview of FSMD	376
11.2.1 Basic RT operation	376
11.2.2 Multiple RT operations and data path	378
11.2.3 FSM as the control path	379
11.2.4 ASMD chart	379
11.2.5 Basic FSMD block diagram	380
11.3 FSMD design of a repetitive-addition multiplier	382
11.3.1 Converting an algorithm to an ASMD chart	382
11.3.2 Construction of the FSMD	385
11.3.3 Multi-segment VHDL description of an FSMD	386
11.3.4 Use of a register value in a decision box	389
11.3.5 Four- and two-segment VHDL descriptions of FSMD	391
11.3.6 One-segment coding style and its deficiency	394
11.4 Alternative design of a repetitive-addition multiplier	396
11.4.1 Resource sharing via FSMD	396
11.4.2 Mealy-controlled RT operations	400
11.5 Timing and performance analysis of FSMD	404
11.5.1 Maximal clock rate	404
11.5.2 Performance analysis	407
11.6 Sequential add-and-shift multiplier	407
11.6.1 Initial design	408
11.6.2 Refined design	412
11.6.3 Comparison of three ASMD designs	417
11.7 Synthesis of FSMD	417
11.8 Synthesis guidelines	418
11.9 Bibliographic notes	418
Problems	418

12 Register Transfer Methodology: Practice   **421**

12.1   Introduction   421
12.2   One-shot pulse generator   422
    12.2.1   FSM implementation   422
    12.2.2   Regular sequential circuit implementation   424
    12.2.3   Implementation using RT methodology   425
    12.2.4   Comparison   427
12.3   SRAM controller   430
    12.3.1   Overview of SRAM   430
    12.3.2   Block diagram of an SRAM controller   434
    12.3.3   Control path of an SRAM controller   436
12.4   GCD circuit   445
12.5   UART receiver   455
12.6   Square-root approximation circuit   460
12.7   High-level synthesis   469
12.8   Bibliographic notes   470
    Problems   470

13 Hierarchical Design in VHDL   **473**

13.1   Introduction   473
    13.1.1   Benefits of hierarchical design   474
    13.1.2   VHDL constructs for hierarchical design   474
13.2   Components   475
    13.2.1   Component declaration   475
    13.2.2   Component instantiation   477
    13.2.3   Caveats in component instantiation   480
13.3   Generics   481
13.4   Configuration   485
    13.4.1   Introduction   485
    13.4.2   Configuration declaration   486
    13.4.3   Configuration specification   488
    13.4.4   Component instantiation and configuration in VHDL 93   488
13.5   Other supporting constructs for a large system   489
    13.5.1   Library   489
    13.5.2   Subprogram   491
    13.5.3   Package   492
13.6   Partition   495
    13.6.1   Physical partition   495
    13.6.2   Logical partition   496
13.7   Synthesis guidelines   497
13.8   Bibliographic notes   497

Problems   497

14 Parameterized Design: Principle   499

14.1   Introduction   499

14.2   Types of parameters   500

    14.2.1   Width parameters   500

    14.2.2   Feature parameters   501

14.3   Specifying parameters   501

    14.3.1   Generics   501

    14.3.2   Array attribute   502

    14.3.3   Unconstrained array   503

    14.3.4   Comparison between a generic and an unconstrained array   506

14.4   Clever use of an array   506

    14.4.1   Description without fixed-size references   507

    14.4.2   Examples   509

14.5   For generate statement   512

    14.5.1   Syntax   513

    14.5.2   Examples   513

14.6   Conditional generate statement   517

    14.6.1   Syntax   517

    14.6.2   Examples   518

    14.6.3   Comparisons with other feature-selection methods   525

14.7   For loop statement   528

    14.7.1   Introduction   528

    14.7.2   Examples of a simple for loop statement   528

    14.7.3   Examples of a loop body with multiple signal assignment statements   530

    14.7.4   Examples of a loop body with variables   533

    14.7.5   Comparison of the for generate and for loop statements   536

14.8   Exit and next statements   537

    14.8.1   Syntax of the exit statement   537

    14.8.2   Examples of the exit statement   537

    14.8.3   Conceptual implementation of the exit statement   539

    14.8.4   Next statement   540

14.9   Synthesis of iterative structure   541

14.10 Synthesis guidelines   542

14.11 Bibliographic notes   542

    Problems   542

15 Parameterized Design: Practice   545

15.1   Introduction   545

15.2 Data types for two-dimensional signals   546
    15.2.1 Genuine two-dimensional data type   546
    15.2.2 Array-of-arrays data type   548
    15.2.3 Emulated two-dimensional array   550
    15.2.4 Example   552
    15.2.5 Summary   554
15.3 Commonly used intermediate-sized RT-level components   555
    15.3.1 Reduced-xor circuit   555
    15.3.2 Binary decoder   558
    15.3.3 Multiplexer   560
    15.3.4 Binary encoder   564
    15.3.5 Barrel shifter   566
15.4 More sophisticated examples   569
    15.4.1 Reduced-xor-vector circuit   570
    15.4.2 Multiplier   572
    15.4.3 Parameterized LFSR   586
    15.4.4 Priority encoder   588
    15.4.5 FIFO buffer   591
15.5 Synthesis of parameterized modules   599
15.6 Synthesis guidelines   599
15.7 Bibliographic notes   600
    Problems   600

16 Clock and Synchronization: Principle and Practice   **603**

16.1 Overview of a clock distribution network   603
    16.1.1 Physical implementation of a clock distribution network   603
    16.1.2 Clock skew and its impact on synchronous design   605
16.2 Timing analysis with clock skew   606
    16.2.1 Effect on setup time and maximal clock rate   606
    16.2.2 Effect on hold time constraint   609
16.3 Overview of a multiple-clock system   610
    16.3.1 System with derived clock signals   611
    16.3.2 GALS system   612
16.4 Metastability and synchronization failure   612
    16.4.1 Nature of metastability   613
    16.4.2 Analysis of $MTBF(T_r)$ 614
    16.4.3 Unique characteristics of $MTBF(T_r)$ 616
16.5 Basic synchronizer   617
    16.5.1 The danger of no synchronizer   617
    16.5.2 One-FF synchronizer and its deficiency   617
    16.5.3 Two-FF synchronizer   619
    16.5.4 Three-FF synchronizer   620

16.5.5 Proper use of a synchronizer   621

16.6 Single enable signal crossing clock domains   623

16.6.1 Edge detection scheme   623

16.6.2 Level-alternation scheme   627

16.7 Handshaking protocol   630

16.7.1 Four-phase handshaking protocol   630

16.7.2 Two-phase handshaking protocol   637

16.8 Data transfer crossing clock domains   639

16.8.1 Four-phase handshaking protocol data transfer   641

16.8.2 Two-phase handshaking data transfer   650

16.8.3 One-phase data transfer   651

16.9 Data transfer via a memory buffer   652

16.9.1 FIFO buffer   652

16.9.2 Shared memory   660

16.10 Synthesis of a multiple-clock system   661

16.11 Synthesis guidelines   662

16.11.1 Guidelines for general use of a clock   662

16.11.2 Guidelines for a synchronizer   662

16.11.3 Guidelines for an interface between clock domains   662

16.12 Bibliographic notes   663

Problems   663

References   665

Topic Index   667

PREFACE

With the maturity and availability of hardware description language (HDL) and synthesis software, using them to design custom digital hardware has become a mainstream practice. Because of the resemblance of an HDL code to a traditional program (such as a C program), some users believe incorrectly that designing hardware in HDL involves simply writing syntactically correct software code, and assume that the synthesis software can automatically derive the physical hardware. Unfortunately, synthesis software can only perform transformation and local optimization, and cannot convert a poor description into an efficient implementation. Without an understanding of the hardware architecture, the HDL code frequently leads to unnecessarily complex hardware, or may not even be synthesizable.

This book provides in-depth coverage on the systematical development and synthesis of efficient, portable and scalable register-transfer-level (RT-level) digital circuits using the VHDL hardware description language. RT-level design uses intermediate-sized components, such as adders, comparators, multiplexers and registers, to construct a digital system. It is the level that is most suitable and effective for today's synthesis software.

RT-level design and VHDL are two somewhat independent subjects. VHDL code is simply one of the methods to describe a hardware design. The same design can also be described by a schematic or code in other HDLs. VHDL and synthesis software will not lead automatically to a better or worse design. However, they can shield designers from low-level details and allow them to explore and research better architectures.

The emphasis of the book is on *hardware* rather than *language*. Instead of treating synthesis software as a mysterious black box and listing "recipe-like" codes, we explain the relationship between the VHDL constructs and the underlying hardware structure and illustrate how to explore the design space and develop codes that can be synthesized into efficient cell-level implementation. The discussion is independent of technology and can be applied to both ASIC and FPGA devices. The VHDL codes listed in the book largely follow the IEEE 1076.6 RTL synthesis standard and can be accepted by most synthesis software. Most codes can be synthesized without modification by the free "demo-version" synthesis software provided by FPGA vendors.

Scope   The book focuses primarily on the design and synthesis of RT-level circuits. A subset of VHDL is used to describe the design. The book is not intended to be a comprehensive ASIC or FPGA book. All other issues, such as device architecture, placement and routing, simulation and testing, are discussed exclusively from the context of RT-level design.

Unique features   The book is a hardware design text. VHDL and synthesis software are used as tools to realize the intended design. Several unique features distinguish the book:

- Suggest a coding style that shows a clear relationship between VHDL constructs and hardware components.
- Use easy-to-understand conceptual diagrams, rather than cell-level netlists, to explain the realization of VHDL codes.
- Emphasize the reuse aspect of the codes throughout the book.
- Consider RT-level design as an integral part of the overall development process and introduce good design practices and guidelines to ensure that an RT-level description can accommodate future simulation, verification and testing needs.
- Make the design "technology neutral" so that the developed VHDL code can be applied to both ASIC and FPGA devices.
- Follow the IEEE 1076.6 RTL synthesis standard to make the codes independent of synthesis software.
- Provide a set of synthesis guidelines at the end of each chapter.
- Contain a large number of non-trivial, practical examples to illustrate and reinforce the design concepts, procedures and techniques.
- Include two chapters on realizing sequential algorithms in hardware (known as "register transfer methodology") and on designing control path and data path.
- Include two chapters on the scalable and parameterized designs and coding.
- Include a chapter on the synchronization and interface between multiple clock domains.

Book organization   The book is basically divided into three major parts. The first part, Chapters 1 to 6, provides a comprehensive overview of VHDL and the synthesis process, and examines the hardware implementation of basic VHDL language constructs. The second part, Chapters 7 to 12, covers the core of the RT-level design, including combinational circuits, "regular" sequential circuits, finite state machine and circuits designed by register transfer methodology. The third part, Chapters 13 to 16, covers the system issues, including the hierarchy, parameterized and scalable design, and interface between clock domains. More detailed descriptions of the chapters follow.

- Chapter 1 presents a "big picture" of digital system design, including an overview on device technologies, system representation, development flow and software tools.
- Chapter 2 provides an overview on the design, usage and capability of a hardware description language. A series of simple codes is used to introduce the basic modeling concepts of VHDL.
- Chapter 3 provides an overview of the basic language constructs of VHDL, including lexical elements, objects, data types and operators. Because VHDL is a strongly typed language, the data types and operators are discussed in more detail.

- Chapter 4 covers the syntax, usage and implementation of concurrent signal assignment statements of VHDL. It shows how to realize these constructs by multiplexing and priority routing networks.
- Chapter 5 examines the syntax, usage and implementation of sequential statements of VHDL. It shows the realization of the sequential statements and discusses the caveats of using these statements.
- Chapter 6 explains the realization of VHDL operators and data types, provides an in-depth overview on the synthesis process and discusses the timing issue involved in synthesis.
- Chapter 7 covers the construction and VHDL description of more sophisticated combinational circuits. Examples show how to transform conceptual ideas into hardware, and illustrate resource-sharing and circuit-shaping techniques to reduce circuit size and increase performance.
- Chapter 8 introduces the synchronous design methodology and the construction and coding of synchronous sequential circuits. Basic "regular" sequential circuits, such as counters and shift registers, in which state transitions exhibit a regular pattern, are examined.
- Chapter 9 explores more sophisticated regular sequential circuits. The design examples show the implementation of a variety of counters, the use of registers as fast, temporary storage, and the construction of pipelined combinational circuits.
- Chapter 10 covers finite state machine (FSM), which is a sequential circuit with "random" transition patterns. The representation, timing and implementation issues of FSMs are studied with an emphasis on its use as the control circuit for a large, complex system.
- Chapter 11 introduces the register transfer methodology, which describes system operation by a sequence of data transfers and manipulations among registers, and demonstrates the construction of the data path (a regular sequential circuit) and the control path (an FSM) used in this methodology.
- Chapter 12 uses a variety of design examples to illustrate how the register transfer methodology can be used in various types of problems and to highlight the design procedure and relevant issues.
- Chapter 13 features the design hierarchy, in which a system is gradually divided into smaller parts. Mechanisms and language constructs of VHDL used to specify and configure a hierarchy are examined.
- Chapter 14 introduces parameterized design, in which the width and functionality of a circuit are specified by explicit parameters. Simple examples illustrate the mechanisms used to pass and infer parameters and the language constructs used to describe the replicated structures.
- Chapter 15 provides more sophisticated parameterized design examples. The main focus is on the derivation of efficient parameterized RT-level modules that can be used as building blocks of larger systems.
- Chapter 16 covers the effect of a non-ideal clock signal and discusses the synchronization of an asynchronous signal and the interface between two independent clock domains.

Audience The intended audience for the book is students in advanced digital system design course and practicing engineers who wish to sharpen their design skills or to learn the effective use of today's synthesis software. Readers need to have basic knowledge of digital systems. The material is normally covered in an introductory digital design course, which is a standard part in all electrical engineering and computer engineering curricula. No prior experience on HDL or synthesis is needed.

Verilog is another popular HDL. Since the book emphasizes hardware and methodology rather than language constructs, readers with prior Verilog experience can easily follow the discussion and learn VHDL along the way. Most VHDL codes can easily be translated into the Verilog language.

Web site An accompanying web site (http://academic.csuohio.edu/chu_p/rtl) provides additional information, including the following materials:

- Errata.
- Summary of coding guidelines.
- Code listing.
- Links to demo-version synthesis software.
- Links to some referenced materials.
- Frequently asked questions (FAQ) on RTL synthesis.
- Lecture slides for instructors.

Errata The book is "self-prepared," which means the author has prepared all materials, including the illustrations, tables, code listing, indexing and formatting, by himself. As the errors are always bound to happen, the accompanying web site provides an updated errata sheet and a place to report errors.

P. P. CHU

Cleveland, Ohio
January 2006

ACKNOWLEDGMENTS

The author would like to express his gratitude to Professor George L. Kramerich for his encouragement and help during the course of this project. The work was partially supported by educational material development grant 0126752 from the National Science Foundation and a Teaching Enhancement grant from Cleveland State University.

P. P. Chu

CHAPTER 1

INTRODUCTION TO DIGITAL SYSTEM DESIGN

Developing and producing a digital system is a complicated process and involves many tasks. The design and synthesis of a register transfer level circuit, which is the focus of this book, is only one of the tasks. In this chapter, we present an overview of device technologies, system representation, development flow and software tools. This helps us to better understand the role of the design and synthesis task in the overall development and production process.

1.1 INTRODUCTION

Digital hardware has experienced drastic expansion and improvement in the past 40 years. Since its introduction, the number of transistors in a single chip has grown exponentially, and a silicon chip now routinely contains hundreds of thousands or even hundreds of millions of transistors. In the past, the major applications of digital hardware were computational systems. However, as the chip became smaller, faster, cheaper and more capable, many electronic, control, communication and even mechanical systems have been "digitized" internally, using digital circuits to store, process and transmit information.

As applications become larger and more complex, the task of designing digital circuits becomes more difficult. The best way to handle the complexity is to view the circuit at a more abstract level and utilize software tools to derive the low-level implementation. This approach shields us from the tedious details and allows us to concentrate and explore high-level design alternatives. Although software tools can automate certain tasks, they are capable of performing only limited transformation and optimization. They cannot, and

RTL Hardware Design Using VHDL: Coding for Efficiency, Portability, and Scalability. By Pong P. Chu
Copyright © 2006 John Wiley & Sons, Inc.

will not, do the design or convert a poor design to a good one. The ultimate efficiency still comes from human ingenuity and experience. The goal of this book is to show how to systematically develop an efficient, portable design description that is both abstract, yet detailed enough for effective software synthesis.

Developing and producing a digital circuit is a complicated process, and the design and synthesis are only two of the tasks. We should be aware of the "big picture" so that the design and synthesis can be efficiently integrated into the overall development and production process. The following sections provide an overview of device technologies, system representation, abstraction, development flow, and the use and limitations of software tools.

1.2 DEVICE TECHNOLOGIES

If we want to build a custom digital system, there are varieties of device technologies to choose, from off-the-shelf simple field-programmable components to full-custom devices that tailor the application down to the transistor level. There is no single best technology, and we have to consider the trade-offs among various factors, including chip area, speed, power and cost.

1.2.1 Fabrication of an IC

To better understand the differences between the device technologies, it is helpful to have a basic idea of the fabrication process of an integrated circuit (IC). An IC is made from layers of doped silicon, polysilicon, metal and silicon dioxide, built on top of one another, on a thin silicon wafer. Some of these layers form transistors, and others form planes of connection wires.

The basic step in IC fabrication is to construct a layer with a customized pattern, a process known as *lithography*. The pattern is defined by a *mask*. Today's IC device technology typically consists of 10 to 15 layers, and thus the lithography process has to be repeated 10 to 15 times during the fabrication of an IC, each time with a unique mask.

One important aspect of a device technology is the silicon area used by a circuit. It is expressed by the length of a smallest transistor that can be fabricated, usually measured in *microns* (a millionth of a meter). As the device fabrication process improved, the transistor size continued to shrink and now approaches a tenth of a micron.

1.2.2 Classification of device technologies

There is an array of device technologies that can be used to construct a custom digital circuit. One major characteristic of a technology is how the customization is done. In certain technologies, all the layers of a device are predetermined, and thus the device can be prefabricated and manufactured as a standard off-the-shelf part. The customization of a circuit can be performed "in the field," normally by downloading a connection pattern to the device's internal memory or by "burning the internal silicon fuses." On the other hand, some device technologies need one or more layers to be customized for a particular application. The customization involves the creation of tailored masks and fabrication of the patterned layers. This process is expensive and complex and can only be done in a fabrication plant (known as a *foundry* or a *fab*). Thus, whether a device needed to be fabricated in a fab is the most important characteristic of a technology. In this book, we use the term *application-specific IC (ASIC)* to represent device technologies that require a fab to do customization.

With an understanding of the difference between ASIC and non-ASIC, we can divide the device technologies further into the following types:

- Full-custom ASIC
- Standard-cell ASIC
- Gate array ASIC
- Complex field-programmable logic device
- Simple field-programmable logic device
- Off-the-shelf small- and medium-scaled IC (SSI/MSI) components

Full-custom ASIC   In *full-custom ASIC* technology, all aspects of a digital circuit are tailored for one particular application. We have complete control of the circuit and can even craft the layout of a transistor to meet special area or performance needs. The resulting circuit is fully optimized and has the best possible performance. Unfortunately, designing a circuit at the transistor level is extremely complex and involved, and is only feasible for a small circuit. It is not practical to use this approach to design a complete system, which now may contain tens and even hundreds of millions of transistors. The major application of full-custom ASIC technology is to design the basic logic components that can be used as building blocks of a larger system. Another application is to design special-purpose "bit-slice" typed circuits, such as a 1-bit memory or 1-bit adder. These circuits have a regular structure and are constructed through a cascade of identical slices. To obtain optimal performance, full-custom ASIC technology is frequently used to design a single slice. The slice is then replicated a number of times to form a complete circuit.

The layouts of a full-custom ASIC chip are tailored to a particular application. All layers are different and a mask is required for every layer. During fabrication, all layers have to be custom constructed, and nothing can be done in advance.

Standard-cell ASIC   In *standard-cell ASIC* (also simply known as *standard-cell*) technology, a circuit is constructed by using a set of predefined logic components, known as *standard cells*. These cells are predesigned and their layouts are validated and tested. Standard-cell ASIC technology allows us to work at the gate level rather than at the transistor level and thus greatly simplifies the design process. The device manufacturer usually provides a library of standard cells as the basic building blocks. The library normally consists of basic logic gates, simple combinational components, such as an and-or-inverter, 2-to-1 multiplexer and 1-bit full adder, and basic memory elements, such as a D-type latch and D-type flip-flop. Some libraries may also contain more sophisticated function blocks, such as an adder, barrel shifter and random access memory (RAM).

In standard-cell technology, a circuit is made of cells. The types of cells and the interconnection depend on the individual application. Whereas the layout of a cell is predetermined, the layout of the complete circuit is unique for a particular application and nothing can be constructed in advance. Thus, fabrication of a standard-cell chip is identical to that of a full-custom ASIC chip, and all layers have to be custom constructed.

Gate array ASIC   In *gate array* ASIC (also simply known as *gate array*) technology, a circuit is built from an array of predefined cells. Unlike standard-cell technology, a gate array chip consists of only one type of cell, known as a *base cell*. The base cell is fairly simple, resembling a logic gate. Base cells are prearranged and placed in fixed positions, aligned as a one- or two-dimensional array. Since the location and type are predetermined, the base cells can be prefabricated. The customization of a circuit is done by specifying the interconnect between these cells. A gate array vendor also provides a library of predesigned components, known as *macro cells*, which are built from base cells. The macro cells have a predefined interconnect and provide the designer with more sophisticated logic blocks.

Compared to standard-cell technology, the fabrication of a gate array device is much simpler, due to its fixed array structure. Since the array is common to all applications, the cell (and transistors) can be fabricated in advance. During construction of a chip, only the masks of metal layers, which specify the interconnect, are unique for an application and therefore must be customized. This reduces the number of custom layers from 10 to 15 layers to 3 to 5 layers and simplifies the fabrication process significantly.

Complex field-programmable device   We now examine several non-ASIC technologies. The most versatile non-ASIC technology is the complex field-programmable device. In this technology, a device consists of an array of generic logic cells and general interconnect structure. Although the logic cells and interconnect structure are prefabricated, both are *programmable*. The programmability is obtained by utilizing semiconductor "fuses" or "switches," which can be set as open- or short-circuit. The customization is done by configuring the device with a specific fuse pattern. This process can be accomplished by a simple, inexpensive device programmer, normally constructed as an add-on card or an adaptor cable of a PC. Since the customization is done "in the field" rather than "in a fab," this technology is known as *field programmable*. (In contrast, ASIC technologies are "programmed" via one or more tailored masks and thus are *mask programmable*.)

The basic structures of gate array ASICs and complex field-programmable devices are somewhat similar. However, the interconnect structure of field-programmable devices is predetermined and thus imposes more constraints on signal routing. To reduce the amount of connection, more functionality is built into the logic cells of a field-programmable device, making a logic cell much more complex than a base cell or a standard cell of ASIC. According to the complexity and structure of logic cells, complex field-programmable devices can be divided roughly into two broad categories: *complex programmable logic device (CPLD)* and *field programmable gate array (FPGA)*.

The logic cell of a CPLD device is more sophisticated, normally consisting of a D-type flip-flop and a PAL-like unit with configurable product terms. The interconnect structure of a CPLD device tends to be more centralized, with few groups of concentrated routing lines. On the other hand, the logic cell of an FPGA device is usually smaller, typically including a D-type flip-flop and a small look-up table or a set of multiplexers. The interconnect structure between the cells tends to be distributed and more flexible. Because of its distributive nature, FPGA is better suited for large, high-capacity complex field-programmable devices.

Simple field-programmable device   Simple field-programmable logic devices, as the name indicates, are programmable devices with simpler internal structure. Historically, these devices are generically called *programmable logic devices (PLDs)*. We add the word *simple* to distinguish them from FPGA and CPLD devices. Simple field-programmable devices are normally constructed as a two-level array, with an *and plane* and an *or plane*. The interconnect of one or both planes can be programmed to perform a logic function expressed in sum-of-product format. The devices include *programmable read only memory (PROM)*, in which the or plane can be programmed; *programmable array logic (PAL)*, in which the and plane can be programmed; and *programmable logic array (PLA)*, in which both planes can be programmed.

Unlike FPGA and CPLD devices, simple field-programmable logic devices do not have a general interconnect structure, and thus their functionality is severely limited. They are gradually being phased out. ROM, PAL and PLA are now used as internal components of an ASIC or CPLD device rather than as an individual chip.

Off-the-shelf SSI/MSI components   Before the emergence of field-programmable devices, the only alternative to ASIC was to utilize the prefabricated off-the-shelf SSI/MSI components. These components are small parts with fixed, limited functionality. One example is the 7400 series transistor transistor logic (TTL) family, which contains more than 100 parts, ranging from simple nand gates to a 4-bit arithmetic unit. A custom system can be designed by a bottom-up approach, building the circuit gradually from the small existing parts. A tailored printed circuit board is needed for each application. The major disadvantage of this approach is that the most resources (power, board area and manufacturing cost) are consumed by the "package" but not by the "silicon," which performs the actual computation. Furthermore, none of today's synthesis software can utilize off-the-shelf SSI/MSI components, and thus automation is virtually impossible. As the programmable devices become more capable and less expensive, designing a large custom circuit using SSI/MSI components is no longer a feasible option and should not be considered.

Summary   We have reviewed six device technologies used to implement custom digital systems. Among them, off-the-shelf SSI/MSI components and simple programmable devices are gradually being phased out and full-custom ASIC is feasible only for a small, specialized circuit. Thus, for a large digital system, there are only three viable device technologies: standard-cell ASIC, gate array ASIC and CPLD/FPGA. In the following subsection, we examine the trade-offs among these technologies.

1.2.3 Comparison of technologies

Once deciding to develop custom hardware for an application, we need to choose from the three device technologies. The major criteria for selection are *area*, *speed*, *power* and *cost*. The first three involve the technical aspects of a circuit. Cost concerns the expenditure associated with the design and production of the circuit as well as the potential lost profits. Each technology has its strengths and weaknesses, and the "best" technology depends on the needs of a particular application.

Area   Chip area (or size) corresponds to the required silicon real estate to implement a particular application. A smaller chip needs fewer resources, simplifies the testing and provides better yield. The chip size depends on the architecture of the circuit and the device technology. The same function can frequently be realized by different architectures, with different areas and speeds. For example, an addition circuit can be realized by a ripple adder (simple but slow), a parallel adder (complex but fast) or a carry-look-ahead adder (somewhere in-between). Once the architecture of a circuit is determined, the area depends on the device technology. In standard-cell technology, the cells and interconnects are customized to this particular application and no silicon is wasted in irrelevant functionality. Thus, the resulting chip is fully optimized and the area is minimal. In gate array technology, the circuit has to be constructed by predefined, prearranged base cells. Since functionality and the placement of the base cells are not tailored to a specific application, silicon use is not optimal. The area of the resulting circuit is normally larger than that of a standard-cell chip. In FPGA technology, a significant portion of the silicon is dedicated to achieving programmability, which introduces a large overhead. Furthermore, the functionalities of logic cells and the interconnect are fixed in advance and it is unlikely that an application can be an exact match for the predetermined structure. A certain percentage of the capacity will be left unutilized. Because of the overhead and relatively low utilization, the area of the resulting FPGA chip is much larger than that of an ASIC chip.

Due to the drastic difference between the device fabrication process and the diversity of applications, it is difficult to determine the exact silicon areas in three technologies. However, it is important to recognize that the difference between standard-cell and gate array technologies is much smaller than that of FPGA and ASIC. In general, a gate array chip may need 20% to 100% larger silicon area than that of a standard-cell chip, but an FPGA chip frequently requires two to five times the area of an ASIC chip.

Speed   The speed of a digital circuit corresponds to the time required to perform a function, frequently represented by the worst-case propagation delay between input and output signals. A faster circuit is always desirable and is essential for computation-intensive applications. At the architecture level, faster operation can be achieved by using a more sophisticated design, which requires a larger area. However, if the identical architecture is used, a chip with a larger area is normally slower, due to its large parasitic capacitance. Since a standard-cell chip has tailored interconnect and utilizes a minimal amount of silicon area, it has the smallest propagation delay and best speed. On the other hand, an FPGA chip has the worst propagation delay. In addition to its large size, the programmable interconnect has a relatively large resistance and capacitance, which introduces even more delay. As with chip area, the speed difference between standard-cell and gate array technologies is much less significant than that between FPGA and ASIC.

Power   Power concerns the energy consumed by a part. In certain applications, such as battery-operated handheld equipment, a low power circuit is of primary importance. At the architecture level, a system can be redesigned to reduce the use of power. If the identical architecture is used, a smaller chip, which consists of fewer transistors, usually consumes less power. Thus, a standard-cell chip consumes the least amount of power and an FPGA chip uses the most power.

Standard-cell technology is clearly the best choice from a technical perspective. A chip constructed using standard-cell ASIC is small and fast, and consumes less power. This should not come as a surprise since the chip is highly optimized and wastes no resources on unnecessary overhead. The price associated with customization is the complexity. Designing and fabricating a standard-cell chip is more involved and time consuming than for the other two technologies.

Cost   The design of a custom digital circuit is seldom a goal in itself. It is an economic activity, and the cost is an important, if not the deciding, factor. We consider three major expenses: production cost, development cost, and time-to-market cost.

Production cost is the expense to produce a single unit. It includes two segments: nonrecurring engineering (NRE) cost and part cost. *NRE cost* (C_{nre}) is the expense that occurs only once (and thus is not recurring) during the production process, regardless of the number of units sold. Thus, it is on a "per design" basis. *Part cost* (C_{per_part}), on the other hand, is on a "per unit" basis, covering the expense required for each individual unit, such as the expense of materials, assembly and manufacturing. Note that the NRE cost is shared by all the units and that the share of each part becomes smaller as the volume increases. The per unit production cost (C_{per_unit}) can be expressed as

$$C_{per\_unit} = C_{per\_part} + \frac{C_{nre}}{\text{units produced}}$$

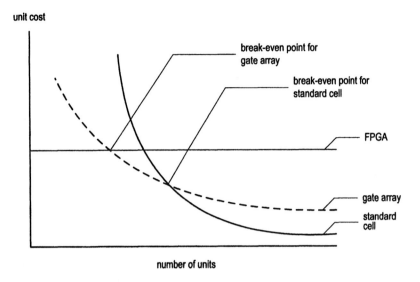

Figure 1.1   Comparison of per unit cost.

The NRE cost of a custom ASIC chip includes the creation of the tailored masks, the development of tests and the fabrication of initial sample chips. The charge is high and can range from several hundred thousand dollars to several million dollars or more. A major factor in the NRE cost is the number of custom masks needed. A standard-cell chip may need 15 or more tailored masks and thus is much more expensive than a gate array chip, which needs only three to five tailored metal layers. On the contrary, an FPGA-based design needs only an inexpensive device programmer to do customization. The NRE cost of creating a mask is negligible and can be considered as zero.

The part cost of an ASIC chip is smaller than that of an FPGA chip since the ASIC chip requires less silicon real estate and has better yield. By the same token, the part cost of a standard-cell chip is smaller than that of a gate array chip since the standard-cell chip is further optimized. If we consider both part cost and NRE cost, the per unit production cost depends on the volume of units, as shown by the previous equation. The volume versus per unit cost plots of three technologies is shown in Figure 1.1. As the volume increases, first the gate array and then the standard-cell technologies, become cost-effective. The intersections of the curves are the break-even points for the FPGA and gate array technologies, and for the gate array and standard-cell technologies.

The second major expense is the *development cost*. The process of transforming an idea to a custom circuit is by no means a simple task. The expense involved in this process is the development cost. It includes the compensation for engineering time as well as the expense of the computing facility and software tools. Although the synthesis procedure is somewhat similar for all device technologies, developing ASIC requires more effort, including physical design, placement and routing, verification and testing. Since the development process is more complex for ASIC, the development cost of an ASIC chip is much higher than that of an FPGA chip. Similarly, due to the high-level optimization, the development cost for a standard-cell chip is much higher than that for a gate array chip.

The third major expense is the *time-to-market cost*. It is actually not a cost, but the lost revenue. In many applications, such as PC peripherals, the life cycle of a product is

Table 1.1   Comparison of device technologies

	FPGA	Gate array	Standard cell
Tailored masks	0	3 to 5	15 or more
Area			best (smallest)
Speed			best (fastest)
Power			best (minimal)
NRE cost	best (smallest)		
Per part cost			best (smallest)
Development cost	best (easiest)		
Time to market	best (shortest)		
Per unit cost		depends on volume	

very short. Eighteen months, the time required to double the chip density, is sometimes considered as the life cycle of the product. Thus, it is very important to introduce the product in a timely manner, and a shipping delay can mean a significant loss in sales. The standard-cell technology requires the most lead time to validate, test and manufacture, ranging from a few months to a year. The gate array technology requires less lead time, from a few weeks to a few months. For FPGA technology, customization involves the programming of a prefabricated chip and can be done in a few minutes.

Summary   The major characteristics of the three device technologies are summarized in Table 1.1. In general, the trade-off is between the optimal use of hardware resources (in terms of chip area, speed and power) and the ease of design (in terms of NRE cost, development cost and manufacturing lead time).

The choice of technology is not necessarily mutual exclusive. For example, ASIC and FPGA developments can be done in parallel to get the benefits of both technologies. The FPGA devices are used as prototypes and in initial shipments to cut the manufacturing lead time. When the ASIC devices become available later, they are used for volume production to reduce cost.

1.3   SYSTEM REPRESENTATION

A large digital system is quite complex. During the development and production process, each task may require a specific kind of information about the system, ranging from system specification to physical component layout. The same system is frequently described in different ways and is examined from different perspectives. We call these perspectives the *representations* or *views* of a system. There are three views:

- Behavioral view
- Structural view
- Physical view

A *behavioral view* describes the functionality (i.e., "behavior") of a system. It treats the system as a black box and ignores its internal implementation. The view focuses on the relationship between the input and output signals, defining the output response when a particular set of input values is applied. The description of a behavioral view is seldom unique. Normally, there are a wide variety of ways to specify the same input–output characteristics.

A *structural view* describes the internal implementation (i.e., structure) of a system. The description is done by explicitly specifying what components are used and how these components are connected. It is more or less the schematic or the diagram of a system. In computer software, we use the term *net* to represent a set of wires that are connected to the same node, and use the term *netlist*, which is a collection of nets, to represent the schematic.

A *physical view* describes the physical characteristics of the system and adds additional information to the structural view. It specifies the physical sizes of components, the physical locations of the components on a board or a silicon wafer, and the physical path of each connection line. An example of a physical view is the printed circuit board layout of a system.

Clearly, the physical view of a system provides the most detailed information. It is the final specification for the system fabrication. On the other hand, the behavioral view imposes fewest constraints and is the most abstract form of description.

1.4 LEVELS OF ABSTRACTION

As chip density reaches hundreds of millions of transistors, it is impossible for a human being, or even a computer, to process this amount of data directly. A key method of managing complexity is to describe a system in several levels of abstraction. An *abstraction* is a simplified model of the system, showing only the selected features and ignoring the associated details. The purpose of an abstraction is to reduce the amount of data to a manageable level so that only the critical information is presented. A high-level abstraction is focused and contains only the most vital data. On the other hand, a low-level abstraction is more detailed and takes account of previously ignored information. Although it is more complex, the low-level abstraction model is more accurate and is closer to the real circuit. In the development process, we normally start with a high-level abstraction and concentrate on the most vital characteristics. As the system is better understood, we then include more details and develop a lower-level abstraction.

Four levels of abstraction are considered in digital system development:

- Transistor level
- Gate level
- Register transfer (RT) level
- Processor level

The division of these levels is based primarily on the size of basic building blocks, which are the transistors, logic gates, function modules and processors respectively.

The level of abstraction and the view are two independent dimensions of a system, and each level has its own views. The levels of abstraction and views can be combined in a *Y-chart*, which is shown in Figure 1.2. In this chart, each axis represents a view and the levels of abstraction increase from the center to the outside.

The following subsections discuss the four levels of abstraction. In the discussion, we examine the five main characteristics at each level of abstraction:

- Basic building blocks
- Signal representation
- Time representation
- Behavioral representation
- Physical representation

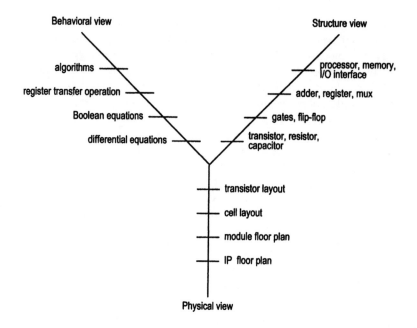

Figure 1.2   Y-chart.

Basic building blocks are the most commonly used parts at the level. These parts are the components used in the structure view. Behavioral and physical representations are the descriptions for the behavioral and physical views.

Signal and timing representations concern how to express a signal's value and how the value changes over time. While the physical signal remains the same, the interpretation of its value and timing is different at each abstraction level. As we expect, more detailed information will be provided at lower levels.

1.4.1   Transistor-level abstraction

The lowest level of abstraction is the transistor level. At this level, the basic building blocks are transistors, resistors, capacitors and so on. The behavior description is usually done by a set of differential equations or even by some type of current–voltage diagram. Analog system simulation software, such as SPICE, can be used to obtain the desired input–output characteristics.

At the transistor level, a digital circuit is treated as an analog system, in which signals are time-varying and can take on any value of a continuous range. For example, the output response of an inverter is plotted at the top of Figure 1.3.

The physical description of the transistor level comprises the detailed layout of components and their interconnections. It essentially defines the masks of various layers and is the final result of the design process.

1.4.2   Gate-level abstraction

The next level of abstraction is the gate level. Typical building blocks include simple logic gates, such as and, or, xor and 1-bit 2-to-1 multiplexer, and basic memory elements, such

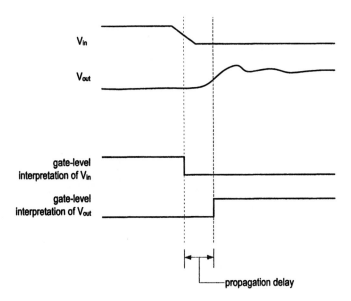

Figure 1.3   Timing characteristic of an inverter.

as latch and flip-flop. Instead of using continuous values, we consider only whether a signal's voltage is above or below a threshold, which is interpreted as logic 1 or logic 0 respectively. Since there are only two values, the input–output behavior is described by Boolean equations. The abstraction essentially converts a continuous system to a discrete system and discards the complex differential equations. Note that logic 0 and logic 1 are only our interpretation, depending on whether a signal's voltage level exceeds a predefined threshold, and the real signal is still the same continuous signal.

The timing information is also simplified at this level. A single discrete number, known as the *propagation delay*, which is defined as the time interval for a system to obtain a stable output response, is used to specify the timing of a gate. The plot at the bottom of Figure 1.3 shows a gate-level interpretation of the corresponding transistor-level signal.

The physical description at this level is the placement of the gates (or cells) and the routing of the interconnection wires.

So far, we use the term *area* or *size* to describe the silicon real estate used to construct a circuit. Alternatively, we can count the number of gates in this circuit (known as *gate count*) and make the measurement independent of the underlying device technology. The area of the two-input nand gate is used as the base unit since it is frequently the simplest physical logic circuit. Instead of using the physical area, we express the size or the complexity of a circuit in terms of the number of equivalent nand gates in that particular device technology.

1.4.3 Register-transfer-level (RT-level) abstraction

At the register-transfer (RT) level, the basic building blocks are modules constructed from simple gates. They include functional units, such as adders and comparators, storage components, such as registers, and data routing components, such as multiplexers. A reasonable name for this level would be *module-level* abstraction. However, the term *register transfer* is normally used in digital design and we follow the general convention.

Register transfer is a somewhat confusing term. It is used in two contexts. Originally, the term was used to describe a design methodology in which the system operation is specified by how the data are manipulated and transferred between storage registers. Since the main components used in the register transfer methodology are the intermediate-size modules, the term has been borrowed to describe module-level abstraction. As the title indicates, the coverage and discussion of this book focus on the RT level. We use the term *RT level* for module-level abstraction and *RT methodology* for the specific design methodology. RT methodology is discussed in Chapters 11 and 12.

The data representation at the RT level becomes more abstract. Signals are frequently grouped together and interpreted as a special kind of data type, such as an unsigned integer or system state. The behavioral description at this level uses general expressions to specify the functional operation and data routing, and uses an extended finite state machine (FSM) to describe a system designed using RT methodology.

A major feature of the RT-level description is the use of a common *clock signal* in the storage components. The clock signal functions as a sampling and synchronizing pulse, putting data into the storage component at a particular point, normally the rising or falling edge of the clock signal. In a properly designed system, the clock period is long enough so that all data signals are stabilized within the clock period. Since the data signals are sampled only at the clock edge, the difference in propagation delays and glitches have no impact on the system operation. This allows us to consider timing in terms of number of clock cycles rather than by keeping track of all the propagation delays.

The physical layout at this level is known as the *floor plan*. It is helpful for us to find the slowest path between the storage components and to determine the clock period.

1.4.4   Processor-level abstraction

Processor-level abstraction is the highest level of abstraction. The basic building blocks at this level, frequently known as *intellectual properties (IPs)*, include processors, memory modules, bus interfaces and so on. The behavioral description of a system is more like a program coded in a conventional programming language, including computation steps and communication processes. The signals are grouped and interpreted as various data types. Time measurement is expressed in terms of a computation step, which is composed of a set of operations defined between two successive synchronization points. A collection of computations may run concurrently in parallel hardware and exchange data through a predefined communication or bus protocol. The physical layout of a processor-level system is also known as the floor plan. Of course, the components used in a floor plan are much larger than those of an RT-level system.

Table 1.2 summarizes the main characteristics at each level. It lists the typical building blocks, signal representation, time representation, representative behavioral description and representative physical description.

1.5   DEVELOPMENT TASKS AND EDA SOFTWARE

Developing a custom digital circuit is essentially a refining and validating process. A system is gradually transformed from an abstract high-level description to final mask layouts. Along with each refinement, the system's function should be validated to ensure that the final product works correctly and meets the specification and performance goals. The major design tasks of developing a digital system are:

Table 1.2   Characteristics of each abstraction level

	Typical blocks	**Signal representation**	**Time representation**	**Behavioral description**	**Physical description**
Transistor	transistor, resistor	voltage	continuous function	differential equation	transistor layout
Gate	and, or, xor, flip-flop	logic 0 or 1	propagation delay	Boolean equation	cell layout
RT	adder, mux, register	integer, system state	clock tick	extended FSM	RT-level floor plan
Processor	processor, memory	abstract data type	event sequence	algorithm in C	IP-level floor plan

- Synthesis
- Physical design
- Verification
- Testing

1.5.1 Synthesis

Synthesis is a refinement process that realizes a description with components from the lower abstraction level. The original description can be in either a behavioral view or a structural view, and the resulting description is a structural view (i.e., netlist) in the lower abstraction level. In the Y-chart, the process either moves the system from behavioral view to structural view or moves it from a high-level abstraction to a low-level abstraction. Thus, synthesis either derives a structural implementation from a behavioral description or realizes an upper level description using finer components. As the synthesis process progresses, more details are added. The final result is a gate-level structural representation using the primitive cells from the chosen device technology. To make the process manageable, synthesis is usually divided into several smaller steps, each performing a specific transformation. The major steps are:

- High-level synthesis
- RT-level synthesis
- Gate-level synthesis
- Technology mapping

High-level synthesis transforms an algorithm into an RT-level description, which is specified explicitly in terms of register transfer operations. Due to the complexity of transformation, it can only be applied to relatively simple algorithms in a narrowly defined application domain.

RT-level synthesis analyzes an RT-level behavioral description and derives the structural implementation using RT-level components. It may also perform a limited degree of optimization to reduce the number of components.

Gate-level synthesis is similar to RT-level synthesis except that gate-level components are used in structural implementation. After the initial circuit is derived, two-level or multilevel optimization is used to minimize the size of the circuit or to meet the timing constraint. In general, generic components are used in gate-level synthesis, and thus the synthesis process is independent of device technology.

Each device technology includes a set of predesigned primitive gate-level components, which can be cells of a standard-cell library or a generic logic cell of an FPGA device. To implement the gate-level circuit in a particular device technology, the generic components have to map into the cells of the chosen technology. The transforming process is known as *technology mapping*. It is the last step in synthesis, and clearly the process is technology dependent.

The synthesis procedure is discussed in detail in Section 6.4.

1.5.2   Physical design

Physical design includes two major parts. The first part is the refinement process between the structural and physical views, which derives a layout for a netlist. The second part involves the analysis and tuning of a circuit's electrical characteristics. The main tasks in physical design include floor planning, placement and routing and circuit extraction.

Floor planning derives layouts at the processor and RT levels. It partitions the system into large function blocks and places these blocks in proper locations to reduce future routing congestion or to achieve certain timing objectives. Furthermore, floor planning may also provide a global plan for the power and clock distribution schemes. *Placement and routing* derives a layout at the gate level. The layout involves the detailed placement of cells and the routing of interconnecting wires.

After the placement and routing are complete, the exact length and location of each interconnect are known, and the associated parasitic capacitance and resistance can be calculated. This process is known as *circuit extraction*. The extracted data are used to construct a resistance and capacitance network, which in turn is used to compute the propagation delays.

In addition to the foregoing tasks, the physical design also includes design rule checking, derivation of the power grid, derivation of the clock distribution network, estimation of power use and assurance of signal integrity.

1.5.3   Verification

Verification is the process of checking whether a design meets the specification and performance goals. It concerns the correctness of the initial design as well as the correctness of refinement processes during synthesis and physical design. Verification has two aspects: *functionality* and *performance*. *Functional verification* checks whether a system generates the desired output response. Performance is represented as certain *timing constraints*. *Timing verification* checks whether the response is generated within the given time constraint. Verification is done in different phases of the design and at different levels of abstraction.

Functional verification   The design of a custom system usually begins with a high-level behavioral description. When it is first created, the primary concern is whether the design functions according to the specifications. We need to check its operation and compare its responses to those desired. Once the functionality of the initial design is verified, we can start the refinement process and gradually convert it to a gate-level structural description. In general, if the initial design does not depend on the internal propagation delay (i.e., is not *delay-sensitive*), the functionality should be maintained through the refinement processes.

In the ideal situation, the design should be "correct by construction" and require no further functional verification. In reality, subtle errors may be introduced in a refinement process, and thus functional verification is still performed after each process to ensure that the new, refined description works correctly.

Timing verification Timing verification checks whether a system meets its performance goals, which are normally expressed in terms of maximal propagation delay or minimal clock frequency. At the processor or RT level, the propagation delay of an input–output path can be calculated by identifying the components in the path and summating the individual delays. However, since these components will be further refined and synthesized, the information is just a rough estimation.

At the gate level, the propagation delay of a path is affected by the delays of the components as well as the interconnection wires. The wiring delay depends on the locations and the lengths of wires. Although they can be estimated during synthesis, the exact values can be obtained only after the placement and routing process. As the size of a transistor continues to shrink, the effect of a wiring delay becomes more dominant. This makes timing verification more difficult since accurate delay information is not available during the synthesis process.

Timing issues and propagation delay are discussed in more detail in Section 6.5.1.

Methods of verification The most commonly used verification method is *simulation*, which is the process of constructing a model of a system, executing the model with input test patterns in a computer, and examining and analyzing the output responses. The model can be an actual or a hypothetical circuit that incorporates functionality and timing information. Simulation is a versatile process that be applied at any level of abstraction, and in behavioral as well as structural views. Utilizing simulation allows us to examine a system's operation in a computer and to detect errors without actually constructing the system.

Simulation essentially provides a sequence of snapshots of system operation, defined by a set of input stimuli. However, there is no guarantee that the selected stimuli can exercise every part of the system and verify the correctness of the entire design. Whereas simulation can do spot checks and detect major design mistakes, it cannot guarantee the absence of errors.

Another limitation of simulation comes from its computation complexity. Hardware operation is concurrent and parallel in nature, and it is time consuming to model its operation in a computer, which performs computational steps sequentially. It becomes a serious problem when we want to simulate low-level models, which may consist of hundreds of thousands or even millions of components.

In addition to simulation, several other methods are used for verification, including timing analysis, formal verification and hardware emulation. *Timing analysis* focuses only on the timing aspects of a circuit. It analyzes the structure of a circuit, determines all possible input–output paths, calculates the propagation delays of these paths and determines the relevant timing parameters, such as worst-case propagation delay and maximal clock frequency. Simulation can provide the relevant timing information for the selected test patterns. However, since these test patterns do not always exercise the critical paths, timing analysis is needed to verify that the system meets the timing specifications.

Formal verification applies formal mathematical techniques to analyze a circuit and determine its property. A popular method in formal verification is *equivalence checking*, which compares two representations of a system and determines whether the two representations perform the same function. It is frequently applied in synthesis to verify that the functionality of a synthesized circuit is identical to the original one. Unlike simulation, formal verification is based on rigorous mathematical reasoning and can ensure that the synthesis is completely error-free.

Hardware emulation physically constructs a prototyping circuit that mimics operation of the system. A common application is to construct an FPGA circuit to emulate a complex ASIC design. Although the FPGA-based system is normally larger and slower than the ASIC system, it is much faster than simulation and it can be physically interfaced with other circuits and studied in detail.

1.5.4   Testing

The meanings of verification and testing are somewhat similar in a dictionary sense. However, they are two very different tasks in digital system development. *Verification* is the process of determining whether a design meets the specification and performance goals. It concerns the correctness of the initial design as well as the refinement processes. On the other hand, *testing* is the process of detecting the physical defects of a die or a package that occurred during manufacturing. When a device is being tested, we already know that the design is correct and the purpose of testing is simply to ensure that this particular part was properly fabricated.

At first glance, testing appears to be easy. All we need to do is simply to apply all possible input combinations and check the output responses. However, because of the large number of input combinations, this approach is not feasible. Instead, we have to utilize special algorithms to obtain a small set of test patterns. This process is known as *test pattern generation*.

For a small circuit, we can develop the testing procedure after completing the initial design and synthesis. However, as a digital circuit becomes larger and more complex, this approach becomes more difficult. Instead of as an afterthought, we have to consider the testing procedure in the initial design and frequently need to add auxiliary circuitry, such as a *scan chain* or *built-in-self-test circuit*, to facilitate the future requirements. This is known as *design-for-test*.

1.5.5   EDA software and its limitations

Developing a large digital circuit is a complicated process that involves complex algorithms and a large amount of data. Computer software is used to automate some tasks. This is known as *electronic design automation (EDA)*. As computers become more powerful, we may ask if it possible to develop a suite of software and automate the development process completely. The ideal scenario would be that human designers only need to develop a high-level behavioral description, and EDA software will perform the synthesis and placement and routing and derive the optimal circuit implementation automatically. The answer is, unfortunately, negative. This is due to the theoretical limitations that cannot be overcome by faster computers or smart software codes.

The synthesis software should be treated as a tool to perform transformation and local optimization. It cannot alter the original architecture or convert a poor design into a good one. The efficiency of the final circuit depends mainly on the initial description.

The limitations and effective use of the EAD software are elaborated in Section 6.1.

1.6   DEVELOPMENT FLOW

Developing a digital circuit is essentially a refining and validating process, gradually transforming an abstract high-level description into a detailed low-level structural description. While all developments follow the basic refinement–validation process, detailed flows depend on the size of the circuit and the target device technology.

The optimization algorithms used in synthesis software are complex. The needed computation time and memory space increase drastically as the circuit size grows. Thus, size is a limiting factor in many synthesis software tools. The software is most effective for an intermediate-sized circuit, which ranges between 2000 and 50,000 gates. For a larger system, we must first partition the circuit into smaller blocks and then process each block individually.

Another factor is the target device technology. The fabrication processes of FPGA and ASIC are very different. Whereas an FPGA chip is an off-the-shelf part that has been prefabricated and pretested, an ASIC design must go through a lengthy, complex fabrication process. Many extra steps are needed to ensure the correctness of the final physical circuit.

The following subsections show the typical development flow of three different types of designs and explain the extra steps needed as the complexity increases. Three types of designs are:

- Medium-sized design targeting FPGA
- Large design targeting FPGA
- Large design targeting ASIC

1.6.1   Flow of a medium-sized design targeting FPGA

The term *medium-sized* here means a design that requires no partition and does not need predesigned IP cores. It is a circuit with up to about 50,000 gates. Current synthesis software and placement-and-routing software can effectively process a circuit of this complexity. This size is not trivial. It corresponds to that of a moderately complex circuit, such as a simple processor or bus interface. The development flow is depicted in Figure 1.4. It is shown in three columns, representing a synthesis track, physical design track and verification track respectively.

The flow starts with the design file, which is normally an RT-level description of the circuit. It may be accompanied by a set of constraints that specify the timing requirements. A separate file, known as a *testbench*, provides a virtual experiment bench for simulation and verification. It incorporates the code to generate input stimuli and to monitor the output responses. Once these files are created, the circuit can be constructed and verified accordingly. The steps in an ideal flow are detailed below.

1. Develop the design file and testbench.
2. Use the design file as the circuit description, and perform a simulation to verify that the design functions as desired.
3. Perform a synthesis.
4. Use the output netlist file of the synthesizer as the circuit description, and perform a simulation and timing analysis to verify the correctness of the synthesis and to check preliminary timing.
5. Perform placement and routing.

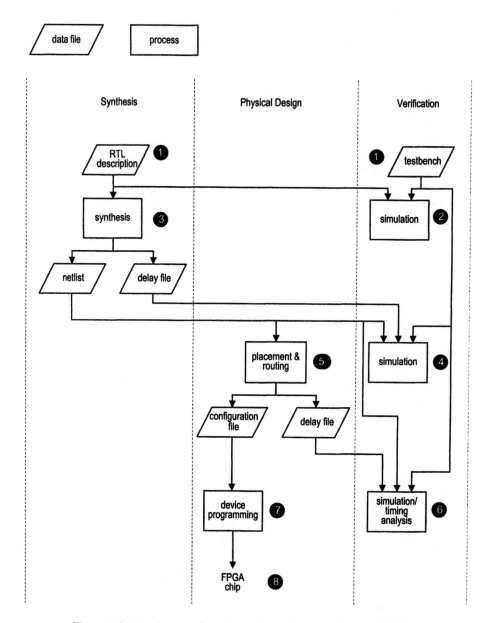

Figure 1.4 Development flow of a medium-sized design targeting FPGA.

6. Annotate the accurate timing information to the netlist, and perform a simulation and timing analysis to verify the correctness of the placement and routing and to check whether the circuit meets the timing constraints.
7. Generate the configuration file and program the device.
8. Verify operation of the physical part.

The flow described above represents an ideal process since it assumes that the initial design description follows the functional specification and meets the timing constraints. In reality, the development flow may consist of several iterations to correct the functional errors or timing problems. We may need to revise the original design file or to fine-tune parameters in synthesis and placement-and-routing software.

1.6.2 Flow of a large design targeting FPGA

A large, complex digital circuit may contain several hundred thousand or even a few million gates. Synthesis tools are not able to perform transformation and optimization effectively in this range. It is necessary to partition the circuit into smaller blocks and to process the blocks individually. The partition process also allows us to use previously designed subsystems or commercial IP cores.

To accommodate a larger design, additional processes must be added to the flow of Figure 1.4. The initial design description tends to be an abstract, high-level behavioral description of the circuit. In the synthesis track, a *partition process* is needed to divide the systems into blocks of adequate size and functionality. The output of the partition process can be considered as a netlist of large blocks. Some blocks may be already designed and verified subsystems, either from a previous project design or from a commercial IP vendor. The other blocks must be designed and synthesized individually as medium-sized circuits, following the development flow of the previous subsection.

In the verification track, an extra step is needed to verify the correctness of the partition results and to check the initial timing. Because of the large number of components, the gate-level netlist becomes very involved, and simulation consumes a significant amount of time. Formal verification techniques and cycle-based simulation are frequently used as an alternative to verify the functionality.

In the physical design track, a floor planning process may be needed. It performs initial placement for the processor-level blocks.

1.6.3 Flow of a large design targeting ASIC

Due to the complexity of ASIC fabrication, the development flow becomes more involved. The additional requirements are the inclusion of a testing track and the expansion of the physical design track.

The purpose of testing is to detect defects in the fabrication process. FPGA devices are tested by vendors before being shipped, and thus we don't need to worry about physical defects of the device. On the other hand, the testing is the integral part of the ASIC design and plays an important role. At the RT level, additional built-in-self-test circuits and special scanning control circuits are frequently added to aid the final testing. These circuits become an integral part of the design and have to be synthesized and verified. At the gate level, scan registers will be strategically inserted around circuit blocks or I/O boundaries. The scan circuit also needs to be synthesized and verified along with the regular design. Finally, test vectors have to be generated for combinational circuit blocks, and simulation has to be performed to ensure that the vectors provide proper fault coverage.

In FPGA-based flow, the physical design track involves only the floor planning and place-ment and routing, which is accomplished by configuring the FPGA device's programmable interconnect structure. The physical design process of an ASIC device is much more complicated since it involves development and verification of the masks. After placement and routing, several additional steps are needed, including design rule checking, physical verification and circuit extraction.

Due to the high NRE cost of an ASIC-based device, it is important that the circuit is simulated and checked thoroughly before fabrication. Thus, the verification track of ASIC-based design flow has to be more comprehensive and more exhaustive.

1.7   OVERVIEW OF THE BOOK

1.7.1   Scope

This book focuses primarily on the design and synthesis of RT-level circuits. A subset of the VHDL hardware description language is used to describe the design. The book is not intended to be a comprehensive ASIC or FPGA book. All other issues, such as device architecture, placement and routing, simulation and testing, are discussed only from the context of RT-level design.

After completing this book, readers should be able to develop and design efficient RT-level systems or subsystem blocks. A physical chip for a medium-sized FPGA design or a large, manually partitioned FPGA design can be obtained with a general synthesis and placement and routing software package. Additional knowledge and more specialized software tools are needed to cover the other tasks for an ASIC design.

1.7.2   Goal

The goal of the book is to learn how to *systematically develop efficient, portable RT-level designs that can easily be integrated into a larger system.* The goal includes three major parts:
- Design for efficiency
- "Design for large"
- Design for portability

Design for efficiency   Availability of HDL and synthesis software relieves us from many tedious, repetitive implementation details and allows us to explore the design at a more abstract level. However, algorithms used in synthesis software can only do transformation and perform local search and optimization. They cannot, and will not, create a good design description or convert a poor design description to a good one. The quality of the circuit lies primarily in the initial description.

The book shows the relationship between VHDL constructs and the hardware compo-nents as well as the effective use of synthesis software tools, and introduces a disciplined way to develop the initial description that leads to efficient implementation.

"Design for large"   We use the term *design for large* loosely to cover three aspects:
- Design of a large module
- Design to be incorporated in a larger system
- Design to facilitate the overall development process

The main purpose of most digital system books, including this one, is to illustrate basic concepts and procedures. For clarity, the design examples are normally explained by circuits with small input size. However, the design and description for a system with a small number of inputs (e.g., a 2-bit multiplier) and a system with a larger number of inputs (e.g., a 32-bit multiplier) can be very different. Although small-input-size examples are used in this book, the design approach and coding style are aimed at a large input size, and thus the design can easily be expanded to a larger, more practical system.

As the digital system becomes more complex, an RT-level description is likely to be a part of a larger system. Although a large processor-level system is not the focus of this book, the coding and development take this into consideration so that the RT-level design can easily be incorporated into a larger system when needed.

The discussion in Section 1.6 shows that the development of a large digital system involves many tasks. RT-level design and synthesis are not an isolated part. A poorly constructed RT-level circuit makes simulation, verification and testing processes unnecessarily difficult or even impossible, and sometimes may need to be revised at a later stage of the development process. While the book focuses on RT-level design and synthesis, it treats this task as an integral part of the development process and uses methodology that can facilitate and even simplify other tasks.

Design for portability   Portability means that the same design description can be used in different applications. We can examine design for portability from three perspectives:

- Device independent
- Software independent
- Design reuse

Device independent means that the same design description can be synthesized to different device technologies. From time to time, the same design may need to migrate to a different technology. It can be from one FPGA vendor to another or from FPGA to ASIC for volume production. The design descriptions of this book carefully avoid any device-dependent feature so that the code can be used for multiple device technologies.

Software independent means that the design description can be accepted by most synthesis software. Since synthesis is a very complex process, software packages from different vendors have different capabilities, support different subsets of hardware description language and may have different interpretations on some subtle language constructs. We try to use the minimal common denominator of the synthesis software so that a design description can be accepted by most software tools and its function will be interpreted in a similar manner.

Design reuse means that the whole or part of the design description can be used again in a different application or project. We interpret the term *reuse* in a broad sense, from the copying of a few lines of code to a complete IP core. While developing an IP core is not the primary goal, we try to make the code modular and scalable when possible so that the same code can be reused in different applications with minimal or no revision.

1.8 BIBLIOGRAPHIC NOTES

This book includes a short bibliographic section at the end of each chapter. The purpose of the section is to provide several of the most relevant references for further exploration. A complete comprehensive bibliography is provided at the end of the book.

Developing a large digital system is a complex process. The text, *Methodology Manual for System-on-a-Chip Designs, 3rd edition* by M. Keating and P. Bricaud, provides an overview and guidelines for the process. The text, *The Design Warrior's Guide to FPGAs* by C. M. Maxfield, introduces relevant issues on FPGAs. Two texts, *FPGA-Based System Design* and *Modern VLSI Design: System-on-Chip Design, 3rd edition*, both by W. Wolf, provide more in-depth reviews of the FPGA and ASIC technologies.

Problems

1.1   An engineer claims the following about the digital format: "In a digital system, logic 0 and logic 1 are represented by two voltage levels. Since there is a significant voltage difference between the two levels, noise will not affect the logic value, and thus digitized information is immune to noise." Is the statement correct? Explain.

1.2   Volume of sale (i.e., the number of parts sold) is a factor when determining which device technology is to be used. Assume that a system can be implemented by FPGA, gate array or standard-cell technology. The per part cost is $15, $3 and $1 for FPGA, gate array and standard cell respectively. Gate array and standard-cell technologies also involve a one-time mask generation cost of $20,000 and $100,000 respectively.

(a) Assume that the number of parts sold is N. Derive the equation of per unit cost for the three technologies.
(b) Plot the three equations with N as the x-axis.
(c) Determine the range of N for which FPGA technology has the minimal per unit cost.
(d) Determine the range of N for which gate array technology has the minimal per unit cost.
(e) Determine the range of N for which standard-cell technology has the minimal per unit cost.

1.3   What is the view (behavioral, structural or physical) of the following illustration?

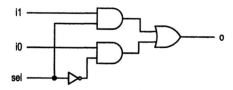

1.4   What is abstraction? Why is it important for digital system design?

1.5   What is the difference between testing and verification?

1.6   In Figure 1.4, the synthesized circuit is simulated in steps 4 and 6. Is the simulation in step 6 necessary? Explain.

CHAPTER 2

OVERVIEW OF HARDWARE DESCRIPTION LANGUAGES

A digital system can be described at different levels of abstractions and from different points of view. As the design process progresses, the level and view are changed, either by human designers or by software tools. It is desirable to have a common framework to exchange information among the designers and various software tools. *Hardware description languages (HDLs)* serve this purpose. In this chapter we provide an overview of the design, use and capability of HDLs. The basic concept and essential modeling features are introduced by a series of codes to show the "big picture" of HDLs. The detailed syntax, language constructs and associated semantics are discussed in subsequent chapters.

2.1 HARDWARE DESCRIPTION LANGUAGES

A digital system can be described at different levels of abstraction and from different points of view. An HDL should faithfully and accurately model and describe a circuit, whether already built or under development, from either the structural or behavioral views, at the desired level of abstraction. Because HDLs are modeled after hardware, their semantics and use are very different from those of traditional programming languages. The following subsections discuss the need, use and design of an HDL.

2.1.1 Limitations of traditional programming languages

There are wide varieties of computer programming languages, from Fortran to C to Java. Unfortunately, they are not adequate to model digital hardware. To understand their limitations, it is beneficial to examine the development of a language. A programming language is characterized by its syntax and semantics. The *syntax* comprises the grammatical rules used to write a program, and the *semantics* is the "meaning" associated with language constructs. When a new computer language is developed, the designers first study the characteristics of the underlying processes and then develop syntactic constructs and their associated semantics to model and express these characteristics.

Most traditional general-purpose programming languages, such as C, are modeled after a sequential process. In this process, operations are performed in sequential order, one operation at a time. Since an operation frequently depends on the result of an earlier operation, the order of execution cannot be altered at will. The sequential process model has two major benefits. At the abstract level, it helps the human thinking process to develop an algorithm step by step. At the implementation level, the sequential process resembles the operation of a basic computer model and thus allows efficient translation from an algorithm to machine instructions.

The characteristics of digital hardware, on the other hand, are very different from those of the sequential model. A typical digital system is normally built by smaller parts, with customized wiring that connects the input and output ports of these parts. When a signal changes, the parts connected to the signal are activated and a set of new operations is initiated accordingly. These operations are performed concurrently, and each operation will take a specific amount of time, which represents the propagation delay of a particular part, to complete. After completion, each part updates the value of the corresponding output port. If the value is changed, the output signal will in turn activate all the connected parts and initiate another round of operations. This description shows several unique characteristics of digital systems, including the connections of parts, concurrent operations, and the concept of propagation delay and timing. The sequential model used in traditional programming languages cannot capture the characteristics of digital hardware, and there is a need for special languages (i.e., HDLs) that are designed to model digital hardware.

2.1.2 Use of an HDL program

To better understand HDL, it is helpful to examine the use of an HDL program. In a traditional programming language, a program is normally coded to solve a specific problem. It takes certain input values and generates the output accordingly. The program is first compiled to machine instructions and then run on a host computer. On the other hand, the application of an HDL program is very different. The program plays three major roles:

- *Formal documentation.* A digital system normally starts with a word description. Unfortunately, since human language is not precise, the description is frequently incomplete and ambiguous, and the same description may be subject to different interpretations. Because the semantics and syntax of an HDL are defined rigorously, a system specified in an HDL program is explicit and precise. Thus, an HDL program can be used as a formal system specification and documentation among various designers and users.

- *Input to a simulator.* As we discussed in Chapter 1, simulation is used to study and verify the operation of a circuit without constructing the system physically. An HDL simulator provides a framework to model the concurrent operations in a sequential host computer, and has specific knowledge of the language's syntactic constructs and the associated semantics. An HDL program, combined with test vector generation and a data collection code, forms a testbench, which becomes the input to the

HDL simulator. During execution, the simulator interprets HDL code and generates responses accordingly.

- *Input to a synthesizer.* The modern development flow is based on the refinement process, which gradually converts a high-level behavioral description to a low-level structural description. Some refinement steps can be performed by synthesis software. The synthesis software takes an HDL program as its input and realizes the circuit from the components of a given library. The output of the synthesizer is a new HDL program that represents the structural description of the synthesized circuit.

2.1.3 Design of a modern HDL

The fundamental characteristics of a digital circuit are defined by the concepts of entity, connectivity, concurrency and timing. *Entity* is the basic building block, modeling after a part of a real circuit. It is self-contained and independent, and has no implicit information about other entities. *Connectivity* models the connecting wires among the parts. It is the way that entities interact with one another. Since the connections of a system are seldom formed as a single thread, many entities may be active at the same time and many operations are performed in parallel. *Concurrency* describes this type of behavior. *Timing* is related to concurrency. It specifies the initiation and completion of each operation and implicitly provides a schedule and order of multiple operations.

The goal of an HDL is to describe and model digital systems faithfully and accurately. To achieve this, the cornerstone of the language should be based on the model of hardware operation, and its semantics should be able to capture the fundamental characteristics of the circuits.

As we discussed in Chapter 1, a digital system can be described at four different levels of abstraction and from three different points of view. Although these descriptions have similar fundamental characteristics, their detailed representations and models vary significantly. Ideally, we wish to develop a single HDL to cover all the levels and all the views. However, this is hardly feasible because the vast differences between abstraction levels and views will make the language excessively complex. Modern HDLs normally cover descriptions in structural and behavior views, but not in physical view. They provide constructs to support modeling at the gate and RT levels, and to a limited degree, at processor and transistor levels. The highlights of modern HDLs are as follows:

- The language semantics encapsulate the concepts of entity, connectivity, concurrency, and timing.
- The language can effectively incorporate propagation delay and timing information.
- The language consists of constructs that can explicitly express the structural implementation (i.e., a block diagram) of a circuit.
- The language incorporates constructs that can describe the behavior of a circuit, including constructs that resemble the sequential process of traditional languages, to facilitate abstract behavioral description.
- The language can efficiently describe the operations and structures at the gate and RT levels.
- The language consists of constructs to support a hierarchical design process.

2.1.4 VHDL

VHDL and Verilog are the two most widely used HDLs. Although the syntax and "appearance" of the two languages are very different, their capabilities and scopes are quite similar.

Both are industrial standards and are supported by most software tools. VHDL is used in this book since it has better support for parameterized design.

VHDL stands for VHSIC (very high speed integrated circuit) HDL. The development of VHDL was sponsored initially by the US Department of Defense as a hardware documentation standard in the early 1980s and then was transferred to the IEEE (Institute of Electrical and Electronics Engineers). IEEE ratified it as IEEE standard 1076 in 1987, which is referred to as VHDL-87. Each IEEE standard is reviewed every few years and is revised as needed. IEEE revised the VHDL standard in 1993, which is referred to as VHDL-93, and made minor modifications and bug fixes in 2001, which is referred to as VHDL-2001. Since no new language construct is added in the new version, there is no significant difference between VHDL-93 and VHDL-2001. A suffix is sometimes added to the IEEE standard to indicate the year the standard was released. For example, VHDL-87 and VHDL-2001 are known as IEEE standards 1076-1987 and IEEE 1076-2001 respectively.

After the initial release, various extensions were developed to facilitate various design and modeling requirements. These extensions are documented in several IEEE standards:

- IEEE standard 1076.1-1999, *VHDL Analog and Mixed Signal Extensions (VHDL-AMS)*: defines the extension for analog and mixed-signal modeling.
- IEEE standard 1076.2-1996, *VHDL Mathematical Packages*: defines extra mathematical functions for real and complex numbers.
- IEEE standard 1076.3-1997, *Synthesis Packages*: defines arithmetic operations over a collection of bits.
- IEEE standard 1076.4-1995, *VHDL Initiative Towards ASIC Libraries (VITAL)*: defines a mechanism to add detailed timing information to ASIC cells.
- IEEE standard 1076.6-1999, *VHDL Register Transfer Level (RTL) Synthesis*: defines a subset that is suitable for synthesis.
- IEEE standard 1164-1993 *Multivalue Logic System for VHDL Model Interoperability (std_logic_1164)*: defines new data types to model multivalue logic.
- IEEE standard 1029.1-1998, *VHDL Waveform and Vector Exchange to Support Design and Test Verification (WAVES)*: defines how to use VHDL to exchange information in a simulation environment.

Standards 1076.3, 1076.6 and 1164 are related to synthesis and are discussed in Chapter 3.

2.2 BASIC VHDL CONCEPT VIA AN EXAMPLE

As its name indicates, HDL describes hardware. Thus, it is essential to read or write HDL code from hardware's perspective. A simple example in this section shows the basic modeling concepts used in HDL and demonstrates the semantic differences between HDLs and traditional programming languages. The example is coded in VHDL and the language constructs are mostly self-explanatory. The purpose of the example is to provide a big picture of HDL and VHDL. The detailed syntax and language constructs are studied in subsequent chapters.

The example is a circuit that detects even parity. There are one output, even, and three inputs, a(2), a(1) and a(0), which are grouped as a bus. The output is asserted when there are even numbers (i.e., 0 or 2) of 1's from the inputs. The truth table of this circuit is shown in Table 2.1.

The VHDL codes for a general description, pure structural description, pure behavioral description and testbench are discussed in the following subsections.

Table 2.1   Truth table of an even-parity detector circuit

a(2)	a(1)	a(0)	even
0	0	0	1
0	0	1	0
0	1	0	0
0	1	1	1
1	0	0	0
1	0	1	1
1	1	0	1
1	1	1	0

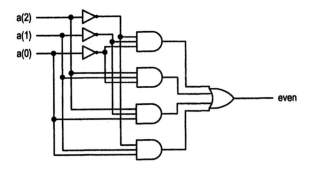

Figure 2.1   Two-level and–or implementation of an even-parity detector circuit.

2.2.1   General description

From Boolean algebra, we know that each row of a truth table represents a product term and the output can be written as the sum-of-products expression

$$even = a(2)' \cdot a(1)' \cdot a(0)' + a(2)' \cdot a(1) \cdot a(0) + a(2) \cdot a(1)' \cdot a(0) + a(2) \cdot a(1) \cdot a(0)'$$

The expression can be realized by a two-level and–or circuit, as shown in Figure 2.1.

The first VHDL description is based on this expression and the code is shown in Listing 2.1. In this book, the reserved words are in boldface font, as in **library**, and comments are in italic font, as in $--$ *this is a comment.*

Listing 2.1   Even-parity detector based on a sum-of-products expression

```
library ieee;
use ieee.std_logic_1164.all;

-- entity declaration
5 entity even_detector is
    port(
        a: in std_logic_vector(2 downto 0);
        even: out std_logic
    );
10 end even_detector;
```

```
      -- architecture body
      architecture sop_arch of even_detector is
15        signal p1, p2, p3, p4 : std_logic;
      begin
          even <= (p1 or p2) or (p3 or p4) after 20 ns;
          p1 <= (not a(2)) and (not a(1)) and (not a(0)) after 15 ns;
          p2 <= (not a(2)) and a(1) and a(0) after 12 ns;
20        p3 <= a(2) and (not a(1)) and a(0) after 12 ns;
          p4 <= a(2) and a(1) and (not a(0)) after 12 ns;
      end sop_arch ;
```

The code consists of two major units: entity declaration and architecture body. The *entity declaration* is:

```
      entity even_detector is
        port(
            a: in std_logic_vector(2 downto 0);
            even: out std_logic
        );
      end even_detector;
```

It specifies the input and output ports of this circuit. There are one output port, even, and one input port, a, which is a three-element array, representing a(2), a(1) and a(0).

The *architecture body* specifies the internal operation or organization of a circuit. The first line of the architecture body shows the name of the body, sop_arch (for sum-of-products architecture), and the corresponding entity, even_detector:

```
      architecture sop_arch of even_detector is
```

The next line is the signal declaration:

```
      signal p1, p2, p3, p4: std_logic;
```

The p1, p2, p3 and p4 signals here can be interpreted as wires that connect the internal parts. The declaration is visible inside this architecture.

The actual architectural description is encompassed within **begin** and **end** sop_arch:

```
          even <= (p1 or p2) or (p3 or p4) after 20 ns;
          p1 <= (not a(2)) and (not a(1)) and (not a(0)) after 15 ns;
          p2 <= (not a(2)) and a(1) and a(0) after 12 ns;
          p3 <= a(2) and (not a(1)) and a(0) after 12 ns;
          p4 <= a(2) and a(1) and (not a(0)) after 12 ns;
```

The fundamental building block inside the architecture body is a concurrent statement. For example, the first line is a *concurrent statement*:

```
      even <= (p1 or p2) or (p3 or p4) after 20 ns;
```

A concurrent statement can be thought of as a circuit part. The left-hand-side signal or port is the output, and all the signals and ports appearing in the right-hand-side expression are the input signals. The right-hand-side expression can be considered as the operation performed by this circuit. The result is available after a specific amount of propagation delay, which is specified by the **after** clause. This particular concurrent statement can be interpreted as a circuit with inputs, p1, p2, p3 and p4, and with an output, even. It performs the or operation among the four inputs, and the operation takes 20 ns. The other four statements can be interpreted in a similar fashion.

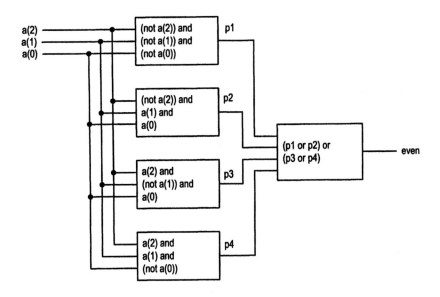

Figure 2.2   Conceptual diagram of sop_arch architecture.

This architecture body consists of five concurrent statements, which can be interpreted as a collection of five circuit parts. These concurrent statements are linked through common signals (or *nets*). When a signal appears on both the right- and left-hand sides, it implies that there is a wire connecting the two parts. Thus, a larger circuit is constructed implicitly through these connections. The conceptual diagram described by this code is shown in Figure 2.2.

Note that since each concurrent statement represents a circuit part and its interconnection, the order of these concurrent statements does not matter. For example, we can rearrange the code as

```
p2 <= (not a(2)) and a(1) and a(0) after 12 ns;
p3 <= a(2) and (not a(1)) and a(0) after 12 ns;
even <= (p1 or p2) or (p3 or p4) after 20 ns;
p1 <= (not a(2)) and (not a(1)) and (not a(0)) after 15 ns;
p4 <= a(2) and a(1) and (not a(0)) after 12 ns;
```

Unlike sequential execution of statements in traditional programming language, concurrent statements are independent and can be activated in parallel. When a concurrent statement's input changes, it is "awakened" and evaluates the expression accordingly. The result will be available after the specific propagation delay, and the new value will be assigned to the output signal. The change of output signals, in turn, may activate other statements and invoke a new round of evaluations.

The incorporation of propagation delay with each concurrent statement is the key ingredient to model the operation of hardware and to ensure the proper interpretation of VHDL code. Sometimes the after clause is omitted because the delay information is not available, as in

```
even <= (p1 or p2) or (p3 or p4);
```

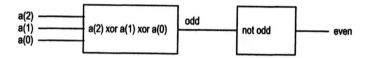

Figure 2.3   Conceptual diagram of the xor_arch architecture body.

In this case, VHDL semantics specifies that there is an implicit δ-delay (delta delay) associated with the operation. A *δ-delay* is an infinitesimal delay that is greater than zero but smaller than any physical number. The previous line can be interpreted as

```
even <= (p1 or p2) or (p3 or p4) after δ;
```

Thus, regardless whether there is an after clause, there is always a propagation delay associated with a concurrent statement.

The truth table is just one method to realize the even-parity detector circuit. An alternative is to use an xor ($\oplus$) operation. Recall that the xor operation can be used to detect odd parity since the $a \oplus b$ expression becomes '1' only when there is a single '1' from the inputs. Thus, the even-parity detector circuit can be implemented by an xor network followed by an inverter and the expression can be written as

$$even = (a(2) \oplus a(1) \oplus a(0))'$$

The architecture body based on this description is shown in Listing 2.2.

Listing 2.2   Even-parity detector based on an xor network

```
architecture xor_arch of even_detector is
   signal odd: std_logic;
begin
   even <= not odd;
s   odd <= a(2) xor a(1) xor a(0);
end xor_arch;
```

Again, the two concurrent statements represent two circuit parts, and the conceptual diagram is shown in Figure 2.3. Since no explicit after clause is used, both statements take a δ-delay to operate.

2.2.2 Structural description

In a structural view, a circuit is constructed of smaller parts. The description specifies what types of parts are used and how these parts are connected. The description is essentially a schematic, representing a block diagram or circuit diagram. Although we treat a concurrent statement of the preceding section as a circuit part, it is our interpretation and the code is not considered as a real structural description. Formal VHDL structural description is done by using the concept of *component*. A component can be either an existing or a hypothetical part. It first has to be *declared* (make known) and then can be *instantiated* (actually used) in the architecture body as needed.

Let us consider the even-parity detector circuit again. Assume that there is a library with predesigned parts, xor2 and not1, which perform the xor and inverting functions respectively. The even-parity detector circuit can be realized by the two parts, as shown in the circuit diagram of Figure 2.4. Based on the schematic, a structural description can be derived accordingly. The code of the architecture body is shown in Listing 2.3.

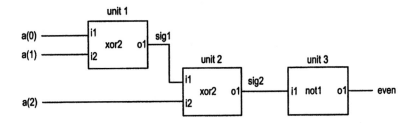

Figure 2.4 Structural diagram of the str_arch architecture.

Listing 2.3 Even-parity detector based on a structural description

```
architecture str_arch of even_detector is
   -- declaration for xor gate
   component xor2
      port(
5         i1, i2: in std_logic;
          o1: out std_logic
      );
   end component;
   -- declaration for invertor
10    component not1
      port(
          i1: in std_logic;
          o1: out std_logic
      );
15    end component;
   signal sig1,sig2: std_logic;

begin
   -- instantiation of the 1st xor instance
20    unit1: xor2
      port map (i1 => a(0), i2 => a(1), o1 => sig1);
   -- instantiation of the 2nd xor instance
   unit2: xor2
      port map (i1 => a(2), i2 => sig1, o1 => sig2);
25    -- instantiation of invertor
   unit3: not1
      port map (i1 => sig2, o1 => even);
end str_arch;
```

Inside the architecture, the components are declared first. For example, the declaration for xor2 is

```
component xor2
   port(
       i1, i2: in std_logic;
       o1: out std_logic
   );
end component;
```

The information contained inside the declaration is similar to that of entity declaration, which specifies the input and output ports of a circuit. In addition to component declaration, there is also a declaration for two internal signals, sig1 and sig2.

The architecture body consists of three statements, each representing a *component instantiation*. The first one is

```
unit1: xor2
    port map (i1=>a(0), i2=>a(1), o1=>sig1);
```

There are three elements in this statement. The first is the label, unit1, which serves as a unique id for this part. The second is the initiated component, xor2. The last is **port map** ···, which specifies the mapping between the formal signals (the I/O ports used in component declaration) and actual signals (the signals used in the architecture body). The mapping indicates that i1, i2 and o1 are connected to a(0), a(1) and sig1 respectively. The code is essentially the textual description of the circuit diagram in Figure 2.4. The three component instantiations together describe the complete circuit. The connections between the components are done implicitly by using the same signal names.

Component instantiation is one type of concurrent statement and can be mixed with other types of concurrent statements. When an architecture body consists only of component instantiations, as in this example, it is just a textual description of a schematic. This is a clumsy way for humans to conceptualize and comprehend this kind of representation. However, textual description put everything into a single VHDL framework so that the description can be handled by the same software tools. There is special design entry software that can convert a schematic to structural VHDL code and vice versa.

Component declaration contains only I/O port information, as in entity declaration. The components can be treated as empty sockets, which provide no clues about their internal functions. A component can be an existing, predesigned circuit or a hypothetical system that is still under construction. An architecture body will be bound with the component at the time of simulation or synthesis. In this example, the components may already be coded, compiled and stored in a library earlier. Their VHDL descriptions are shown in Listing 2.4.

Listing 2.4 Predesigned component

```
  -- 2-input xor gate
  library ieee;
  use ieee.std_logic_1164.all;
  entity xor2 is
5     port(
          i1, i2: in std_logic;
          o1: out std_logic
      );
  end xor2;
10
  architecture beh_arch of xor2 is
  begin
      o1 <= i1 xor i2;
  end beh_arch;
15
  -- invertor
  library ieee;
  use ieee.std_logic_1164.all;
  entity not1 is
20    port(
```

```
        i1: in std_logic;
        o1: out std_logic
    );
  end not1;
25 architecture beh_arch of not1 is
  begin
        o1 <= not i1;
  end beh_arch;
```

Structural description and the use of components help the design in several ways. First, they facilitate hierarchical design. A complex system can be divided into several smaller subsystems, each represented by a component and designed individually. The subsystem, if needed, can be further divided into even smaller modules. Second, they provide a method to use predesigned circuits. These circuits, including complex IP cores and certain specialized library cells, can be instantiated in the description and treated as black boxes. Finally, structural description can be used to represent the result of synthesis: a gate- or cell-level netlist.

2.2.3 Abstract behavioral description

In a large design, the implementation can be very complex and the construction can be a time-consuming process. In the beginning, we frequently just want to study system operation rather than focusing on construction of the actual circuit, and prefer an abstract description. Since human reasoning and algorithms resemble a sequential process, the sequential semantics of traditional language is more adequate. VHDL provides language constructs that resemble the sequential semantics, including the use of variable and sequential execution. These features are considered as exceptions to the regular VHDL semantics, and they are encapsulated in a special construct, known as a *process*. This kind of code is sometimes referred to as *behavioral description*. However, there is no precise definition for the term behavioral description. According to VHDL, all codes, except for pure component instantiation, are considered as behavioral.

The basic skeleton of a process is

```
process(sensitivity_list)
    variable declaration;
begin
    sequential statements;
end process;
```

A process has a *sensitivity list*, which is composed of a set of signals. When a signal in the sensitivity list changes, the process is activated. Inside the process, the semantic is similar to that of a traditional programming language. Variables can be used and execution of the statements is sequential. The use of process is shown in two examples, both describing the even-parity detector circuit. The first example is based on the xor network, as in the xor_arch architecture. The architecture body is shown in Listing 2.5.

Listing 2.5   Even-parity detector based on a behavioral description

```
  architecture beh1_arch of even_detector is
  signal odd: std_logic;
  begin
      -- invertor
5     even <= not odd;
```

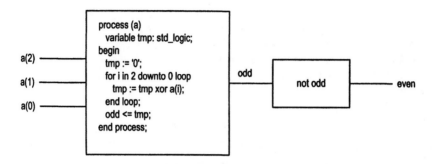

Figure 2.5 Conceptual diagram of the beh1_arch architecture.

```
      —  xor  network  for  odd  parity
     process(a)
         variable tmp: std_logic;
     begin
10       tmp := '0';
         for i in 2 downto 0 loop
             tmp := tmp xor a(i);
         end loop;
         odd <= tmp;
15     end process;
     end beh1_arch;
```

The xor network is described by a process that utilizes a variable and a for loop statement. Unlike signal and signal assignment in a concurrent statement, the variable and loop do not have direct hardware counterparts. We treat a process as one indivisible part whose behavior is specified by the sequential statements. The graphic interpretation of the beh1 architecture is shown in Figure 2.5.

The second example uses a single process to describe the desired operation in an algorithm. The algorithm first sums up the number of 1's from input, performs a modulo-2 operation to find the remainder, and then uses an if statement to check the value of the remainder to generate the final result. The VHDL code is shown in Listing 2.6.

Listing 2.6 Even-parity detector based on another behavioral description

```
     architecture beh2_arch of even_detector is
     begin
         process(a)
             variable sum, r: integer;
5        begin
             sum := 0;
             for i in 2 downto 0 loop
                 if a(i)='1' then
                     sum := sum +1;
10               end if;
             end loop ;
             r := sum mod 2;
             if (r=0) then
                 even <= '1';
15           else
```

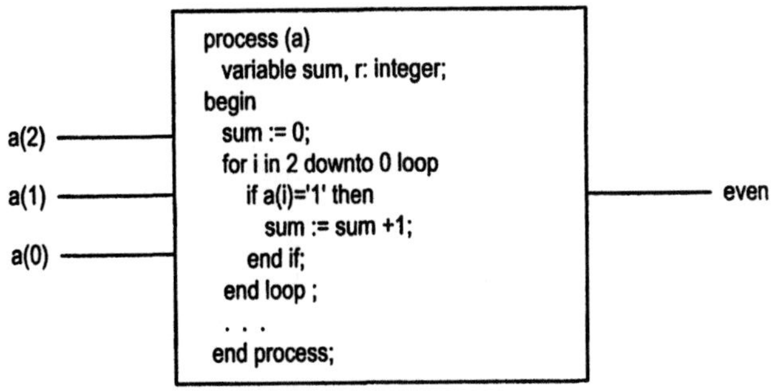

```
             process (a)
                variable sum, r: integer;
             begin
a(2) ───────    sum := 0;
                for i in 2 downto 0 loop
a(1) ───────       if a(i)='1' then                    ─────── even
                      sum := sum +1;
a(0) ───────       end if;
                end loop ;
                . . .
             end process;
```

Figure 2.6   Conceptual diagram of the beh2_arch architecture.

```
        even <='0';
    end if;
  end process;
end beh2_arch;
```

Since there is only one process, the graphic interpretation has only one part, as in Figure 2.6. While the code is very straightforward and easy to understand, it provides no clues about the underlying structure or how to realize the code in hardware.

2.2.4   Testbench

One major use of a VHDL program is simulation, which is used to study the operation of a circuit or to verify the correctness of a design. Performing simulation is similar to doing an experiment with a physical circuit, in which we connect the circuit's input to a stimulus (e.g., a function generator) and observe the output (e.g., by logic analyzer). Simulating a VHDL description is like doing a virtual experiment, in which the physical circuit is replaced by the corresponding VHDL description. Furthermore, we can develop VHDL utility routines to imitate the stimulus generator (which is known as a *test vector generator*) and to collect and compare the output responses. The framework is known as a *testbench*.

A simple VHDL testbench for the previous even detection circuit is shown in Listing 2.7. The testbench includes a test vector generator that generates a stimulus and a verifier that verifies the correctness of the output response. The testbench consists of an entity declaration and architecture body. Since the testbench is self-contained, no port is specified in the entity declaration. There are three concurrent statements in the architecture body, including one component instantiation and two processes. The component instantiation specifies that the even_detector is used and its I/O pins are connected to the internal test generator and verifier. The first process is the stimulus generator. It produces all possible test vector combinations, from "000" to "111". These vectors are generated in sequential order, each lasting for 200 ns. The second process is the verifier. It takes the input test vector, waits for 100 ns to let the output settle down, checks the output value with the known value and reports the results. The two processes are for demonstration purposes only, and we don't need to worry about the syntax detail.

Listing 2.7   Simple testbench

```
library ieee;
use ieee.std_logic_1164.all;
```

```
     entity even_detector_testbench is
   5 end even_detector_testbench;

     architecture tb_arch of even_detector_testbench is
        component even_detector
           port(
  10           a: in std_logic_vector(2 downto 0);
              even: out std_logic
           );
        end component;
        signal test_in: std_logic_vector(2 downto 0);
  15    signal test_out: std_logic;

     begin
        -- instantiate the circuit under test
        uut: even_detector
  20       port map( a=>test_in, even=>test_out);
        -- test vector generator
        process
        begin
           test_in <= "000";
  25       wait for 200 ns;
           test_in <= "001";
           wait for 200 ns;
           test_in <= "010";
           wait for 200 ns;
  30       test_in <= "011";
           wait for 200 ns;
           test_in <= "100";
           wait for 200 ns;
           test_in <= "101";
  35       wait for 200 ns;
           test_in <= "110";
           wait for 200 ns;
           test_in <= "111";
           wait for 200 ns;
  40    end process;
        --verifier
        process
           variable error_status: boolean;
        begin
  45       wait on test_in;
           wait for 100 ns;
           if ((test_in="000" and test_out = '1') or
              (test_in="001" and test_out = '0') or
              (test_in="010" and test_out = '0') or
  50          (test_in="011" and test_out = '1') or
              (test_in="100" and test_out = '0') or
              (test_in="101" and test_out = '1') or
              (test_in="110" and test_out = '1') or
              (test_in="111" and test_out = '0'))
  55       then
```

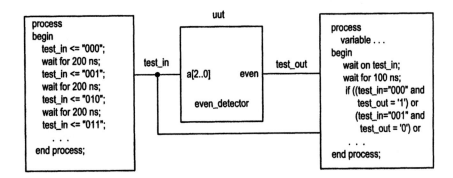

Figure 2.7 Conceptual diagram of an even_detector testbench.

```
              error_status := false;
         else
              error_status := true;
         end if;
 60           — error reporting
         assert not error_status
              report "test failed."
              severity note;
      end process;
 65 end tb_arch;
```

The graphic interpretation of this VHDL code is shown in Figure 2.7. Most of today's simulation software can keep track of the execution of a VHDL program and display the relevant information in a tabular or graphic format.

2.2.5 Configuration

The VHDL intentionally separates the entity declaration and architecture body into two independent design units. We can associate multiple architecture bodies with a single entity declaration. For example, the even_detector entity of this section has about a half dozen architecture bodies. At the time of simulation or synthesis, we can choose a specific architecture body to *bind* with the entity.

An analogy of the entity and architecture is the socket and IC chip. An entity declaration can be thought of as a socket of a printed circuit board, which is empty but has fixed input and output pins. Architecture bodies can be thought of as IC chips with the same outline. While the input and output pins of these chips are identical, their internal circuitry and performances may be very different. We can select a chip and insert it into the socket according to our particular need.

VHDL provides a mechanism, known as *configuration*, to specify the binding information. In the previous example, the even_detector entity has five different architecture bodies. The component declaration and component instantiation of test_bench does not specify which body is to be used. The test_bench is like a printed circuit board with an empty socket, and one of the five possible chips can be inserted into the circuit. A simple configuration declaration unit is shown in Listing 2.8, in which the sop_arch architecture is bound with the even_detector entity.

Listing 2.8   Simple configuration

```
configuration demo_config of even_detector_testbench is
    for tb_arch
        for uut: even_detector
            use entity work.even_detector(sop_arch);
 5      end for;
    end for;
end demo_config;
```

In a VHDL program, a configuration unit is not always needed. If there is no configuration unit, the entity is automatically bound with the last compiled architecture body. The configuration is particularly helpful for the development and verification of large systems.

2.3 VHDL IN DEVELOPMENT FLOW

The examples from the previous sections show the basic language constructs and capabilities of VHDL. The choice of these constructs is not accidental. They are carefully selected to facilitate system development. In the following subsections, we discuss the use of VHDL in the development flow and the difference between coding for modeling and coding for synthesis.

2.3.1 Scope of VHDL

The scope and coverage of VHDL in a simplified development flow is illustrated in Figure 2.8. The design of a complex system normally begins with an abstract high-level description, which describes the desired behavior of the system, and a testbench, which includes a set of test vectors to exercise various functions of the system. The description and testbench allow designers to study the system operation in detail, discover any misconception or inconsistency, clarify and finalize the specification, and eventually establish the desired I/O behavior for future verification. The beh2_arch architecture (in Listing 2.6) of even_detector and the corresponding test_bench (in Listing 2.7) resemble these kinds of codes. In a large system, the abstract description is normally not suitable for synthesis. It either leads to unnecessarily complex circuitry or cannot be synthesized at all.

Once the specification and behavior of a system are completely understood, a synthesis-oriented code can be developed. This code is normally an RT-level description and provides a "sketch" of the underlying hardware organization so that synthesis software can derive an efficient implementation. The xor_arch and beh1_arch architecture bodies in Listings 2.2 and 2.5 resemble this kind of description. A synthesis-oriented description needs to be verified first. By utilizing the VHDL configuration, we can bind the new architecture body to the entity and use the same testbench and the previously established test vectors. After comparing the simulation responses with the known results, we can easily determine whether the new description meets the specification.

Once verified, the synthesis-oriented description can be synthesized. The result is a gate-level netlist, represented by a structural VHDL description. The code will be similar to the str_arch architecture body Listing 2.3. In a large design, the description is normally too tedious for humans to comprehend. Instead, it is usually plugged into the testbench via a new configuration unit. The testbench will be simulated to verify the correctness of synthesis and to study the system timing. The netlist description can then be passed to placement and routing software for further processing. The placement and routing tool

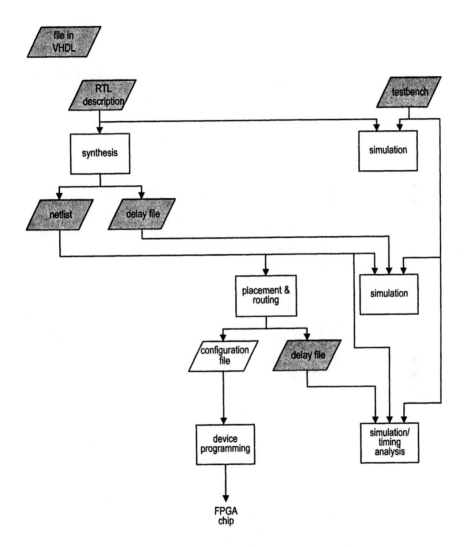

Figure 2.8 Coverage of VHDL in development flow.

will generate the layout or configuration files, which are not in VHDL. However, additional timing information will be augmented to the previous structural description. The new description will again be plugged into the testbench for final timing verification.

In summary, VHDL provides a unified environment for the entire development flow. It not only contains constructs to describe the design at various stages of the design, from the abstraction behavior to the post placement-and-routing cell-level netlist, but also provides a framework for simulation and verification.

2.3.2 Coding for synthesis

VHDL is used to model all aspects of digital hardware and to facilitate the entire design process. After a VHDL code is developed, it can be "executed" in a simulator or synthesizer. The natures of the two executions are quite different.

In simulation, the design is realized only in a virtual environment: the software simulator. The host computer utilizes its instruction set to mimic operation of the circuit. Since the host computer normally contains one processing unit, the circuit simulation is done sequentially, in which all constructs and operators of the VHDL code implicitly shared a single resource in a time-multiplexing fashion. In synthesis, on the other hand, all constructs and operators of the VHDL code are mapped to hardware. Let us consider a task that consists of 10 addition operations. In simulation, the number of addition operators, +, in VHDL code does not play a significant role since only one addition can be simulated at a time. In synthesis, each addition operator is mapped to a hardware adder, which is fairly complex, and thus it is desirable to share the hardware and to reduce the number of addition operators in VHDL description. Similarly, sophisticated control structures, such as loop or conditional branch, can be easily simulated in a sequential host but cannot be efficiently mapped to hardware.

For synthesis, only a subset of VHDL can be used. Many modeling language constructs, such as file operations and assertion statements, are not meaningful for hardware implementation. The others, such as floating-point number or complicated operators, are too complex to be synthesized automatically. IEEE defines a subset of VHDL that is suitable for RT-level synthesis in IEEE standard 1076.6. Even though the scope of the synthesizable subset is restricted, it still contains a rich collection of language constructs and is very flexible. The same circuit can be coded in a wide variety of descriptions, ranging from abstract high-level behavioral-like specification to detailed gate-level structural description. Although all these descriptions can be synthesized, there is no guarantee that the synthesized circuit is an efficient implementation. The synthesis software can perform only local search and local optimization, and the resulting circuit depends heavily on the initial description. An inadequate description consumes a large amount of CPU time during synthesis, introduces excessively complex circuitry and even fails to be synthesized.

This book focuses on RT-level design and synthesis, not VHDL. We are using VHDL as a vehicle to describe our intended hardware implementation. Our emphasis is on coding for synthesis, which means to develop VHDL code that accurately describes the underlying hardware structure and to provide adequate information to guide the synthesis software to generate an efficient implementation.

2.4 BIBLIOGRAPHIC NOTES

HDL is very different from a traditional programming language. The book, *Hardware Description Languages: Concepts and Principles* by S. Ghosh, discusses general issues in designing HDL. Both VHDL and Verilog are IEEE standards. They are documented by *IEEE Standard VHDL Language Reference Manual* and *IEEE Standard for Verilog Hardware Description Language* respectively. The other relevant VHDL standards are also documented in the IEEE publications. The standards themselves are difficult to read. The text, *The Designer's Guide to VHDL* by P. J. Ashenden, provides a detailed and comprehensive discussion of VHDL. The texts, *Starter's Guide to Verilog 2001* by M. D. Ciletti, and *Verilog HDL, 2nd edition*, by S. Palnitkar, provide good coverage on Verilog.

The verification of a design and the derivation of the testbench are two of the major tasks in the development flow. The text, *Writing Testbenches: Functional Verification of HDL Models, 2nd edition*, by J. Bergeron, discusses this topic in detail.

Problems

2.1 What are the syntax and semantics of a programming language?

2.2 List three major differences between an HDL and a traditional programming language, such as C.

2.3 In a traditional programming language, such as C, we can write the statement a=!a, and in VHDL, we can write a concurrent statement as a **<= not a after** 10 ns;.
   (a) Draw the circuit diagram for the VHDL statement.
   (b) Describe the operation of the circuit in part (a).
   (c) Discuss the differences between the VHDL and C statements.

2.4 For the even-parity detector circuit, rewrite the expression in product-of-sums format. Revise the code of the sop_arch architecture body according to the new expression.

2.5 For the VHDL code shown below, treat each concurrent statement as a circuit part and draw the conceptual block diagram accordingly.

```
y <= e1 and e0;
e0 <= (a0 and b0) or ((not a0) and (not b0));
e1 <= (a1 and b1) or ((not a1) and (not b1));
```

2.6 A circuit diagram consisting of the xor2 component is shown below. Follow the code of the str_arch architecture body to derive a structural VHDL description for this circuit.

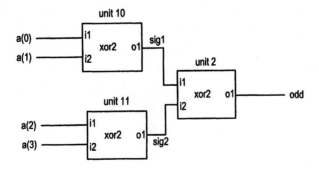

2.7   The VHDL structural description of a circuit is shown below. Derive the block diagram according to the code.

```
library ieee;
use ieee.std_logic_1164.all;
entity hundred_counter is
   port(
      clk, reset: in std_logic;
      en: in std_logic;
      q_ten, q_one: out std_logic_vector(3 downto 0);
      p_ten: out std_logic
   );
end hundred_counter;

architecture str_arch of hundred_counter is
   component dec_counter
      port(
         clk, reset: in std_logic;
         en: in std_logic;
         q: out std_logic_vector(3 downto 0);
         pulse: out std_logic
      );
   end component;
   signal p_one, p_ten: std_logic;
begin
   one_digit: dec_counter
      port map (clk=>clk, reset=>reset, en=>en,
                pulse=>p_one, q=>q_one);
   ten_digit: dec_counter
      port map (clk=>clk, reset=>reset, en=>p_one,
                pulse=>p_ten, q=>q_ten);
end str_arch;
```

2.8   From the description of the VHDL process in Section 2.2.3, discuss the differences between the VHDL process and the traditional programming languages' procedure and function.

2.9   We want to change the input of the even-parity detector circuit from 3 bits to 4 bits, i.e., from a(2 downto 0) to a(3 downto 0). Revise the VHDL codes of the five architecture bodies to accommodate the change.

2.10   If we want to change the input of the even-parity detector circuit from 3 bits to 10 bits, discuss the amount of code modifications needed in each architecture body.

2.11   Explain why VHDL treats the entity declaration and architecture body as two separate design units.

2.12   Think of two applications that can use the configuration construct of the VHDL.

CHAPTER 3

BASIC LANGUAGE CONSTRUCTS OF VHDL

To use a programming language, we first have to learn its syntax and language constructs. In this chapter, we illustrate the basic skeleton of a VHDL program and provide an overview of the basic language constructs, including lexical elements, objects, data types and operators. VHDL is a *strongly typed language* and imposes rigorous restriction on data types and operators. We discuss this aspect in more detail.

3.1 INTRODUCTION

VHDL is a complex language. It is designed to describe both the structural and behavioral views of a digital system at various levels of abstraction. Many of the language constructs are intended for modeling and for abstract, behavioral description. Only a small portion of VHDL can be synthesized and realized physically in hardware. The IEEE 1076.6 RTL synthesis standard tries to define a subset that can be accepted by most synthesis tools. The focus of this book is synthesis, and thus the discussion is limited primarily to this subset.

VHDL was revised twice by IEEE and there are three versions: VHDL-87, VHDL-93 and VHDL-2001. Since only simple, primitive language constructs can be synthesized, the revisions do not have a significant impact on synthesis except for some differences in the syntactical appearances. Since IEEE 1076.6 mainly follows the syntax of VHDL-87, we use the syntax of VHDL-87 in the book in general and highlight the difference if any VHDL-93 feature is used.

This chapter discusses only the basic, most commonly used language constructs in VHDL and some extensions defined in IEEE standards 1076.3 and 1164. In subsequent chapters, more specialized features are covered within the context.

3.2 SKELETON OF A BASIC VHDL PROGRAM

3.2.1 Example of a VHDL program

A VHDL program is composed of a collection of *design units*. A synthesizable VHDL program needs at least two design units: an entity declaration and an architecture body associated with the entity. The skeleton of a typical VHDL program can best be explained by an example. Let us consider the even_detector circuit of Chapter 2. The VHDL code is shown in Listing 3.1. It uses implicit δ-delays in signal assignment statements. Note that we use the boldface font for the VHDL's reserved words.

Listing 3.1   Even-parity detector

```
library ieee;
use ieee.std_logic_1164.all;
entity even_detector is
    port(
5       a: in std_logic_vector(2 downto 0);
        even: out std_logic
    );
end even_detector;

10 architecture sop_arch of even_detector is
    signal p1, p2, p3, p4 : std_logic;
    begin
        even <= (p1 or p2) or (p3 or p4);
        p1 <= (not a(0)) and (not a(1)) and (not a(2));
15      p2 <= (not a(0)) and a(1) and a(2);
        p3 <= a(0) and (not a(1)) and a(2);
        p4 <= a(0) and a(1) and (not a(2));
    end sop_arch ;
```

3.2.2 Entity declaration

The entity declaration describes the external interface, or "outline" of a circuit, including the name of the circuit and the names and basic characteristics of its input and output ports. In the example, the entity declaration indicates that the name of the circuit is even_detector and the circuit has a 3-bit input port, a, and a 1-bit output port, even.

The simplified syntax of an entity declaration is

```
entity entity_name is
    port(
        port_names: mode data_type;
        port_names: mode data_type;
        ...
        port_names: mode data_type
    );
end entity_name;
```

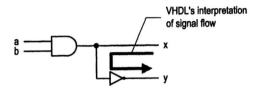

Figure 3.1 Demonstration circuit for mode.

Note that there is no semicolon (;) in the last port declaration.

A *port declaration* is composed of the port_names, mode and data_type terms. The port_names and data_type terms are self-explanatory. The mode term indicates the direction of the signal, which can be **in, out** or **inout**. The **in** and **out** keywords indicate that the signal flows "into" and "out of" the circuit respectively. They represent the fact that the corresponding port is an input or an output of the circuit. The **inout** keyword indicates that the signal flows in both directions and that the corresponding port is a bidirectional port. The mode term can also be **buffer**. It can cause a subtle compatibility problem and is not used in the book.

In the example, the port declaration shows that there are two ports. The a port is an input signal and its data type is std_logic_vector(2 **downto** 0), which represents a 3-bit bus, and the even port is an output port and its data type is std_logic.

Note that a port with the **out** mode cannot be used as an input signal. For example, consider the simple circuit shown in Figure 3.1. We may be tempted to use the following code to describe the circuit:

```
library ieee;
use ieee.std_logic_1164.all;
entity mode_demo is
    port(
        a, b: in std_logic;
        x, y: out std_logic
    );
end mode_demo;
architecture wrong_arch of mode_demo is
begin
    x <= a and b;
    y <= not x;
end wrong_arch;
```

Since the x signal is used to obtain the y signal, VHDL considers it as an external signal that "flows into" the circuit, as shown in Figure 3.1. This violates the **out** mode and leads to a syntax error. One way to fix the problem is to change the mode of the x port to the **inout** mode. It is a poor solution since the x port is not actually a bidirectional port. A better alternative is to use an internal signal to represent the intermediate result, as shown in the revised code:

```
architecture ok_arch of mode_demo is
    signal ab: std_logic;
begin
    ab <= a and b;
    x <= ab;
    y <= not ab;
end ok_arch;
```

3.2.3   Architecture body

The architecture body specifies the internal operation or organization of a circuit. In VHDL, we can develop multiple architecture bodies for the same entity declaration and later choose one body to bind with the entity for simulation or synthesis. The simplified syntax of an architecture body is

```
architecture arch_name of entity_name is
   declarations;
begin
   concurrent statement;
   concurrent statement;
   concurrent statement;
   . . .
end arch_name;
```

The first line of the architecture body shows the name of the body and the corresponding entity. An architecture body may include an optional declarative section, which consists of the declarations of some objects, such as signals and constants, which are used in the architecture description. The example includes a declaration of internal signals:

```
signal p1, p2, p3, p4: std_logic;
```

The main part of the architecture body consists of the concurrent statements that describe the operation or organization of the circuit. As we discussed in Chapter 2, each concurrent statement describes an individual part and the architecture can be thought of as a collection of interconnected circuit parts. There are a variety of concurrent statements, which are discussed in subsequent chapters.

3.2.4   Design unit and library

Design units are the fundamental building blocks in a VHDL program. When a program is processed, it is broken into individual design units and each unit is analyzed and stored independently. There are five kinds of design units:

- Entity declaration
- Architecture body
- Package declaration
- Package body
- Configuration

We have just studied the entity declaration and architecture body. A *package* of VHDL normally contains a collection of commonly used items, such as data types, subprograms and components, which are needed by many VHDL programs. As the name suggests, a *package declaration* consists of the declaration of these items. A *package body* normally contains the implementation and code of the subprograms.

In VHDL, multiple architecture bodies can be associated with an entity declaration. A *configuration* specifies which architecture body is to be bound with the entity declaration. The package and configuration are discussed in Chapter 13.

A VHDL *library* is a place to store the design units. It is normally mapped into a directory in the computer's hard disk storage. The software defines mapping between the symbolic VHDL library name and the physical directory. By VHDL default, the design units will be stored in a library named work.

To facilitate the synthesis, IEEE has developed several VHDL packages, including the `std_logic_1164` package and the `numeric_std` package, which are defined in IEEE standards 1164 and 1076.3. These packages are discussed in Sections 3.5.2 and 3.5.4. To use a predefined package, we must include the *library* and *use statements* before the entity declaration. The first two lines of the example are for this purpose:

```
library ieee;
use ieee.std_logic_1164.all;
```

The first line invokes a library named `ieee`, and the second line makes the `std_logic_1164` package visible to the subsequent design unit. We must invoke this library because we want to use some predefined data types, `std_logic` and `std_logic_vector`, of the `std_logic_1164` package.

3.2.5 Processing of VHDL code

A VHDL program is normally processed in three stages:

1. Analysis
2. Elaboration
3. Execution

During the *analysis stage*, the software checks the syntax and some static semantic errors of the VHDL code. The analysis is performed on a design unit basis. If there is no error, the software translates the code of the design unit into an intermediate form and stores it in the designated library. A VHDL file can contain multiple design units, but a design unit cannot be split into two or more files.

In a complex design, the system is normally described in a hierarchical manner. The top level may include subsystems as instantiated components, as in the example in Section 2.2.2. During the *elaboration stage*, the software starts from the designated top-level entity declaration and binds its architecture body according to the configuration specification. If there are instantiated components, the software replaces each instantiated component with the corresponding architecture body description. The process may repeat recursively until all instantiated components are replaced. The elaboration process essentially selects and combines the needed architectural descriptions, and creates a single "flattened" description.

During the *execution stage*, the analyzed and elaborated description is usually fed to simulation or synthesis software. The former simulates and "runs" the description in a computer, and the latter realizes the description by physical circuits.

3.3 LEXICAL ELEMENTS AND PROGRAM FORMAT

3.3.1 Lexical elements

The lexical elements are the basic syntactical units in a VHDL program. They include comments, identifiers, reserved words, numbers, characters and strings.

Comments A comment starts with two dashes, `--`, followed by the comment text. Anything after the `--` symbol in the line will be ignored. The comment is for documentation purposes only and has no effect on the code. For example, we have added comments to the previous VHDL code:

```
—*********************************************************
— example  to  show  the  caveat  of  the  out  mode
—*********************************************************
architecture ok_arch of mode_demo is
   signal ab: std_logic;  — ab is the internal signal
begin
   ab <= a and b;
   x <= ab;                — ab connected to the x output
   y <= not ab;
end ok_arch;
```

For clarity, we use *italic type* for comments.

Identifiers   An identifier is the name of an object in VHDL. The basic rules to form an identifier are:

- The identifier can contain only alphabetic letters, decimal digits and underscores.
- The first character must be a letter.
- The last character cannot be an underscore.
- Two successive underscores are not allowed.

For example, the following identifiers are valid:

```
A10, next_state, NextState, mem_addr_enable
```

On the other hand, the following identifiers violate one of the rules and will cause a syntax error during analysis of the program:

```
sig#3, _X10, 7segment, X10_, hi__there
```

Since VHDL is not case sensitive, the following identifiers are the same:

```
nextstate, NextState, NEXTSTATE, nEXTsTATE
```

It is good practice to be consistent with the use of case. In this book, we use capital letters for symbolic constants and use a special suffix, such as _n, to represent a special characteristics of an identifier. For example, the _n suffix is used to indicate an active-low signal. If we see a signal with a name like oe_n, we know it is an active-low signal.

It is also a good practice to use descriptive identifier for better readability. For example, consider the name for a signal that enables the memory address buffer. The mem_addr_en is good, mae is too short, and memory_address_enable is probably too cumbersome.

Reserved words   Some words are reserved in VHDL to form the basic language constructs. These reserved words are:

```
abs access after alias all and architecture array assert
attribute begin block body buffer bus case component
configuration constant disconnect downto else elsif end
entity exit file for function generate generic guarded
if impure in inertial inout is label library linkage
literal loop map mod nand new next nor not null of on
open or others out package port postponed procedure
process pure range record register reject rem report
return rol ror select severity shared signal sla sll
sra srl subtype then to transport type unaffected units
until use variable wait when while with xnor xor
```

Numbers, characters and strings A *number* in VHDL can be integer, such as 0, 1234 and 98E7, or real, such as 0.0, 1.23456 or 9.87E6. It can be represented in other number bases. For example, 45 can be represented as 2#101101# and 16#2D# in base 2 and base 16 respectively. We can also add an underscore to enhance readability. For example, 12_3456 is the same as 123456, and 2#0011_1010_1101# is the same as 2#001110101101#.

A *character* in VHDL is enclosed in single quotation marks, such as 'A', 'Z' and '3'. Note that 1 and '1' are different since the former is a number and the latter is a character.

A *string* is a sequence of characters enclosed in double quotation marks, such as "Hello" and "10000111". Again, note that that 2#10110010# and "10110010" are different since the former is a number and the latter is a string. Unlike the number, we cannot arbitrarily use an underscore inside a string. The "10110010" and "1011_0010" strings are different.

3.3.2 VHDL program format

VHDL is a case-insensitive free-format language, which means that the letter case does not matter, and "white space" (space, tab and new-line characters) can be inserted freely between lexical elements. For example, the VHDL program

```vhdl
library ieee;
use ieee.std_logic_1164.all;
entity even_detector is
   port(
      a: in std_logic_vector(2 downto 0);
      even: out std_logic
   );
end even_detector;

architecture eg_arch of even_detector is
   signal p1, p2, p3, p4 : std_logic;
begin
   even <= (p1 or p2) or (p3 or p4);
   p1 <= (not a(0)) and (not a(1)) and (not a(2));
   p2 <= (not a(0)) and a(1) and a(2);
   p3 <= a(0) and (not a(1)) and a(2);
   p4 <= a(0) and a(1) and (not a(2));
end eg_arch;
```

is the same as

```vhdl
library ieee; use ieee.std_logic_1164.all;entity
even_detector is port(a: in std_logic_vector(2
downto 0);even: out std_logic);end even_detector;
architecture eg_arch of even_detector is signal p1,
p2, p3, p4: std_logic; begin even <= (p1 or p2) or
(p3 or p4); p1 <= (not a(0)) and (not a(1)) and
(not a(2)); p2 <= (not a(0)) and a(1) and a(2);
p3 <= a(0) and (not a(1)) and a(2); p4 <= a(0) and
a(1) and (not a(2)); end eg_arch;
```

This extreme example demonstrates the importance of proper formatting. Although the program format does not affect the content or efficiency of a design, it has a significant impact on human users. An adequately documented and formatted program makes the code easier to comprehend and helps us to locate potential design errors. It will save a tremendous amount of time for future code revision and maintenance. This is perhaps the easiest way to enhance the reusability of the code. Section 3.6.2 lists the basic guidelines for code formatting and documentation.

It is a good idea to include a short "header" comment in the beginning of the file. The header should provide general information about the design and the "design environment." A representative header of the previous VHDL program is shown below.

```
--*********************************************************
---
-- Author : p chu
---
-- File : even_det . vhd
---
-- Design units :
--        entity even_detector
--            function : check even # of 1's from input
--            input : a
--            output : even
--        architecture sop_arch :
--            truth-table-based sum-of-products
--            implementation
---
-- Library / package :
--        ieee . std_logic_1164 : to use std_logic
---
-- Synthesis and verification :
--        Synthesis software : . . .
--        Options / script : . . .
--        Target technology : . . .
--        Testbench : even_detector_tb
---
-- Revision history
--        Version 1.0 :
--        Date : 9/2005
--        Comments : Original
---
--*********************************************************
```

The first two parts list the author and file name. The "Design units" part provides a brief description about the design units in the file. The description includes the input and output ports and the function of the entity, and the implementation method of the architecture body. The final "Revision history" part provides general information about the development. The "Library/package" and "Synthesis and verification" parts describe the design environment. The idea here is to provide the necessary information for users to reconstruct or duplicate the implementation. The "Library/package" part lists the packages and libraries that are referred to in the design file, and explains briefly the use of these packages. It is especially essential when a nonstandard or custom package is involved. The "Synthesis and verification" part lists the EDA software and the script or relevant options used in the synthesis, the original targeting device technology, as well as, if available, the testbench used to verify the design. Since synthesis software from different manufacturers supports different subsets of the VHDL and may interpret certain VHDL constructs differently, this information allows future users to duplicate the original implementation.

3.4 OBJECTS

An *object* in VHDL is a named item that holds the value of a specific data type. There are four kinds of objects: **signal, variable, constant** and **file**. A construct known as **alias** is somewhat like an object. We do not discuss the **file** object in this book since it cannot be synthesized.

Signals The signal is the most common object and we already used it in previous examples. A signal has to be declared in the architecture body's declaration section. The simplified syntax of signal declaration is

```
signal signal_name , signal_name , ... : data_type;
```

For example, the following line declares the a, b and c signals with the std_logic data type:

```
signal a, b, c: std_logic;
```

According to the VHDL definition, we can specify an optional initial value in the signal declaration. For example, we can assign an initial value of '0' to the previous signals:

```
signal a, b, c: std_logic := '0';
```

While this is sometimes handy for simulation purposes, it should not be used in synthesis since not many physical devices can implement the desired effect.

The simplified syntax of signal assignment is

```
signal_name <= projected_waveform;
```

We examined the concept of projected_waveform in Section 2.2.1 and discuss it in more detail in Chapter 4. From the synthesis point of view, a signal represents a wire or "a wire with memory" (i.e., a register or latch).

The input and output ports of the entity declaration are also considered as signals.

Variables A variable is a concept found in a traditional programming language. It can be thought of as a "symbolic memory location" where a value can be stored and modified. There is no direct mapping between a variable and a hardware part. A variable can only be declared and used in a process and is local to that process (the exception is a shared variable, which is difficult to use and is not discussed). The main application of a variable is to describe the abstract behavior of a system.

The syntax of variable declaration is similar to that of signal declaration:

```
variable variable_name , variable_name , ... : data_type
```

An optional initial value can be assigned to variables as well.

The simplified syntax of variable assignment is

```
variable_name := value_expression;
```

Note that no timing information is associated with a variable, and thus only a value, not a waveform, can be assigned to a variable. Since there is no delay, the assignment is known as an *immediate assignment* and the notion := is used. We examine variables in detail when the VHDL process is discussed in Chapter 5.

Constants A constant holds a value that cannot be changed. The syntax of constant declaration is

```
constant constant_name: data_type := value_expression;
```

The value_expression term specifies the value of the constant. A simple example is

```
constant BUS_WIDTH: integer := 32;
constant BUS_BYTES: integer := BUS_WIDTH / 8;
```

Note that we use capital letters for constants in this book.

Since an identifier name and data type convey more information than does a literal alone, the proper use of constants can greatly enhance readability of the VHDL code and make the code more descriptive. Consider the behavioral description of even_detector in Section 2.2.3:

```
architecture beh1_arch of even_detector is
    signal odd: std_logic;
begin
    . . .
    tmp := '0';
    for i in 2 downto 0 loop
        tmp := tmp xor a(i);
    end loop;
    . . .
```

The code uses a "hard literal," 2, to specify the upper boundary of the loop's range. It becomes much clearer if we replace it with a symbolic constant:

```
architecture beh1_arch of even_detector is
    signal odd: std_logic;
    constant BUS_WIDTH: integer := 3;
begin
    . . .
    tmp := '0';
    for i in (BUS_WIDTH-1) downto 0 loop
        tmp := tmp xor a(i);
    end loop;
    . . .
```

Alias Alias is not a data object. It is the alternative name for an existing object. As a constant, the purpose of an alias is to enhance code clarity and readability. One form of the signal alias is especially helpful for synthesis. Consider a machine instruction of a processor that is 16 bits wide and consists of fields with an operation code and three registers. The instruction is stored in memory as a 16-bit word. After it is read from memory, we can use an alias to identify the individual field:

```
signal word: std_logic_vector(15 downto 0);
alias op: std_logic_vector(6 downto 0) is word(15 downto 9);
alias reg1: std_logic_vector(2 downto 0) is word(8 downto 6);
alias reg2: std_logic_vector(2 downto 0) is word(5 downto 3);
alias reg3: std_logic_vector(2 downto 0) is word(2 downto 0);
```

Clearly, a name like reg1 is more descriptive than word(8 **downto** 6). Unfortunately, some synthesis software does not support this language construct. We can achieve this in a somewhat cumbersome way by declaring four new signals in the architecture body and assigning them with the proper portions of the word signal.

3.5   DATA TYPES AND OPERATORS

In VHDL, each object has a *data type*. A data type is defined by:

- A set of values that an object can assume.
- A set of operations that can be performed on objects of this data type.

VHDL is a known as *strongly typed language*, which means that an object can only be assigned a value of its type, and only the operations defined with the data type can be performed on the object. If a value of a different type has to be assigned to an object, the value must be converted to the proper data type by a *type conversion function* or *type casting*.

The motivation behind a strongly typed language is to catch errors in the early stage. For example, if a Boolean value is assigned to a signal of integer type or an arithmetic operation is applied to a signal of character type, the software can detect the error during the analysis stage. On the downside, the rigid type requirement may introduce many type-conversion functions and make the code cumbersome and difficult to understand.

To facilitate modeling and simulation, VHDL is rich in data types. In theory, any data type with a finite number of values can be mapped into a set of binary representations and thus can be realized in hardware. However, we refrain from doing this since the mapping introduces another dimension of uncertainty in synthesis and may lead to compatibility problems in larger designs. Our focus is on a small set of predefined data types that are relevant to synthesis. For a signal, we are mainly confined to the std_logic, std_logic_vector, signed and unsigned data types. A few user-defined data types will be used for specific applications and they will be discussed as needed.

The following subsections examine the relevant data types, operators and type conversions in VHDL and two synthesis-related IEEE packages.

3.5.1   Predefined data types in VHDL

Commonly used data types   There are about a dozen predefined data types in VHDL. Only the following data types are relevant to synthesis:

- integer: VHDL does not define the exact range of the integer type but specifies that the minimal range is from $-(2^{31} - 1)$ to $2^{31} - 1$, which corresponds to 32 bits. Two related data types (formally known as *subtypes*) are the natural and positive data types. The former includes 0 and the positive numbers and the latter includes only the positive numbers.
- boolean: defined as (false, true).
- bit: defined as ('0', '1').
- bit_vector: defined as a one-dimensional array with elements of the bit data type.

The original intention of the bit data type is to represent the two binary values used in Boolean algebra and digital logic. However, in a real design, a signal may assume other values, such as the high impedance of a tri-state buffer's output or a "fighting" value because of a conflict (e.g., two outputs are wired together, forming a short circuit). To solve the problem, a set of more versatile data types, std_logic and std_logic_vector, are introduced in the IEEE std_logic_1164 package. To achieve better compatibility, we should avoid using the bit and bit_vector data types. The std_logic and std_logic_vector data types are discussed in Section 3.5.2.

In VHDL, data types similar to the boolean and bit types are known as the *enumeration* data types since their values are enumerated in a list.

Table 3.1   Operators and applicable data types of VHDL-93

Operator	Description	Data type of operand a	Data type of operand b	Data type of result
a ** b	exponentiation	`integer`	`integer`	`integer`
abs a	absolute value	`integer`		`integer`
not a	negation	`boolean, bit, bit_vector`		`boolean, bit, bit_vector`
a * b	multiplication	`integer`	`integer`	`integer`
a / b	division			
a **mod** b	modulo			
a **rem** b	remainder			
+ a	identity	`integer`		`integer`
− a	negation			
a + b	addition	`integer`	`integer`	`integer`
a − b	subtraction			
a & b	concatenation	1-D array, element	1-D array, element	1-D array
a **sll** b	shift-left logical	`bit_vector`	`integer`	`bit_vector`
a **srl** b	shift-right logical			
a **sla** b	shift-left arithmetic			
a **srl** b	shift-right arithmetic			
a **rol** b	rotate left			
a **ror** b	rotate right			
a = b	equal to	any	same as a	`boolean`
a /= b	not equal to			
a < b	less than	scalar or 1-D array	same as a	`boolean`
a <= b	less than or equal to			
a > b	greater than			
a >= b	greater than or equal to			
a **and** b	and	`boolean, bit, bit_vector`	same as a	same as a
a **or** b	or			
a **xor** b	xor			
a **nand** b	nand			
a **nor** b	nor			
a **xnor** b	xnor			

Operators   About 30 operators are defined in VHDL. In a strongly typed language, the definition of data type includes the operations that can be performed on the object of this data type. It is important to know which data types can be used with a particular operator.

Descriptions of these operators and the applicable data types are summarized in Table 3.1. Only synthesis-related data types are listed. Most operators and data types are self-explanatory. Relational operators and the concatenation operator (&) can be applied to arrays and are discussed in Section 3.5.2.

Note that the operators in the table are defined in VHDL-93. The shift operators and the **xnor** operator are not defined in VHDL-87 and are not supported by IEEE 1076.6 RTL synthesis standard either.

Table 3.2 Precedence of the VHDL operators

Precedence	Operators
Highest	**** abs not**
	* / mod rem
	+ − (identity and negation)
	& + − (addition and subtraction)
	sll srl sla sra rol ror
	= /= < <= > >=
Lowest	**and or nand nor xor xnor**

During synthesis, operators in the VHDL code will be realized by physical components. Their hardware complexities vary significantly, and many operators, such as multiplication and division, cannot be synthesized automatically. This issue is discussed in Chapter 6.

The *precedence* of the operators is shown in Table 3.2, which is divided into seven groups. The operators in the same group have the same precedence. The operators in the upper group have higher precedence over the operators in the lower group. For example, consider the expression

```
a + b > c or a < d
```

The + operator will be evaluated first, and then the > and < operators, and then the **or** operator.

If an expression consists of several identical operators, evaluation begins at the leftmost operator and progresses toward the right (known as *left-associative*). For example, consider the expression

```
a + b + c + d
```

The a + b expression will be evaluated first, and then c is added, and then d is added.

Parentheses can be used in an expression. They have the highest preference and thus can alter the order of evaluation. For example, we can use parentheses to make the previous expression be evaluated from right to left:

```
a + (b + (c + d))
```

Unlike the logic expression used in Boolean algebra, the **and** and **or** operators have the same precedence in VHDL, and thus we must use parentheses to specify the desired order, as in

```
(a and b) or (c and d)
```

It is a good practice to use parentheses to make the code clear and readable, even when they are not needed. For example, the expression

```
a + b > c or a < d
```

can be written as

```
((a + b) > c) or (a < d)
```

This is more descriptive and reduces the chance for error or misinterpretation.

Table 3.3   VHDL operators versus conventional Boolean algebra notations

VHDL operator	Boolean algebra notation
not	$'$
and	$\cdot$
or	$+$
xor	$\oplus$
+	$+$

Notation   To make an expression compact, we use the conventional symbols $'$, $\cdot$ and $+$ of Boolean algebra for *not*, *and* and *or* operations in our discussion. They are expressed as **not**, **and** and **or** in VHDL code. We also assume that $\cdot$ has precedence over $+$. For example, in our discussion, we may write

$$y = a \cdot b + a' \cdot b'$$

When coded in VHDL, This expression becomes

```
y <= (a and b) or ((not a) and (not b));
```

The notations used in our discussion and VHDL code are summarized in Table 3.3. Note that the $+$ notation is used as both or and addition operations in our discussion. Since they are used in different contexts, it should not introduce confusion.

3.5.2   Data types in the IEEE std_logic_1164 package

To better reflect the electrical property of digital hardware, several new data types were developed by IEEE to serve as an extension to the `bit` and `bit_vector` data types. Theses data types are defined in the `std_logic_1164` package of IEEE standard 1164. In this subsection, we discuss the new data types, the operations defined over these data types and the conversion between these data types and the predefined VHDL data types.

std_logic and std_logic_vector data types   The two most useful data types defined in the `std_logic_1164` package are `std_logic` and `std_logic_vector`. Formally speaking, the `std_logic` data type is actually a subtype of the `std_ulogic` data type. Since the `std_ulogic` data type is "unresolved," it has some limitations and will not be used in this book.

To use the new data types, we must include the necessary library and use statements before the entity declaration:

```
library ieee;
use ieee.std_logic_1164.all;
```

The `std_logic` data type consists of nine possible values, which are shown in the following list:

```
('U', 'X', '0', '1', 'Z', 'W', 'L', 'H', '-')
```

These values are interpreted as follows:

- '0' and '1': stand for "forcing logic 0" and "forcing logic 1," which mean that the signal is driven by a circuit with a regular driving current.

- 'Z': stands for high impedance, which is usually encountered in a tri-state buffer.
- 'L' and 'H': stand for "weak logic 0" and "weak logic 1," which means that the signal is obtained from wired-logic types of circuits, in which the driving current is weak.
- 'X' and 'W': stand for "unknown" and "weak unknown." The unknown represents that a signal reaches an intermediate voltage value that can be interpreted as neither logic 0 or logic 1. This may happen because of a conflict in output (such as a logic-0 signal and a logic-1 signal being tied together). They are used in simulation for an erroneous condition.
- 'U': stands for uninitialized. It is used in simulation to indicate that a signal or variable has not yet been assigned a value.
- '-': stands for don't-care.

Among these values, only '0', '1' and 'Z' are used in synthesis. The 'L' and 'H' values are seldom used now since current design practice rarely utilizes a wired-logic circuit. The use of 'Z' and the potential problem of '-' are discussed in Chapter 6.

A VHDL array is defined as a collection of elements with the same data type. Each element in the array is identified by an index. The std_logic_vector data type is an array of elements with the std_logic data type. It can be thought of as a group of signals or a bus in a logic circuit.

The use of std_logic_vector can best be explained by a simple example. Let us consider an 8-bit signal, a. Its declaration is

```
signal a: std_logic_vector(7 downto 0);
```

It indicates that the a signal has 8 bits, which are indexed from 7 down to 0. The most significant bit (MSB, the leftmost bit) has the index 7, and the least significant bit (LSB, the rightmost bit) has the index 0. We can access a single bit by using an index, such as a(7) or a(2), and access a portion of the index by using a range, such as a(7 **downto** 3) or a(2 **downto** 0).

Another form of std_logic_vector is using an ascending range, as in

```
signal a: std_logic_vector(0 to 7);
```

Since its MSB is associated with index 0 and may cause some confusion if the array is interpreted as a binary number, we don't use this form in the book.

Overloaded operators   Recall that the definition of a data type includes a set of values and a set of operations to be performed on this data type. In VHDL, we can use the same function or operator name for operands of different data types. There may exist multiple functions with the same name, each for a different data type. This is known as *overloading* of a function or operator.

In the std_logic_1164 package, all logical operators, which include **not, and, nand, or, nor, xor** and **xnor**, are overloaded with the std_logic and std_logic_vector data types. In other words, we can perform the logical operations over the objects with the std_logic or std_logic_vector data types. The overloaded operators are summarized in Table 3.4. Note that the arithmetic operators are not overloaded, and thus these operations cannot be applied.

Type conversion   The std_logic_1164 package also defines several type conversion functions for conversion between the bit and std_logic data types as well as between the bit_vector and std_logic_vector data types. The relevant functions are summarized in Table 3.5.

Table 3.4   Overloaded operators in the IEEE `std_logic_1164` package

Overloaded operator	Data type of operand a	Data type of operand b	Data type of result
not a	std_logic_vector std_logic		same as a
a **and** b a **or** b a **xor** b a **nand** b a **nor** b a **xnor** b	std_logic_vector std_logic	same as a	same as a

Table 3.5   Functions in the IEEE `std_logic_1164` package

Function	Data type of operand a	Data type of result
to_bit(a)	std_logic	bit
to_stdulogic(a)	bit	std_logic
to_bitvector(a)	std_logic_vector	bit_vector
to_stdlogicvector(a)	bit_vector	std_logic_vector

Use of the conversion function is shown below. Assume that the s1, s2, s3, b1 and b2 signals are defined as

```
signal s1, s2, s3: std_logic_vector(7 downto 0);
signal b1, b2: bit_vector(7 downto 0);
```

The following statements are wrong because of data type mismatch:

```
s1 <= b1;            -- bit_vector assigned to std_logic_vector
b2 <= s1 and s2;     -- std_logic_vector assigned to bit_vector
s3 <= b1 or s2;      -- or is undefined between bit_vector
                     -- and std_logic_vector
```

We can use the conversion functions to correct these problems:

```
s1 <= to_stdlogicvector(b1);
b2 <= to_bitvector(s1 and s2);
s3 <= to_stdlogicvector(b1) or s2;
```

The last statement can also be written as

```
s3 <= to_stdlogicvector(b1 or to_bitvector(s2));
```

3.5.3   Operators over an array data type

Several operations are defined over the one-dimensional array data types in VHDL, including the concatenation and relational operators and the array aggregate. In this subsection, we demonstrate the use of these operators with the std_logic_vector data type. Note that these operators can be applied in any array data types, and thus no overloading is needed.

Relational operators for an array   In VHDL, the relational operators can be applied to the one-dimensional array data type. The two operands must have the same element type, but their lengths may differ. When an operator is applied, the two arrays are compared element by element. The comparison procedure starts from the leftmost element and continues until a result can be established. If one array reaches the end before another, that array is considered to be "smaller" and the two arrays are considered to be not equal. For example, all following operations return `true`:

```
"011"="011",  "011">"010",  "011">"00010",  "0110">"011"
```

Arrays with unequal lengths can sometimes introduce subtle, unexpected results. For example, assume that the `sig1` and `sig2` signals are with an array data type of different lengths and we accidentally write

```
if (sig1=sig2) then
    . . .
else
    . . .
```

Because of the different lengths, the comparison expression is always evaluated as false, and thus the then branch will never be taken. This kind of error is difficult to debug since the code is syntactically correct. In this book, we always use operands of identical length.

Concatenation operator   The concatenation operator, `&`, is very useful for array manipulation. We can combine segments of elements and smaller arrays to form a larger array. For example, we can shift the elements of the array to the right by two positions and append two 0's to the front:

```
y <= "00" & a(7 downto 2);
```

or append the MSB to the front (known as an arithmetic shift):

```
y <= a(7) & a(7) & a(7 downto 2);
```

or rotate the elements to the right by two positions:

```
y <= a(1 downto 0) & a(7 downto 2);
```

Array aggregate   *Array aggregate* is not an operator. It is a VHDL language construct to assign a value to an object of array data type. For the `std_logic_vector` data type, the simplest way to express an aggregate is to use a collection of `std_logic` values inside double quotation marks. For example, if we want to assign a value of `"10100000"` to the a signal, it can be written as

```
a <= "10100000";
```

Another way is to list each value of the element in the corresponding position, which is known as *positional association*. The previous assignment becomes

```
a <= ('1','0','1','0','0','0','0','0');
```

We can also use the form of `index => value` to explicitly specify the value for each index, known as *named association*. The statement can be written as

```
a <= (7=>'1', 6=>'0', 0=>'0', 1=>'0', 5=>'1',
      4=>'0', 3=>'0', 2=>'0');
```

It means that the value associated with index 7 (i.e., a(7)) is '1', the value associated with index 6 is '0', and so on. Note that the order of the index => value pairs does not matter. We can combine the index, as in

```
a <= (7|5=>'1', 6|4|3|2|1|0=>'0');
```

or use a reserved word, **others**, to cover all the unused indexes, as in

```
a <= (7|5=>'1', others=>'0');
```

One frequently encountered array aggregate is all 0's, which is used in the initialization of a counter or a memory element. For example, if we want to assign "00000000" to the a signal, we can write

```
a <= (others=>'0');
```

It is more compact than

```
a <= "00000000";
```

The code remains the same even when the width of the a signal is later revised.

3.5.4   Data types in the IEEE numeric_std package

In addition to logical operations, digital hardware frequently involves arithmetic operation as well. If we examine VHDL and the std_logic_1164 package, the arithmetic operations are defined only over the integer data type. To perform addition of the a and b signals, we must use the integer data type, as in

```
signal a, b, sum: integer;
 . . .
sum <= a + b;
```

It is difficult to realize this statement in hardware since the code doesn't indicate the range (number of bits) of the a and b signals. Although this does not matter for simulation, it is important for synthesis since there is a huge difference between the hardware complexity of an 8-bit adder and that of a 32-bit adder.

A better alternative is to use an array of 0's and 1's and interpret it as an unsigned or signed number. We can define the width of the input and the size of the adder precisely, and thus have better control over the underlying hardware. The IEEE numeric_std package was developed for this purpose.

Signed and unsigned data types   The IEEE numeric_std package is a part of IEEE standard 1176.3. Two new data types, signed and unsigned, are defined in the package. Both data types are an array of elements with the std_logic data type. For the unsigned data type, the array is interpreted as an unsigned binary number, with the leftmost element as the MSB of the binary number. For the signed data type, the array is interpreted as a signed binary number in 2's-complement format. The leftmost element is the MSB of the binary number, which represents the sign of the number.

Note that the std_logic_vector, unsigned and signed data types are all defined as an array of elements with the std_logic data type. Since VHDL is a strongly typed language, they are considered as three independent data types. It is reasonable since the three data types are interpreted differently. For example, consider a 4-bit binary representation "1100". It represents the number 12 if it is interpreted as an unsigned number and represents the number

−4 if it is interpreted as a signed number. It may also just represent four independent bits (e.g., four status signals) if it is interpreted as a collection of bits.

Since the `signed` and `unsigned` data types are arrays, their declarations are similar to that of the `std_logic_vector` data type, as in

```
signal x, y: signed(15 downto 0);
```

To use the `signed` and `unsigned` data types, we must include the library statement before the entity declaration. Furthermore, we must include the `std_logic_1164` package since the `std_logic` data type is used in the `numeric_std` package. These statements are

```
library ieee;
use ieee.std_logic_1164.all;
use ieee.numeric_std.all;
```

Overloaded operators   Since the goal of the `numeric_std` package is to support the arithmetic operations, the relevant arithmetic operators, which include **abs**, *****, **/**, **mod**, **rem**, **+** and **−**, are overloaded. These operators can now take two operands, with data types `unsigned` and `unsigned`, `unsigned` and `natural`, `signed` and `signed` as well as `signed` and `integer`. For example, the following are valid assignment statements:

```
signal a, b, c, d: unsigned(7 downto 0);
. . .
a <= b + c;
d <= b + 1;
e <= (5 + a + b) - c;
```

The overloading definition of addition and subtraction follows the model of a physical adder. The sum automatically "wraps around" when overflow occurs.

The relational operators, which include **=**, **/=**, **<**, **>**, **<=** and **>=**, are also overloaded. The overloading serves two purposes. First, it makes the operator take two operands with data types `unsigned` and `natural` as well as `signed` and `integer`. Second, for two operands with the `unsigned` or `signed` data types, the overloading overrides the original left-to-right element-by-element comparison procedure and treats the two arrays as two binary numbers. For example, consider the expression `"011" > "1000"`. If the data type of the two operands is `std_logic_vector`, the expression returns `false` because the first element of `"011"` is smaller than the first element of `"1000"`. If the data type of the two operands is `unsigned`, the `>` operator is overloaded and the two operands are interpreted as 3 and 8 respectively. The expression returns `false` again. However, if the data type is `signed`, they are interpreted as 3 and −8, and thus the expression returns `true`.

A summary of the overloaded operators is given in Table 3.6.

Functions   The `numeric_std` package defines several new functions. The new functions include:

- `shift_left`, `shift_right`, `rotate_left`, `rotate_right`: used for shifting and rotating operations. Note that these are new functions, not the overloaded VHDL operators.
- `resize`: used to convert an array to different sizes.
- `std_match`: used to compare objects with the `'-'` value.
- `to_unsigned`, `to_signed`, `to_integer`: used to do type conversion between the two new data types and the `integer` data type.

Table 3.6 Overloaded operators in the IEEE numeric_std package

Overloaded operator	Description	Data type of operand a	Data type of operand b	Data type of result
abs a - a	absolute value negation	signed		signed
a * b a / b a mod b a rem b a + b a - b	arithmetic operation	unsigned unsigned, natural signed signed, integer	unsigned, natural unsigned signed, integer signed	unsigned unsigned signed signed
a = b a /= b a < b a <= b a > b a >= b	relational operation	unsigned unsigned, natural signed signed, integer	unsigned, natural unsigned signed, integer signed	boolean boolean boolean boolean

Table 3.7 Functions in the IEEE numeric_std package

Function	Description	Data type of operand a	Data type of operand b	Data type of result
shift_left(a,b) shift_right(a,b) rotate_left(a,b) rotate_right(a,b)	shift left shift right rotate left rotate right	unsigned, signed	natural	same as a
resize(a,b) std_match(a,b)	resize array compare '-'	unsigned, signed unsigned, signed std_logic_vector, std_logic	natural same as a	same as a boolean
to_integer(a) to_unsigned(a,b) to_signed(a,b)	data type conversion	unsigned, signed natural integer	natural natural	integer unsigned signed

The functions are summarized in Table 3.7. The shift functions are similar to the VHDL shift operators but with different data types. Note that the IEEE 1076.6 RTL synthesis standard supports the shift functions of the numeric_std package but not the shift operators of VHDL. The synthesis issues of the shift functions and the use of the std_match function are discussed in Chapter 6.

Type conversion Conversion between two different data types can be done by a *type conversion function* or *type casting*. There are three type conversion functions in the numeric_std package: to_unsigned, to_signed, and to_integer. The to_intger function takes an object with an unsigned or signed data type and converts it to the integer data type. The to_unsigned and to_signed functions convert an integer into an object with the unsigned or signed data type of a specific number of bits. It takes

Table 3.8 Type conversions of numeric data types

Data type of a	To data type	Conversion function/type casting
unsigned, signed	std_logic_vector	std_logic_vector(a)
signed, std_logic_vector	unsigned	unsigned(a)
unsigned, std_logic_vector	signed	signed(a)
unsigned, signed	integer	to_integer(a)
natural	unsigned	to_unsigned(a, size)
integer	signed	to_signed(a, size)

two parameters. The first is the integer number to be converted, and the other specifies the desired number of bits (or size) in the new unsigned or signed data type.

The std_logic_vector, unsigned and signed data types are all defined as an array with elements of the std_logic data type. They are known as *closely related data types* in VHDL. Conversion between these types is done by a procedure known as *type casting*. To do type casting, we simply put the original object inside parentheses prefixed by the new data type. This can best be explained by an example:

```
signal u1, u2: unsigned(7 downto 0);
signal v1, v2: std_logic_vector(7 downto 0);
. . .
u1 <= unsigned(v1)
v2 <= std_logic_vector(u2);
```

Table 3.8 summarizes all the type conversions in the numeric_std package. Note that the std_logic_vector data type is not interpreted as a number and thus cannot be directly converted to an integer and vice versa.

Type conversion between various numeric data types is frequently confusing to new VHDL users. The following examples of signal assignment statements demonstrate and clarify the use of these data types and data conversions. Assume that some signals are declared as follows:

```
library ieee;
use ieee.std_logic_1164.all;
use ieee.numeric_std.all;
. . .
signal s1, s2, s3, s4, s5, s6: std_logic_vector(3 downto 0);
signal u1, u2, u3, u4, u5, u6, u7: unsigned(3 downto 0);
signal sg: signed(3 downto 0);
. . .
```

The following assignments to the signals u3 and u4 are valid since the + operator is overloaded with the unsigned and natural types:

```
u3 <= u2 + u1;   — ok, both operands unsigned
u4 <= u2 + 1;    — ok, operands unsigned and natural
```

On the other hand, the following two assignments are invalid due to type mismatch:

```
u5 <= sg;   — not ok, type mismatch
u6 <= 5;    — not ok, type mismatch
```

We must use type casting and the conversion function to covert the expressions to the proper type:

```
u5 <= unsigned (sg);      -- ok, type casting
u6 <= to_unsigned (5,4);  -- ok, conversion function
```

The arithmetic operators are not overloaded with the mixed data types signed and unsigned, and thus the following statement is invalid:

```
u7 <= sg + u1;    -- not ok, + undefined over the types
```

We must convert the data type of the operand as follows:

```
u7 <= unsigned (sg) + u1;  -- ok, but be careful
```

We need to be aware of the different interpretations of the signed and unsigned types. For example, "1111" is -1 for the signed type but is 15 for the unsigned type. This kind conversion should proceed with care.

Two assignments for signals with std_logic_vector data type are

```
s3 <= u3;   -- not ok, type mismatch
s4 <= 5;    -- not ok, type mismatch
```

Both of them are invalid because of type mismatch. We must use type casting and a conversion function to correct the problem:

```
s3 <= std_logic_vector (u3);   -- ok, type casting
s4 <= std_logic_vector (to_unsigned (5,4));  -- ok
```

Note that two type conversions are needed for the second statement.

Arithmetic operations cannot be applied to the std_logic_vector data type since no overloading is defined for this type. Thus, the following statements are invalid:

```
s5 <= s2 + s1;  -- not ok, + undefined over the types
s6 <= s2 + 1;   -- not ok, + undefined over the types
```

To fix the problem, we must convert the operands to the unsigned (or signed) data type, perform addition, and then convert the result back to the std_logic_vector data type. The revised code becomes

```
s5 <= std_logic_vector (unsigned (s2) + unsigned (s1));  -- ok
s6 <= std_logic_vector (unsigned (s2) + 1);              -- ok
```

3.5.5 The std_logic_arith and related packages

For historical reasons, several packages similar to the IEEE numeric_std package are used in some EDA software and existing VHDL codes. The packages are:

- std_logic_arith
- std_logic_unsigned
- std_logic_signed

They are not a part of the IEEE standards, but many software vendors store these packages in the ieee library. They can be invoked by

```
library ieee;
use ieee.std_logic_1164.all;
use ieee.std_logic_arith.all;
```

Because of the use of the ieee term, these packages sometimes cause confusion. They are not used them in this book. For reference, we explain briefly the use of these packages in this subsection.

The purpose of the std_logic_arith package is similar to that of the numeric_std package. It defines two new data types, unsigned and signed, and overloads the +, – and * operators with these data types. The package also includes similar shifting, sizing and type conversion functions although the names of these functions are different.

Instead of defining new data types, the std_logic_unsigned and std_logic_signed packages define overloaded arithmetic operators for the std_logic_vector data type. In other words, the std_logic_vector data type is interpreted as unsigned and signed binary numbers in the std_logic_unsigned and std_logic_signed packages respectively. The two packages clearly cannot be used at the same time.

With one of the packages, the previous code segments becomes valid and no type conversion is needed:

```
library ieee;
use ieee.std_logic_1164.all;
use ieee.std_logic_arith.all;
use ieee.std_logic_unsigned.all;
. . .
signal s1, s2, s3, s4, s5, s6: std_logic_vector(3 downto 0);
. . .
s5 <= s2 + s1;  -- ok, + overloaded with std_logic_vector
s6 <= s2 + 1;   -- ok, + overloaded with std_logic_vector
```

The overloading means that we can treat the std_logic_vector data type as "a collection of bits" as well as an unsigned binary number. This package actually beats the motivation behind a strongly typed language. The IEEE 1076.6 RTL synthesis standard states explicitly that the unsigned and signed data types defined in IEEE 1076.3 are the only array types that can be used to represent unsigned and signed numbers.

3.6 SYNTHESIS GUIDELINES

In this and subsequent chapters, we summarize the good design and coding practices mentioned in the chapter and present them as a set of guidelines at the end of the chapter. Since the book focuses on synthesis, these guidelines are applied only to synthesis, not to general modeling or simulation. These suggested guidelines help us to avoid some common mistakes and to increase the compatibility, portability and efficiency of VHDL codes.

3.6.1 Guidelines for general VHDL

- Use the std_logic_vector and std_logic data types instead of the bit_vector or bit data types.

- Use the numeric_std package and the unsigned and signed data types for synthesizing arithmetic operations.

- Use only the descending range (i.e., **downto**) in the array specification of the unsigned, signed and std_logic_vector data types.

- Use parentheses to clarify the intended order of evaluation.

- Don't use user-defined data types unless there is a compelling reason.

- Don't use immediate assignment (i.e., :=) to assign an initial value to a signal.

- Use operands with identical lengths for the relational operators.

3.6.2 Guidelines for VHDL formatting

- Include an information header for each file.

- Be consistent with the use of case.

- Use proper spaces, blank lines and indentations to make the code clear.

- Add necessary comments.

- Use symbolic constant names to replace hard literals in VHDL code.

- Use meaningful names for the identifiers.

- Use a suffix to indicate a signal's special property, such as _n for the active-low signal.

- Keep the line width within 72 characters so that the code can be displayed and printed properly by various editors and printers without wrapping.

3.7 BIBLIOGRAPHIC NOTES

VHDL is a complex language. It is formally specified by IEEE standard 1076. The most recent version, VHDL-2001, is specified by IEEE standard 1076-2001, and VHDL-87 is specified by IEEE standard 1076-1987. The standard is documented in *IEEE Standard VHDL Language Reference Manual*, which sometimes known simply as *LRM*. Since *LRM* gives the formal definition of VHDL, it is difficult to read. The book, *The Designer's Guide to VHDL, 2nd edition*, by P. J. Ashenden, provides a detailed and comprehensive discussion of the VHDL language. It has several chapters on basic VHDL concepts, data types and alias. The book, *VHDL for Logic Synthesis* by A. Rushton, has a chapter on numeric_std package and provides a detailed discussion on functions.

After synthesis software is installed, we can normally find the files that contain the source codes of IEEE std_logic_1164 and numeric_std packages as well as std_logic_arith, std_logic_unsigned and std_logic_signed packages. These packages provide detailed information about operator overloading and function definitions.

Although formatting is not real design, good coding style and documentation are essential for a project, especially for a large project that involves many design teams. Many organizations set and enforce their own coding and documentation standards. An example is *VHDL Modeling Guideline* from the European Space Agency.

The text, *Reuse Methodology Manual* by M. Keating and P. Bricaud, also provides some rules and guidelines for the use and formatting of VHDL.

Problems

3.1 Write an entity declaration for a memory circuit whose input and output ports are shown below. Use only the std_logic or std_logic_vector data types.

- addr: 12-bit address input
- wr_n: 1-bit write-enable control signal
- oe_n: 1-bit output-enable control signal
- data: 8-bit bidirectional data bus

3.2   What is the difference between a variable and a signal?

3.3   What is a strongly typed language?

3.4   What is the limitation of using the bit data type to represent a physical signal?

3.5   Assume that a is a 10-bit signal with the std_logic_vector(9 **downto** 0) data type. List the 10 bits assigned to the a signal.

   (a) a <= (**others**=>'1');
   (b) a <= (1|3|5|7|9=>'1', **others**=>'0');
   (c) a <= (9|7|2=>'1', 6=>'0', 0=>'1', 1|5|8=>'0', 3|4=>'0');

3.6   Assume that a and y are 8-bit signals with the std_logic_vector(7 **downto** 0) data type. If the signals are interpreted as unsigned numbers, the following assignment statement performs a / 8. Explain.

```
y <= "000" & a(7 downto 3);
```

3.7   Assume the same a and y signals in Problem 3.6. We want to perform a **mod** 8 and assign the result to y. Rewrite the previous signal assignment statement using only the & operator.

3.8   Assume that the following double-quoted strings are with the std_logic_vector data type. Determine whether the relational operation is syntactically correct. If yes, what is the result (i.e., true or false)?

   (a) "0110" > "1001"
   (b) "0110" > "0001001"
   (c) 2#1010# > "1010"
   (d) 1010 > "1010"

3.9   Repeat Problem 3.8, but assume that the data type is unsigned.

3.10   Repeat Problem 3.8, but assume that the data type is signed.

3.11   Determine whether the following signal assignment is syntactically correct. If not, use the proper conversion function and type casting to correct the problem.

```
library ieee;
use ieee.std_logic_1164.all;
use ieee.numeric_std.all;
 . . .
signal s1, s2, s3, s4, s5, s6, s7: std_logic_vector(3 downto 0);
signal u1, u2, u3, u4, u5, u6, u7: unsigned(3 downto 0);
signal sg: signed(3 downto 0);
 . . .
u1 <= 2#0001#;
u2 <= u3 and u4;
u5 <= s1 + 1;
u6 <= u3 + u4 + 3;
u7 <= (others=>'1');
```

```
s2  <=  s3 + s4 -1;
s5  <=  (others=>'1');
s6  <=  u3 and u4;
sg  <=  u3 - 1;
s7  <=  not sg;
```

3.12 For the following VHDL segment, correct the type mismatch with proper conversion function(s).

```
library ieee;
use ieee.std_logic_1164.all;
use ieee.numeric_std.all;
. . .
signal src, dest: std_logic_vector(15 downto 0);
signal amount: std_logic_vector(3 downto 0);
. . .
dest <= shift_left(src, amount);
```

3.13 For the following VHDL segment, correct the type mismatch with proper conversion function(s).

```
library ieee;
use ieee.std_logic_1164.all;
use ieee.numeric_std.all;


. . .
signal src, dest: std_logic_vector(15 downto 0);
signal amount: std_logic_vector(3 downto 0);
. . .
dest <= src sll amount;
```

3.14 For the following VHDL segment, correct the type mismatch with proper conversion function(s).

```
library ieee;
use ieee.std_logic_1164.all;
use ieee.std_logic_arith.all;
use ieee.std_logic_unsigned.all;
. . .
signal src, dest: std_logic_vector(15 downto 0);
signal amount: std_logic_vector(3 downto 0);
. . .
dest <= src sll amount;
```

CHAPTER 4

CONCURRENT SIGNAL ASSIGNMENT STATEMENTS OF VHDL

Concurrent signal assignment statements are simple, yet powerful VHDL statements. Since there is a clear mapping between the language constructs of an assignment statement and hardware components, we can easily visualize the conceptual diagram of the VHDL description. This helps us to develop a more efficient design. According to the VHDL definition, concurrent signal assignment statement has two basic forms: the *conditional signal assignment statement* and the *selected signal assignment statement*. For discussion purposes, we add an additional one, the *simple signal assignment statement*, which is a conditional assignment statement without any condition expression.

4.1 COMBINATIONAL VERSUS SEQUENTIAL CIRCUITS

A digital circuit can be broadly classified as combinational or sequential. A *combinational circuit* has no internal memory or state and its output is *a function of inputs only*. Thus, the same input values will always produce an identical output value. In a real circuit, the output may experience a short transient period after an input signal changes. However, the identical output value will be obtained when the signal is stabilized. In term of implementation, a combinational circuit is a circuit without memory elements (latches or flip-flops) or a closed feedback loop. A *sequential circuit*, on the other hand, has an internal state, and its output is *a function of inputs as well as the internal state*.

Although concurrent signal assignment statements can be used to describe sequential circuits, this is not the preferred method. We limit the discussion to combinational circuits in this chapter. We use the VHDL *process* to specify sequential circuits and study them in Chapter 8.

4.2   SIMPLE SIGNAL ASSIGNMENT STATEMENT

4.2.1   Syntax and examples

A simple signal assignment statement is a conditional signal assignment statement without the condition expression and thus is a special case of a conditional signal assignment statement. In VHDL definition, the simplified syntax of the simple signal assignment statement can be written as

```
signal_name <= projected_waveform;
```

The projected_waveform clause consists of two kinds of specifications: the expression of a new value for the signal and the time when the new value takes place. For example, consider the statement

```
y <= a + b + 1 after 10 ns;
```

which indicates that whenever the a or b signal changes, the expression a+b+1 will be evaluated, and its result will be assigned to the y signal after 10 ns.

The time aspect of projected_waveform normally corresponds to the internal propagation delay to complete the computation of the expression. However, since the propagation delay depends on the components, device technology, routing, fabrication process and operation environment, it is impossible to synthesize a circuit with an exact amount of delay. Therefore, for synthesis, explicit timing information is not specified in VHDL code. The default δ-delay is used in the projected waveform. The syntax becomes

```
signal_name <= value_expression;
```

The value_expression clause can be a constant value, logical operation, arithmetic operation and so on. Following are a few examples:

```
status <= '1';
even <= (p1 and p2) or (p3 and p4);
arith_out <= a + b + c - 1;
```

Note that the timing aspect is not dropped. It is just specified implicitly as a δ-delay. The previous statements implicitly imply

```
status <= '1' after δ;
even <= (p1 and p2) or (p3 and p4) after δ;
arith_out <= <= a + b + c - 1 after δ;
```

4.2.2   Conceptual implementation

Deriving the conceptual hardware block diagram for a simple signal assignment statement is straightforward. The entire statement can be thought of as a circuit block. The output of the circuit is the signal in the left-hand side of the statement, and the inputs are all the signals that appear in the right-hand-side value expression. We then map each operator of the value expression into a smaller circuit block and connect their inputs and outputs accordingly. The conceptual diagrams of three previous statements are shown in Figure 4.1.

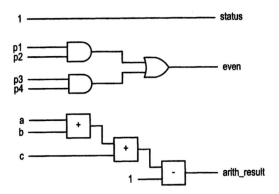

Figure 4.1 Conceptual diagrams of three simple signal assignment statements.

Note that these diagrams are only conceptual sketches. They will be transformed and simplified during synthesis. The circuit sizes of different VHDL operators vary significantly, and some of them, like the division operator, cannot be synthesized automatically. We examine this issue in detail in Chapter 6.

4.2.3 Signal assignment statement with a closed feedback loop

According to VHDL definition, it is syntactically correct for a signal to appear on both sides of a concurrent signal assignment statement. When an output signal is used as an input in the value expression, a closed feedback loop is formed. This may lead to the creation of an internal state or even oscillation. Consider the following VHDL statement:

```
q <= (q and (not en)) or (d and en);
```

In this example, the q signal is the output but also appears in the right-hand-side expression. The q output takes the value of the d signal if the en signal is '1' and it keeps its previous value if the en signal is '0'. Note that the output (i.e., q) now depends on input (i.e., en and d) as well as internal state (the previous value of q), and thus the circuit is no longer a combinational circuit. If we modify the previous statement by inverting q:

```
q <= ((not q) and (not en)) or (d and en);
```

the q output oscillates between '0' and '1' when the en signal is '0'.

When a signal assignment statement contains a closed feedback loop, it becomes sensitive to internal propagation delay and may exhibit race or oscillation. This kind of circuit confuses synthesis software and complicates verification and testing processes. It is a really bad coding practice and should be avoided completely in VHDL synthesis. The shortfall of delay-sensitive design and the disciplined derivation of sequential circuits are discussed in detail in Chapters 8 and 9.

Table 4.1   Function table of a 4-to-1 multiplexer

Input	Output
s	x
0 0	a
0 1	b
1 0	c
1 1	d

4.3   CONDITIONAL SIGNAL ASSIGNMENT STATEMENT

4.3.1   Syntax and examples

The simplified syntax of conditional signal assignment statement is shown below. As in Section 4.2.2, we assume that a timing specification is embedded implicitly in δ-delay and use value_expression to substitute the projected_waveform clause:

```
signal_name <= value_expr_1 when boolean_expr_1 else
               value_expr_2 when boolean_expr_2 else
               value_expr_3 when boolean_expr_3 else
                    . . .
               value_expr_n;
```

The boolean_expr_i (i= 1, 2, 3, ..., n) term is a Boolean expression that returns true or false. These Boolean expressions are evaluated successively in turn until one is found to be true, and the corresponding value expression is assigned to the output signal. In other words, the first Boolean expression, boolean_expr_1, is checked first. If it is true, the first value expression, value_expr_1, will be assigned to the output signal. If it is false, the second Boolean expression, boolean_expr_2, will be checked next. This process continues until all Boolean expressions are checked. The last value expression, value_expr_n, will be assigned to the signal if none of the Boolean expressions is true.

In the remaining subsection, we use several simple examples to illustrate the use of conditional signal assignment statements. The circuits include a multiplexer, a decoder, a priority encoder and a simple arithmetic logic unit (ALU).

Multiplexer   A multiplexer is essentially a virtual switch that routes a selected input signal to the output. The function table of an 8-bit 4-to-1 multiplexer is show in Table 4.1. In this circuit, the a, b, c and d signals can be considered as input data, and the s signal is a 2-bit selection signal that specifies which input data will be routed to the output. The VHDL code for this circuit is shown in Listing 4.1.

Listing 4.1   4-to-1 multiplexer based on a conditional signal assignment statement

```
library ieee;
use ieee.std_logic_1164.all;
entity mux4 is
   port(
      a,b,c,d: in std_logic_vector(7 downto 0);
      s: in std_logic_vector(1 downto 0);
      x: out std_logic_vector(7 downto 0)
   );
```

```
      end mux4;
10
      architecture cond_arch of mux4 is
      begin
          x <= a when (s="00") else
               b when (s="01") else
15             c when (s="10") else
               d;
      end cond_arch;
```

The first two lines are used to invoke the IEEE std_logic_1164 package so that the std_logic data type can be used. The next part is the entity declaration, which specifies the input and output ports of this circuit. The input ports include a, b, c and d, which are four 8-bit input data, and s, which is the 2-bit control signal. The output port is the 8-bit x signal. The architecture part uses one conditional signal assignment statement. The Boolean condition s="00" is evaluated first. If it is true, the first value expression, a, is assigned to x. If it is false, the next Boolean condition, s="01", will be evaluated. If it is true, b is assigned to x or the next Boolean expression, s="10", will be evaluated. If all three Boolean expressions are false, the last value expression, d, is assigned to x.

There is an issue about the use of the std_logic data type. At first glance, it seems that s is implied to be "11" when the first three Boolean expressions are false, and thus d is assigned to x. However, there are nine possible values in std_logic data type and, for the 2-bit s signal, there are 81 (i.e., 9*9) possible combinations, including the expected "00", "01", "10" and "11" as well as the metavalue combinations, such as "0Z", "UX", "0-" and so on. Therefore, d is assigned to x for the "11" condition, as well as other 77 metavalue combinations. However, these 77 combinations can exist only in simulation. In a real circuit, comparison of metavalues, as in s="0Z", cannot be implemented, and sometimes is meaningless, as in s="UX". In general, except for the limited use of 'Z', the metavalues of the std_logic data type will be ignored by synthesis software, and thus the final circuit will be synthesized as we originally expected. Some synthesis software also accepts VHDL code using 'X' for the unused metavalue combinations:

```
x <= a when (s="00") else
     b when (s="01") else
     c when (s="10") else
     d when (s="11") else
     'X';
```

The code leads to the same physical implementation.

Binary decoder   A binary decoder is an n-to-2^n decoder, which has an n-bit input and a 2^n-bit output. Each bit of the output represents an input combination. Based on the value of the input, the circuit activates the corresponding output bit. The function table of a simple 2-to-2^2 decoder is shown in Table 4.2. The VHDL code for this circuit is shown in Listing 4.2.

Listing 4.2   2-to-2^2 binary decoder based on a conditional signal assignment statement

```
library ieee;
use ieee.std_logic_1164.all;
entity decoder4 is
    port(
5       s: in  std_logic_vector(1 downto 0);
```

Table 4.2 Function table of a 2-to-2^2 binary decoder

Input	Output
s	x
0 0	0001
0 1	0010
1 0	0100
1 1	1000

Table 4.3 Function table of a 4-to-2 priority encoder

Input	Output	
r	code	active
1 – – –	11	1
0 1 – –	10	1
0 0 1 –	01	1
0 0 0 1	00	1
0 0 0 0	00	0

```
        x: out std_logic_vector(3 downto 0)
      );
   end decoder4;

10 architecture cond_arch of decoder4 is
   begin
        x <= "0001" when (s="00") else
             "0010" when (s="01") else
             "0100" when (s="10") else
15           "1000";
   end cond_arch;
```

Again, the first two lines are used to invoke the IEEE std_logic_1164 package. The entity declaration shows the circuit with a 2-bit input, a, and a 4-bit output, x. The architecture body uses one conditional signal assignment statement, which evaluates the Boolean conditions s="00", s="01" and s="10" one after another. The value expressions are constants that reflect the desired output patterns.

Priority encoder A priority encoder checks the input requests and generates the code of the request with highest priority. The function table of a 4-to-2 priority encoder is shown in Table 4.3. There are four input requests, r(3), r(2), r(1) and r(0). The outputs include a 2-bit signal, code, which is the binary code of the highest-priority request, and a 1-bit signal, active, which indicates whether there is an active request. The r(3) request has the highest priority. When it is asserted, the other three requests are ignored and the code signal becomes "11". If r(3) is not asserted, the second highest request, r(2), is examined. If it is asserted, the code signal becomes "10". The process repeats until all the requests are checked. The code signal returns "00" when only r(0) is asserted or no request is asserted. The active signal can be used to distinguish the two conditions. The VHDL code for this circuit is shown in Listing 4.3. The requests are grouped together and

Table 4.4 Function table of a simple ALU

Input	Output
ctrl	result
0 - -	src0 + 1
1 0 0	src0 + src1
1 0 1	src0 - src1
1 1 0	src0 **and** src1
1 1 1	src0 **or** src1

represented by a 4-bit signal, r. Individual bits of the r signal are checked in descending order, starting with r(3). Since operation of the priority encoder is similar to the definition of the conditional signal assignment statement, it is a good way to code this type of circuit (note the simple Boolean expressions in the code). A separate simple signal assignment statement is used to describe the active output.

Listing 4.3 4-to-2 priority encoder based on a conditional signal assignment statement

```
library ieee;
use ieee.std_logic_1164.all;
entity prio_encoder42 is
    port(
5       r: in std_logic_vector(3 downto 0);
        code: out std_logic_vector(1 downto 0);
        active: out std_logic
    );
end prio_encoder42;

10
    architecture cond_arch of prio_encoder42 is
    begin
        code <= "11" when (r(3)='1') else
                "10" when (r(2)='1') else
15              "01" when (r(1)='1') else
                "00";
        active <= r(3) or r(2) or r(1) or r(0);
    end cond_arch;
```

Simple ALU An ALU performs a set of arithmetic and logical operations. The function table of a simple ALU is shown in Table 4.4. The inputs include two 8-bit data sources, scr0 and src1, and a control signal, ctrl, which specifies the function to be performed. The output is the 8-bit result signal, which is the computed result. There are five functions, including three arithmetic operations, which are incrementing, addition and subtraction, and two logical operations, which are bitwise and and or operations. Furthermore, we assume that the input and output are interpreted as signed integers when an arithmetic function is selected.

For this circuit, the input data are interpreted as a collection of bits for the logical operation and as a signed number for the arithmetic operation. To achieve better portability, we normally use the std_logic_vector data type in the port declaration and then convert it to the desired data type in architecture body. The VHDL code is shown in Listing 4.4. The

IEEE numeric_std package and its signed data type are used to facilitate the arithmetic operation. When an addition, subtraction or incrementing operation is specified, we first convert the input to the signed data type, perform the operation and then convert the result back to the std_logic_vector data type. To make the code clear, we introduce three separate simple signal assignment statements and the sum, diff, and inc signals for the intermediate results of arithmetic operations.

Listing 4.4   Simple ALU based on a conditional signal assignment statement

```
   library ieee;
   use ieee.std_logic_1164.all;
   use ieee.numeric_std.all;
   entity simple_alu is
5     port(
         ctrl: in   std_logic_vector(2 downto 0);
         src0, src1: in std_logic_vector(7 downto 0);
         result: out std_logic_vector(7 downto 0)
      );
10 end simple_alu;

   architecture cond_arch of simple_alu is
      signal sum, diff, inc: std_logic_vector(7 downto 0);
   begin
15    inc <= std_logic_vector(signed(src0)+1);
      sum <= std_logic_vector(signed(src0)+signed(src1));
      diff <= std_logic_vector(signed(src0)-signed(src1));
      result <= inc   when ctrl(2)='0' else
                sum   when ctrl(1 downto 0)="00" else
20              diff  when ctrl(1 downto 0)="01" else
                src0 and src1 when ctrl(1 downto 0)="10" else
                src0 or src1;
   end cond_arch;
```

4.3.2 Conceptual Implementation

Recall that the syntax of the simplified conditional signal assignment statement is

```
signal_name <= value_expr_1 when boolean_expr_1 else
               value_expr_2 when boolean_expr_2 else
               value_expr_3 when boolean_expr_3 else
                       . . .
               value_expr_n;
```

Its semantics specifies that the Boolean expressions are evaluated in descending order until a condition is true, and then the corresponding value expression is assigned to the output signal. The key to implementing this construct is to achieve the desired descending order of evaluations. In a traditional programming language, descending order is implicitly observed because of the sequential execution of a single, shared CPU. In synthesis, we must use hardware to achieve this task.

The structure of conditional signal assignment statement implies a priority routing network since the Boolean expressions are evaluated in an orderly manner and the one evaluated earlier assumes a higher priority. Once the evaluation of a Boolean expression is true, the

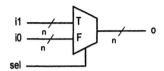

Figure 4.2   Conceptual diagram of an abstract multiplexer.

result of the corresponding value expression is routed to output. Unlike the *temporal* execution of the traditional programming language, the priority routing network is done on a *spatial* basis. Furthermore, since we cannot create hardware dynamically, dedicated hardware is needed for each Boolean expression and each value expression.

In summary, constructing the conditional signal assignment statement requires three groups of hardware:

- Value expression circuits
- Boolean expression circuits
- Priority routing network

Value expression circuits realize the value expressions, `value_expr_1, ···, value_expr_n`, and one of the results is routed to the output. *Boolean expression circuits* realize the Boolean expressions, `boolean_expr_1, ···, boolean_expr_n`, and their values are used to control the priority routing network. The *priority routing network* is the structure that routes and controls the desired value to the output signal.

A priority network can be implemented by a sequence of 2-to-1 multiplexers. To better illustrate the conceptual implementation, we utilize an "abstract multiplexer." Recall that a multiplexer is like a switch and uses a selection signal to select an input port and connect it to the output port. Any signal appearing in that input port will be routed to the output port. In an abstract multiplexer, the selection and input port designation are specified around the data type of the selection signal. Each input port is designated to a value of the data type of the selection signal, and one input port is selected according to the current value of the selection signal. For example, if the selection signal has a data type of `boolean`, there will be two input ports, designated as T (for `true`) and F (for `false`) respectively. If the selection signal has a value of `true`, the data from the T port will be routed to output. On the other hand, if the selection signal has a value of `false`, the data from the F port will be selected. The block diagram of this multiplexer is shown in Figure 4.2. The number of bits of the inputs and output may vary, and the symbol, n, is used to designate the width of the buses. During synthesis, the symbolic values can easily be mapped into binary representations of a physical multiplexer.

With the 2-to-1 abstract multiplexer, we can start to construct a priority network. Let us first consider a simple conditional signal assignment statement that has only one when clause:

```
sig <= value_expr_1 when boolean_expr_1 else
       value_expr_2;
```

The conceptual realization of this statement is shown in Figure 4.3. The three "clouds" represent the implementations of `value_expr_1`, `value_expr_2` and `boolean_expr_1` respectively. The result of `boolean_expr_1` is connected to the selection signal of the multiplexer. If it is `true`, the result from `value_expr_1` will be routed to the output port of the multiplexer. Otherwise, the result from `value_expr_2` will be routed to the output port.

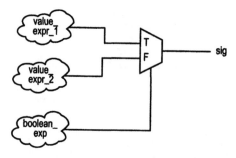

Figure 4.3   Conceptual diagram of a simple conditional signal assignment statement.

When there are more when clauses, we can perform the previous process repetitively and build the routing network in stages. Consider a statement with three when clauses:

```
sig <= value_expr_1 when boolean_expr_1 else
       value_expr_2 when boolean_expr_2 else
       value_expr_3 when boolean_expr_3 else
       value_expr_4;
```

The construction sequence is shown in Figure 4.4. First we construct the first when clause, which corresponds to the highest-priority condition. If the result of `boolean_expr_1` is `true`, the result of the corresponding value expression, `value_expr_1`, is routed to output, as shown in Figure 4.4(a). On the other hand, if the result of `boolean_expr_1` is `false`, the result from the remaining part of the statement, which is shown as a single cloud, will be used. This cloud can be constructed using a multiplexer similar to the first when clause, with its output connected to the F port of the rightmost multiplexer, as shown in Figure 4.4(b). After repeating this process one more time, we construct the third when clause and complete the conceptual implementation, as shown in Figure 4.4(c).

The construction process can be applied repeatedly to any number of when clauses. Since each clause will introduce one extra stage of multiplexer network, the depth of the network grows as the number of clauses increases. Although the conceptual construction is straightforward, it is difficult for synthesis software to transform an extremely deep multiplexer network to an efficient implementation. Thus, we should be aware of the impact on the number of when clauses. Discussion in Chapter 6 provides more insight on this issue.

4.3.3   Detailed implementation examples

Obtaining the conceptual diagram is only the first step in synthesis. We must derive the more detailed implementation for the multiplexers and "clouds" and eventually construct everything by using cells of the given technology library. Many of these tasks can be done in synthesis software, which is discussed in Chapter 6. In this section, we manually derive some simple circuits from VHDL segments to illustrate the basic synthesis process.

Implementation of a 2-to-1 multiplexer   An abstract 2-to-1 multiplexer has two symbolic ports, T and F. We can map it directly to a regular 2-to-1 multiplexer. The schematic of a 1-bit 2-to-1 multiplexer is shown in Figure 4.5(a). The two abstract ports, T and F, are mapped to the $i1$ and $i0$ ports respectively. In this circuit, the and cells can be interpreted as "passing gates," controlled by separate enable signals. When the enable signal is '1',

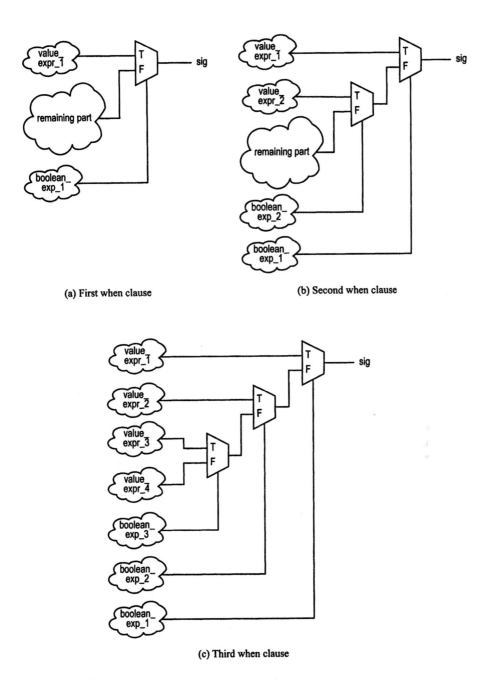

(a) First when clause

(b) Second when clause

(c) Third when clause

Figure 4.4   Construction of a multi-condition conditional signal assignment statement.

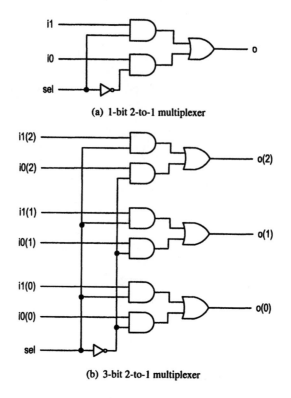

(a) 1-bit 2-to-1 multiplexer

(b) 3-bit 2-to-1 multiplexer

Figure 4.5   Gate-level implementation of a multiplexer.

the gate is open and the input signal is passed to output. When it is '0', the gate is closed and the output is set to '0'. The two enable signals are *sel* and *sel'* respectively, and thus one of the inputs will be passed to output. In terms of a logic expression, the output can be expressed as

$$o = sel' \cdot i0 + sel \cdot i1$$

For an n-bit 2-to-1 multiplexer, the control signals remain the same, but the gating structure will be duplicated n times. The schematic of a 3-bit 2-to-1 multiplexer is shown in Figure 4.5(b).

Example 1   Consider the following VHDL segment:

```
. . .
signal a,b,y: std_logic;
. . .
y <= '0' when a=b else
       '1';
. . .
```

This is a simple conditional signal assignment statement that contains one when clause. The conceptual diagram is shown in Figure 4.6(a). Let us consider the implementation of a=b, which is a 1-bit comparison circuit. According to VHDL definition, the input data type is std_logic, which has nine values, and the output data type is boolean, whose value can be true or false. During synthesis, we only consider the '0' and '1' of the std_logic

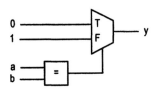 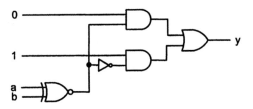

(a) Conceptual diagram   (b) Gate-level diagram

Figure 4.6 Synthesis of example 1.

Table 4.5 Truth table of a 1-bit comparator.

input a b	output a=b
0 0	1
0 1	0
1 0	0
1 1	1

data type since the other seven values are meaningless for a physical circuit. We also map the true and false to logic 1 and logic 0 of the physical circuit. Now the operation a=b can be represented in a traditional truth table, as shown in Table 4.5. The function can be expressed as $a' \cdot b' + a \cdot b$, or simply $(a \oplus b)'$, which is an xnor gate. We can now refine the conceptual diagram into the gate-level implementation, and the new diagram is shown in Figure 4.6(b). We can derive the logic expression of this circuit. Based on the expression of the multiplexer, the output can be expressed as

$$y = sel' \cdot i0 + sel \cdot i1$$

The sel, $i0$ and $i1$ are connected to $(a \oplus b)'$, '1' and '0' respectively, and thus the expression becomes

$$y = sel' \cdot i0 + sel \cdot i1 = (a \oplus b)'' \cdot 1 + (a \oplus b)' \cdot 0$$

which can be simplified to

$$y = a \oplus b$$

Thus, the final simplified circuit is a single xor gate.

Example 2 Consider the following VHDL segment:

```
. . .
signal r: std_logic_vector(2 downto 1);
signal y: std_logic_vector(1 downto 0);
. . .
y <= "10" when r(2)='1' else
     "01" when r(1)='1' else
     "00";
```

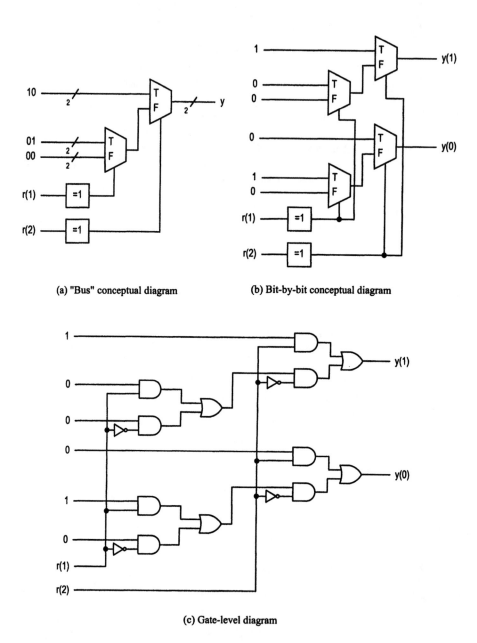

(a) "Bus" conceptual diagram   (b) Bit-by-bit conceptual diagram

(c) Gate-level diagram

Figure 4.7   Synthesis of example 2.

The conceptual diagram of this segment is shown in Figure 4.7(a). The next step is to derive gate-level implementation. Since the output has two bits, we have to split the conceptual diagram into two single-bit diagrams, as in Figure 4.7(b). Note that the implementation of the Boolean expressions, `r(2)='1'` and `r(1)='1'`, consists simply of the `r(2)` and `r(1)` signals themselves, and no additional logic is needed. After we substitute the multiplexer with its gate-level implementation, the resulting circuits are shown in Figure 4.7(c). We can derive the logic expressions for `y(0)` and `y(1)` using a procedure similar to that in example 1. After simplification, these logic expressions become

$$y(1) = r(2)$$

$$y(0) = r(2)' \cdot r(1)$$

Example 3   Consider the following VHDL segment:

```
.  .  .
signal  a,b,c,x,y,r:  std_logic;
.  .  .
r  <=  a  when  x=y  else
       b  when  x>y  else
       c;
.  .  .
```

The conceptual diagram of this segment is shown in Figure 4.8(a). By using the procedure to realize the a=b expression of example 1, we can derive the implementation of the x>y expression, which is $x \cdot y'$. The corresponding gate level circuit is shown in Figure 4.8(b). We can also derive the logic expression for the output and perform simplification to reduce the circuit size. The logic expression for this circuit is more involved and manually simplifying this circuit becomes a tedious task. This task is better left for software, which is good for a mechanical and repetitive procedure.

Example 4   Consider the following VHDL segment:

```
.  .  .
signal  a,b,r:  unsigned (7  downto  0);
signal  x,y:  unsigned (3  downto  0);
.  .  .
r  <=  a+b  when  x+y>1  else
       a-b-1  when  x>y  and  y!=0  else
       a+1;
.  .  .
```

The initial block diagram of this segment is shown in Figure 4.9(a). While the initial block diagram is similar to the previous examples, the value expressions and Boolean expressions are more involved. More complex components, such as an adder and comparator, are needed for implementation. After we implement the clouds, the block diagram is shown in Figure 4.9(b). We can continue to refine the circuit by replacing these components with their gate-level implementations and eventually derive the logic expressions. With these components, performing gate-level simplification becomes much more difficult and good coding practice at the RT level can improve the circuit efficiency significantly. These issues are discussed in Chapter 7.

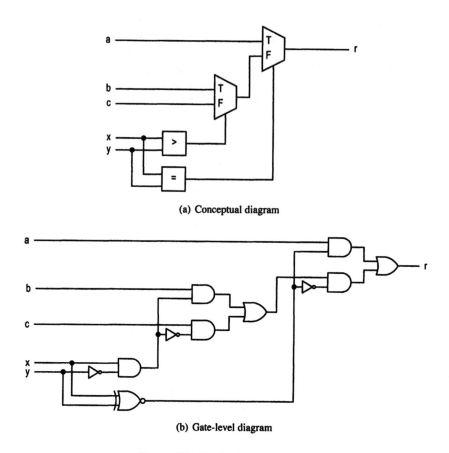

(a) Conceptual diagram

(b) Gate-level diagram

Figure 4.8   Synthesis of example 3.

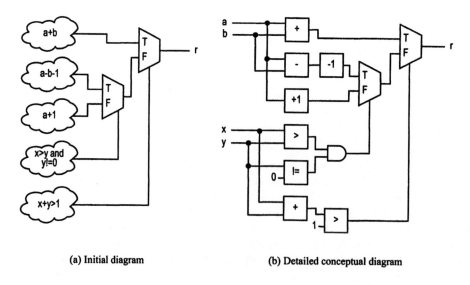

(a) Initial diagram   (b) Detailed conceptual diagram

Figure 4.9   Refinement of example 4.

4.4 SELECTED SIGNAL ASSIGNMENT STATEMENT

4.4.1 Syntax and examples

The simplified syntax of the selected signal assignment statement is shown below. As in conditional signal assignment statement, we assume that the timing specification is embedded in δ-delay and substitute `value_expression` for the `projected_waveform` clause.

```
with select_expression select
    signal_name <= value_expr_1 when choice_1,
                   value_expr_2 when choice_2,
                   value_expr_3 when choice_3,
                      . . .
                   value_expr_n when choice_n;
```

The selected signal assignment statement assigns an expression to a signal according to the value of `select_expression`. It is somewhat like a case statement in a traditional programming language. The `select_expression` term is used as the key for selection and it must result in a value of a discrete type or one-dimensional array. In other words, the evaluated result of `select_expression` can have only a finite number of possibilities. For example, a signal of the `bit_vector(1 downto 0)` data type can be used as `select_expression` since it contains only four possible values: `"00"`, `"01"`, `"10"` or `"11"`. A choice (i.e., `choice_i`) must be a valid value or a set of valid values of `select_expression`. The values of choices have to be *mutually exclusive* (i.e., no value can be used more than once) and *all inclusive* (i.e., all values have to be used). In other words, all possible values of `select_expression` must be covered by one and only one choice. The reserved word, **others**, can be used in the last choice (i.e., `choice_n`) to represent all the previously unused values.

We use the same multiplexer, binary decoder, priority encoder and ALU circuit of Section 4.3.1 to illustrate use of the selected signal assignment statement. Since this statement is a natural match to implement a truth table, an additional example is included for this purpose.

Multiplexer Let us consider the 8-bit 4-to-1 multiplexer of Section 4.3.1. The VHDL code for this circuit is shown in Listing 4.5. The entity declaration is identical and thus is omitted.

Listing 4.5 4-to-1 multiplexer based on a selected signal assignment statement

```
architecture sel_arch of mux4 is
begin
    with s select
      x <= a when "00",
           b when "01",
           c when "10",
           d when others;
    s
end sel_arch;
```

We need to be cautious about the metavalues of the `std_logic` and `std_logic_vector` data types. There is an issue about the use of these data types for `select_expression`. Recall that there are nine possible values in `std_logic` data type and there are 81 (i.e., $9*9$) possible combinations for the 2-bit s signal, including the expected `"00"`, `"01"`, `"10"` and `"11"` as well as 77 other metavalue combinations, such as `"ZZ"`, `"UX"` and `"0-"`, which are not meaningful in synthesis and will be ignored accordingly. In the code, the **when others** clause covers the "11" choice as well as the metavalue combinations. We cannot simply list the last choice as "11":

```
with s select
   x <= a when "00",
        b when "01",
        c when "10",
        d when "11";
```

This causes a syntax error since only 4 of 81 values are covered, and thus the choices are not all-inclusive. Some synthesis software may accept the following form:

```
with s select
   x <= a when "00",
        b when "01",
        c when "10",
        d when "11",
        'X' when others;  -- may also use '-'
```

The last line will be ignored during synthesis and the same physical circuit will be derived.

Binary decoder   The VHDL code for the 2-to-2^2 binary decoder of Section 4.3.1 is shown in Listing 4.6. Again, it is necessary to use **others** as the last choice to cover all metavalue combinations.

Listing 4.6   2-to-2^2 binary decoder based on a selected signal assignment statement

```
architecture sel_arch of decoder4 is
begin
   with s select
      x <= "0001" when "00",
           "0010" when "01",
           "0100" when "10",
           "1000" when others;
end sel_arch;
```

Priority encoder   The VHDL code for the 4-to-2 priority encoder is shown in Listing 4.7. Recall that "11" will be assigned to code if r(3) is '1'. This consists of eight possible input combinations of the r signal, which are "1000", "1001", "1010", . . . , "1111". All of them are listed in the first choice. Note that the symbol | is used for specifying multiple values.

Listing 4.7   4-to-2 priority encoder based on a selected signal assignment statement

```
architecture sel_arch of prio_encoder42 is
begin
   with r select
      code <= "11" when "1000"|"1001"|"1010"|"1011"|
                        "1100"|"1101"|"1110"|"1111",
              "10" when "0100"|"0101"|"0110"|"0111",
              "01" when "0010"|"0011",
              "00" when others;
      active <= r(3) or r(2) or r(1) or r(0);
end sel_arch;
```

Intuitively, we may wish to use the '-' (don't-care) value of the st
make the code compact:

```
with r select
    code <= "11" when "1---",
            "10" when "01--",
            "01" when "001-",
            "00" when others;
```

While this is syntactically correct, the code does not describe the intended circuit. In VHDL, the '-' value is treated just as an ordinary value of std_logic. Since the '-' value will never occur in the physical circuit, the "1---", "01--" and "001-" choices will never be met and the code is the same as

```
code <= "00";
```

This, of course, is not the intended priority encoding circuit. We discuss this issue in more detail in Chapter 6.

A simple ALU The VHDL code of the simple ALU specified in Table 4.4 is shown in Listing 4.8. Note that all four possible combinations of the ctrl signal, "000", "001", "010" and "011", are listed in the first choice.

Listing 4.8   Simple ALU based on a selected signal assignment statement

```
architecture sel_arch of simple_alu is
    signal sum, diff, inc: std_logic_vector(7 downto 0);
begin
    inc <= std_logic_vector(signed(src0)+1);
5   sum <= std_logic_vector(signed(src0)+signed(src1));
    diff <= std_logic_vector(signed(src0)-signed(src1));
    with ctrl select
        result <= inc              when "000"|"001"|"010"|"011",
                  sum              when "100",
10              diff             when "101",
                  src0 and src1    when "110",
                  src0 or src1     when others;  -- "111"
end sel_arch;
```

Truth Table Implementation   A truth table can be used to specify any combinational function. It is a simple and useful way to describe a small, random combinational circuit. Because the choices list all the possible combinations, the selected signal assignment statement is a natural match for the truth table description. A simple two-input truth table is shown in Table 4.6.

The corresponding VHDL code is shown in Listing 4.9. The a and b signals are concatenated as tmp, which is then used as the select expression. Each row of the truth table now becomes a choice in the selected signal assignment statement and the truth table is implemented accordingly.

Listing 4.9   Truth table based on selected signal assignment statement

```
library ieee;
use ieee.std_logic_1164.all;
entity truth_table is
```

Table 4.6   Truth table of a two-input function

Input	Output
a b	y
0 0	0
0 1	1
1 0	1
1 1	1

```
      port(
5         a,b: in    std_logic;
          y: out     std_logic
      );
   end truth_table;

10 architecture a of truth_table is
       signal tmp: std_logic_vector(1 downto 0);
   begin
       tmp <= a & b;
       with tmp select
15         y <= '0' when "00",
              '1' when "01",
              '1' when "10",
              '1' when others;  --  "11"
   end a;
```

4.4.2   Conceptual implementation

Recall that the syntax of the selected signal assignment is

```
with select_expression select
    signal_name <= value_expr_1 when choice_1,
                   value_expr_2 when choice_2,
                   value_expr_3 when choice_3,
                        . . .
                   value_expr_n when choice_n;
```

Conceptually, the selected signal assignment statement can be thought as an abstract multi-plexing circuit that utilizes a selection signal to route the result of the designated expression to output. In this multiplexing circuit, each possible value of select_expression has a designated input port in the multiplexer, and select_expression works as the selection signal of this multiplexer. Once its value is determined, the result of the designated value expression is passed to the output port of the multiplexer. In Section 4.3.2, we utilized an abstract 2-to-1 multiplexer with a selection signal of the boolean data type. The multiplexer can be generalized for other kinds of selection signals. For example, consider a selection signal with $k + 1$ different possible values, c0, c1, ..., ck. The abstract multiplexer has $k + 1$ ports, each corresponding to a value, as shown in Figure 4.10.

It is possible that the input and output have multiple bits and the symbol n is used to designate the width of the buses. The conceptual implementation of the selected signal

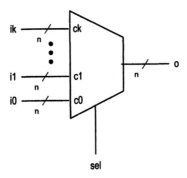

Figure 4.10   Abstract $(k + 1)$-to-1 multiplexer.

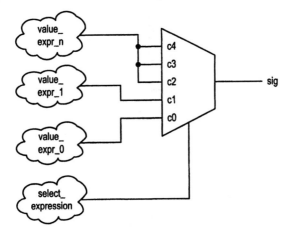

Figure 4.11   Conceptual diagram of a selected signal assignment statement.

assignment statement involves a single abstract multiplexer and is straightforward. Consider the following statement:

```
with select_expression select
    sig <= value_expr_0 when c0,
           value_expr_1 when c1,
           value_expr_n when others;
```

We assume that select_expression may result in one of five possible values: c0, c1, c2, c3 and c4. Note that the last choice, **when others**, of this statement implicitly covers c2, c3 and c4. The conceptual realization of this statement is shown in Figure 4.11.

The clouds represent the implementation of the three value expressions, value_expr_0, value_expr_1 and value_expr_n, and select_expression respectively. The evaluated results of the value expressions are fed into the designated input ports of the multiplexer. The result of select_expression is connected to the selection port of the multiplexer and its value determines which data will be routed to the output port.

All selected signal assignment statements have a similar conceptual diagram. The main difference is in the number of values that select_expression can assume, which in turn determines the size of the multiplexer. Despite the simple conceptual construction, certain

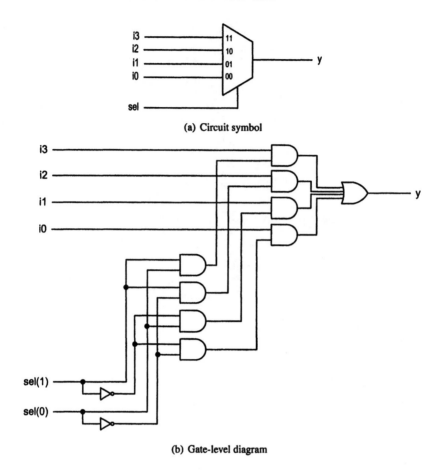

(a) Circuit symbol

(b) Gate-level diagram

Figure 4.12   Circuit symbol and gate-level diagram of a 4-to-1 multiplexer.

device technologies may have difficulty supporting an extremely wide multiplexing circuit. Thus, we should be aware of the number of values in a selection expression.

4.4.3 Detailed implementation examples

As in the implementation of a conditional signal assignment statement, we continue the refining process and realize the conceptual diagram using gate-level components. Following examples illustrate the derivation.

k-to-1 multiplexer   An abstract multiplexer with k symbolic ports can easily be mapped to a physical k-to-1 multiplexer with a $\log_2 k$-bit selection signal. The symbol and gate-level diagram of a 1-bit 4-to-1 multiplexer are shown in Figure 4.12. We use the binary representations, "00", "01", "10" and "11", as the names of the ports. The upper and cells can be thought of as "passing gates," each controlled by an enable signal. The corresponding input will be passed to output when the enable signal is '1'. The bottom part is a 2-to-4 binary decoder that generates the enable signal, in which only one bit is activated. In term of a logic expression, the output can be expressed as

$$y = (sel(1)' \cdot sel(0)') \cdot i0 + (sel(1)' \cdot sel(0)) \cdot i1 + (sel(1) \cdot sel(0)') \cdot i2 + (sel(1) \cdot sel(0)) \cdot i3$$

For a multiple-bit 4-to-1 multiplexer, the enable signals remain the same, but the gating structure will be duplicated multiple times.

In VHDL code, the selection signal frequently has a data type of std_logic_vector, which includes many meaningless combinations. During synthesis, only '0' and '1' of nine values will be used, as we discussed in Section 4.3.1.

Example 1   Consider the following VHDL segments:

```
  . . .
signal s: std_logic_vector (1 downto 0);
  . . .
  with s select
      x <= (a and b) when "11",
           (a or b)  when "01"|"10",
           '0'       when others;
  . . .
```

This is a simple selected signal assignment statement. The selection expression has a data type of std_logic_vector(1 **downto** 0). Again, although there are 81 possible values, only "00", "01", "10" and "11" are meaningful for synthesis. Thus, only a 4-to-1 multiplexer is needed. The conceptual diagram and the refined gate-level diagram are shown in Figure 4.13. The logic expression for this circuit is

$$x = (s(1)' \cdot s(0)') \cdot 0 + (s(1)' \cdot s(0)) \cdot (a+b) + (s(1) \cdot s(0)') \cdot (a+b) + (s(1) \cdot s(0)) \cdot (a \cdot b)$$

Example 2   Consider the truth table in Table 4.6 and the corresponding VHDL segment:

```
tmp <= a & b;
with tmp select
    y <= '0' when "00",
         '1' when "01",
         '1' when "10",
         '1' when others;
```

The conceptual diagram is shown in Figure 4.14. The logic expression for this circuit is

$$y = (a' \cdot b') \cdot 0 + (a' \cdot b) \cdot 1 + (a \cdot b') \cdot 1 + (a \cdot b) \cdot 1$$

The expression can be simplified to $a + b$, which is the or function specified in the truth table.

Example 3   Consider the following VHDL segment:

```
  . . .
signal a,b,r: unsigned (7 downto 0);
signal s: std_logic_vector (1 downto 0);
  . . .
  with s select
      r <= a+1   when "11",
           a-b-1 when "10",
           a+b   when others;
  . . .
```

This segment contains more sophisticated expressions. After we realized the value expression clouds, the block diagram is shown in Figure 4.15.

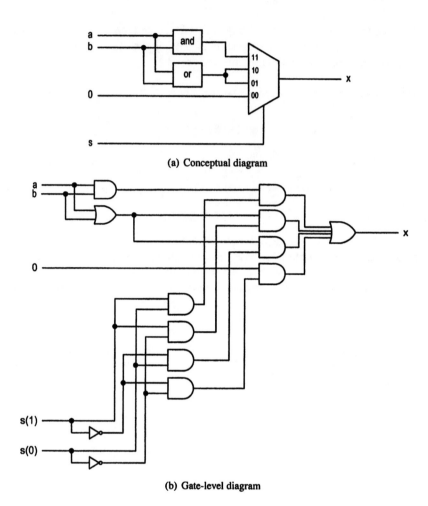

(a) Conceptual diagram

(b) Gate-level diagram

Figure 4.13   Synthesis of example 1.

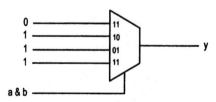

Figure 4.14   Conceptual diagram of truth table–based description.

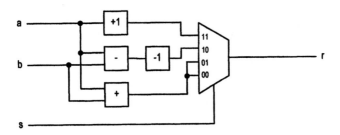

Figure 4.15   Block diagram of example 3.

4.5   CONDITIONAL SIGNAL ASSIGNMENT STATEMENT VERSUS SELECTED SIGNAL ASSIGNMENT STATEMENT

4.5.1   Conversion between conditional signal assignment and selected signal assignment statements

From the synthesis point of view, the conditional signal assignment statement and the selected signal assignment statement imply two different routing structures. The examples presented in the previous sections show that we can describe the same circuit using either a conditional or a selected signal assignment statement. Actually, the conversion between the two forms of assignment statements is always possible.

Converting a selected signal assignment statement to a conditional signal assignment statement is straightforward. Consider a general selected signal assignment statement in which there are eight possible choices: $c_7, c_6, \ldots, c_1, c_0$.

```
with sel select
    sig <= value_expr_0 when c0,
           value_expr_1 when c1|c3|c5,
           value_expr_2 when c2|c4,
           value_expr_n when others;
```

We can describe the choices of a when clause as a Boolean expression. For example, when $c_2|c_4$ can be expressed as (sel=c2) or (sel=c4). We can then use these Boolean expressions and convert the selected signal assignment statement to a new format:

```
sig <=
    value_expr_0 when (sel=c0) else
    value_expr_1 when (sel=c1) or (sel=c3) or (sel=c5) else
    value_expr_2 when (sel=c2) or (sel=c4) else
    value_expr_n;
```

Converting a conditional signal assignment statement to a selected statement needs more manipulation. Let us consider a general conditional signal assignment statement with three Boolean expressions:

```
sig <= value_expr_0 when bool_exp_0 else
       value_expr_1 when bool_exp_1 else
       value_expr_2 when bool_exp_2 else
       value_expr_n;
```

We need a 3-bit auxiliary selection signal, `sel`, in which each bit represents a Boolean expression. By specifying proper choices, we can preserve the desired priority. The converted code is

```
sel(2) <= '1' when bool_exp_0 else '0';
sel(1) <= '1' when bool_exp_1 else '0';
sel(0) <= '1' when bool_exp_2 else '0';
with sel select
    sig <= value_expr_0 when "100"|"101"|"110"|"111",
           value_expr_1 when "010"|"011",
           value_expr_2 when "001",
           value_expr_n when others;
```

Note that the pattern of the selected signal assignment statement is very similar to a priority encoder except that the request signals are replaced by the auxiliary selection signals generated from the Boolean expressions.

4.5.2 Comparison between conditional signal assignment and selected signal assignment statements

In the selected signal assignment statement, each choice can be considered as a row in a table. Thus, this statement is a good match for a circuit described by a truth table or a truth table–like function table, such as the decoder, truth table and multiplexer examples discussed in Section 4.4. On the other hand, it is less effective when certain input conditions are given preferential treatment. For example, if we examine the priority encoder example of Section 4.4, eight of the 16 ports of the multiplexer are connected to an identical expression.

The conditional signal assignment statement implicitly enforces the order of the operation and is a natural match for a circuit that needs to give preferential treatment for certain conditions or to prioritize the operations. The priority encoder is a good example of this kind of circuit. The conditional signal assignment statement can also handle complicated conditions. For example, we can write

```
pc_next <=
    pc_reg + offset when (state=jump and a=b) else
    pc_reg + 1 when (state=skip and flag='1') else
    . . .
```

A conditional signal assignment statement is less effective to describe a truth table since it may "overspecify" the circuit and thus add unnecessary constraints. For example, consider the multiplexer of Section 4.3.1. The original VHDL segment is

```
x <= a when (s="00") else
     b when (s="01") else
     c when (s="10") else
     d;
```

The code can also be written as

```
x <= c when (s="10") else
     a when (s="00") else
     b when (s="01") else
     d;
```

or

```
x <= c  when  (s="10")  else
       b  when  (s="01")  else
       a  when  (s="00")  else
       d ;
```

or many other possible variations. These codes give priority to the condition in the first when clause, although it is not part of the original specification. While this type of code is not wrong, the extra constraint may introduce additional circuitry and make synthesis and optimization more difficult.

Ideally, the synthesis software should automatically determine the optimal structure and derive identical gate-level implementation, regardless of the language constructs used in VHDL descriptions. In reality, this is possible only for small, trivial designs. For a general design, we have to be aware of the effect of the statements on the routing and the "layout" of the final implementation. These aspects are illustrated by examples in Chapter 7.

4.6 SYNTHESIS GUIDELINES

- Avoid a closed feedback loop in a concurrent signal assignment statement.

- Think of the conditional signal assignment and selected signal assignment statements as routing structures rather than sequential control constructs.

- The conditional signal assignment statement infers a priority routing structure, and a larger number of when clauses leads to a long cascading chain.

- The selected signal assignment statement infers a multiplexing structure, and a large number of choices leads to a wide multiplexer.

4.7 BIBLIOGRAPHIC NOTES

Since the focus of the book is on synthesis, only synthesis-related aspects of the concurrent signal assignment statement are discussed. The complete discussion on these constructs can be found in *The Designer's Guide to VHDL, 2nd edition*, by P. J. Ashenden.

The discussion in this chapter illustrates the general schemes to realize concurrent signal assignment statements in various routing structures. Individual synthesis software may map certain language constructs to specific hardware architectures. The software vendors sometimes include a "style guide" in their documentation. It shows the mapping between hardware architecture and the VHDL language constructs.

Problems

4.1   Add an enable signal, en, to a 2-to-4 decoder. When en is '1', the decoder functions as usual. When en is '0', the decoder is disabled and output becomes "0000". Use the conditional signal assignment statement to derive this circuit. Draw the conceptual diagram.

4.2   Repeat Problem 4.1, but use the selected signal assignment statement instead.

4.3   Consider a 2-by-2 switch. It has two input data ports, x(0) and x(1), and a 2-bit control signal, ctrl. The input data are routed to output ports y(0) and y(1) according to the ctrl signal. The function table is specified below.

Input	Output		Function
ctrl	y1	y0	
00	x1	x0	pass
01	x0	x1	cross
10	x0	x0	broadcast x0
11	x1	x1	broadcast x1

(a) Use concurrent signal assignment statements to derive the circuit.

(b) Draw the conceptual diagram.

(c) Expand it into gate-level circuit and derive the simplified logic expression in sum-of-products format.

4.4   Consider a comparator with two 8-bit inputs, a and b. The a and b are with the std_logic_vector data type and are interpreted as unsigned integers. The comparator has an output, agtb, which is asserted when a is greater than b. Assume that only a single-bit comparator is supported by synthesis software. Derive the circuit with concurrent signal assignment statement(s).

4.5   Repeat Problem 4.4, but assume that a and b are interpreted as signed integers.

4.6   We wish to design a shift-left circuit manually. The inputs include a, which is an 8-bit signal to be shifted, and ctrl, which is a 3-bit signal specifying the amount to be shifted. Both are with the std_logic_vector data type. The output y is an 8-bit signal with the std_logic_vector data type. Use concurrent signal assignment statements to derive the circuit and draw the conceptual diagram.

CHAPTER 5

SEQUENTIAL STATEMENTS OF VHDL

As the name suggests, sequential statements are executed in sequence. The semantics of these statements is more like that of a traditional programming language. Since they are not compatible with the general concurrent execution model of VHDL, sequential statements have to be enclosed inside a construct known as a *process*. The main purpose of sequential statements is to describe and model a circuit's "abstract behavior." Unlike concurrent signal assignment statements, there is no clear mapping between sequential statements and hardware components. Some statements and coding styles are difficult or even impossible to synthesize. To use processes and sequential statements for synthesis, the VHDL description has to be coded in a disciplined way so that the code can be faithfully mapped into the intended hardware configuration.

5.1 VHDL PROCESS

5.1.1 Introduction

A *process* is a VHDL construct that contains a set of actions to be executed sequentially. These actions are known as *sequential statements*. The process itself is a concurrent statement. It can be interpreted as a circuit part enclosed inside a black box whose behavior is described by the sequential statements. We may or may not be able to construct physical hardware that exhibits the desired behavior.

Sequential statements include a rich variety of constructs, and they can exist only inside a process. The execution inside a process is sequential, and thus the order of the statements is important. Many sequential constructs don't have clear counterparts in hardware implementation, and are difficult, if not impossible, to synthesize. We examine the use and synthesis of the following sequential statements in this chapter:

- *wait statement*
- *sequential signal assignment statement*
- *variable assignment statement*
- *if statement*
- *case statement*
- simple *for loop statement*

More sophisticated loop statements as well as two other sequential statements, the *exit* and *next statements*, are discussed in Chapter 14. Note that we should not confuse sequential statements with sequential circuits. *Sequential statements* are VHDL statements inside a process, and *sequential circuits* are circuits with internal states. A process and its internal sequential statements can be used to describe a combinational or sequential circuit. As in Chapter 4, our discussion in this chapter is limited to combinational circuits.

The process has two basic forms. The first form has a *sensitivity list* but no wait statement inside the process. The second form has one or more wait statements but no sensitivity list. Because of its clarity, we use mainly the first form in this book. The second form is examined briefly in Section 5.1.3.

5.1.2 Process with a sensitivity list

The syntax of a process with a sensitivity list is

```
process (sensitivity_list)
   declarations;
begin
   sequential statement;
   sequential statement;
   . . .
end process;
```

The sensitivity_list is a list of signals to which the process responds (i.e., is "sensitive to"). The declarations part consists of various declarations that are local to the process.

Whereas the appearance of a VHDL process is like a function or procedure of a traditional programming language, the behavior of the process is very different. A VHDL process is not invoked (or called) by another routine. It acts like a circuit part, which is either active (known as *activated*) or inactive (known as *suspended*). A VHDL process is activated when a signal in the sensitivity list changes its value, like a circuit responding to an input signal. Once a process is activated, its statements will be executed sequentially until the end of the process. The process is then suspended until the next change of signal. A simple process with a single sequential signal assignment is

```
signal a,b,c,y: std_logic;  -- in architecture declaration
. . .
process (a,b,c)
begin
   y <= a and b and c;
end process;
```

When any input (i.e., a, b or c) changes, the process is activated and its statement is executed. The statement evaluates the expression and assigns the result to the y signal. This process simply describes a three-input and circuit with a, b and c inputs and y output.

One tricky issue about the process is the *incomplete sensitivity list*, which is a list with one or more input signals missing. For example, the b and c signals are omitted from the sensitivity list of the previous example:

```
signal a,b,c,y: std_logic;
. . .
process(a)
begin
    y <= a and b and c;
end process;
```

When the a signal changes, the process is activated and the circuit acts as expected. On the other hand, when the b or c signal changes, the process remains suspended and the y signal keeps its previous value. This implies that the circuit has some sort of memory element that is triggered at both positive and negative edges of the a signal. When the a signal changes, the expression is evaluated and the result is stored in the memory element. This is not the circuit behavior we expected, and it cannot be synthesized by regular hardware components.

For a combinational circuit, the output is a function of input. This implies that the circuit responds to any input change. Thus, *all* input signals of a combinational circuit should be included in the sensitivity list. A process with incomplete sensitivity can be used to describe a circuit with internal memory and thus infer a memory element. This is discussed in Chapter 8.

5.1.3   Process with a wait statement

A process with wait statements has one or more wait statements but no sensitivity list. The wait statement has several forms:

```
wait on signals;
wait until boolean_expression;
wait for time_expression;
```

Use of the wait statement can best be explained by an example. The code segment of Section 5.1.2 can be rewritten as

```
process
begin
    y <= a and b and c;
    wait on a, b, c;
end process;
```

Note that there is no sensitivity list. The process starts automatically after the system initialization. It continues the execution until a wait statement is reached and then becomes suspended. The statement **wait on a, b, c** means that the process waits for a change in the a or b or c signal. When one of them changes value, the process is activated. It executes to the end of the process, then returns to the beginning of the process and continues execution. It becomes suspended again when reaching the wait statement. The overall effect of this process describes a three-input and gate, as in the previous example.

The behavior of the two other types of wait statements is similar except that the process waits until a special Boolean condition is asserted or waits for a specific amount of time.

Since multiple wait statements are allowed, a process with wait statements can be used to model complex timing behavior and sequential events. However, in synthesis, only few well-defined forms of wait statements can be used, and normally only one wait statement is allowed in a process. Since a process with a sensitivity list can clearly show the input signals and make the code clearer and more descriptive, we prefer this form and normally don't use the wait statement in this book.

5.2   SEQUENTIAL SIGNAL ASSIGNMENT STATEMENT

The syntax of a sequential signal assignment is identical to that of the simple concurrent signal assignment of Chapter 4 except that the former is inside a process. It can be written as

```
signal_name <= projected_waveform;
```

The projected_waveform clause consists of a value expression and a time expression, which is generally used to represent the propagation delay. As in the concurrent signal assignment statement, the delay specification cannot be synthesized and we always use the default δ-delay. The syntax becomes

```
signal_name <= value_expression;
```

Note that the concurrent conditional and selected signal assignment statements cannot be used inside the process.

For a signal assignment with δ-delay, the behavior of a sequential signal assignment statement is somewhat different from that of its concurrent counterpart. If a process has a sensitivity list, the execution of sequential statements is treated as a "single abstract evaluation," and the actual value of an expression will not be assigned to a signal until the end of the process. This is consistent with the black box interpretation of the process; that is, the entire process is treated as one indivisible circuit part, and the signal is assigned a value only after the completion of all sequential statements.

Inside a process, a signal can be assigned multiple times. If all assignments are with δ-delays, only the last assignment takes effect. Because the signal is not updated until the end of the process, it *never* assumes any "intermediate" value. For example, consider the following code segment:

```
a,b,c,d,y: std_logic;
 . . .
process(a,b,c,d)
begin
    y <= a or c;
    y <= a and b;
    y <= c and d;
end process;
```

It is the same as

```
process(a,b,c,d)
begin
    y <= c and d;
end process;
```

Although this segment is easy to understand, multiple assignments may introduce subtle mistakes in a more complex code and make synthesis very difficult. Unless there is a

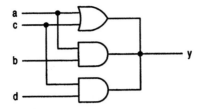

Figure 5.1   Conceptual diagram of multiple concurrent signal assignments.

compelling reason, it is a good idea to avoid assigning a signal multiple times. The only exception is the assignment of a default value in the if and case statements. This is discussed in Sections 5.4.3 and 5.5.3.

The result will be very different if the multiple assignments are the concurrent signal assignment statements. Assume that the previous three assignment statements are concurrent signal assignment statements (i.e., not inside a process). The code segment becomes

```
a,b,c,d,y: std_logic;
. . .
-- the statements are not inside a process
y <= a or c;
y <= a and b;
y <= c and d;
```

The code is syntactically correct since multiple assignments are allowed for a signal with the std_logic data type (since it is a "resolved" data type). The corresponding circuit is shown in Figure 5.1. Although the syntax is fine, the design is incorrect because of the potential output conflict. The y signal may get a value of 'X' in simulation if any two of the output values of the three gates are different.

5.3 VARIABLE ASSIGNMENT STATEMENT

The syntax of a variable assignment statement is

```
variable_name := value_expression;
```

The immediate assignment notion, :=, is used for the variable assignment. There is no time dimension (i.e., no propagation delay) and the assignment takes effect *immediately*. The behavior of the variable assignment is just like that of a regular variable assignment used in a traditional programming language. For example, consider the code segment

```
signal a, b, y: std_logic;
. . .
process(a,b)
   variable tmp: std_logic;
begin
   tmp := '0';
   tmp := tmp or a;
   tmp := tmp or b;
   y <= tmp;
end process;
```

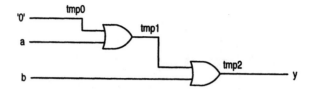

Figure 5.2   Conceptual implementation of simple variable assignments.

The tmp variable assumes the value immediately in each sequential statement and assigns its value, which is equal to $a + b$, to the y signal. Note that the variables are "local" to the process and have to be declared inside the process.

Although the behavior of a variable is easy to understand, mapping it into hardware is difficult. For example, to realize the previous process in hardware, we have to rename the variables tmp0, tmp1 and tmp2 and change the process to

```
process(a,b)
   variable tmp0, tmp1, tmp2: std_logic;
begin
   tmp0 := '0';
   tmp1 := tmp0 or a;
   tmp2 := tmp1 or b;
   y <= tmp2;
end process;
```

For synthesis purposes, we can now interpret the variables as signals or nets. The corresponding diagram is shown in Figure 5.2. Because of the lack of clear hardware mapping, we should try to use signals in code in general and resort to variables only for the characteristics that cannot be described by signals.

For comparison purposes, let us repeat the previous segment by replacing the variables with signals:

```
signal a, b, y, tmp: std_logic;
. . .
process(a,b,tmp)
begin
   tmp <= '0';
   tmp <= tmp or a;
   tmp <= tmp or b;
   y <= tmp;
end process;
```

Note that the signals have to be "global" and declared outside the process, and the tmp signal has to be included in the sensitivity list. This code is the same as

```
process(a,b,tmp)
begin
   tmp <= tmp or b;
   y <= tmp;
end process;
```

This code implies a combinational loop with an or gate, as shown in Figure 5.3.

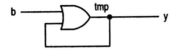

Figure 5.3   Conceptual implementation of erroneous signal assignments.

5.4   IF STATEMENT

5.4.1   Syntax and examples

The simplified syntax of an if statement is

```
if boolean_expr_1 then
    sequential_statements;
elsif boolean_expr_2 then
    sequential_statements;
elsif boolean_expr_3 then
    sequential_statements;
 . . .
else
    sequential_statements;
end if;
```

An if statement has one *then branch*, one or more optional *elsif branches* and one optional *else branch*. The boolean_expr_i term is a Boolean expression that returns true or false. These Boolean expressions are evaluated sequentially. When an expression is evaluated as true, the statements in the corresponding branch will be executed and the remaining branches will be skipped. If none of the expressions is true and the else branch exists, the statements in the else branch will be executed.

We use the same circuit examples as in Chapter 4, which include a multiplexer, a decoder, a priority encoder and a simple ALU, to illustrate use of an if statement. The if statement description of an 8-bit 4-to-1 multiplexer is shown in Listing 5.1. Since the multiplexer is a combinational circuit, all input signals, including a, b, c, d and s, are in the sensitivity list. Note that the signals used in the Boolean expressions are also the input signals.

Listing 5.1   4-to-1 multiplexer based on an if statement

```
architecture if_arch of mux4 is
begin
    process(a,b,c,d,s)
    begin
5       if (s="00") then
            x <= a;
        elsif (s="01")then
            x <= b;
        elsif (s="10")then
10          x <= c;
        else
            x <= d;
        end if;
    end process;
15 end if_arch;
```

The if statement versions of binary decoder, priority encoder and simple ALU are shown in Listings 5.2, 5.3 and 5.4 respectively.

Listing 5.2   2-to-4 decoder based on an if statement

```
architecture if_arch of decoder4 is
begin
    process(s)
    begin
5       if (s="00") then
            x <= "0001";
        elsif (s="01")then
            x <= "0010";
        elsif (s="10")then
10          x <= "0100";
        else
            x <= "1000";
        end if;
    end process;
15 end if_arch;
```

Listing 5.3   4-to-2 priority encoder based on an if statement

```
architecture if_arch of prio_encoder42 is
begin
    process(r)
    begin
5       if (r(3)='1') then
            code <= "11";
        elsif (r(2)='1')then
            code <= "10";
        elsif (r(1)='1')then
10          code <= "01";
        else
            code <= "00";
        end if;
    end process;
15  active <= r(3) or r(2) or r(1) or r(0);
    end if_arch;
```

Listing 5.4   Simple ALU based on an if statement

```
architecture if_arch of simple_alu is
    signal src0s, src1s: signed(7 downto 0);
begin
    src0s <= signed(src0);
5   src1s <= signed(src1);
    process(ctrl,src0,src1,src0s,src1s)
    begin
        if (ctrl(2)='0') then
            result <= std_logic_vector(src0s + 1);
10      elsif (ctrl(1 downto 0)="00")then
            result <=   std_logic_vector(src0s + src1s);
        elsif (ctrl(1 downto 0)="01")then
```

```
              result <= std_logic_vector(src0s - src1s);
           elsif (ctrl(1 downto 0)="10")then
15             result <= src0 and src1 ;
           else
              result <= src0 or src1;
           end if;
        end process;
20 end if_arch;
```

5.4.2 Comparison to a conditional signal assignment statement

An if statement is somewhat like a concurrent conditional signal assignment statement. If the sequential statements inside an if statement consist of only the signal assignment of a single signal, as in previous examples, the two statements are equivalent. Consider the following conditional signal assignment statement:

```
sig <= value_expr_1 when boolean_expr_1 else
       value_expr_2 when boolean_expr_2 else
       value_expr_3 when boolean_expr_3 else

           . . .

       value_expr_n;
```

It can be written as

```
process(...)
begin
   if boolean_expr_1 then
      sig <= value_expr_1;
   elsif boolean_expr_2 then
      sig <= value_expr_2;
   elsif boolean_expr_3 then
      sig <= value_expr_3;

       . . .

   else
      sig <= value_expr_n;
   end if;
end process;
```

Thus, our discussion in Chapter 4 regarding the conditional signal assignment statement can also be applied to the if statement.

The equivalency, however, is true only for this simple scenario. An if statement is much more general since a branch of the if statement can be a *sequence* of sequential statements. Proper and disciplined use of an if statement can make code more descriptive and sometimes even more efficient. For example, an if statement is a sequential statement, and thus it can be nested in a branch of another if statement. Assume that we want to find the maximum value of three signals, a, b and c. One way to do it is by using nested if statements:

```
process(a,b,c)
begin
   if (a > b) then
      if (a > c) then
         max <= a;    --- a>b and a>c
      else
```

```
        max <= c;   --- a>b and c>=a
      end if;
   else
      if (b > c) then
         max <= b;   --- b>=a and b>c
      else
         max <= c;   --- b>=a and c>=b
      end if;
   end if;
end process;
```

We have to use three conditional signal assignment statements to achieve the same task:

```
signal ac_max, bc_max: std_logic;
. . .
ac_max <= a when (a > c) else c;
bc_max <= b when (b > c) else c;
max <= ac_max when (a > b) else bc_max;
```

We can also convert code using one conditional signal assignment statement. Since it cannot be nested, we have to "flatten" the Boolean conditions of the if statements. The following code follows the pattern of the previous nested Boolean conditions:

```
max <= a when ((a > b) and (a > c)) else
       c when (a > b) else
       b when (b > c) else
       c;
```

Although the code is shorter, it is not very descriptive and is difficult to understand.

Another situation suitable for the if statement is when many operations are controlled by the same Boolean conditions. For example, consider the following code segment:

```
process(a,b)
begin
   if (a > b and op="00") then
      y <= a - b;
      z <= a - 1;
      status <= '0';
   else
      y <= b - a;
      z <= b - 1;
      status <= '1';
   end if;
end process;
```

The Boolean conditions and the if-then-else structure is shared by three signals. On the other hand, we need three conditional signal assignment statements to describe the same circuit:

```
y <= a-b when (a > b and op="00") else
        b-a;
z <= a-1 when (a > b and op="00") else
        b-1;
status <= '0' when (a > b and op="00") else
             '1';
```

5.4.3 Incomplete branch and Incomplete signal assignment

In Section 5.1.2, we learned that an incomplete sensitivity list, in which one or more input signals are omitted, may lead to unexpected circuit behavior. This may also happen to the *incomplete branch* and *incomplete signal assignment*. According to VHDL semantics, the else branch is optional and a signal does not need to be assigned in all branches. Although syntactically correct, the omissions introduce unwanted memory elements (i.e., latches).

Incomplete branch In VHDL, only the then branch is mandatory and the other branches can be omitted. For example, the following statement is an attempt to code a comparator that compares the a and b inputs and asserts the eq output when a and b are equal:

```
process(a,b)
begin
   if (a=b) then
        eq <= '1';
   end if;
end process;
```

The code is syntactically correct. When a is equal to b, the eq signal becomes '1'. When a is not equal to b, there is no else branch and thus no action is taken. VHDL semantics specifies that the eq signal does not change and *keeps its previous value*. Thus, the previous statement is the same as

```
process(a,b)
begin
   if (a=b) then
        eq <= '1';
   else
        eq <= eq;
   end if;
end process;
```

This implies a circuit with a closed feedback loop, which constitutes internal states or memory. Clearly, this description does not meet the intended specification. The correct code should be

```
process(a,b)
begin
   if (a=b) then
        eq <= '1';
   else
        eq <= '0';
   end if ;
end process;
```

For a combinational circuit, the else branch should always be included to avoid the unwanted memory or latch.

Incomplete signal assignment An if statement has several branches. It is possible that a signal is assigned only in some, but not all branches. Although syntactically correct, the incomplete signal assignment infers unwanted memory. For example, the following statement attempts to describe a comparator with three outputs, gt, lt and eq, which indicate the conditions "a is greater than b," "a is less than b" and "a is equal to b" respectively:

```
process (a,b)
begin
   if  (a>b) then
       gt <= '1';
   elsif (a=b) then
       eq <= '1';
   else
       lt <= '1';
   end if;
end process;
```

The VHDL semantics specifies that a signal will *keep its previous value* if it is not assigned. When a is greater than b, the first branch is taken and the gt signal becomes '1'. The eq and lt signals keep their previous values since they are not assigned. A similar situation occurs in two other branches since only one output signal is assigned. This implies that three unwanted memory elements are inferred from the code. The correct code should have the signals assigned in all branches:

```
process (a,b)
begin
   if  (a>b) then
       gt <= '1';
       eq <= '0';
       lt <= '0';
   elsif (a=b) then
       gt <= '0';
       eq <= '1';
       lt <= '0';
   else
       gt <= '0';
       eq <= '0';
       lt <= '1';
   end if;
end process;
```

One way to make the code compact and clear is to assign a default value for each signal in the beginning of the process:

```
process (a,b)
begin
   gt <= '0';
   eq <= '0';
   lt <= '0';
   if  (a>b) then
       gt <= '1';
   elsif (a=b) then
       eq <= '1';
   else
       lt <= '1';
   end if;
end process;
```

Recall that, in a process, only the last signal assignment takes effect. If a signal is assigned in a branch of the if statement, that assignment takes effect. If it is not assigned in any branch, the default assignment takes effect. The output signals are therefore always assigned. We can treat the assignment of a default value as shorthand for the previous code segment.

For a combinational circuit, an output signal should be assigned in all branches of an if statement. It is a good practice to assign a default value at the beginning of the process to cover the unassigned branches.

5.4.4 Conceptual Implementation

An if statement evaluates a set of Boolean expressions in sequential order and takes action when the first Boolean condition is met. To achieve this in hardware, we need a priority routing network similar to that of a conditional signal assignment statement.

Discussion in Section 5.4.2 shows that a simple one-output-signal if statement is equivalent to a conditional signal assignment statement. We can apply the same procedure as that described in Section 4.3.2 to derive the conceptual block diagram for the simple if statement. Consider an if statement with four branches:

```
if boolean_expr_1 then
    sig <= value_expr_1;
elsif boolean_expr_2 then
    sig <= value_expr_2;
elsif boolean_expr_3 then
    sig <= value_expr_3;
else
    sig <= value_expr_4;
end if;
```

We first derive the circuit for the first branch by constructing the rightmost 2-to-1 multiplexing circuit and the boolean_expr_1 and value_expr_1 circuits. We can then repeat the process and complete the implementation branch by branch. The finished diagram is identical to Figure 4.4.

An if statement is more flexible and can accommodate more than one statement in each branch. The following examples illustrate the construction of two more complex forms. The first form is an if statement with multiple statements in each branch. The following code shows an if statement with two output signals:

```
if boolean_expr then
    sig_a <= value_expr_a_1;
    sig_b <= value_expr_b_1
else
    sig_a <= value_expr_a_2;
    sig_b <= value_expr_b_2;
end if;
```

Since there are two signals, two separating routing networks are needed. When each routing network has its own multiplexer, the two networks use the same Boolean expressions to control the selection signals of the multiplexers. Thus, the boolean_exp circuit is actually shared. The conceptual diagram is shown in Figure 5.4. We can apply the same idea to derive a conceptual diagram for an if statement with more output signals.

The second form is a nested if statement; that is, one or more if statements is used inside the branches of an if statement. The following code shows a two-level nested if statement:

```
if boolean_expr_1 then
    if boolean_expr_2 then
```

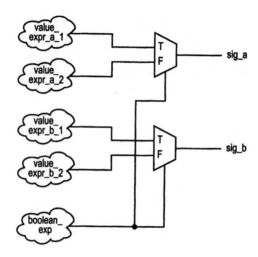

Figure 5.4   Conceptual implementation of an if statement with multiple signal assignments.

```
      signal_a <= value_expr_1;
   else
      signal_a <= value_expr_2;
   end if;
else
   if boolean_expr_3 then
      signal_a <= value_expr_3;
   else
      signal_a <= value_expr_4;
   end if;
end if;
```

The conceptual diagram can be constructed in a hierarchal manner, and the derivation process is shown in Figure 5.5. We first derive the routing structure for the outer if statement, as in Figure 5.5(a), and then realize the two inner if statements inside the then and else branches of the outer if statement, as in Figure 5.5(b). We can apply this procedure repeatedly if the code consists of more nested levels.

5.4.5   Cascading single-branched if statements

Because of the sequential semantics, a signal can be assigned multiple times inside a process and only the last assignment takes effect. We can use this property to construct a priority circuit using a sequence of single-branched if statements (i.e., if statements with only a then branch). For example, the previous priority encoder can be rewritten using three if statements, as shown in Listing 5.5. The code signal is first assigned with "00". If the r(1) request is asserted, the code signal will be reassigned with "01". This procedure continues until the end of the process. Clearly, the Boolean conditions in the later if statements have the higher priority and can override the earlier conditions. Thus, this sequence of if statements implicitly forms a priority circuit.

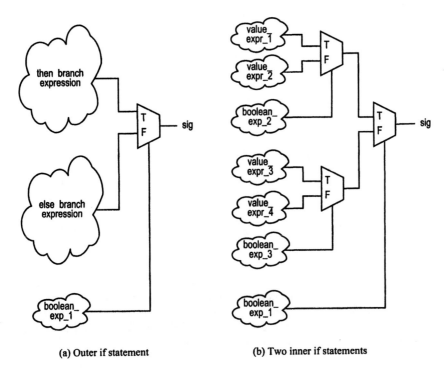

(a) Outer if statement   (b) Two inner if statements

Figure 5.5   Conceptual implementation of a nested if statement.

Listing 5.5   Priority encoder based on cascading if statements

```
architecture cascade_if_arch of prio_encoder42 is
begin
    process(r)
    begin
        code <="00";
        if (r(1)='1') then
            code <= "01";
        end if;
        if (r(2)='1') then
            code <= "10";
        end if;
        if (r(3)='1') then
            code <= "11";
        end if;
    end process;
    active <= r(3) or r(2) or r(1) or r(0);
end cascade_if_arch;
```

We can generalize this idea and replace an if statement with multiple elsif branches with a sequence of simple cascading single-branched if statements. For example, consider the following code segment:

```
if boolean_expr_1 then
    sig <= value_expr_1;
```

```
    elsif boolean_expr_2 then
        sig <= value_expr_2;
    elsif boolean_expr_3 then
        sig <= value_expr_3;
    else
        sig <= value_expr_4;
    end if;
```

It can be rewritten as

```
    sig <= value_expr_4;
    if boolean_expr_3 then
        sig <= value_expr_3;
    end if;
    if boolean_expr_2 then
        sig <= value_expr_2;
    end if;
    if boolean_expr_1 then
        sig <= value_expr_1;
    end if;
```

Although inferring the same priority configuration, VHDL code in this form is less clear and more difficult for software to synthesize. We should avoid this style in general. However, due to its repetitive nature, this form is sometimes useful to describe a replicated structure of parameterized design. This aspect is discussed in detail in Chapter 14.

5.5 CASE STATEMENT

5.5.1 Syntax and examples

The simplified syntax of a case statement is

```
    case case_expression is
        when choice_1 =>
            sequential statements;
        when choice_2 =>
            sequential statements;
        . . .
        when choice_n =>
            sequential statements;
    end case;
```

A case statement uses the value of case_expression to select a set of sequential statements. The case_expression term functions just as the select_expression term of the concurrent selected signal assignment statement. Its data type must be a discrete type or one-dimensional array. The choice_i term is a value or a set of values that can be assumed by case_expression. The choices have to be *mutually exclusive* (i.e., no value can be used more than once) and *all-inclusive* (i.e., all values must be included). The keyword **others** can be used in choice_n in the end to cover all unused values.

Again, we use the same multiplexer, binary decoder, priority encoder and ALU circuits to show the use of the case statement. The VHDL code of the multiplexer is shown in Listing 5.6. Note that there are 81 (9*9) possible combinations for the 2-bit s signal, including the normal "00", "01", "10" and "11" combinations as well as 77 other metavalue combinations. This issue was examined in the selected signal assignment statement in Section 4.4.1, and the discussion can be applied to the case statement as well. In the code, we use the **when others** clause to cover "11" and all unused combinations. The signals used in case_expression are the inputs to the circuit and thus should be included in the sensitivity list.

Listing 5.6   4-to-1 multiplexer based on a case statement

```
architecture case_arch of mux4 is
begin
    process(a,b,c,d,s)
    begin
5       case s is
            when "00" =>
                x <= a;
            when "01" =>
                x <= b;
10          when "10" =>
                x <= c;
            when others =>
                x <= d;
        end case;
15  end process;
end case_arch;
```

The VHDL codes for the other three examples are shown in Listings 5.7, 5.8 and 5.9.

Listing 5.7   2-to-4 decoder based on a case statement

```
architecture case_arch of decoder4 is
begin
    process(s)
    begin
5       case s is
            when "00" =>
                x <= "0001";
            when "01" =>
                x <= "0010";
10          when "10" =>
                x <= "0100";
            when others =>
                x <= "1000";
        end case;
15  end process;
end case_arch;
```

Listing 5.8   4-to-2 priority encoder based on a case statement

```
architecture case_arch of prio_encoder42 is
begin
    process(r)
    begin
5       case r is
            when "1000"|"1001"|"1010"|"1011"|
                 "1100"|"1101"|"1110"|"1111" =>
```

```
                      code <= "11";
                when "0100"|"0101"|"0110"|"0111" =>
 10                   code <= "10";
                when "0010"|"0011" =>
                      code <= "01";
                when others =>
                      code <= "00";
 15        end case;
        end process;
        active <= r(3) or r(2) or r(1) or r(0);
    end case_arch;
```

Listing 5.9   Simple ALU based on a case statement

```
architecture case_arch of simple_alu is
    signal src0s, src1s: signed(7 downto 0);
begin
    src0s <= signed(src0);
 5  src1s <= signed(src1);
    process(ctrl,src0,src1,src0s,src1s)
    begin
        case ctrl is
            when "000"|"001"|"010"|"011" =>
 10             result <=  std_logic_vector(src0s + 1);
            when "100" =>
                result <=  std_logic_vector(src0s + src1s);
            when "101" =>
                result <= std_logic_vector(src0s - src1s);
 15         when "110" =>
                result <= src0 and src1;
            when others =>    -- "111"
                result <= src0 or src1;
        end case;
 20 end process;
   end case_arch ;
```

5.5.2 Comparison to a selected signal assignment statement

A case statement is somewhat like a concurrent selected signal assignment statement. If each when branch of a case statement consists only of the assignment of a single signal, the two statements are equivalent. Consider a selected signal assignment statement:

```
with sel_exp select
    sig <= value_expr_1 when choice_1,
           value_expr_2 when choice_2,
           value_expr_3 when choice_3,
           . . .
           value_expr_n when choice_n;
```

It can be rewritten as

```
process (...)
begin
```

```
    case sel_exp is
       when choice_1 =>
           sig <= value_expr_1;
       when choice_2 =>
           sig <= value_expr_2;
       when choice_3 =>
           sig <= value_expr_3;
         . . .
       when choice_n =>
           sig <= value_expr_n;
    end case;
  end process;
```

Thus, the discussion in Chapter 4 regarding the selected signal assignment statement can also be applied to a case statement. Again, the equivalence is limited to this simple scenario. The case statement is much more flexible and general since each when branch can consist of a sequence of sequential statements. The comparison between the if statement and the conditional signal assignment statement in Section 5.4.2 can be applied here as well.

5.5.3 Incomplete signal assignment

Unlike an if statement, the choices of a case statement have to be inclusive, and thus no omitted when clause is allowed. Any "incomplete when clause" will lead to a syntax error and thus be detected when the VHDL code is analyzed. However, incomplete signal assignment can still occur and infer unwanted memory. For example, the following statement attempts to describe a priority encoder with a 3-bit input request signal, a, and three output signals, high, middle and low. The a(3) signal has the highest priority. When it is '1', the high signal will be asserted. The two other output signals are for two other lower requests. The code uses a case statement:

```
  process(a)
  begin
    case a is
       when "100"|"101"|"110"|"111" =>
           high <= '1';
       when "010"|"011" =>
           middle <= '1';
       when others =>
           low <='1';
    end case;
  end process;
```

Again, the VHDL semantics specifies that a signal will keep its previous value if it is unassigned. If the a signal is "111", the first when clause is taken and the high signal is assigned a '1'. Since the middle and low signals are unspecified, they keep their previous values. A similar situation occurs in other when clauses, and therefore three unwanted memory elements are inferred. To fix the problem, we must make sure to have the signals assigned in all when clauses:

```
  process(a)
  begin
    case a is
       when "100"|"101"|"110"|"111" =>
```

```
            high <= '1';
            middle <= '0';
            low <= '0';
         when "010"|"011" =>
            high <= '0';
            middle <= '1';
            low <= '0';
         when others =>
            high <= '0';
            middle <= '0';
            low <= '1';
      end case;
   end process;
```

As in the if statement discussion, we can also use a default assignment to make the code clearer and more compact:

```
process(a)
begin
   high <= '0';
   middle <= '0';
   low <= '0';
   case a is
      when "100"|"101"|"110"|"111" =>
         high <= '1';
      when "010"|"011" =>
         middle <= '1';
      when others =>
         low <='1';
   end case;
end process;
```

5.5.4 Conceptual implementation

A case statement utilizes the value of case_expression to select a set of sequential statements to execute. Conceptually, it can be thought of as an abstract multiplexing circuit that utilizes case_expression as the selection signal to route the results of designated expressions to output signals. A case statement with a single output signal can be implemented by an abstract multiplexer identical to the one used in the selected signal assignment statement in Section 4.4.2. Consider the following case statement:

```
case case_exp is
   when c0 =>
      sig <= value_expr_0;
   when c1 =>
      sig <= value_expr_1;
   when others =>
      sig <= value_expr_n;
end case;
```

We assume that case_exp may result in one of five possible values: c0, c1, c2, c3 and c4. The **when others** clause implicitly covers c2, c3 and c4. The conceptual diagram is identical to Figure 4.11 except that the selection_expression circuit is replaced by the case_exp circuit.

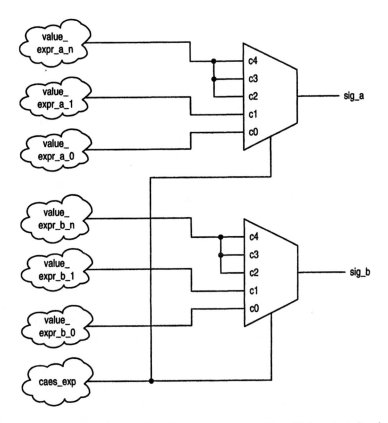

Figure 5.6   Conceptual implementation of a case statement with multiple output signals.

The previous scheme can easily be extended for a case statement with multiple output signals. We simply duplicate the abstract multiplexer for each signal and connect the case_exp to the selection signals of all multiplexers. For example, the following case statement has two output signals:

```
case case_exp is
   when c0 =>
      sig_a <= value_expr_a_0;
      sig_b <= value_expr_b_0;
   when c1 =>
      sig_a <= value_expr_a_1;
      sig_b <= value_expr_b_1;
   when others =>
      sig_a <= value_expr_a_n;
      sig_b <= value_expr_b_n;
end case;
```

The corresponding conceptual diagram is shown in Figure 5.6. As an if statement, a case statement is very general. Any valid sequence of sequential statements can be included inside a when branch. We can derive the conceptual diagram from the outermost level and iterate through the inner levels, as in the nested if statement example in Section 5.4.4.

5.6   SIMPLE FOR LOOP STATEMENT

5.6.1   Syntax

VHDL provides a variety of loop constructs, including *simple infinite loop*, *for loop* and *while loop*, as well as mechanisms to terminate a loop, including the *exit* statement, which skips the remaining iterations of the loop, and the *next* statement, which skips the remaining part of the current iteration. These constructs are mainly for modeling. Only few, very restricted forms of loop can be realized by hardware and synthesized automatically. In this section, we limit the discussion to the simple for loop statement and use it as shorthand for repetitive statements. A more general application of loops is discussed in Chapter 14.

The simplified syntax of the for loop statement is:

```
for index in loop_range loop
    sequential statements;
end loop;
```

The for loop repeats the loop body of sequential statements for a fixed number of iterations. The loop_range term specifies a range of values between the left and right bounds. A loop index, index, is used to keep track of the iteration and takes a successive value from loop_range in each iteration, starting with the leftmost value. The loop index automatically takes the data type of loop_range's element and does not need to be declared. For synthesizable code, loop_range must be determined at the time of synthesis (i.e., be *static*) and cannot change with the input signal. The loop body is a sequence of sequential statements. It is very flexible and versatile but can be difficult or impossible to synthesize. In this chapter, we limit it to sequential signal assignment statements.

5.6.2   Examples

Use of the for loop statement is demonstrated by two examples. The first example is a 4-bit xor circuit and its code is shown in Listing 5.10. The for loop performs bitwise xor operation on two 4-bit signals. The operation is done one bit at a time. The loop range is WIDTH-1 **downto** 0. We use a symbolic constant here to make the code more readable and to facilitate future modification. The loop index is i. It is local to the loop and does not need to be declared. The index assumes a value of 3, the leftmost value in the range, in the first iteration, and then assumes a value of 2 in the second iteration. The iteration continues until the value of the rightmost value, 0, is used.

Listing 5.10   Bitwise xor operation using a for loop statement

```
library ieee;
use ieee.std_logic_1164.all;
entity bit_xor is
    port(
5       a, b: in std_logic_vector(3 downto 0);
        y: out std_logic_vector(3 downto 0)
    );
end bit_xor;

10 architecture demo_arch of bit_xor  is
    constant WIDTH: integer := 4;
begin
```

```
      process(a,b)
      begin
15        for i in (WIDTH-1) downto 0 loop
              y(i) <= a(i) xor b(i);
          end loop;
      end process;
  end demo_arch;
```

The code here is just for demonstration purposes. The same operation can actually be achieved by a single statement:

```
y <= a xor b;
```

The second example is a *reduced-xor* circuit, which performs the xor operation over a group of signals. For example, consider a group of four signals, a_3, a_2, a_1 and a_0. The reduced-xor operation of the four signals is $a_3 \oplus a_2 \oplus a_1 \oplus a_0$. The for-loop VHDL description of this circuit is shown in Listing 5.11.

Listing 5.11   Reduced-xor operation using a for loop statement

```
   library ieee;
   use ieee.std_logic_1164.all;
   entity reduced_xor_demo is
       port(
5          a: in std_logic_vector(3 downto 0);
           y: out std_logic
       );
   end reduced_xor_demo;

10 architecture demo_arch of reduced_xor_demo is
       constant WIDTH: integer := 4;
       signal tmp: std_logic_vector(WIDTH-1 downto 0);
   begin
       process(a,tmp)
15     begin
           tmp(0) <= a(0);    -- boundary bit
           for i in 1 to (WIDTH-1) loop
               tmp(i) <= a(i) xor tmp(i-1);
           end loop;
20     end process;
       y <= tmp(WIDTH-1);
   end demo_arch;
```

5.6.3 Conceptual Implementation

The basic way to realize a for loop in hardware is to *unroll* or *flatten* the loop and convert it into code that contains no loop constructs. The flattened code can then be constructed accordingly. This implies that we replicate the hardware described by the loop body for each iteration. To unroll a loop, the range has to be constant and has to be known at the time of synthesis. That is why the range has to be static. We cannot, for example, use the value of an input signal to set the range's right boundary.

Let us first consider the bitwise xor code. The for loop can be unrolled by manually substituting index i into the loop body for four iterations. The flattened code becomes

```
y(3) <= a(3) xor b(3);
y(2) <= a(2) xor b(2);
y(1) <= a(1) xor b(1);
y(0) <= a(0) xor b(0);
```

We can derive the conceptual implantation accordingly. Similarly, the reduced-xor code can be unrolled and the flattened code is

```
tmp(0) <= a(0);
tmp(1) <= a(1) xor tmp(0);
tmp(2) <= a(2) xor tmp(1);
tmp(3) <= a(3) xor tmp(2);
y <= tmp(3);
```

Since we limit the loop body to sequential signal assignment statements, the implementation is straightforward. The for loop can be thought of as shorthand for repetitive statements. A unique property of the for loop is that we can use the range to control hardware replication. It is very useful for the development of parameterized design, in which the "width" of the circuit (e.g., the input width of an adder) can be adjusted to match a specific need. For example, we can change the value of the WIDTH constant of the reduced-xor code to accommodate different input widths. The implementation and synthesis of more versatile loop structure and the parameterized design are examined in Chapters 14 and 15.

5.7 SYNTHESIS OF SEQUENTIAL STATEMENTS

The nature of concurrent and sequential statements is very different. Concurrent statements are modeled after hardware, and thus there is a clear, direct mapping between a concurrent statement and a hardware structure. On the other hand, sequential statements are intended to describe the abstract behavior of a system, and some constructs cannot be easily realized by hardware. Sequential statements are more flexible and versatile than concurrent statements. For synthesis, this is a mixed blessing. On the positive side, the flexibility allows us to specify the desired design in a compact, clear and descriptive manner and to explore more design alternatives. On the negative side, the flexibility can easily be abused. It may make us think falsely that we can synthesize hardware directly from sequential descriptions. This usually leads to unnecessarily complex or unsynthesizable implementation.

Our goal is to develop code for synthesis. When we use sequential statements, we should think in terms of hardware rather than treating them as a way to describe a sequential algorithm. This helps us to focus on the underlying hardware complexity and the efficiency of the design. One good way to check sequential statements is to ask ourselves whether we can derive the conceptual diagram manually. If we cannot, the description is probably also too difficult for synthesis software to synthesize.

5.8 SYNTHESIS GUIDELINES

5.8.1 Guidelines for using sequential statements

- Variables should be used with care. A signal is generally preferred. A statement like n:=n+1 can cause great confusion for synthesis.

- Except for the default value, avoid overriding a signal multiple times in a process.

- Think of the if and case statements as routing structures rather than as sequential control constructs.

- An if statement infers a priority routing structure, and a larger number of elsif branches leads to a long cascading chain.

- A case statement infers a multiplexing structure, and a large number of choices leads to a wide multiplexer.

- Think of a for loop statement as a mechanism to describe the replicated structure.

- Avoid "innovative" use of language constructs. We should be innovative about the hardware architecture but not about the description of the architecture. Synthesis software may not be able to interpret the intention of the code.

5.8.2 Guidelines for combinational circuits

- For a combinational circuit, include all input signals in the sensitivity list to avoid unexpected behavior.

- For a combinational circuit, include all branches of an if statement to avoid unwanted latch.

- For a combinational circuit, an output signal should be assigned in every branch of the if and case statements to avoid unwanted latch.

- For a combinational circuit, it is a good practice to assign a default value to each signal at the beginning of the process.

5.9 BIBLIOGRAPHIC NOTES

The bibliographic information for this chapter is similar to that of Chapter 4.

Problems

5.1 Consider a circuit described by the following code segment:

```
process(a)
begin
   q <= d;
end process;
```

    (a) Describe the operation of this circuit.
    (b) Does this circuit resemble any real physical component?

5.2 Consider the following code segment:

```
process(a,b)
begin
   if a='1' then
      q <= b;
   end if;
end process;
```

(a) Describe the operation of this circuit.

(b) Draw the conceptual diagram of this circuit.

5.3 Add an enable signal, en, to the 2-to-4 decoder discussed in Section 5.4.1. When en is '1', the decoder functions as usual. When en is '0', the decoder is disabled and the output becomes "0000". Use an if statement to derive this circuit and draw the conceptual diagram.

5.4 Repeat Problem 5.3, but use a case statement to derive the circuit.

5.5 Derive the conceptual diagram for the following code segment:

```
if (a > b and op="00") then
    y <= a - b;
    z <= a - 1;
    status <= '0';
else
    y <= b - a;
    z <= b - 1;
    status <= '1';
end if;
```

5.6 Consider the 2-by-2 switch discussed in Problem 4.3. Its inputs are x1, x0 and ctrl, and its outputs are y1 and y0. The functional table is shown below. Use one if statement to derive the circuit.

ctrl	y1	y0
0 0	x1	x0
0 1	x0	x1
1 0	x0	x0
1 1	x1	x1

5.7 Repeat Problem 5.6, but use a case statement to derive the circuit.

5.8 Consider the following code segment:

```
if (a > b) then
    y <= a - b;
else
    if (a > c) then
        y <= a - c;
    else
        y <= a + 1;
    end if;
end if;
```

(a) Draw the conceptual diagram.

(b) Rewrite the code using two concurrent conditional signal assignment statements.

(c) Rewrite the code using one concurrent conditional signal assignment statement.

(d) Rewrite the code using one case statement.

5.9   Consider the following code segment:

```
y <= (others=>'0');
if (a > b) then
    y <= a - b;
end if;
if (crtl='1') then
    y <= c;
end if;
```

(a) Rewrite the code using one if statement.

(b) Draw the conceptual diagram.

5.10   Assume that op is a 2-bit signal with the std_logic_vector data type. Consider the following code segment:

```
case op is
    when "00" =>
        y <= (others => '0');
    when "01" =>
        if (a > 0) then
            y <= a - 1;
        else
            y <= a + 1;
        end if;
    when others =>
        y <= a + b;
end case;
```

(a) Draw the conceptual diagram.

(b) Rewrite the code using concurrent conditional and selected signal assignment statements.

5.11   Consider the shift-left circuit discussed in Problem 4.6. The inputs include a, which is an 8-bit signal to be shifted, and ctrl, which is a 3-bit signal specifying the amount to be shifted. Both are with the std_logic_vector data type. The output y is an 8-bit signal with the std_logic_vector data type. Use an if statement to derive the circuit and draw the conceptual diagram.

5.12   Repeat Problem 5.11, but use a case statement.

CHAPTER 6

SYNTHESIS OF VHDL CODE

Synthesizing VHDL code is the process of realizing the VHDL description using the primitive logic cells from the target device's library. In Chapters 4 and 5, we discussed how to derive a conceptual diagram from VHDL statements. The conceptual diagram can be considered as the first step in realizing the code. The diagram is refined further during synthesis. The synthesis process involves complex algorithms and a large amount of data, and computers are needed to facilitate the process. Although today's synthesis software appears to be sophisticated and capable, there are fundamental limitations. Understanding the capability and limitation of synthesis software will help us better utilize this tool and derive more efficient designs. This chapter explains the realization of VHDL operators and data types, provides an in-depth overview on the synthesis process, and discusses the timing issue involved in synthesis.

6.1 FUNDAMENTAL LIMITATIONS OF EDA SOFTWARE

Developing a large digital circuit is a complicated process and involves many difficult tasks. We have to deal with complex algorithms and procedures and handle a large amount of data. Computers are used to facilitate the process. As computers become more powerful, we may ask if it is possible to develop a suite of software and completely automate the synthesis process. The ideal scenario is that human designers would only need to develop a high-level behavioral description and EDA software would perform the synthesis and placement and routing and automatically derive the optimal circuit implementation. The is unfortunately not possible. The limitation comes from the theoretical study of computational algorithms.

Although this book does not cover EDA algorithms, it will be helpful to know the capability and limitation of EDA software tools so that they can be used effectively.

For the purposes of discussion, we can separate an EDA software tool into a core and a shell. The *core* is the algorithms that perform the transformation or optimization, and the *shell* wraps the algorithm, including data conversion, memory and file management and user interface. Although the shell is important, the core algorithms ultimately determine the quality and efficiency of the software tool. The problems encountered in EDA are not unique. In fact, they are formulated and transformed into optimization problems in other fields, especially in the study of graph theory. This section provides a layperson's overview of computability and computation complexity, which helps us understand the fundamental limitation of EDA software.

6.1.1 Computability

Computability concerns whether a problem can be solved by a computer algorithm. If an algorithm exists, the problem is *computable* (or *decidable*). Otherwise, the problem is *uncomputable* (or *undecidable*). An example of an uncomputable problem is the "halting problem." Some programs, such as a compiler, take another program as input and check certain properties (e.g., syntax) of that program. The halting problem asks whether we can develop a program that takes any program and its input and determines whether computation of that program will eventually halt (e.g., no infinite loop). It can be proven mathematically that no such program can be developed, and thus the halting problem is uncomputable. Informally speaking, any attempt to examine the "meaning" of a program is uncomputable.

Equivalence checking discussed in Section 1.5.3 essentially compares whether two programs perform the same function, which goes further than the halting problem. Therefore, equivalence checking is uncomputable; i.e., it is not possible to develop an EDA tool that determines the equivalence of *any* two descriptions. However, it is possible to use some clever techniques to determine the equivalence of some descriptions, which are coded following certain guidelines. Thus, while equivalence checking cannot guarantee to work all of the time, it can be useful some of the time.

6.1.2 Computation complexity

If a problem is computable, an algorithm can be derived to solve the problem. The *computation complexity* concerns the efficiency of an algorithm. The computation complexity can be further divided into *time complexity*, which is a measure of the time needed to complete the computation, and *space complexity*, which is a measure of hardware resources, such as memory, needed to complete the computation. Since most statements on time complexity can be applied to space complexity as well, in the remaining section we focus on time complexity.

Big-O notation   The computation time of an algorithm depends on the size of the input as well as on the type of processor, programming language, compiler and even personal coding style. It is difficult to determine the exact time needed to complete execution of an algorithm. To characterize an algorithm, we normally focus on the impact of input size and try to filter out the effect of the "interferences" on measurement. Instead of determining the exact function for computation time, we usually consider only the *order* of this function.

The order is defined as follows. Given two functions, $f(n)$ and $g(n)$, we say that $f(n)$ is $O(g(n))$ (pronounced as $f(n)$ is big-O of $g(n)$ or $f(n)$ is of order $g(n)$) if two constants,

Table 6.1   Scaling of some commonly used big-O functions

Input size	Big-O function					
n	n	$\log_2 n$	$n \log_2 n$	n^2	n^3	2^n
2	2 μs	1 μs	2 μs	4 μs	8 μs	4 μs
4	4 μs	2 μs	8 μs	16 μs	64 μs	16 μs
8	8 μs	3 μs	24 μs	64 μs	512 μs	256 μs
16	16 μs	4 μs	64 μs	256 μs	4 ms	66 ms
32	32 μs	5 μs	160 μs	1 ms	33 ms	71 min
48	48 μs	5.5 μs	268 μs	2 ms	111 ms	9 years
64	64 μs	6 μs	384 μs	4 ms	262 ms	600,000 years

n_0 and c, can be found to satisfy

$$f(n) < cg(n) \text{ for any } n, n > n_0$$

The $g(n)$ function is normally a simple function, such as n, $n \log_2 n$, n^2, n^3 or 2^n. For example, all the following functions are $O(n^2)$:

- $0.1n^2$
- $n^2 + 5n + 9$
- $500n^2 + 1000000$

The purpose of big-O notation is twofold. First, it drops the less important, secondary terms since the highest-order term becomes the dominant factor as n becomes large. Second, it concentrates on the rate of change and ignores the constant coefficient in a function. After removing the constant coefficients and lower-order terms, we eliminate the effect of coding style, instruction set and hardware speed, and can concentrate on the effectiveness of an algorithm. Big-O notation is essentially a scaling factor or growth rate, indicating the resources needed as input size increases.

Commonly encountered orders are $O(1)$, $O(\log_2 n)$, $O(n)$, $O(n \log_2 n)$, $O(n^2)$, $O(n^3)$ and $O(2^n)$. $O(n)$ indicates the linear growth rate, in which the required computation resources increase in proportion to the input size. $O(1)$ means that the required computation resources are constant and do not depend on input size. $O(\log_2 n)$ indicates the logarithmic growth rate, which changes rather slowly. For a problem with $O(1)$ or $O(\log_2 n)$, the input size has very little impact on the resources. $O(n^2)$ and $O(n^3)$ have faster growth rates and the required computation resources become more significant as the input size increases. All of the orders discussed so far are considered as being of *polynomial order* since they have the form of $O(n^k)$, where k is a constant. On the other hand, $O(2^n)$ indicates the exponential growth rate and the computation time increases geometrically. Note that an increment of 1 in input size doubles the computation time. $O(2^n)$ grows faster than does any polynomial order.

An example using these functions is shown in Table 6.1, which lists the required computation times of algorithms of varying computation complexity. For comparison, we assume that it takes 2 μs for an $O(n)$ algorithm to perform a computation of input size 2. The table shows the required times as the input size increases from 2 to 64 under different big-O functions.

One example of $O(2^n)$ complexity is the exhaustive testing of a combinational circuit. One way to test a combinational circuit is to apply all possible input combinations exhaustively and examine their output responses. For a circuit with n inputs, there are

2^n possible input combinations. If we assume that the testing equipment can check 1 million patterns per second, exhaustively testing a 64-bit circuit takes about 600,000 years (i.e., $\frac{2^{64}}{10^6*60*60*24*365}$) to complete. Thus, although simple and straightforward, this method is not practical in reality.

Intractable and tractable problems   In most problems, if a polynomial order ($O(n^k)$) algorithm can be found, the exponent k is normally very small (say, 1, 2, or 3). Even though the growth rate is much worse than the linear rate, we can tolerate applying the algorithm to problems with nontrivial input sizes. We call these problems *tractable*. On the other hand, computation theory has shown that a polynomial-order solution cannot be found or is "unlikely" to be found for some problems. The only existing solutions are the algorithms with nonpolynomial order, such as $O(2^n)$. We call these problems *intractable*. As we have seen in Table 6.1, the computation time for the $O(2^n)$ algorithm simply grows too fast and the algorithm is not practical even for a moderate-sized n. Improvement in hardware speed will not change the situation significantly.

The situation is not completely hopeless for an intractable problem. An intractable problem usually means that it takes $O(2^n)$ computation time to find the *optimal* answer for *any* given input. It is frequently possible to find a polynomial-order algorithm, based on some smart tactics and *heuristics* (an educated guess), that permits us to obtain a valid, *suboptimal* answer or the optimal solution for *some* input patterns.

Synthesis as an intractable problem   The focus of this book is on describing a design in textual HDL code and then using synthesis software to realize the circuit. From the computation complexity point of view, the synthesis consists of several intractable problems, and thus no polynomial-time algorithm exists. We can treat the synthesis process as a searching procedure. For a given specification, there are possibly $O(2^n)$ valid circuit configurations. Finding the optimal configuration corresponds to a global search, exhaustively checking and comparing all $O(2^n)$ possible configurations. Real synthesis software must limit the search space. It normally performs the search on a local basis and applies some smart tactics and heuristics to guide the direction of the search. The starting point of the search corresponds to the configuration described in our HDL code. Since the search is local, the initial starting point plays a key role. A good initial description will put the starting point in a good location, and an efficient configuration can be obtained accordingly. On the other hand, if the initial description is poor, the good configurations will be far away. Since synthesis software doesn't perform a global search, it is unlikely that software can obtain an efficient configuration.

6.1.3 Limitations of EDA software

Like synthesis, other design tasks contain intractable or even undecidable computation problems. This is the inherent, theoretical limitation of EDA software and cannot be overcome by fast hardware, smart software code or human talents. Heuristics and tricks of software algorithms can sometimes find good solutions for certain types of inputs. There is no guarantee that the solutions are optimal or that the algorithm will work for all types of inputs. Therefore, it is impossible to use EDA software to completely automate the design process. This limitation is real and here to stay. The quality and efficiency of a design still rely on a human designer's experience, insight, ingenuity and imagination, which, to some degree, can be considered as the ultimate heuristics that cannot be coded into software.

6.2 REALIZATION OF VHDL OPERATORS

When we develop VHDL code for synthesis, language constructs in the code are eventually mapped to hardware. In the previous chapters, we illustrated the realization (i.e., the conceptual diagram) of basic concurrent and sequential statements. VHDL operators are used as building components in these diagrams. In a conventional programming language, we don't pay too much attention to the operators since most operations, including integer arithmetic operations, logical operations and shift operations, take the same amount of resources: one instruction cycle of the CPU. This is totally different in synthesis. Hardware complexities and operation speed of VHDL operators vary significantly and are processed differently during synthesis. To derive an efficient design, we have to be aware of the implications of VHDL operators on hardware implementation.

Only a subset of VHDL operators can be synthesized automatically. The subset normally includes the logical operators, relational operators as well as addition and subtraction operators. Some software may also include more complicated operators, such as shift or multiplication operators. Software can rarely automatically synthesize division (/), **mod**, **rem** and exponential (******) operators or any operators associated with floating-point datatype operands. The following subsections provide an overview of the realization of VHDL operators.

6.2.1 Realization of logical operators

Logical operators can be mapped directly to logic gates, and their synthesis is straightforward. The **and, nand, or** and **nor** operators have similar area and delay characteristics. The **xor** and **xnor** operators are slightly more involved and their implementation requires more silicon area and experiences a larger propagation delay.

In VHDL, a logical operation can be applied over operands with multiple bits. For example, let a and b be 8-bit signals with a data type of std_logic_vector(7 **downto** 0). The expression a **xor** b means that the **xor** operation is applied to eight individual bits in parallel. Since each bit of the input operates independently, the area of the circuit grows linearly with the number of input bits (i.e., on the order of $O(n)$), and the propagation delay is a constant (i.e., on the order of $O(1)$).

6.2.2 Realization of relational operators

There are six relational operators in VHDL: =, /=, <, <=, > and =>. According to their hardware implementation, these operators can be divided into the equality group, which includes the = and /= operators, and the greater–less group, which includes the other four operators.

In the equality group, operators can easily be implemented by a tree-like structure. For this implementation, the circuit area grows linearly with the number of input bits (i.e., $O(n)$), and the delay grows at a relatively slow $O(\log_2 n)$ rate. In the greater–less group, the operation exhibits a strong data dependency of input bits. For example, to determine the "greater than" relationship, we first have to compare the most significant bits of two operands and, if they are equal, the next lower bits and so on. This leads to larger area and propagation delay. Because of the circuit complexity, these operators can be implemented in a variety of ways, each with a different area–delay characteristic. In the minimal-area implementation, both area and delay grow linearly (i.e., $O(n)$) with the number of input bits. There are several different ways to improve the performance (i.e., reduce the delay), all at the expense of extra hardware.

6.2.3 Realization of addition operators

The addition operator (+) is the most basic arithmetic operator. Several other operators, including subtraction (-), negation (- with one operand) and absolute value (**abs**), can easily be derived from the addition operator.

The addition operation has an even stronger data dependency of individual bits since the least significant bit of input may affect the most significant bit of the result. It is normally the most complex operator that can be synthesized automatically. Since the adder is the basis of other arithmetic operations, its implementation has been studied extensively and a wide range of circuits that exhibit different area–delay characteristics has been developed. The minimal-area circuit, sometimes known as a *serial* or *ripple* adder, can easily be implemented by cascading a series of 1-bit full adders. In this implementation, both area and delay grow linearly (i.e., $O(n)$).

6.2.4 Synthesis support for other operators

Synthesis support for other more complicated operators is sporadic. It depends on individual synthesis software, the width of the input operands as well as the targeted device technology. Some high-end synthesis software can automatically derive multiplication operator (∗) and shift operators (**sll**, **srl**, **sla**, **sra**, **rol** and **ror** of VHDL, and `shift_left`, `shift_right`, `rotate_left` and `rotate_right` of the IEEE `numeric_std` library). Because of the hardware complexity, we must be extremely careful if these operators are used in a VHDL code. Synthesis software rarely supports division-related operators (/, **mod** and **rem**) or the exponential operator (∗∗) or any operators associated with floating-point data-type operands.

Since the emphasis in this book is on portable description, we will not use these operators in our VHDL codes. Examples in Chapters 8 and 15 show how to design and derive VHDL code for some of these operators.

6.2.5 Realization of an operator with constant operands

The operands of VHDL operators can sometimes be a constant expression, which does not depend on the value of any input signal. Such constant operands have a significant implication in the synthesis process.

Operator with all constant operands   If all the operands of an expression are constants, we can evaluate the expression in advance and replace it with a constant value. However, it is good practice to use constant symbols and constant expressions in VHDL code. They make the code more descriptive. For example, consider the following code segment:

```
constant OFFSET: integer := 8;
signal boundary: unsigned(8 downto 0);
signal overflow: std_logic;
. . .
overflow <= '1' when boundary > (2**OFFSET-1) else
            '0';
```

The operands of operators ** and - are constants, and the 2**OFFSET-1 expression can be replaced by a constant, 255. Although we can use 255 in VHDL code, it is less clear about how the value is obtained. In a large, complex VHDL program that involves many constant values, keeping track of the meaning of all constants becomes difficult. It is advisable to use constant symbols and constant expressions.

During synthesis, software can easily detect constant expressions and replace them with constants during preprocessing (in the elaboration phase of VHDL code). Since no physical hardware will be inferred from constant expressions, we can use them freely in VHDL code.

Operator with partial constant operands   Most VHDL operators have two operands. Sometimes one of the operands is a constant, as in count+1. Instead of using a full-fledged operator implementation, synthesis software can "propagate" and "embed" the constant value into the circuit implementation. From a synthesis point of view, a constant operand actually decreases the number of inputs of the circuit by half and thus can significantly reduce the circuit complexity. For example, if a and b are two 8-bit signals and op is a VHDL operator, implementing the a op b expression requires a combination circuit with 16 inputs. On the other hand, if one operand is a constant, say "0001001", implementing the a op "00010001" expression only requires a combination circuit with eight inputs.

The following three examples further depict the difference between a full-fledged circuit and the simplified implementation. The first example is of a rotation operator. Assume that x and y are 8-bit signals and consider the following rotation operation:

```
y <= rotate_right(x, 3);
```

Since the shifting amount is a constant of 3, no actual shifting circuit is needed. This operation can be implemented by properly connecting the input signals to the output signals, which requires no logic at all. It is the same as

```
y <= x(2 downto 0) & x(7 downto 3);
```

The second example is of an equality operator. Let us consider a 4-bit equality comparator with inputs of $x_3x_2x_1x_0$ and $y_3y_2y_1y_0$. The logic expression of this operation is

$$(x_3 \oplus y_3)' \cdot (x_2 \oplus y_2)' \cdot (x_1 \oplus y_1)' \cdot (x_0 \oplus y_0)'$$

If one operand is a constant, say, $y_3y_2y_1y_0 = 0000$, the expression can be simplified to

$$x_3' \cdot x_2' \cdot x_1' \cdot x_0'$$

The comparator is reduced to a 4-input nor gate. Thus, there is a significant difference between a full-fledged comparator and a reduced comparator.

The last example is of an addition operator. A frequently used operation in VHDL is incrementing: adding 1 to a signal, as in count+1. A minimal-area implementation of the addition operator is done by cascading 1-bit full adders. On the other hand, a minimal-area incrementor can be implemented by half adders, whose size is about one half that of full adders. Thus, the circuit area of an incrementor is only about one half that of a regular addition operator.

6.2.6   An example implementation

It will be helpful to have a comprehensive table that lists the areas and delays of synthesizable operators. However, because of the complexity of the synthesis process and device

Table 6.2 Circuit area and delay of some commonly used VHDL operators

Width	VHDL operator									
	nand	xor	$>_a$	$>_d$	=	$+1_a$	$+1_d$	$+_a$	$+_d$	mux
Area (gate count)										
8	8	22	25	68	26	27	33	51	118	21
16	16	44	52	102	51	55	73	101	265	42
32	32	85	105	211	102	113	153	203	437	85
64	64	171	212	398	204	227	313	405	755	171
Delay (ns)										
8	0.1	0.4	4.0	1.9	1.0	2.4	1.5	4.2	3.2	0.3
16	0.1	0.4	8.6	3.7	1.7	5.5	3.3	8.2	5.5	0.3
32	0.1	0.4	17.6	6.7	1.8	11.6	7.5	16.2	11.1	0.3
64	0.1	0.4	35.7	14.3	2.2	24.0	15.7	32.2	22.9	0.3

technology, a small variation in VHDL code, synthesis algorithm, or device parameters will lead to different results. Table 6.2 shows one synthesis result for several representative operators of different input widths in a 0.55-micron CMOS standard-cell technology. The subscripts a and d indicates that the circuit is optimized for area and for delay respectively.

The unit of area is a gate count, which is the equivalent number of 2-input nand gates used to implement the circuit, and the unit of propagation delay is the nanosecond (ns). We need to be cautious about the data in the table. The data is valid only for a particular version of a particular software on a particular device technology and should not be overly interpreted or analyzed. However, this data does show a general trend and provide a rough idea about the relative complexity of different operators. The information for a 2-to-1 multiplexer, which is the basic component for routing, is also included in the table for reference.

There are several important observations to be made from the table. First, as we expect, the area and propagation delay vary significantly among the different operators. For example, the area of a 32-bit fast addition operator is more than 10 times larger than that of a 32-bit nand operator, and the propagation delay of the adder is more than 100 times longer than that of the nand operator.

The second observation is about the trade-off between area and delay. In digital system design, it is generally not possible to find an optimal implementation, which has both minimal area and minimal delay. We normally have to invest more resources (a larger area) for better performance (less delay). Except for the trivial implementation of logical operators, other operators have multiple implementations with different area–delay characteristics. Table 6.2 shows the area and delay characteristics of two implementations, in which one is optimized for a smaller area and the other is optimized for less delay.

The third observation is about scaling, the impact of increasing the size of the input of an operator (e.g., from 8 bits to 16 bits to 32 bits). The growth rates of area and delay are not always linear (i.e., $O(n)$). In general, the growth rate of delay is on the order of $O(1)$, $O(\log_2 n)$ or $O(n)$, while the growth rate of area is between the orders $O(n)$ and $O(n^2)$. Since the commercial synthesis software normally does not reveal its internal algorithms, the growth rate observation is true only for this particular software and device. Chapter 15 provides an in-depth discussion of the design of some operators.

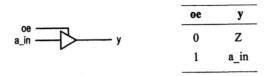

oe	y
0	Z
1	a_in

Figure 6.1 Tri-state buffer.

6.3 REALIZATION OF VHDL DATA TYPES

6.3.1 Use of the `std_logic` data type

VHDL supports a rich set of data types. During synthesis, these data types must be mapped into binary representations so that they can be realized in a physical circuit. The VHDL standard itself does not define the mapping mechanism, and thus the mapping is left for synthesis software. To have better control of the final implementation, we limit our use of data types primarily to the `std_logic` data type and its derivatives, the `std_logic_vector`, `signed` and `unsigned` data types. The only exception is the user-defined enumeration data type, which is used for the description of a finite state machine and is discussed in Chapter 9.

Recall that there are nine possible values in the `std_logic` data type. Among them, `'0'` and `'1'` are interpreted as logic 0 and logic 1 and are used in regular synthesis. `'L'` and `'H'` are interpreted as weak 0 and weak 1, as in wired logic. Since modern device technologies no longer use this kind of circuitry, the two values should not be used. `'U'`, `'X'` and `'W'` are meaningful only in modeling and simulation, and they cannot be synthesized. The two remaining values, `'Z'` and `'-'`, which represent high impedance and "don't-care" respectively, have some impact on synthesis. Their use is discussed in the following subsections.

6.3.2 Use and realization of the `'Z'` value

The `'Z'` value means high impedance or an open circuit. It is not a value in Boolean algebra but a special electrical property exhibited in a physical circuit. Only a special kind of component, known as a *tri-state buffer*, can have an output of this value. The symbol and function table of a tri-state buffer are shown in Figure 6.1. When the oe (for "output enable") signal is `'1'`, the buffer acts as a short circuit and the input is passed to output. On the other hand, when the oe signal is `'0'`, the y output appears to be an open circuit.

VHDL description of a tri-state buffer High impedance cannot be handled by regular logic and can exist only in the output of a tri-state buffer. The VHDL description of the tri-state buffer of Figure 6.1 is

```
y <= a_in when oe='1' else
     'Z';
```

We cannot use a value of `'Z'` as an input or manipulate it as a logic value. For example, the following statements cannot be realized and are meaningless in synthesis:

```
f <= 'Z' and a;
y <= data_a when in_bus='Z' else
     data_b;
```

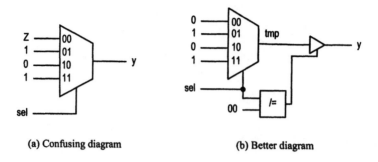

(a) Confusing diagram   (b) Better diagram

Figure 6.2   Use of 'Z' as an output value.

Since a tri-state buffer is not an ordinary logic value, it is a good idea to code it in a separate statement. For example, consider the following VHDL description:

```
with sel select
   y <= 'Z' when "00",
        '1' when "01"|"11",
        '0' when others;
```

Although the code is correct, direct transformation to a conceptual diagram, as shown in Figure 6.2(a), cannot be synthesized. To clarify the intended structure, the code should be modified as

```
with sel select
   tmp <= '1' when "01"|"11",
          '0' when others;
y <= tmp when sel/="00" else
     'Z';
```

Following the description, we can easily derive the intended block diagram, as shown in Figure 6.2(b).

The major application of a tri-state buffer is to implement a bidirectional I/O port to save the pin count and to form a bus.

VHDL description of a bidirectional I/O port   As a silicon device packs more circuitry into a chip, the number of I/O signals increases accordingly. A bidirectional I/O pin can be used as either an input or an output and thus makes more efficient use of an I/O pin. Most FPGA and memory devices utilize bidirectional I/O pins.

The schematic of a simple circuit with bidirectional I/O port, bi, is shown in Figure 6.3. The dir signal controls the direction of the I/O port. When it is '0', the port is used as an input port. The tri-state buffer is in a high-impedance state, and thus the sig_out signal is blocked. The external signal connected to the bi port is routed to the sig_in signal. When the dir signal is '1', the port is used as an output port and the sig_out signal is connected to an external circuit. Note that the sig_out signal is implicitly routed back to the sig_in signal when the dir signal is '1'. If this causes a problem, we can add an additional tri-state buffer to break the return path, as shown in Figure 6.4. Since the control signals of tri-state buffers are connected to a complementary enable signal, only one tri-state buffer is enabled at a time.

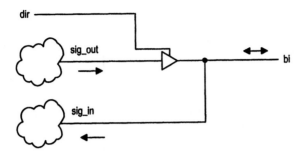

Figure 6.3 Single-buffer bidirectional I/O port.

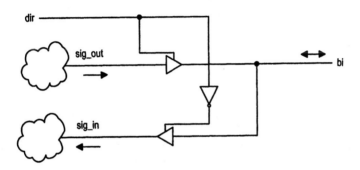

Figure 6.4 Dual-buffer bidirectional I/O port.

The VHDL description for a bidirectional port is straightforward. We first specify the mode as **inout** in port declaration and then describe the tri-state buffer accordingly. The VHDL segment for the single-buffer diagram of Figure 6.3 is

```
entity bi_demo is
    port(
        bi: inout std_logic;
        . . .

begin
    sig_out <= output_expression;
    . . .
    some_signal <= expression_with_sig_in;
    . . .
    bi <= sig_out when dir='1' else 'Z';
    sig_in <= bi;
    . . .
```

To accommodate the dual-buffer configuration of Figure 6.4, we just need to modify the last statement to reflect the change:

```
    sig_in <= bi when dir='0' else 'Z';
```

Tri-state buffer–based bus   Another application of the tri-state buffer is to form a bus. The diagram of a simple tri-state buffer–based bus (or simply tri-state bus) is shown in Figure 6.5, in which four sources are connected to the bus. The signal src_select specifies

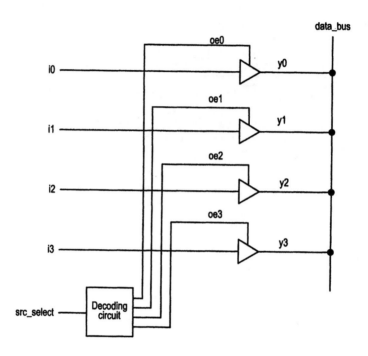

Figure 6.5 Tri-state bus.

which input source is to be placed on the bus. It is connected to a decoding circuit that generates four *non-overlapping* control signals, oe(0), oe(1), oe(2) and oe(3). Only one can be activated at a time, and the input connected to the activated buffer is placed on the bus. The VHDL code for this circuit is

```
-- binary decoder
with src_select select
    oe <= "0001" when "00",
          "0010" when "01",
          "0100" when "10",
          "1000" when others; -- "11"
-- tri-state buffers
y0 <= i0 when oe(0)='1' else 'Z';
y1 <= i1 when oe(1)='1' else 'Z';
y2 <= i2 when oe(2)='1' else 'Z';
y3 <= i3 when oe(3)='1' else 'Z';
data_bus <= y0;
data_bus <= y1;
data_bus <= y2;
data_bus <= y3;
```

Despite its simple appearance, the internal tri-state buses presents a serious problem in the development flow. Since the theoretical models of most EDA algorithms are based on Boolean algebra, which is defined according to two logic values, the software tools cannot handle the high-impedance state. The tri-state bus thus imposes a problem in optimization, timing analysis, verification and testing. Furthermore, internal tri-state bus is technology dependent, and thus the design is less portable.

Table 6.3   Function tables of a 3-to-2 priority encoder

Input	Output		Input	Output
req	code		req	code
1 0 0	10		1 – –	10
1 0 1	10		0 1 –	01
1 1 0	10		0 0 1	00
1 1 1	10		0 0 0	00
0 1 0	01			
0 1 1	01			
0 0 1	00			
0 0 0	00			

Table 6.4   Don't-care used as an output value

input	output
a b	f
0 0	0
0 1	1
1 0	1
1 1	–

A tri-state bus essentially performs multiplexing. For example, the previous design can be replaced by a 4-to-1 multiplexer:

```
with src_select select
    data_bus <= i0 when "00",
                i1 when "01",
                i2 when "10",
                i3 when others;    —— "11"
```

This scheme is more robust and portable and thus is the preferred choice. The major application of the tri-state bus is to construct the external back-plan bus of a printed circuit board. An add-on card can easily be added to or removed from the bus without affecting subsystems residing on other cards.

6.3.3 Use of the '–' value

Don't-care is not a valid logic value in Boolean algebra but is used to facilitate the design process. Don't-care can be used as an input value to make a function table clear and compact. For example, the original function table of a 3-input priority encoder is shown on the left of Table 6.3. When $req(2)$ is '1', the output should be "10" regardless of the values of other requests. Instead of using four rows, we can use 1-- to indicate the condition. The revised table, as on the right of Table 6.3, is more compact and more descriptive.

When used as an output value, don't-care indicates that the exact value is not important. This happens when some of the input combinations are not used. During the synthesis process, we can assign a value that helps to reduce the circuit complexity. A simple example is shown in Table 6.4, in which the output value for the input pattern "11" is don't-care. If don't-care is assigned to '0' during synthesis, f becomes $a' \cdot b + a \cdot b'$. On the other hand, when it is assigned to '1', f can be simplified to $a + b$, which requires much less hardware.

According to the definition of the std_logic data type, the '-' value is designated as "don't-care." However, VHDL treats '-' as an independent symbolic value of the std_logic data type rather than "0 or 1." This definition is somewhat different from our conventional use and may lead to unexpected behaviors and subtle mistakes. The following paragraphs discuss the use of this value.

Use of '-' as an input value   Let us first examine the issues related to using '-' as an input value. Consider the priority function of Table 6.3. We may be tempted to code the circuit as follows:

```
y <= "10" when req="1--" else
     "01" when req="01-" else
     "00" when req="001" else
     "00";
```

The code is syntactically correct. However, in a physical circuit, an input signal can only assume a value of '0' or '1' but never '-', and thus the req="1--" and req="01-" expressions will always be false. If the value of the req signal is "111", none of the Boolean expression is true and "00" will be assigned to y accordingly. To correct the problem, we have to eliminate the comparison of '-' in Boolean expressions:

```
y <= "10" when req(2)='1' else
     "01" when req(2 downto 1)="01" else
     "00" when req(2 downto 0)="001" else
     "00";
```

The code is just for demonstration purposes and is not very efficient. Better code for priority encoding circuit was illustrated in Section 4.3.1.

In the IEEE numeric_std package, there is a function, std_match(), which performs don't-care comparisons according to the traditional interpretation. The function compares two inputs of std_logic_vector data type and interprets '-' as a don't-care in a conventional sense. The previous code can be written as

```
. . .
use ieee.numeric_std.all;
. . .
y <= "10" when std_match(req,"1--") else
     "01" when std_match(req,"01-") else
     "00" when std_match(req,"001") else
     "00";
```

Our discussion of '-' is also applied to the choice expression in a selected signal assignment statement and case statement. For example, the following code seems to be the direct implementation of the compact function table of Table 6.3:

```
with req select
  y <= "10" when "1--",
       "01" when "01-",
       "00" when "001",
       "00" when others;
```

The code is syntactically correct. Again, since a physical input signal can never assume a value of '-', the choices "1--" and "01-" will never occur. If the value of the req signal is "111", there is no match and "00" will be assigned to y. There is no easy fix in this case. We must explicitly specify choice expressions in terms of '0' and '1', as in the original left function table of Table 6.3. The correct VHDL code is

```
with req select
    y <= "10" when "100"|"101"|"110"|"111",
         "00" when "010"|"011",
         "00" when others;
```

Use of `'-'` as an output value Don't-care can also be used as an output value and assigned to a signal. For example, the function table of Table 6.3 can easily be translated to VHDL code:

```
sel <= a & b;
with sel select
    y <= '0' when "00",
         '1' when "01",
         '1' when "10",
         '-' when others;
```

The code is syntactically correct. According to the VHDL definition, `'-'`, not "0 or 1," will be assigned to y if sel is "11". Since a real `'-'` does not exist in a physical implementation, this symbol cannot be synthesized. During synthesis, some software flags an error, and others treat it as a conventional don't-care and perform optimization accordingly.

6.4   VHDL SYNTHESIS FLOW

Synthesizing VHDL code is the process of realizing a VHDL description using the primitive logic cells from the target device's library. It is a complex process. To make it manageable, we normally divide VHDL synthesis into steps, including *high-level synthesis*, *RT-level synthesis*, gate-level synthesis (commonly known as *logic synthesis*) and cell-level synthesis (commonly known as *technology mapping*). High-level synthesis transforms an algorithm into an architecture consisting of a *data path* and *control path*. It is substantially different from the other three steps and is done by specialized software tools. It is reviewed in Section 12.7.

RT-level synthesis, logic synthesis and technology mapping generate structural netlists utilizing generic RT-level components, generic gate-level components and device-dependent cells respectively. The detailed flow is shown in Figure 6.6. Basically, the entire circuit is transformed and optimized level by level, from an RT-level netlist to a gate-level netlist and then to a cell-level netlist, as shown in the left column of the flowchart. Some RT-level components, such as adder and comparator, can be quite complex. They are normally handled by a *module generator*, as shown in the right column of the flowchart. Our current discussion is limited to the synthesis flow of combinational circuits. It can easily be expanded to include sequential circuits, which are discussed in Chapter 8.

6.4.1   RT-level synthesis

RT-level synthesis transforms a behavioral VHDL description into a circuit constructed by components from a generic RT-level library. The term *generic* implies that the components are common to all technologies and thus the library is not technology dependent. The components can be classified into three categories: functional units, routing units and storage units. *Functional units* are used to implement the logic, relational and arithmetic operators encountered in VHDL code. *Routing units* are various multiplexers used to construct the routing structure of a VHDL description, as discussed in Chapters 4 and 5.

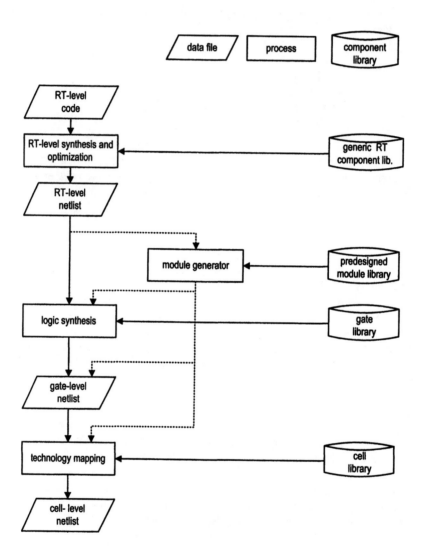

Figure 6.6   Synthesis flow.

Storage units are registers and latches, which are used only in sequential circuits and are discussed in Chapter 8.

RT-level synthesis includes the derivation and optimization of an RT-level netlist. During the process, VHDL statements are converted into corresponding structural implementation, somewhat similar to the derivation of conceptual diagrams discussed in Chapters 4 and 5. Some optimization techniques, such as operator sharing, common code elimination and constant propagation, can be applied to reduce circuit complexity or to enhance performance. Unlike gate- and cell-level synthesis, optimization at the RT level is performed in an ad hoc way and its scope is very limited. Good design can drastically alter the RT-level structure and help software to derive a more effective implementation.

6.4.2   Module generator

After the RT-level synthesis, the initial description is converted to a netlist of generic RT-level components. These components have to be transformed into lower-level implementation for further processing. Some RT-level components, such as logical operators and multiplexers, are simple and can be mapped directly into gate-level implementation. They are known as *random logic* since they show less regularity and can be optimized later in logic synthesis. The other components are quite complex and need special software, known as a *module generator*, to derive the gate-level implementation. These components include adder, subtractor, incrementor, decrementor, comparator and, if supported, shifter and multiplier as well. They usually show some kind of repetitive structure and sometimes are known as *regular logic*. Regular logic is usually designed in advance. A module generator can produce modules in different levels of detail:

- Gate-level behavioral description.
- Presynthesized gate-level netlist.
- Presynthesized cell-level netlist.

A gate-level behavioral description can be thought of as VHDL code that uses only simple signal assignment and logical operators, which can easily be mapped to a gate-level netlist. The description is general and independent of underlying device technology. The description will be flattened and combined with the random logic to form a single gate-level netlist. The merged netlist will be synthesized together later in logic synthesis. Chapter 15 discusses the generation of some frequently used components.

Because of the regular and repetitive nature of these components, it is possible to further explore their properties and manually derive and synthesize the netlist at the gate level or even at the cell level. Manual design can explore this regularity and derive a more efficient implementation. The resulting circuit is more efficient than a circuit obtained from logic synthesis. When a presynthesized gate- or cell-level netlist is used, it will not be flattened and merged with the random logic. The random logic will be independently processed through logic synthesis and even technology mapping. The netlist of random logic and the netlists of regular components will be merged after these processes. The right column in the synthesis flow of Figure 6.6 shows the various possibilities for module generation.

There are two advantages to the non-flattened approach. First, it can utilize predesigned, highly optimized modules. Second, since these modules are extracted from the original circuit, the remaining part is smaller and thus is easier to process and optimize. On the other hand, the non-flattened modules may isolate the random logic and thus reduce the chance for further optimization. For example, the adder of Figure 6.7 separates the random logic circuits into two parts and forces them to be processed independently. It may introduce

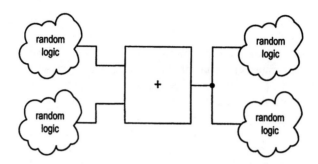

Figure 6.7   Random logic with a regular component.

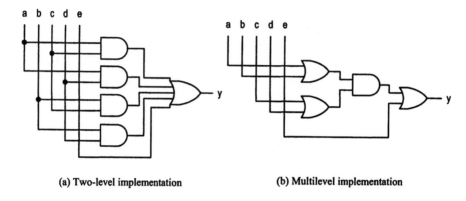

(a) Two-level implementation   (b) Multilevel implementation

Figure 6.8   Two-level versus multilevel implementation.

more optimization opportunities if we flatten the adder, merge it with the four random logic circuits, and then process and optimize them together. There is no clear-cut rule as to which approach is more effective. Some synthesis software allows users to specify the desired option.

6.4.3 Logic synthesis

Logic synthesis is the process of generating a structural view using an optimal number of generic primitive gate-level components, such as a not gate, and gate, nand gate, or gate and nor gate. Again, the term *generic* means that the components are not tied to a particular device technology and there is no detailed information about the components' size or propagation delay. At this level, a circuit can be expressed by a Boolean function, and these generic components are essentially the operators of Boolean algebra. Logic synthesis can be divided into *two-level synthesis* and *multilevel synthesis*.

The most commonly used two-level form is the sum-of-products form, in which the first level of logic corresponds to and gates and the second level to or gates. An example is shown in Figure 6.8(a). Other two-level forms can easily be derived from the sum-of-products form. Two-level synthesis is to derive an optimal sum-of-products form for a Boolean function. The goal of optimization is to reduce the number of product terms (i.e., the number of and gates) and the number of input literals (i.e., the total fan-ins of and gates). The well-known Karnaugh map technique is a method to manually obtain the optimal two-level implementation for a circuit with up to four or five inputs. A more realistic circuit may contain dozens or even several hundred inputs and cannot be optimized manually. Obtaining the optimal two-level circuit is actually an intractable problem and thus is not practical. However, this process is well understood, and many efficient algorithms to obtain good, suboptimal circuits have been developed.

Because of the large number of fan-ins for the and and or gates, the two-level sum-of-products form can only be implemented by using a special ASIC structure, known as *programmable logic array (PLA)*, and, with some modification, by using *programmable array logic (PAL)*-based CPLD devices. However, the two-level form is a formal way of expressing Boolean functions and is frequently used as a basis for processing and manipulating logic expressions. Two-level synthesis can reduce the information needed to represent a function and theoretically can serve as a staring point of multilevel processing.

Multilevel representation, as its name indicates, expresses a Boolean function by using multiple levels of gates. Its form is far less stringent than that of the two-level form and provides several degrees of freedom, leading to better efficiency and more flexibility. The implementation may be exploited by optimizing area, by optimizing delay, or even by obtaining an optimal area–delay trade-off point. An example of multilevel implementation of the previous two-level implementation is shown in Figure 6.8(b). It reduces both the number of gates and the number of fan-ins. Modern device technologies are based on small cells whose fan-in is limited to a small number. Thus, multilevel synthesis is more appropriate.

Processing and optimizing a multilevel logic are more difficult. Optimization is normally based on heuristic methods, which exploit various Boolean or algebraic transformations or search and replace circuit patterns according to a rule database. Because of the flexibility of multilevel representation, synthesis results vary significantly, and a minor modification in initial description may lead to a totally different implementation.

6.4.4 Technology mapping

Logic synthesis generates an optimized netlist that utilizes generic components. *Technology mapping* is the process of transforming the netlist using components from the target device's library. These components are commonly referred to as *cells*, and the technology library is normally provided by a semiconductor vendor who manufactured (as in FPGA technology) or will manufacture (as in ASIC technology) the device. Whereas a generic component is defined by its function, a cell is further characterized by a set of physical parameters, such as area, delay, and input and output capacitance load. In the case of ASIC technology, each cell is associated with the physical layout or prediffused patterns.

Although technology mapping can be done by simple translation between generic components and logic cells, the resulting circuit is not very efficient since the translation does not exploit the functionalities, areas and delays of the cells. Obtaining optimal mapping is a very difficult process, which involves intractable problems. Again, heuristic and rule-based algorithms are used to find suboptimal solutions. The following subsections use two simple examples to illustrate the technology mapping process of a hypothetical standard-cell ASIC library and a 5-input look-up table (LUT)–based FPGA.

Standard-cell technology A library from standard-cell technology normally consists of several dozen to several hundred cells, including combinational, sequential and interface cells. Combinational cells consist of simple gates, such as and, or, nand, nor, xor etc., and sometimes slightly complex circuits, such as 1-bit full adder, 1-bit 2-to-1 multiplexer

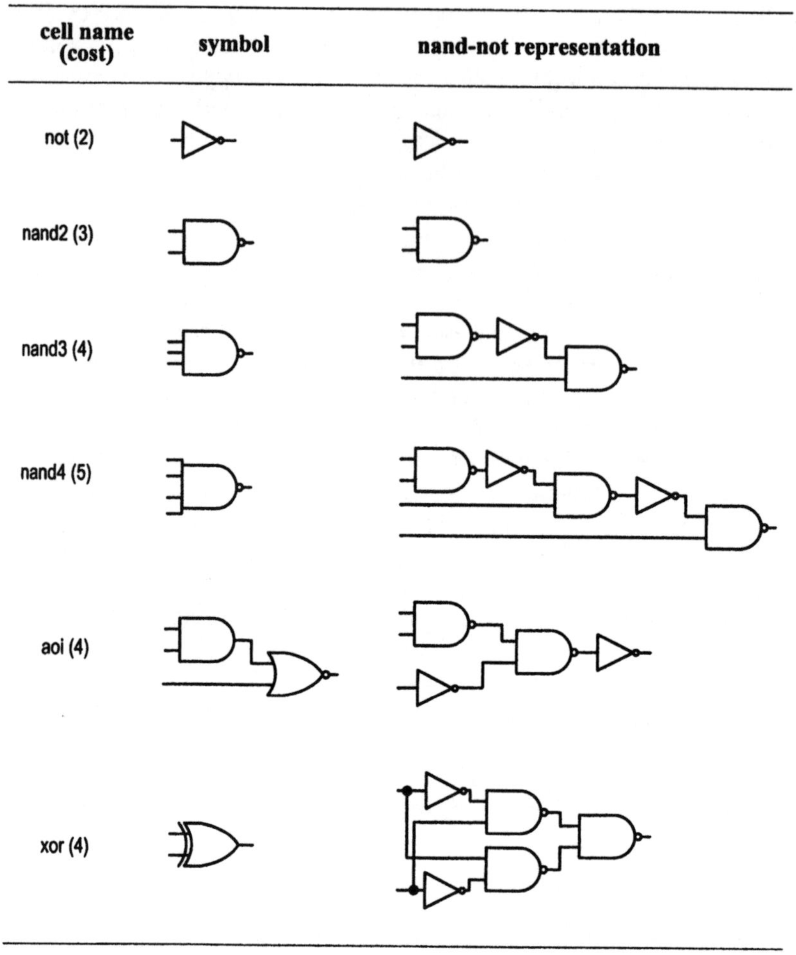

cell name (cost)	symbol	nand-not representation
not (2)		
nand2 (3)		
nand3 (4)		
nand4 (5)		
aoi (4)		
xor (4)		

Figure 6.9   Simple hypothetical ASIC cell library.

etc. A simple hypothetical technology library with seven cells is shown in Figure 6.9. The columns are the name of the cell, its relative area (cost), its symbol and its normal form. The *normal form*, which represents a cell using 2-input nand gates and inverters, is used to facilitate the mapping process.

The cells of a technology library are optimized and tuned for a particular technology. They are manually designed from scratch at the transistor level rather than being based on simple logic gates. For example, if the aoi cell is implemented using the simpler nand2 and not cells, its area is 11, which is about four times the area of the nand2 cell. However, if it is implemented directly at the transistor level, its area is 5, which is about twice the area of the nand2 cell. This explains why there are many different primitive cells in a typical standard-cell library. Furthermore, since fine adjustments can be made at the transistor level, multiple cells of different area–delay trade-offs may exist for the same logic function.

The mapping can best be illustrated by the example shown in Figure 6.10. The initial mapping in Figure 6.10(a) is a trivial one-to-one gate-to-cell translation and its area is 31. The better one, in Figure 6.10(b), is optimized and its area is reduced to 17. Although

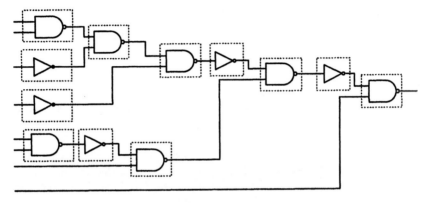

(a) Initial mapping

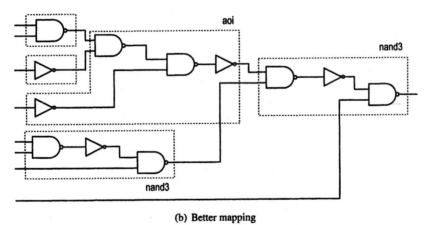

(b) Better mapping

Figure 6.10   Standard-cell technology mapping example.

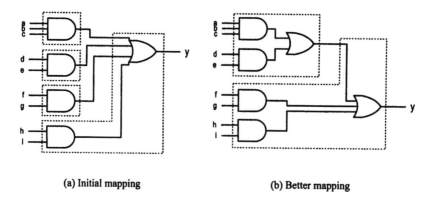

(a) Initial mapping   (b) Better mapping

Figure 6.11   LUT-based FPGA technology mapping example.

this is a simple example, it demonstrates the importance of good mapping as well as the complexity of the technology mapping process.

LUT-based FPGA technology   Because an FPGA device is prefabricated in advance, its technology library normally consists of only a single cell. This cell can, however, be "programmed" or configured to perform different logic functions. The most commonly used construction is based on a small *look-up table* (*LUT*). We can program a LUT by specifying its contents, as in a truth table description of a logic function. If a LUT can accommodate 2^n rows (i.e., n inputs), it can be used to realize any combinational function with n or fewer inputs. A typical FPGA cell consists of a 4-, 5-, or 6-input LUT and a D-type flip-flop.

An example of technology mapping using 5-input LUT cells is shown in Figure 6.11. Since a LUT cell concerns only the number of inputs, the netlist does not need to be converted into normal form. The mapping in Figure 6.11(a) is a trivial one-to-one gate-to-cell translation, and it requires four LUT cells. The mapping in Figure 6.11(b) is more efficient and reduces the number to only two LUT cells.

Precaution with FPGA technology   From technology mapping's point of view, one difference between ASIC and FPGA technologies is the size of the cells. The cell size of an ASIC device is very small, and thus any minor adjustment will be reflected in the implementation. For example, the previous standard-cell library has 2-, 3- and 4-input nand cells. If we can improve our design by eliminating one input of a product term in the logic expression, we can use a smaller nand cell and reduce the circuit area by a small amount.

On the other hand, the cell size of a FPGA device is relatively large. A 5-input LUT-based cell can implement any 1-, 2-, 3-, 4- or 5-input logic function, regardless of the complexity of the function. A wide range of functions can be implemented by this cell, and all of them are considered to have the same area under the FPGA technology. For example, both the $a \cdot b$ and $a \oplus b \oplus c \oplus d \oplus e$ expressions can be mapped into a single LUT cell. Although the internal utilizations of the cells are very different, the two expressions are considered to have the same area. This may cause an unexpected result when we synthesize a circuit using FPGA technology. This phenomenon will be further amplified if we take into consideration the built-in flip-flop within a logic cell. For example, we can construct a 1-bit counter and its area remains a single cell.

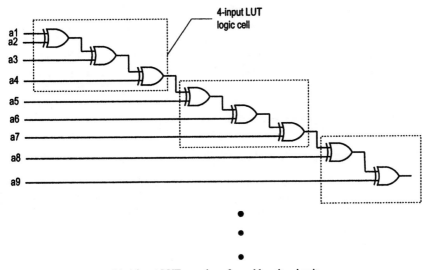

(a) 4-input LUT mapping of an odd-parity circuit

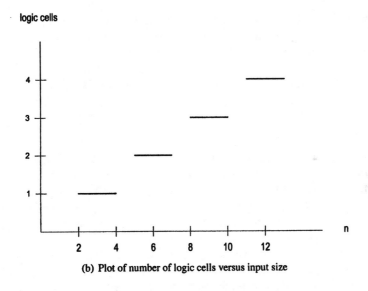

(b) Plot of number of logic cells versus input size

Figure 6.12 Discontinuity of LUT cell-based implementation.

The FPGA-based implementation may also exhibit a "discontinuity" phenomenon. For example, let us use a 4-input LUT logic cell to implement an odd-parity circuit, which has an expression of

$$a_1 \oplus a_2 \oplus a_3 \oplus \cdots \oplus a_n$$

A simple cascading chain implementation and mapping is shown in Figure 6.12(a). The number of logic cells needed for different input size (i.e., n) is plotted in Figure 6.12(b). It looks like a staircase and exhibits discontinuities (i.e., a sudden change) at certain points. For example, if we increase the input size from 6 to 7, there is no change in the number of logic cells, and thus the area remains unchanged. But if we change the input size from 7 to 8, the number of logic cells increases from 2 to 3, and thus the area increases 50%.

For a larger, more complex circuit, we can expect that the cell utilization and discontinuity will average out and the result is more like that of an ASIC device. Nevertheless, occasional fluctuations and randomness are unavoidable, and targeting an FPGA device still introduces a new dimension of complication in synthesis. Although the discussion in the remainder of the book can be applied equally to both ASIC and FPGA devices, we target the design using ASIC devices for the area and performance data.

6.4.5 Effective use of synthesis software

Despite its fundamental limitation, synthesis software is still a powerful and necessary tool, which can automate many design tasks and perform certain tedious and repetitive computations. A good designer should understand the capabilities and limitation of software, and know what this tool can and cannot do as well as when to compromise.

VHDL description of logical operators   In general, synthesis software is very effective in performing logic synthesis and technology mapping for a small to moderate-sized circuit whose complexity is around 5000 to 50,000 equivalent gates. Although optimization involves intractable problems, these problems have been studied thoroughly and many good heuristics and searching procedures have been developed. Furthermore, although a circuit is processed at the gate or cell level, even a very simple design consists of hundreds or thousands of components. It is not practical to manipulate the design manually at this level.

VHDL logical operators can be mapped directly to gate-level components. Their implementations are simple and straightforward. Since synthesis is very effective at this level, we need not worry about the sharing and optimization of logical operators in a VHDL description.

VHDL description of arithmetic and relational operators   Optimization at the RT level involves complex arithmetic and relational operators and routing structure. It is not well developed and is frequently done on an ad hoc basis. Human intervention is required, and we have to specify explicitly the desired design in a VHDL description. Simple modifications on code frequently can improve circuit efficiency significantly.

There is no comprehensive procedure or algorithm to detect sharing and to perform optimization for arithmetic and relational operators. It frequently depends on the designer's insight and knowledge of a circuit. VHDL is a good vehicle to explore design at this level. Sections 7.2 and 7.3 provide a comprehensive array of examples for this topic.

VHDL description of layout and routing structure   *Routing structure* indicates how "data" propagate through various parts of the system, from input ports to output ports. Although a VHDL program cannot explicitly specify the placement of components or the layout of a design, it implicitly describes the routing structure and, to some degree, the shape of the implementation. Recall that each VHDL statement can be considered as a circuit part, and a VHDL program implicitly connects these parts. Although all parts of a combinational circuit operate concurrently, some outputs of these parts are not valid initially. The valid value can be thought of as data that propagates from one part to another and eventually to the circuit output. The data flow forms a routing structure, which, in turn, implicitly determines the shape or layout of the physical circuit.

Regardless of the shape of the initial VHDL description, the placement and routing process will eventually implement the circuit on a two-dimensional silicon chip. If the

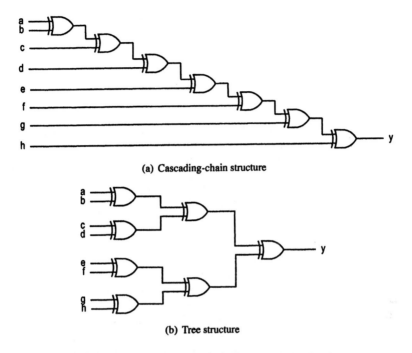

(a) Cascading-chain structure

(b) Tree structure

Figure 6.13   Routing structures of an odd-parity circuit.

shape of the initial description resembles the shape of the chip, the description can help the placement and routing process and make the final implementation smaller and faster. Two routing structures of a simple example of an odd-parity circuit are shown in Figure 6.13. The one in Figure 6.13(a) is a cascading-chain structure described by the statement

```
y<=(((((((a xor b) xor c) xor d) xor e) xor f) xor g) xor h);
```

and the one in Figure 6.13(b) has a tree structure described by the statement

```
y<=((a xor b) xor (c xor d)) xor ((e xor f) xor (g xor h));
```

Both structures use the same number of xor gates, but the propagation delay is much smaller in the tree structure.

Although synthesis software can recognize a few specific patterns and rearrange the routing structure on a local basis, it cannot make any major global change. Good VHDL coding can outline the basic "skeleton" of the implementation and provide a framework for synthesis. It has a greater impact than the local optimization performed by synthesis software. The coding technique is discussed in detail in Section 7.4 .

6.5 TIMING CONSIDERATIONS

A digital circuit cannot respond instantaneously, and the output is actually a function of time. The most important time-domain characteristic is the *propagation delay*, which is the time required for the circuit to generate a valid, stabilized output value after an input change. It is one of the major design criteria for a circuit.

Another time-domain phenomenon, known as a *hazard*, is the possible occurrence of unwanted fluctuations of an output signal before it is stabilized. Although a hazard is a transient response, it may cause circuit malfunction in a poorly conceived design. The following subsections examine the propagation delay and hazard in more detail and discuss several timing issues that have an impact on synthesis.

6.5.1 Propagation delay

It takes a digital circuit a certain amount of time to reach a valid stable output response after an input change. In digital design, we treat this time as the delay required to propagate a signal from the input port to the output port, and call it *propagation delay* or simply *delay*. A digital system normally has multiple input and output ports, and each input–output path may exhibit a different delay. We consider the worst-case scenario and use the largest input–output delay as the system's propagation delay.

The propagation delay reflects how fast a system can operate and is usually considered as the *performance* or the *speed* of the system. Combined with the circuit size (area), they are the two most important design criteria of a digital system.

To compute the delay of a system, we first determine the delays of individual components and identify all possible paths between input and output ports. We then calculate the delay of each path by summing up the individual component delays of the path and eventually determine the system delay.

The system delay calculation clearly depends on the information of its underlying components. The best estimation can be obtained at the cell level since the netlist is final, and the accurate physical and electrical characteristics of cells are provided. The least accurate estimation is at the RT level since the components must be further transformed and optimized.

Propagation delay at the cell level   To determine the exact time-domain behavior of a cell, we have to examine and analyze it at the transistor level, which is modeled by transistors, resistors and capacitors. The delay is due mainly to parasitic capacitance, which occurs at two overlapping layers and thus exists everywhere. When a transistor changes state, these capacitors have to be charged or discharged and thus introduce a delay. Analyzing a cell at this level is extremely complex and can be done only at a small scale. The analysis provides basic data for cell-level modeling.

To manage the complexity, timing analysis at the cell level has to rely on a much simpler model. One commonly used approach is a simplified linear model, in which all parasitic capacitance is lumped as a single capacitor and only the first-order effect is considered. In this model, the delay of a cell is expressed as

$$delay = d_{intrinsic} + r * C_{load}$$

The first term in the expression, $d_{intrinsic}$, is associated with the internal circuit of the cell. It models the time required for transistors to change state (i.e., switch on or off). The second term is associated with the external circuits driven by the cell. The parameter C_{load} is the total capacitive load driven by this cell, which includes the input capacitance of cells connected to the output of current cell and the parasitic capacitance of the interconnect wires. An example is shown in Figure 6.14. The load is the summation of the input capacitance of three cells driven by the and gate (C_{g1}, C_{g2} and C_{g3}) and parasitic capacitances of three interconnect wires (C_{w1}, C_{w2} and C_{w3}).

The r parameter represents the driver capability of the cell and can loosely be considered as the output impedance of the cell. When r is small, the cell can allow more current (i.e., larger driver capability) and thus can charge or discharge the capacitance load in a shorter

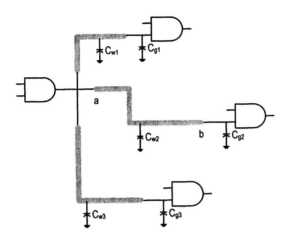

Figure 6.14   Delay estimation at the cell level.

period, leading to a smaller delay. At the transistor level, we can reduce the delay by using a larger transistor to increase the driver capability.

Impact of wiring on cell-level delay estimation   The accuracy of cell-level delay estimation depends on several factors. The first factor is the accuracy of the parameters used in delay calculation. We can obtain fairly accurate values for $d_{intrinsic}$, r, and input capacitance from the manufacturer's data sheet. After technology mapping, fan-out of each cell can be obtained from the netlist, and thus the total input capacitance load can easily be determined. The wire capacitance, on the other hand, depends on the actual length and location of each wire. Since this information is not available at the synthesis stage, software sometimes uses a statistical model to provide a rough estimation. Accurate information can only be extracted after place and routing is performed. This is one reason that the system has to be simulated and verified again after the placement and routing process.

The second factor is the accuracy of the model. The linear cell-level model is only an approximation and ignores higher-order effects. In some circumstances, these effects become more dominant, and more sophisticated models have to be used. For example, a more complex distributed RC model can be used to obtain better estimation than a simple lumped circuit. Some models for a wire between points a and b are shown in Figure 6.15(b) - (d).

When the transistor geometry is relatively large, the wire capacitance and higher-order effects do not contribute much to the overall delay and can safely be ignored. Accurate timing information can be obtained in the synthesis stage. However, as the transistor becomes smaller and submicron technology becomes available, the wiring delay gradually becomes the dominant part and the high-order effects have more impact. This makes the design process harder since we need to do placement and routing to obtain accurate timing information.

In addition to the inherent errors of approximation, the fabrication process and operation environment (such as temperature) affect the delay characteristics as well. In general, there is no way that we can control the exact delay of a cell. A device manufacturer can only guarantee the boundary of operation, normally in terms of the *maximal* propagation delay. While VHDL incorporates the timing aspect in the language, it is primarily for modeling purposes. For example, we can specify an and gate with a 2-ns delay as:

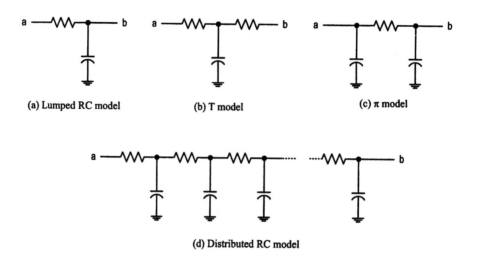

(a) Lumped RC model   (b) T model   (c) π model

(d) Distributed RC model

Figure 6.15   Wiring models.

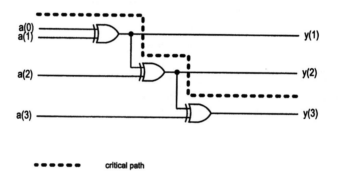

Figure 6.16   Topological critical path.

```
f <= b and c after 2 ns;
```

During synthesis, the timing part will be completely ignored since there is no technology that can produce a gate with an exact 2-ns delay.

System delay   Once cell delays are known, we can calculate the delay of a path by adding the individual cell delays along the path. A digital system typically has many paths between input and output ports, and their delays are different. Since the system has to accommodate the worst-case scenario, the system delay is defined as the longest delay. The corresponding path is considered as the longest path and is known as the *critical path*.

A simple method of determining the critical path is to treat the netlist as a graph, extract all possible paths and then determine the longest path accordingly. An example is shown in Figure 6.16. Since the topology of the system alone determines the critical path, it is also known as the *topologically critical path*.

Using the topologically critical path to determine the system delay may occasionally overestimate the actual value because of a *false path*, a path along which no signal transition can propagate. An example of a false path is shown in Figure 6.17. The topologically critical

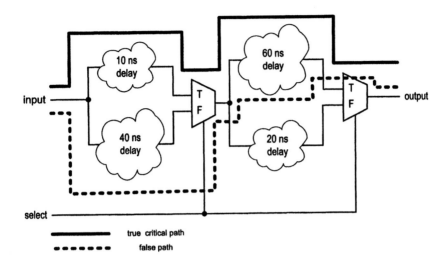

Figure 6.17 False path.

path is the route that includes the circuit with 40- and 60-ns delays. However, in realty, the input signal can propagate through either the top part (when the `select` signal is '1') or the bottom part (when the `select` signal is '0') but never the topologically critical path. Since no signal actually passes through the false paths, they should be excluded from system delay calculation. To determine the true critical path is much harder since the analysis involves not only the topology but also the internal logic operations.

Because of the large number of cells in a system, cell-level timing analysis is always done by software. This feature is normally integrated into the synthesis software. Most software uses the topological critical path to determine the system delay. Some software allows users manually to exclude potential false paths.

Delay estimation at the RT level We can apply the same principle to analyze and calculate the propagation delay at the RT level. The accuracy of the calculation depends on the components used in the RT-level diagram. If an RT-level diagram consists primarily of simple logical operators and is mainly random logic, the circuit is subjected to a significant amount of transformation and optimization during logic synthesis and technology mapping. Since the final circuit may not resemble the original RT-level diagram, the RT-level delay calculation will not faithfully reflect delay in the synthesized circuit.

On the other hand, if an RT-level diagram consists of many complex operators and function blocks, these components become the dominating part of a delay calculation. Furthermore, since these components are presdesigned and optimized, their delay characteristics will not change significantly during synthesis. Thus, the delay calculation will be much more accurate for this type of circuit. Calculating RT-level delay allows us to identify the critical path and thus better understand the performance of the circuit, and eventually helps to derive an efficient design and VHDL code with the desired area–delay characteristics. RT-level delay estimation is shown in many design examples in the subsequent chapters.

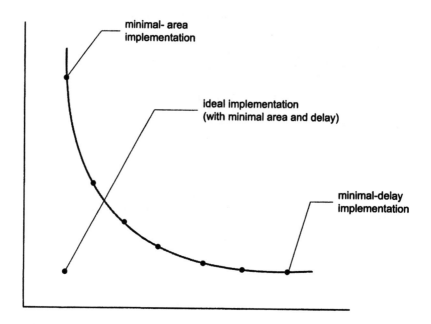

Figure 6.18   Area–delay trade-off curve.

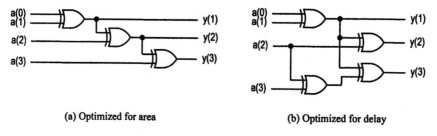

(a) Optimized for area   (b) Optimized for delay

Figure 6.19   Delay constraint implementation.

6.5.2 Synthesis with timing constraints

The circuit area and system delay are two major design criteria. In most applications, we cannot find a design or an implementation that is optimized for both criteria. A faster circuit normally is more complex and needs more silicon real estate, and a smaller circuit normally has to sacrifice some performance. For the same application, there frequently exist multiple implementations that exhibit different area–delay characteristics. A typical area–delay curve is shown in Figure 6.18, in which each point is a possible implementation. Of course, the trade-off can be achieved only in a limited range. We cannot reduce the area or increase the performance indefinitely.

Multilevel logic synthesis is quite flexible, and it is possible to add additional gates to achieve shorter delay. An example is shown in Figure 6.19. The circuit performs three xor operations. The diagram in Figure 6.19(a) is the initial design, which is optimized for area. The critical path is from a(0) or a(1) to y(3), and the system delay is three times the delay of an xor gate. The diagram in Figure 6.19(b) is the revised circuit. It shortens the

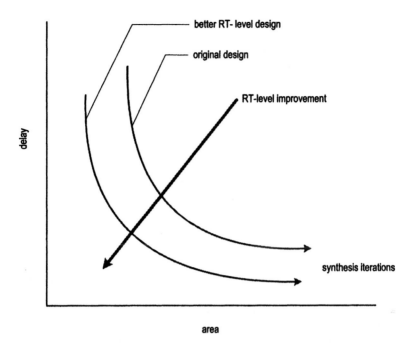

better RT- level design

original design

RT-level improvement

synthesis iterations

delay

area

Figure 6.20   Synthesis iterations and the impact of RT-level change.

critical path by adding an extra xor gate, and the system delay is reduced to twice the delay of an xor gate.

The synthesis procedure discussed in Section 6.4 focuses on minimizing the circuit area. A combinational system is normally part of a larger system. To meet a certain performance goal, we sometimes have to add a specific timing constraint for synthesis. As we discussed earlier, it is impossible to synthesize a circuit with an exact propagation delay. Instead, the timing constraint is specified in terms of maximal allowable propagation delay. Since the system delay depends only on the delay of the critical path, it is not wise to blindly optimize all paths. Synthesis with a timing constraint utilizes an iterative procedure. First, the minimal-area implementation is obtained from regular synthesis. The implementation will be analyzed to determine the critical path and the system delay. If the delay exceeds the constraint, extra gates will be provided to speed up the critical path. The revised implementation will be analyzed again for the critical path (which is the second longest path in the original implementation) and checked to see whether the new system delay is within the constraint. The process may repeat several times until a satisfactory implementation is found. The iteration process in an area–delay space is shown in Figure 6.20.

The previous iteration procedure is done at the gate or cell level and thus is too tedious for human designers. However, it is possible to apply the procedure in at the RT level. A block diagram shows the basic routing structure and the locations of complex RT-level modules. Since the delays of the complex modules constitute the major portion of the system delay, we can identify the paths that contain these modules, estimate the rough delays of these paths, and determine the critical path accordingly. This kind of analysis helps us to explore various architectural alternatives and eventually to derive a more efficient design. Our understanding of the system and insight can lead to "global" optimization, and it is normally much more effective than gate- or cell-level optimization done by synthesis software. The impact of an innovative RT-level architectural change on the area–delay space is shown in Figure 6.20.

6.5.3 Timing hazards

The propagation delay of a system is the time required to generate a valid, steady-state output value. *Timing hazards* are the fluctuations occurring during the transient period. In a digital system, many paths may lead to the same output port. Since each path's delay is different, signals may propagate to the output port at different times. Before the output port produces a steady-state value, it may fluctuate several times. The fluctuations are one or more short undesired pulses, known as *glitches*. We say that a circuit has timing hazards if it can produce glitches. The following subsections discuss the two types of hazards and how to deal with them.

Static hazards   A *static hazard* is the condition that a circuit's output produces a glitch when it should remain at a steady value. It is further divided into static-1 hazard and static-0 hazard. A *static-1 hazard* occurs when a circuit's output produces a '0' glitch. An example is shown in Figure 6.21. The Karnaugh map of a function and its implementation are shown in Figure 6.21(a). The corresponding Boolean function is

$$sh = a \cdot b' + b \cdot c$$

Assume that a and c are '1', and that b changes from '1' to '0'. Regular analysis, which is based on Boolean algebra and deals with steady-state value, predicts that the output should be '1' all the time. However, if we consider transient behavior, there are two converging paths with different delays. Assume that the delay of inverter is T_{not} and the delay of the and gate and or gate is T_{and} and the wire delays are 0. The timing diagram and the sequence of events are shown in Figure 6.21(b). An unwanted '0' glitch of width T_{not} occurs at the output because the signal in the bottom path propagates faster than that in the top path.

Similarly, a *static-0 hazard* is the condition that a circuit's output produces a '1' glitch when Boolean algebra analysis predicts that the output should be a steady '0'.

Dynamic hazards   A *dynamic hazard* is the condition that a circuit's output produces a glitch when it changes from '1' to '0' or '0' to '1'. An example of a circuit with a dynamic hazard is shown in Figure 6.22(a). Assume that a, c and d are '1' and that b changes from '1' to '0'. The timing diagram in Figure 6.22(b) shows that there is a '1' glitch when the dh output changes from '0' to '1'. The glitch is due to the different propagation delays of the converging paths.

Dealing with hazards   There are some techniques to eliminate hazards caused by a single input change. For example, we can add a redundant product term to eliminate the previous static hazard:

$$sh = a \cdot b' + b \cdot c + a \cdot c$$

The revised Karnaugh map and circuit are shown in Figure 6.21(c). Although deriving a hazard-free circuit is possible, this approach is problematic if the design is later processed by synthesis software. The problems are discussed in detail in the next section.

In a real-world application, the hazard situation will become even more complicated because of the possibility of multiple input signal transitions. If the inputs of a combinational

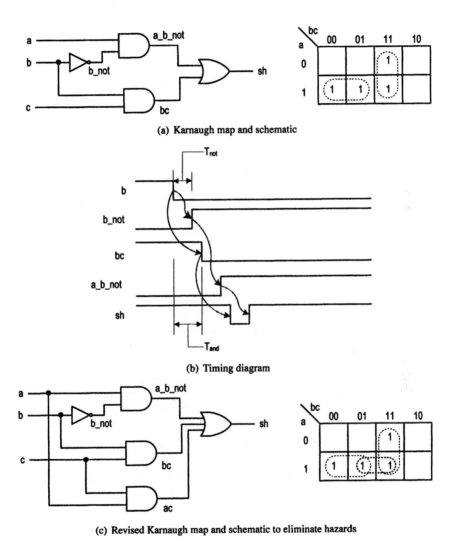

(a) Karnaugh map and schematic

(b) Timing diagram

(c) Revised Karnaugh map and schematic to eliminate hazards

Figure 6.21   Static hazards example.

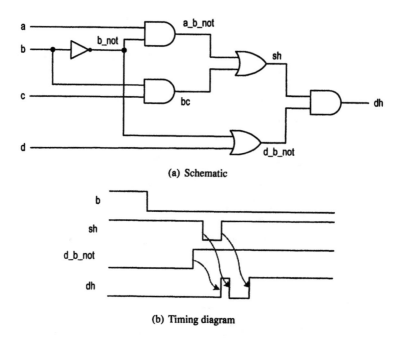

(a) Schematic

(b) Timing diagram

Figure 6.22   Dynamic hazards example.

circuit are connected to the outputs of an edge-triggered register, the register's outputs may change almost simultaneously at the transition edge of the clock signal. For example, when a 4-bit counter circulates from "1111" to "0000", four input bits change almost simultaneously. Multiple changes will activate several paths at the same time and frequently lead to glitches in an output signal. Unless we utilize a specialized counter, which is normally not practical, it is impossible to eliminate hazards.

Since there is no easy way to eliminate hazards, we have to live with them. In a combinational circuit, the most effective way to handle hazards is to ignore the output during the transient period. Recall that the propagation delay is the time for an input signal to propagate through the longest path in a system. If there is a glitch, it will occur within this period of time. After that, the output will always be a valid, steady-state value. As long as we know when to examine the output, the existence of glitches does not matter. This "wait until the output is stabilized" idea is one of the motivations behind the *synchronous design methodology*, in which a clock signal "samples" input signals at the proper time and stores the values in a register. The synchronous design methodology is elaborated in Chapter 8.

6.5.4   Delay-sensitive design and its dangers

In a digital system, most theoretical studies and design methodologies are based on steady-state analysis. Boolean algebra, the theoretical foundation of digital logic, conveys no time-domain information. When we use Boolean algebra to describe a digital circuit, we actually implicitly describe its steady-state behavior. Modeling and analyzing the transient behavior can be very hard, and most of the time we choose not to deal with it directly. Instead, we determine when the transient period ends and ignore the responses within the period. This approach is embedded in the concept of system delay, which specifies the time needed to reach the steady state in the worst-case scenario. Most design methodologies

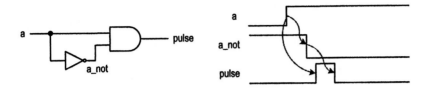

Figure 6.23   Delay-sensitive edge detection circuit.

and synthesis algorithms, such as time-constrained optimization, are based on system delay rather than the exact transient behavior.

In a few circumstances, we need to consider the transient behavior to understand a circuit's function and operation. We use the term *delay-sensitive design* to describe this type of circuit.

One example is the hazard elimination circuit in Figure 6.21. If we examine only the steady-state behavior, Boolean algebra shows that the $a \cdot c$ term does not serve any useful purpose and that the $a \cdot b' + b \cdot c + a \cdot c$ and $a \cdot b' + b \cdot c$ expressions are equivalent. The circuit is meaningful only if the transient behavior is considered.

One old, commonly used delay-sensitive design trick is to use cascading gates to generate a delay. An example is shown in Figure 6.23. The purpose of this circuit is to generate a short pulse when the input a switches from '0' to '1'. The inverter introduces a small delay and causes a monetary '1' pulse, as shown on the timing diagram. If we use steady-state analysis, the $a \cdot a'$ expression can be reduced to '0', and the circuit becomes a wire connected to ground. Again, this circuit makes sense only if we consider its transient behavior.

Although a delay-sensitive design can be useful in a few special situations, we should avoid using VHDL description and synthesis software to construct this kind of circuit. Transformation and optimization algorithms used in synthesis software are based on the model of steady-state value and propagation delay, and cannot interpret or process transient-related information.

Deriving VHDL code for a delay-sensitive circuit is not very difficult. For example, we can revise the VHDL code from

```
sh <= (a and (not b)) or (b and c);
```

to

```
sh <= (a and (not b)) or (b and c) or (a and c);
```

to describe the hazard-free circuit in Figure 6.21, and can use the statement

```
pulse <= a and (not a);
```

to describe the pulse generation circuit in Figure 6.23. However, it is unlikely that the desired effect can be preserved during the synthesis process. The potential complications are as follows:

- During logic synthesis, the logic expressions will be rearranged and optimized. Redundant product terms, if they exist, will be removed during the optimization process. It is unlikely that the original expression can be preserved.
- If we assume that the logic expression remains unchanged after logic synthesis, the netlist may be converted to other cells during technology mapping. Again, the original logic expression will be altered.

- If we assume that the original logic expression survives after technology mapping, wire delays will be changed after the placement and routing process. The change will alter the delay of the path and may invalidate the previous analysis.
- If we assume that the circuit is synthesized according to the specification, the design may hinder other steps in the verification and testing process. For example, the redundant product term used in the logic expression will complicate the test vector generation or even make the circuit untestable.

In summary, VHDL-based synthesis is not feasible for delay-sensitive design. If this kind of circuit is really needed, as in an asynchronous sequential circuit, we should construct the circuit manually using cells from the target device library. We may even need to manually perform the placement and routing to ensure that wire delay is within a tolerable range. Since our focus is on RT-level HDL synthesis, we will not discuss this approach in the remainder of the book.

6.6   SYNTHESIS GUIDELINES

- Be aware of the theoretical limitation of synthesis software.

- Be aware of the hardware complexity of different VHDL operators.

- Isolate tri-state buffers from other logic and code them in a separate segment.

- Unless there is a compelling reason, use a multiplexer instead of an internal tri-state bus.

- Avoid using the '-' value of the std_logic data type as an input value.

- In RT-level description, there is no effective way to eliminate glitches from a combinational circuit. We should deal with the glitches rather than attempting to derive a glitch-free combinational circuit.

- Do not use delay-sensitive design in RT-level description.

6.7   BIBLIOGRAPHIC NOTES

Synthesis is a complicated process and involves many difficult computation problems. The texts, *Synthesis and Optimization of Digital Circuits* by G. De Micheli, and *Logic Synthesis* by S. Devadas et al., provide comprehensive coverage of the theoretical foundations and relevant algorithms.

Because most software vendors do not allow users to publish benchmark information, there is very little documentation on the "behavior" of synthesis tools. The article, *Visualizing the Behavior of Logic Synthesis Algorithms* of *SNUG* (Synopsys Users Group Conference) 1998, by H. A. Landman, presents an interesting study of the relationship between the circuit area and timing constraints.

Problems

6.1   Determine the order (big-O) of the following functions:
   (a) 1.5
   (b) $2^n + 10^3 n^2$

   (c) $3n^2 + 500n + 50$

   (d) $0.01n^3 + 10n^2$

   (e) $2n \log_5 n + 2n^2 + 20n + 45$

6.2 A programmer developed an optimization algorithm with an order of $O(n!)$. Can the algorithm be applied to a large input size? Explain.

6.3 One way to specify a combinational circuit is to describe its function by a truth table, in which we list all possible input combinations and their desired output values. Assume that the circuit has n inputs.

   (a) What is the size (number of the rows) of the table?

   (b) What is the problem with this approach?

6.4 Assume that a and b are 3-bit inputs (let a be $a_2 a_1 a_0$ and b be $b_2 b_1 b_0$).

   (a) Determine the Boolean expression (in terms of a_2, a_1, a_0, b_2, b_1 and b_0) for the relational operation $a > b$.

   (b) Assume that b is a constant and that $b = $ "101". Determine the Boolean expression again.

   (c) Assume that a is a constant and that $a = $ "101". Determine the Boolean expression again.

6.5 Assume that a and b are 16-bit inputs interpreted as unsigned numbers. Write five VHDL programs for the following operations. Synthesize the programs using an ASIC device. Compare their area and propagation delay and discuss the impact of a constant operand.

- a + b
- a + "0000000000000001"
- a + "0000000010000000"
- a + "1000000000000000"
- a + "1010101010101010"

6.6 Repeat Problem 6.5, but use an FPGA device.

6.7 In a tri-state buffer, there are two special timing parameters, T_{zo} and T_{oz}. T_{zo} (known as the *turn-on time*) is the required time for the output port to transit from Z to a regular, valid value after the control signal is activated. T_{oz} (known as the *turn-off time*) is the required time to force the output port to Z after control signal is deactivated. Manufacturers normally guarantee that $T_{zo} > T_{oz}$. Explain why this constraint is necessary.

6.8 Repeat the synthesis of the odd-parity circuit of Section 6.4.4 using the software available to you. Make the range of n between 2 and 20. Choose an FPGA device based on a 5-input LUT cell as the target technology. Plot the circuit size versus n and the propagation delay versus n. Discuss the result.

6.9 Repeat Problem 6.8 but choose an ASIC device as the target technology.

6.10 In the past, the design process was sometimes divided into a "front-end" process, which included the initial RT-level development and synthesis, and a "back-end" process, which included placement and routing and physical synthesis. The front-end and back-end processes were normally handled by two independent design teams without much interaction. Explain why this approach is no longer feasible for design targeting submicron ASIC technology.

6.11   If your software supports synthesis with a timing constraint, obtain the area–delay trade-off curve for the following VHDL code. You can first synthesize the circuit with no constraint to obtain the minimal-area implementation and then gradually impose smaller values on the maximal allowable delay.

```
library ieee;
use ieee.std_logic_1164.all;
use ieee.numeric_std.all;
entity hamming is
   port(
      a,b: in std_logic_vector(7 downto 0);
       y: out std_logic_vector(3 downto 0)
   );
end hamming;

architecture effi_arch of hamming is
   signal diff: unsigned(7 downto 0);
   signal lev0_0, lev0_2, lev0_4, lev0_6: unsigned(1 downto 0);
   signal lev1_0, lev1_4: unsigned(2 downto 0);
   signal lev2: unsigned(3 downto 0);
begin
   diff <= unsigned(a xor b);
   lev0_0 <= ('0' & diff(0)) + ('0' & diff(1));
   lev0_2 <= ('0' & diff(2)) + ('0' & diff(3));
   lev0_4 <= ('0' & diff(4)) + ('0' & diff(5));
   lev0_6 <= ('0' & diff(6)) + ('0' & diff(7));
   lev1_0 <= ('0' & lev0_0) + ('0' & lev0_2);
   lev1_4 <= ('0' & lev0_4) + ('0' & lev0_6);
   lev2 <= ('0' & lev1_0) + ('0' & lev1_4);
   y <= std_logic_vector(lev2);
end effi_arch;
```

6.12   Use your software to synthesize the VHDL code for the hazard elimination circuit of Section 6.5.4. Examine the netlist of the synthesized circuit and determine whether it preserves the redundant product term.

6.13   Use your software to synthesize the VHDL code for the edge detection circuit of Section 6.5.4. Examine the netlist of the synthesized circuit and determine whether it can still generate the desired pulse.

CHAPTER 7

COMBINATIONAL CIRCUIT DESIGN: PRACTICE

After learning the implementation of key VHDL constructs and reviewing the synthesis process in Chapters 4, 5 and 6, we are ready to study the construction and VHDL description of more sophisticated combinational circuits. Examples will show how to transform conceptual ideas into hardware and illustrate resource-sharing and circuit-shaping techniques to reduce circuit size and increase performance. This chapter follows and demonstrates the main theme of the book: to research an efficient design and derive the VHDL code accordingly.

7.1 DERIVATION OF EFFICIENT HDL DESCRIPTION

Although the appearance of VHDL code is very different from a schematic diagram, VHDL code is just another way to describe a circuit. Synthesis software carries out a series of refinements and transforms a textual VHDL description to a cell-level netlist. Although software can perform simplification and local optimization, it does not know the meaning or intention of the code and cannot exploit alternative designs or change the architectural of the circuit.

The quality of a design and its description are two independent factors. We can express the initial design by a schematic diagram or by a textual VHDL program. Similarly, we can realize and synthesize the design either manually by paper and pencil or automatically by synthesis software. Using VHDL and synthesis software does not lead automatically to either a good or a bad design. VHDL description and synthesis software, however, can

RTL Hardware Design Using VHDL: Coding for Efficiency, Portability, and Scalability. By Pong P. Chu **163**
Copyright © 2006 John Wiley & Sons, Inc.

shield tedious implementation details and greatly simplify the realization process. They allow us to have more time to explore and investigate alternative design ideas.

Derivation of an efficient, synthesizable VHDL description requires two major tasks:

- Research to find an efficient design.
- Develop VHDL code that accurately describes the design.

For a problem in digital system development, there is seldom a single unique solution. A large number of possible designs exist. The resulting implementations differ in size and performance and their quality may vary significantly. There is no simple, mechanical way to derive an efficient design. It frequently relies on a designer's experience, insight and understanding of the problem.

After we find a design, the next step is to derive VHDL code that describes the design accurately. Although the VHDL textual code cannot precisely specify the final structural implementation, it describes the "big picture" that establishes the basic skeleton of the circuit. For a complex design, it is useful to draw a rough schematic sketch to help in locating the key components and identifying the critical path.

In addition to faithfully describing the intended design, good VHDL code should be clear and compact, and can be easily "scaled." Scalability concerns the amount of code modification needed when the signal width of a circuit changes. For example, after we develop a VHDL code for an 8-bit barrel shifter, how much modification is required if the input is increased to 16 bits, 32 bits or even 64 bits? The development of scalable and parameterized VHDL code is discussed in detail in Chapters 14 and 15. In this chapter, we need only be aware of this aspect of VHDL code, and discuss it when appropriate.

7.2 OPERATOR SHARING

When a VHDL program is synthesized, all statements and language constructs of the program will be mapped to hardware. One way to reduce the overall size of synthesized hardware is to identify the resources that can be used by different operations. This is known as *resource sharing*. Performing resource sharing normally introduces some overhead and may penalize performance, and thus is worthwhile only for large, complex constructs. Although the exact size depends on the underlying target technology, data from Table 6.2 provides a good estimation of the relative sizes of commonly synthesizable components. Ideally, synthesis software should identify the sharing opportunities and perform the optimization automatically. Unfortunately, in reality, software's capability varies and sometimes is rather limited in this respect. We may need to explicitly describe the desired sharing in VHDL code. This section discusses the operator sharing and the next section illustrates functionality sharing.

In certain VHDL constructs, operations are mutually exclusive; i.e., only one operation is active at a particular time. These constructs include the conditional signal assignment statement (or the equivalent if statement in a process) and the selected signal assignment statement (or the equivalent case statement in a process). Recall that the basic expression of a conditional signal assignment statement is

```
signal_name <= value_expr_1 when boolean_expr_1 else
               value_expr_2 when boolean_expr_2 else
               value_expr_3 when boolean_expr_3 else
               . . .
               value_expr_n;
```

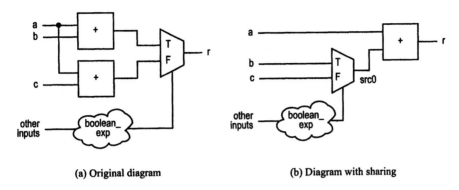

(a) Original diagram   (b) Diagram with sharing

Figure 7.1   Simple operator sharing.

The value expressions `value_expr_1`, `value_expr_2`, ... , `value_expr_n` are mutually exclusive since only one expression needs to be evaluated and passed to output. Similarly, recall that the basic expression of a selected signal assignment statement is

```
with select_expression select
     signal_name <= value_expr_1 when choice_1,
                    value_expr_2 when choice_2,
                    value_expr_3 when choice_3,

                    . . .

                    value_expr_n when choice_n;
```

Since choices are mutually exclusive, only one expression actually has to be evaluated. The same argument can be applied to the if statement and the case statement since their implementations are similar to the conditional and selected signal assignment statements.

If the same operator is used in several different expressions, it can be shared. The sharing is normally done by routing the proper data to or from this particular operator. We demonstrate the coding technique in the following examples and discuss the degree of saving and its potential impact on system performance.

7.2.1 Sharing example 1

Consider the following code segment:

```
r <= a+b when boolean_exp else
     a+c;
```

The block diagram of this code is shown in Figure 7.1(a).

There are two adders and one multiplexer. The adder can be shared because only one addition operation is needed at any time. We can revise the code as follows:

```
src0 <= b when boolean_exp else
        c;
r <= a + src0;
```

The block diagram of the revised code is shown in Figure 7.1(b). Instead of multiplexing the addition results, it multiplexes the desired source operand to the input of the adder. One adder can be eliminated in this new implementation.

Now we compare the propagation delays of the two circuits. Let the propagation delays of the adder, the multiplexer and the `boolean_exp` circuit be T_{adder}, T_{mux} and $T_{boolean}$

respectively. In the first circuit, the adders and `boolean_exp` operate in parallel, and thus the overall propagation delay is $\max(T_{adder}, T_{boolean}) + T_{mux}$. In the second circuit, the propagation delay is $T_{boolean} + T_{mux} + T_{adder}$. This reflects the fact that the `boolean_exp` operation and addition operations are performed concurrently in the first circuit whereas they are done in cascade in the second circuit. If the `boolean_exp` circuit is very simple and its delay is negligible, there will be no performance penalty on the shared design.

7.2.2 Sharing example 2

Consider the following code segment:

```
process (a,b,c,d,...)
begin
   if boolean_exp_1 then
      r <= a+b;
   elsif boolean_exp_2 then
      r <= a+c;
   else
      r <= d+1;
   end if;
end process;
```

The block diagram of this code is shown in Figure 7.2(a).

The implementation needs two adders, one incrementor and two multiplexers. The addition and increment operations can share the same adder because only one branch of the if statement is executed at a time. Assume that the signals are 8 bits wide. The revised code becomes

```
process (a,b,c,d,...)
begin
   if boolean_exp_1 then
      src0 <= a;
      src1 <= b;
   elsif boolean_exp_2 then
      src0 <= a;
      src1 <= c;
   else
      src0 <= d;
      src1 <= "00000001";
   end if;
end process;
r <= src0 + src1;
```

The block diagram of the new code is shown in Figure 7.2(b). We use two multiplexers to route the desired source operands to the inputs of the adder. The new circuit eliminates one adder and one incrementor but requires two additional multiplexers. To determine whether the sharing is worthwhile, we examine the circuit size of the adder, incrementor and multiplexer given in Table 6.2. Since a multiplexer is smaller, especially when compared with an adder, the sharing indeed leads to a smaller size. It is likely that the multiplexing circuit can be further simplified during logic synthesis, due to the duplicated input patterns (the a signal is used twice) and constant input ("00000001"). The saving will become more significant if a high-performance adder (the one optimized for delay) is used.

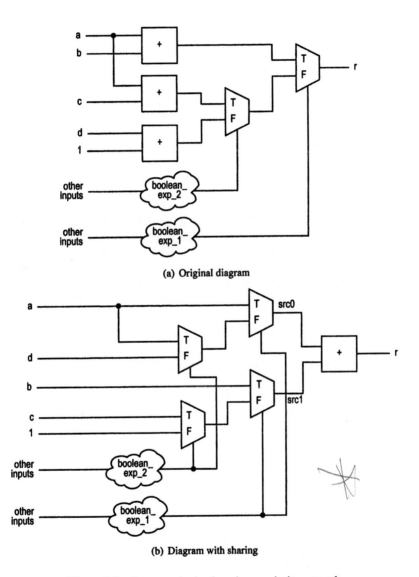

(a) Original diagram

(b) Diagram with sharing

Figure 7.2   Operator sharing based on a priority network.

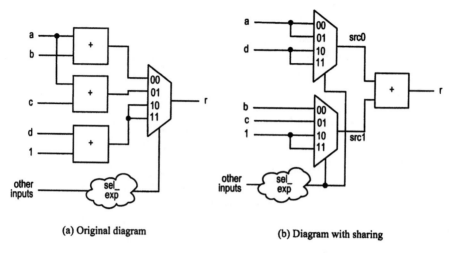

(a) Original diagram   (b) Diagram with sharing

Figure 7.3   Operator sharing based on a multiplexer.

Determining the propagation delays of these circuits is more involved since they depend on the relative values of the delays of the `boolean_exp_1` circuit, the `boolean_exp_2` circuit and the multiplexer. However, observation from the previous example still applies. The two Boolean circuits and three adders operate in parallel in the first circuit whereas the Boolean circuits and the adder operate in cascade in the second circuit. Thus, the first circuit should always have a smaller propagation delay.

7.2.3   Sharing example 3

Assume that the `sel` signal is 2 bits wide. Consider the following code segment:

```
with sel_exp select
    r <= a+b when "00",
         a+c when "01",
         d+1 when others;
```

This example is similar to the previous one but uses the selected signal assignment statement. The block diagram of this code is shown in Figure 7.3(a).

The circuit needs two adders, one incrementor and one 4-to-1 multiplexer. We can revise the code to share the adder:

```
with sel_exp select
    src0 <= a when "00"|"01",
            d when others;
with sel_exp select
    src1 <= b when "00",
            c when "01",
            "00000001" when others;
r <= src0 + src1;
```

The block diagram of the new code is shown in Figure 7.3(b). We use two multiplexers to route the desired source operands to the adder. The new circuit eliminates one adder and one incrementor but requires an additional 4-to-1 multiplexer. Since an adder and an

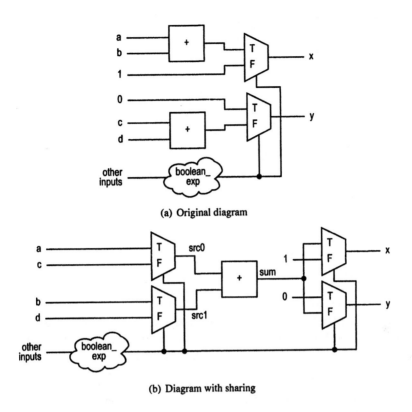

(a) Original diagram

(b) Diagram with sharing

Figure 7.4   Complex operator sharing.

incrementor are more complex than a multiplexer, the revision leads to a significant saving. Again, the second circuit may suffer a longer propagation delay because of the cascaded operations, as in example 1.

7.2.4 Sharing example 4

Consider the following code segment:

```
process (a,b,c,d,...)
begin
    if boolean_exp then
        x <= a + b;
        y <= (others=>'0');
    else
        x <= "00000001";
        y <= c + d;
    end if;
end process;
```

The block diagram of this code is shown in Figure 7.4(a). The implementation needs two adders and two multiplexers. The adder can be shared since the executions of two branches of the if statement are mutually exclusive. The revised code is as follows:

```
process (a,b,c,d,sum,...)
begin
    if boolean_exp then
        src0 <= a;
        src1 <= b;
        x <= sum;
        y <= (others=>'0');
    else
        src0 <= c;
        src1 <= d;
        x <= "00000001";
        y <= sum;
    end if;
end process;
sum <= src0 + src1;
```

The block diagram of this code is shown in Figure 7.4(b). This example illustrates the worst-case scenario of operator sharing, in which the operator has no common sources or destinations. We need a multiplexing structure to route one set of signals to the adder's input and a demultiplexing structure to route the addition result to one of the two output signals. The demultiplexing is done using two 2-to-1 multiplexers. Note that the addition result (the sum signal) is connected to the T port of the upper output multiplexer and the F port of the lower output multiplexer.

The new circuit eliminates one adder but adds two additional multiplexers. The merit of sharing in this circuit is less clear, and it depends on the relative sizes of an adder and two multiplexers. Again, we use the numbers given in Table 6.2 for estimation. If a slow adder ($+_a$, optimized for area) is used, the size of two multiplexers is about the same as that of one adder. On the other hand, if a faster adder ($+_d$, optimized for delay) is used, the saving is significant.

7.2.5 Summary

Operator sharing is done by providing additional multiplexing circuits to route input and output signals into or out of the operator. The merit of sharing and the degree of saving depend on the relative complexity of the multiplexing circuit and the operator. Substantial savings are possible for complex operators. However, sharing normally forces evaluation of the Boolean expressions and evaluation of the operators to be performed in cascade and thus may introduce extra propagation delay.

7.3 FUNCTIONALITY SHARING

In a large, complex digital system, such as a processor, an array of functions is needed. Some functions may be related and have certain common characteristics. It is possible for several functions to share a common circuit or to utilize one function to construct another function. We call this approach *functionality sharing*. Unlike operator sharing, there is no systematic way to identify functionality sharing. This kind of sharing is done in an ad hoc, case-by-case basis and relies on the designer's insight and intimate understanding of the system. It is more difficult for synthesis software to identify functionality sharing.

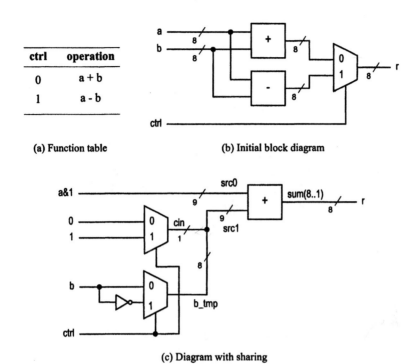

(a) Function table

ctrl	operation
0	a + b
1	a - b

(b) Initial block diagram

(c) Diagram with sharing

Figure 7.5   Addition–subtraction circuit.

7.3.1   Addition–subtraction circuit

Consider a simple arithmetic circuit that performs either addition or subtraction. A control signal, `ctrl`, specifies the desired operation. The function table of this circuit is shown in Figure 7.5(a).

Our first design follows the function table, and the VHDL code is very straightforward, as shown in Listing 7.1. Note that the signals are converted to the `signed` data type internally to accommodate arithmetic operation.

Listing 7.1   Initial description of an addition–subtraction circuit

```
library ieee;
use ieee.std_logic_1164.all;
use ieee.numeric_std.all;
entity addsub is
    port(
        a,b: in std_logic_vector(7 downto 0);
        ctrl: in std_logic;
        r: out   std_logic_vector(7 downto 0)
    );
end addsub;

architecture direct_arch of addsub is
    signal src0, src1, sum: signed(7 downto 0);
begin
    src0 <= signed(a);
```

```
      src1 <= signed(b);
      sum <= src0 + src1 when ctrl='0' else
             src0 - src1;
      r <= std_logic_vector(sum);
20 end direct_arch;
```

The conceptual diagram for this code is shown in Figure 7.5(b), which consists of an adder, a subtractor and a 2-to-1 multiplexer.

Since the adder and subtractor are different operators, we cannot directly apply the earlier operator-sharing technique. In 2's-complement representation, recall that the subtraction, $a - b$, can be calculated indirectly as $a + \bar{b} + 1$, where $\bar{b}$ is the bitwise inversion of b. Therefore, it is possible to share the functionality of the adder. After inverting b and putting a carry-in of 1, we can utilize the same adder to perform subtraction. The VHDL code is shown in Listing 7.2.

Listing 7.2 More efficient description of an addition–subtraction circuit

```
    architecture shared_arch of addsub is
       signal src0, src1, sum: signed(7 downto 0);
       signal cin: signed(0 downto 0); -- carry-in bit
    begin
5      src0 <= signed(a);
       src1 <= signed(b) when ctrl='0' else
              signed(not b);
       cin <= "0" when ctrl='0' else
              "1";
10     sum <= src0 + src1 + cin;
       r <= std_logic_vector(sum);
    end shared_arch;
```

Note that the expression a + src_b + cin has two addition operators. Since cin is either "0" or "1", it can be mapped to the carry-in port of a typical adder. In other words, the + cin operation can be embedded into the a + src_b operation and no separate incrementor is needed. Most synthesis software should be able to derive the correct implementation.

Alternatively, we can manually describe the carry-in operation and use only one addition operator in the VHDL code. The trick is to use an extra bit in the adder to mimic the effect of carry-in operation. The internal adder is extended to 9 bits, in which the original input takes 8 MSBs and the extra bit is the LSB. The LSBs of the two operands are connected to 1 and the carry-in input, c_{in}, respectively. For example, if the two original operands are

$$a_7 a_6 a_5 a_4 a_3 a_2 a_1 a_0 \text{ and } b_7 b_6 b_5 b_4 b_3 b_2 b_1 b_0$$

The extended operands will be

$$a_7 a_6 a_5 a_4 a_3 a_2 a_1 a_0 1 \text{ and } b_7 b_6 b_5 b_4 b_3 b_2 b_1 b_0 c_{in}$$

After the addition, the LSB will be discarded and the higher 8 bits will be used as the output. When c_{in} is 1, a carry will be propagated from the LSB to 8 MSBs, effectively adding 1 to the 8 MSBs of the adder. On the other hand, when c_{in} is 0, no carry occurs. Since the LSB of the sum is discarded, there is no impact on the addition of 8 MSBs. The VHDL code of this design is shown in Listing 7.3.

Listing 7.3   Manual carry-in description of an addition–subtraction circuit

```
architecture manual_carry_arch of addsub is
    signal src0, src1, sum: signed(8 downto 0);
    signal b_tmp: std_logic_vector(7 downto 0);
    signal cin: std_logic; — carry−in bit
5 begin
    src0 <= signed(a & '1');
    b_tmp <= b when ctrl='0' else
             not b;
    cin <= '0' when ctrl='0' else
10            '1';
    src1 <= signed(b_tmp & cin);
    sum <= src0 + src1;
    r <= std_logic_vector(sum(8 downto 1));
end manual_carry_arch;
```

The diagram for this design is shown in Figure 7.5(c).

7.3.2   Signed–unsigned dual-mode comparator

In the IEEE numeric_std package, the signed and unsigned data types are defined to represent an array of bits as signed and unsigned integers respectively. The signed data type is in 2's-complement format. An example of 4-bit binary representations and their signed and unsigned interpretations are shown as a "binary wheel" in Figure 7.6. Note that the addition and subtraction operations are identical for the two data types. The addition and subtraction of a positive amount corresponds to a move clockwise and counterclockwise along the wheel, and thus the same hardware can be used. However, this is not true for relational operators.

This example considers a greater-than comparator in which the input can be interpreted as either unsigned or signed. The input data type (or the operation mode of the comparator) is specified by a control signal, mode. Our first design uses two comparators, one for each data type, and then uses the mode signal to select the desired result. The VHDL code is shown in Listing 7.4. Note that by definition of VHDL, the comparison in std_logic_vector data type (i.e., a > b) and the comparison in unsigned date type (i.e., unsigned(a) > unsigned(b)) implies the same implementation. For clarity, we use the latter in the VHDL code.

Listing 7.4   Initial description of a dual-mode comparator

```
library ieee;
use ieee.std_logic_1164.all;
use ieee.numeric_std.all;
entity comp2mode is
5    port(
        a,b: in std_logic_vector(7 downto 0);
        mode: in std_logic;
        agtb: out std_logic
    );
10 end comp2mode;

architecture direct_arch of comp2mode is
    signal agtb_signed, agtb_unsigned: std_logic;
```

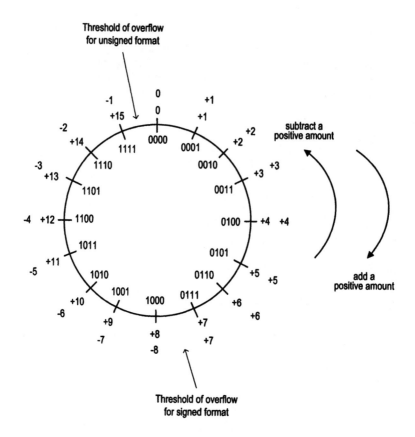

Figure 7.6 Four-bit binary wheel.

```
     begin
15       agtb_signed <= '1' when signed(a) > signed(b) else
                        '0';
         agtb_unsigned <= '1' when unsigned(a) > unsigned(b) else
                        '0';
         agtb <= agtb_unsigned when (mode='0') else
20                  agtb_signed;
     end direct_arch ;
```

To identify a potential sharing opportunity, we must examine the implementation of a comparator for the signed data type. First, if two inputs have different sign bits, the one with '0' is greater than the one with '1' since a positive number or 0 is always greater than a negative number. If two inputs have the same sign, we can ignore the sign bit and compare the remaining bits in a regular fashion (i.e., as the unsigned or std_logic_vector data type). At first glance, this may not be obvious for two negative numbers. We can verify it by checking the binary representations of signed numbers in Figure 7.6. For example, the binary representations of -1, -4 and -7 are "1111" and "1100" and "1001". After discarding the sign bit, we can see that "111" > "100" > "001", which is consistent with $-1 > -4 > -7$. Based on this observation, we can develop the rules for a dual-mode comparator:

- If a and b have the same sign bit, compare the remaining bits in a regular fashion.

- If a's sign bit is '1' and b's sign bit is '0', a is greater than b when in unsigned mode and b is greater than a when in signed mode.
- If a's sign bit is '0' and b's sign bit is '1', reverse the previous result.

The VHDL code for the design is shown in Listing 7.5. The agtb_mag signal is the comparison result of 7 LSBs of a and b, and the a1_b0 signal is a special status indicating that the MSBs (signs) of a and b are '1' and '0' respectively. The last conditional signal assignment statement translates the previous rules into logic expressions.

Listing 7.5   More efficient description of a dual-mode comparator

```
architecture shared_arch of comp2mode is
signal a1_b0, agtb_mag: std_logic;
begin
    a1_b0 <= '1' when a(7)='1' and b(7)='0' else
             '0';
    agtb_mag <= '1' when a(6 downto 0) > b(6 downto 0) else
                '0';
    agtb <= agtb_mag when (a(7)=b(7)) else
            a1_b0 when mode='0' else
            not a1_b0;
end shared_arch;
```

The new design eliminates one comparator and reduces the circuit size of the dual-mode comparator by one half.

7.3.3   Difference circuit

Assume that we want to implement a circuit that takes two unsigned numbers and calculates their difference; i.e., performs the function $|a - b|$. The straightforward design is to compute both a − b and b − a, compare a and b, and then select the proper subtraction result accordingly. The VHDL code is shown in Listing 7.6. Note that the signals are converted to the unsigned data type for arithmetic operation.

Listing 7.6   Initial description of a difference circuit

```
library ieee;
use ieee.std_logic_1164.all;
use ieee.numeric_std.all;
entity diff is
    port(
        a,b: in std_logic_vector(7 downto 0);
        result: out std_logic_vector(7 downto 0)
    );
end diff;

architecture direct_arch of diff is
    signal au, bu, ru, diffab, diffba: unsigned(7 downto 0);
begin
    au <= unsigned(a);
    bu <= unsigned(b);
    diffab <= au - bu;
    diffba <= bu - au;
    ru <= diffab when (au >= bu) else
          diffba;
```

```
20    result <= std_logic_vector(ru);
    end direct_arch;
```

One observation about the initial design is the implementation of the relational operation >=. The result of a >= b can be indirectly obtained from a − b by examining the sign bit of the subtraction result. If the sign bit is '0', the result is positive or 0 and thus a >= b is true. Otherwise, the result is negative and a >= b is false. We consider this scenario as functionality sharing since the operation a >= b indirectly utilizes the functionality of a − b. To apply the idea to this example, we must modify the internal representation since the original inputs a and b are interpreted as unsigned numbers and have no sign bit. We extend the internal signals by one bit and interpret them as a signed number. The VHDL code of the revised design is shown in Listing 7.7. Note that since both extended signals, as and bs, are positive (with '0' in MSB) and subtraction is performed, we need not worry about the overflow condition and can use the sign bit of a − b (i.e., diffab(8) in code) directly.

Listing 7.7   Better description of a difference circuit

```
architecture shared_arch of diff is
    signal as, bs, rs, diffab, diffba: signed(8 downto 0);
begin
    as <= signed('0'&a);
5   bs <= signed('0'&b);
    diffab <= as - bs;
    diffba <= bs - as;
    rs <= diffab when diffab(8)='0' else
            diffba;
10    result <= std_logic_vector(rs(7 downto 0));
    end shared_arch;
```

The revised design can be further optimized by replacing the b − a expression with 0 − diffab (or simply −diffab). Since the 0 − diffab operation has a constant operand (i.e., 0), the circuit size is about half that of a full subtractor. The final code is listed in Listing 7.8.

Listing 7.8   Most efficient description of a difference circuit

```
architecture effi_arch of diff is
    signal as, bs, rs, diffab, diffba: signed(8 downto 0);
begin
    as <= signed('0'&a);
5   bs <= signed('0'&b);
    diffab <= as - bs;
    diffba <= 0 - diffab;
    rs <= diffab when diffab(8)='0' else
            diffba;
10    result <= std_logic_vector(rs(7 downto 0));
    end effi_arch;bout
```

An alternative design approach is to use the operator-sharing technique. The code is shown in Listing 7.9. We first compare the two inputs and route the larger one to src0 and smaller one to src1, and then perform the subtraction. The design requires one subtractor and one comparator, and its size is comparable to that of the effi_arch architecture in Listing 7.8.

Listing 7.9   Alternative description of a difference circuit

```
architecture shared3_arch of diff is
    signal au, bu, ru, src0, src1: unsigned(7 downto 0);
begin
    au <= unsigned(a);
5   bu <= unsigned(b);
    process(au,bu)
    begin
        if au >= bu then
            src0 <= au;
10          src1 <= bu;
        else
            src0 <= bu;
            src1 <= au;
        end if;
15  end process;
    ru <= src0 - src1;
    result <= std_logic_vector(ru);
end shared3_arch;
```

7.3.4 Full comparator

Assume that we need a comparator that has three outputs, indicating the greater-than, equal-to and less-than conditions respectively. The straightforward design is to use three relational operators, each for an output condition. The VHDL code for this design is shown in Listing 7.10. Clearly, three separate relational circuits are needed when it is synthesized.

Listing 7.10   Initial description of a full comparator

```
library ieee;
use ieee.std_logic_1164.all;
entity comp3 is
    port(
5       a,b: in std_logic_vector(15 downto 0);
        agtb, altb, aeqb: out std_logic
    );
end comp3 ;

10 architecture direct_arch of comp3 is
begin
    agtb <= '1' when a > b else
            '0';
    altb <= '1' when a < b else
15          '0';
    aeqb <= '1' when a = b else
            '0';
end direct_arch;
```

If we examine the three operations carefully, we can see that the three conditions are mutually exclusive, and the third one can be derived if the other two are known. Thus, the functionality of the first two relational circuits can be shared to obtain the third output. The code of the revised design is shown in Listing 7.11.

Listing 7.11 Better description of a full comparator

```
architecture share1_arch of comp3 is
   signal gt, lt: std_logic;
begin
   gt <= '1' when a > b else
5          '0';
   lt <= '1' when a < b else
          '0';
   agtb <= gt;
   altb <= lt;
10 aeqb <= not (gt or lt);
end share1_arch;
```

The third statement means "a is equal to b" if the condition "a is greater than b or a is less than b" is not true. This simple revision eliminates the comparison circuit for the equal-to operator.

If we look Table 6.2, the equal-to circuit is smaller and faster than the greater-than circuit (especially compared with the circuit optimized for delay). This is due to the internal implementation of these circuits. We can further optimize the circuit by using the equal-to operator to replace either the greater-than or less-than operator. The code of the final design is shown in Listing 7.12.

Listing 7.12 Most efficient description of a full comparator

```
architecture share2_arch of comp3 is
   signal eq, lt: std_logic;
begin
   eq <= '1' when a = b else
5          '0';
   lt <= '1' when a < b else
          '0';
   aeqb <= eq;
   altb <= lt;
10 agtb <= not (eq or lt);
end share2_arch;
```

Although the observation of mutual exclusiveness of three outputs is trivial for us, it involves the meaning (semantics) of the operators. Most synthesis software is unable to take advantage of this property and optimize the code segment.

7.3.5 Three-function barrel shifter

A barrel shifter is a circuit that can shift input data by any number of positions. Both VHDL standard and the IEEE std_logic_1164 package define a set of shifting and rotating operators. Because of the complexity of the shifting circuit, some synthesis software is unable to synthesize these operators automatically. Shifting operations can be done in either the left or right direction and are divided into *rotate*, *logic shift* and *arithmetic shift*. In this example, we consider an 8-bit shifting circuit that can perform rotate right, logic shift right or arithmetic shift right, in which lower bits, 0's or sign bits are shifted into left positions respectively. In addition to the 8-bit data input, this circuit has a control signal, lar (for logic shift, arithmetic shift and rotate), which specifies the operation to be performed, and a control signal, amt (for amount), which specifies the number of positions to be rotated or shifted.

A straightforward design is to construct a rotate-right circuit, a logic shift-right circuit and an arithmetic shift-right circuit, and then use a multiplexer to select the desired output. The VHDL code of this design is shown in Listing 7.13. The individual shifting circuit is implemented by a selected signal assignment statement.

Listing 7.13   Initial description of a barrel shifter

```
   library ieee;
   use ieee.std_logic_1164.all;
   entity shift3mode is
       port(
 5         a: in std_logic_vector(7 downto 0);
           lar: in std_logic_vector(1 downto 0);
           amt: in std_logic_vector(2 downto 0);
           y: out std_logic_vector(7 downto 0)
       );
10 end shift3mode ;

   architecture direct_arch of shift3mode is
       signal logic_result, arith_result, rot_result:
           std_logic_vector(7 downto 0);
15 begin
       with amt select
           rot_result <=
               a                                when "000",
               a(0) & a(7 downto 1)             when "001",
20             a(1 downto 0) & a(7 downto 2)    when "010",
               a(2 downto 0) & a(7 downto 3)    when "011",
               a(3 downto 0) & a(7 downto 4)    when "100",
               a(4 downto 0) & a(7 downto 5)    when "101",
               a(5 downto 0) & a(7 downto 6)    when "110",
25             a(6 downto 0) & a(7) when others; -- 111
       with amt select
           logic_result <=
               a                                when "000",
               "0"        & a(7 downto 1) when "001",
30             "00"       & a(7 downto 2) when "010",
               "000"      & a(7 downto 3) when "011",
               "0000"     & a(7 downto 4) when "100",
               "00000"    & a(7 downto 5) when "101",
               "000000"   & a(7 downto 6) when "110",
35             "0000000"  & a(7) when others; -- 111
       with amt select
           arith_result <=
               a                                when "000",
               a(7) & a(7 downto 1)             when "001",
40             a(7)&a(7)     & a(7 downto 2) when "010",
               a(7)&a(7)&a(7)& a(7 downto 3) when "011",
               a(7)&a(7)&a(7)&a(7)&
                   a(7 downto 4)             when "100",
               a(7)&a(7)&a(7)&a(7)&a(7)&
45                 a(7 downto 5)             when "101",
```

```
                a(7)&a(7)&a(7)&a(7)&a(7)&a(7)&
                  a(7 downto 6)                    when "110",
                a(7)&a(7)&a(7)&a(7)&a(7)&a(7)&a(7)&
                  a(7)                             when others;
50      with lar select
          y <= logic_result when "00",
               arith_result when "01",
               rot_result   when others;
     end direct_arch;
```

The implementation includes three 8-bit 8-to-1 multiplexers and one 8-bit 3-to-1 multi-plexer.

If we examine the output of three shifting operations, we can see that their patterns are very similar and the only difference is the data being shifted into the left part. It is possible to share the functionality of a shifting circuit. To take advantage of this, we use a preprocessing circuit to modify the left part of the input data to the desired format and then pass it to the shifting circuit. The VHDL code based on this idea is given in Listing 7.14.

Listing 7.14   Better description of a barrel shifter

```
architecture shared_arch of shift3mode is
    signal shift_in: std_logic_vector(7 downto 0);
begin
    with lar select
s       shift_in <= (others=>'0')   when "00",
                    (others=>a(7))  when "01",
                    a               when others;
    with amt select
        y <= a                                          when "000",
10           shift_in(0)          & a(7 downto 1) when "001",
             shift_in(1 downto 0) & a(7 downto 2) when "010",
             shift_in(2 downto 0) & a(7 downto 3) when "011",
             shift_in(3 downto 0) & a(7 downto 4) when "100",
             shift_in(4 downto 0) & a(7 downto 5) when "101",
15           shift_in(5 downto 0) & a(7 downto 6) when "110",
             shift_in(6 downto 0) & a(7)          when others;
    end shared_arch;
```

In this code, one 8-bit 3-to-1 multiplexer is used to preprocess the input. Depending on the lar signal, its output shift_in can be the a input, repetitive 0's or repetitive sign bits. The shift_in signal is then passed to the shifting circuit and becomes the left part of the final output. The improved design consists of one 8-bit 8-to-1 multiplexer and one 8-bit 3-to-1 multiplexer. It has a similar critical path but eliminates two 8-bit 8-to-1 multiplexers.

7.4 LAYOUT-RELATED CIRCUITS

After synthesis, placement and routing will derive the actual physical layout of a digital circuit on a silicon chip. Although we cannot use VHDL code to specify the exact layout, it is possible to outline the general "shape" of the circuit. This will help the synthesis process and the placement and routing process to derive a more efficient circuit. Examples in this section show how to shape the circuit layout in VHDL code.

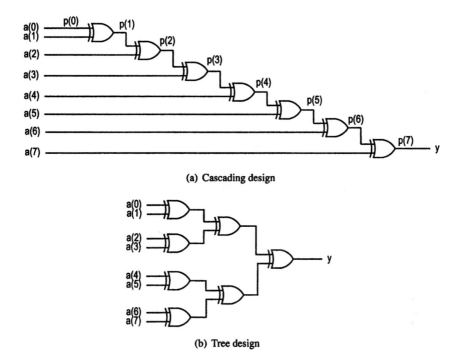

(a) Cascading design

(b) Tree design

Figure 7.7   Reduced-xor circuit.

7.4.1   Reduced-xor circuit

A reduced-xor function is to apply xor operations over all bits of an input signal. For example, let $a_7a_6a_5a_4a_3a_2a_1a_0$ be an 8-bit signal. The reduced-xor of this signal is

$$a_7 \oplus a_6 \oplus a_5 \oplus a_4 \oplus a_3 \oplus a_2 \oplus a_1 \oplus a_0$$

Since this function returns '1' if there are odd number of 1's in its input, it can be used to determine the odd parity of the input signal.

A straightforward design is shown in Figure 7.7(a). This design can be easily transformed into a VHDL code, which is shown in Listing 7.15.

Listing 7.15   Initial description of a reduced-xor circuit

```
library ieee;
use ieee.std_logic_1164.all;
entity reduced_xor is
    port(
        a: in std_logic_vector(7 downto 0);
        y: out std_logic
    );
end reduced_xor;

architecture cascade1_arch of reduced_xor is
begin
    y <= a(0) xor a(1) xor a(2) xor a(3) xor
         a(4) xor a(5) xor a(6) xor a(7);
```

```
end cascade1_arch;
```

By VHDL definition, the xor operator is left associative. Thus, the expression

```
a(0) xor a(1) xor a(2) xor ... xor a(7)
```

is the same as

```
(...((a(0) xor a(1)) xor a(2)) xor ...) xor a(7))
```

We can also use an 8-bit internal signal, p, to represent the intermediate results, as in Figure 7.7(a). The code for the architecture body is shown in Listing 7.16.

Listing 7.16   Alternative description of a reduced-xor circuit

```
architecture cascade2_arch of reduced_xor is
   signal p: std_logic_vector(7 downto 0);
begin
   p(0) <= a(0);
   p(1) <= p(0) xor a(1);
   p(2) <= p(1) xor a(2);
   p(3) <= p(2) xor a(3);
   p(4) <= p(3) xor a(4);
   p(5) <= p(4) xor a(5);
   p(6) <= p(5) xor a(6);
   p(7) <= p(6) xor a(7);
   y <= p(7);
end cascade2_arch;
```

Except for the first statement, a clear pattern exists between the inputs and outputs of these statements. By Boolean algebra, we know that $x = 0 \oplus x$. We can rewrite the first statement as

```
p(0) <= '0' xor a(0);
```

to make it match the pattern. Once this is done, we can use a more compact vector form to replace these statements, as shown in Listing 7.17.

Listing 7.17   Compact description of a reduced-xor circuit

```
architecture cascade_compact_arch of reduced_xor is
   constant WIDTH: integer := 8;
   signal p: std_logic_vector(WIDTH-1 downto 0);
begin
   p <= (p(WIDTH-2 downto 0) & '0') xor a;
   y <= p(WIDTH-1);
end cascade_compact_arch;
```

Although this design uses a minimal number of xor gates, it suffers a long propagation delay. The single cascading chain of xor gates becomes the critical path and the corresponding propagation delay is proportional to the number of xor gates in the chain. As the number of inputs increases, the propagation delay increases proportionally. Thus, the delay has an order of $O(n)$. Because of the associativity of the xor operator, we can arbitrarily change the order of operation. The initial design can be rearranged as a tree to reduce the length of its critical path, as shown in Figure 7.7(b). In VHDL code, we can use parentheses to force the desired order of operation, and the revised architecture body is shown in Listing 7.18.

Listing 7.18   Better description of a reduced-xor circuit

```
architecture tree_arch of reduced_xor is
begin
    y <= ((a(7) xor a(6)) xor (a(5) xor a(4))) xor
         ((a(3) xor a(2)) xor (a(1) xor a(0)));
5 end tree_arch;
```

In this new design, the critical path is reduced to three xor gates while the number of xor gates remains unchanged. Since we achieve better performance without adding extra hardware resource, it is a better design. In general, when we rearrange a cascade structure of n elements into a treelike structure, there will be $\log_2 n$ levels in the tree. The critical path is proportional to the number of levels in the tree and thus has an order of $O(\log_2 n)$.

Since this is a trivial circuit, synthesis software should be able to automatically transform the cascade design into the tree structure either by exploring the associative property or by performing time-constraint optimization. It is likely to obtain the same synthesis results for all codes in this example. However, for a more involved circuit, synthesis software is unable to do this, and we need to manually specify the order of operation to obtain a more efficient circuit.

Finally, let us examine the scalability of these codes. Assume that we want to increase the input to 16 bits. In the cascade1_arch, cascade2_arch and tree_arch architectures, we have to add eight additional **xor** a(i) terms or eight additional p(i+1) <= p(i) **xor** a(i) statements respectively. The number of revisions is proportional to the number of inputs and thus is on the order of $O(n)$. In the cascade_compact_arch architecture, the code remains the same except that the number in the constant statement has to be changed from 8 to 16. The needed revision is $O(1)$, and this code is highly scalable.

7.4.2 Reduced-xor-vector circuit

A reduced-xor-vector function is to apply xor operations over all possible combinations of lower bits of an input signal. It can best be explained by an example. Let $a_7 a_6 a_5 a_4 a_3 a_2 a_1 a_0$ be an 8-bit signal. Applying the reduced-xor-vector function to it returns eight values, and they are defined as

$$
\begin{aligned}
y_0 &= a_0 \\
y_1 &= a_1 \oplus a_0 \\
y_2 &= a_2 \oplus a_1 \oplus a_0 \\
y_3 &= a_3 \oplus a_2 \oplus a_1 \oplus a_0 \\
y_4 &= a_4 \oplus a_3 \oplus a_2 \oplus a_1 \oplus a_0 \\
y_5 &= a_5 \oplus a_4 \oplus a_3 \oplus a_2 \oplus a_1 \oplus a_0 \\
y_6 &= a_6 \oplus a_5 \oplus a_4 \oplus a_3 \oplus a_2 \oplus a_1 \oplus a_0 \\
y_7 &= a_7 \oplus a_6 \oplus a_5 \oplus a_4 \oplus a_3 \oplus a_2 \oplus a_1 \oplus a_0
\end{aligned}
$$

A straightforward design is to follow the definition of this function, which can easily be transformed into the VHDL code shown in Listing 7.19.

Listing 7.19 Initial description of a reduced-xor-vector circuit

```
library ieee;
use ieee.std_logic_1164.all;
entity reduced_xor_vector is
   port(
s         a: in std_logic_vector(7 downto 0);
          y: out std_logic_vector(7 downto 0)
     );
end reduced_xor_vector;

10 architecture direct_arch of reduced_xor_vector is
   begin
      y(0) <= a(0);
      y(1) <= a(1) xor a(0);
      y(2) <= a(2) xor a(1) xor a(0);
15    y(3) <= a(3) xor a(2) xor a(1) xor a(0);
      y(4) <= a(4) xor a(3) xor a(2) xor a(1) xor a(0);
      y(5) <= a(5) xor a(4) xor a(3) xor a(2) xor a(1) xor a(0);
      y(6) <= a(6) xor a(5) xor a(4) xor a(3) xor a(2) xor a(1)
               xor a(0);
20    y(7) <= a(7) xor a(6) xor a(5) xor a(4) xor a(3) xor a(2)
               xor a(1) xor a(0);
   end direct_arch;
```

In this code, each output is described independently, and no sharing is imposed. If no optimization is performed during synthesis, the synthesized circuit needs 28 xor gates. There are lots of common expressions that can be shared to reduce the number of xor gates.

Note that there is a simple relationship between the successive output values:

$$y_{i+1} = a_{i+1} \oplus y_i$$

The design based on this observation is shown in Figure 7.8(a), in which only seven xor gates are needed.

The VHDL code for this design is similar to the cascade2_arch architecture in Listing 7.16 except that all intermediate internal values are used as output. We need to modify the last statement and the VHDL code, as shown in Listing 7.20.

Listing 7.20 Sharing description of a reduced-xor-vector circuit

```
architecture shared1_arch of reduced_xor_vector is
   signal p: std_logic_vector(7 downto 0);
   begin
      p(0) <= a(0);
s     p(1) <= p(0) xor a(1);
      p(2) <= p(1) xor a(2);
      p(3) <= p(2) xor a(3);
      p(4) <= p(3) xor a(4);
      p(5) <= p(4) xor a(5);
10    p(6) <= p(5) xor a(6);
      p(7) <= p(6) xor a(7);
      y <= p;
   end shared1_arch;
```

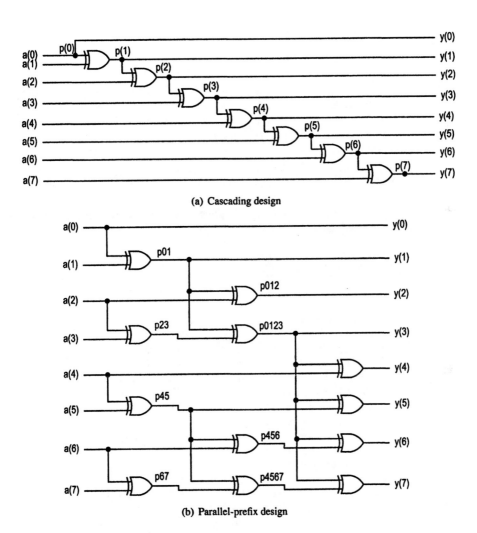

(a) Cascading design

(b) Parallel-prefix design

Figure 7.8 Reduced-xor-vector circuit.

Similarly, the compact cascade_compact_arch architecture in Listing 7.17 can be revised, too, as shown in Listing 7.21.

Listing 7.21 Compact description of reduced-xor-vector circuit

```
architecture shared_compact_arch of reduced_xor_vector is
   constant WIDTH: integer := 8;
   signal p: std_logic_vector(WIDTH-1 downto 0);
begin
   p <= (p(WIDTH-2 downto 0) & '0') xor a;
   y <= p;
end shared_compact_arch;
```

The critical path of this circuit is the path to obtain the $y(7)$ signal, which has the largest number of xor gates along the path. Our earlier discussion shows that the propagation delay is on the order of $O(n)$. To increase the performance, we have to rearrange the cascading chain to a treelike structure. The simple xor tree of tree_arch architecture of the previous example is not adequate since it cannot produce all the needed output values. One straightforward design is to create an independent xor tree for each output value. The design needs 28 xor gates, and the critical path of the circuit is the critical path of the tree used to implement $y(7)$, which is the largest tree and has three levels of xor gates. The VHDL code is similar to the direct_arch architecture except that we use parentheses to force the order of evaluation, as shown in Listing 7.22.

Listing 7.22 Tree description of a reduced-xor-vector circuit

```
architecture direct_tree_arch of reduced_xor_vector is
begin
   y(0) <= a(0);
   y(1) <= a(1) xor a(0);
   y(2) <= a(2) xor a(1) xor a(0);
   y(3) <= (a(3) xor a(2)) xor (a(1) xor a(0));
   y(4) <= (a(4) xor a(3)) xor (a(2) xor a(1)) xor a(0);
   y(5) <= (a(5) xor a(4)) xor (a(3) xor a(2)) xor
           (a(1) xor a(0));
   y(6) <= ((a(6) xor a(5)) xor (a(4) xor a(3))) xor
           ((a(2) xor a(1)) xor a(0));
   y(7) <= ((a(7) xor a(6)) xor (a(5) xor a(4))) xor
           ((a(3) xor a(2)) xor (a(1) xor a(0)));
end direct_tree_arch;
```

A more elegant design is shown in Figure 7.8(b). This design is targeted for performance and limits the critical path within three levels of xor gates. Within this constraint, it tries to share as many common expressions as possible. Instead of 28 xor gates, this design needs only 12 xor gates. We can derive the VHDL code according to the circuit diagram, as shown in Listing 7.23.

Listing 7.23 Parallel-prefix description of a reduced-xor-vector circuit

```
architecture optimal_tree_arch of reduced_xor_vector is
   signal p01, p23, p45, p67, p012,
          p0123, p456, p4567: std_logic;
begin
   p01 <= a(0) xor a(1);
   p23 <= a(2) xor a(3);
```

```
      p45  <= a(4) xor a(5);
      p67  <= a(6) xor a(7);
      p012  <= p01 xor a(2);
10    p0123  <= p01 xor p23;
      p456  <= p45 xor a(6);
      p4567  <= p45 xor p67;
      y(0)  <= a(0);
      y(1)  <= p01;
15    y(2)  <= p012;
      y(3)  <= p0123;
      y(4)  <= p0123 xor a(4);
      y(5)  <= p0123 xor p45;
      y(6)  <= p0123 xor p456;
20    y(7)  <= p0123 xor p4567;
   end optimal_tree_arch;
```

Although the same design principle can be used for a circuit with a larger number of inputs, revising the VHDL code will be very tedious and error-prone. Actually, this design is not just a lucky observation. It is based on *parallel-prefix structure*, and the systematic development of VHDL code for this circuit is discussed in Chapter 15.

There are two important observations for this example. The first is the trade-off between circuit size and performance. In a digital circuit, we normally have to use more hardware resources to improve the performance, as in this example. The cascading design needs a minimal number of xor gates, which is on the order of $O(n)$, but suffers a large propagation delay, which is also on the order of $O(n)$. The parallel-prefix design, on the other hand, requires $0.5n \log_2 n$ xor gates, but its delay is only on the order of $O(\log_2 n)$.

The second observation is about the capability of the synthesis software. Ideally, we hope the synthesis software can automatically derive the desired implementation regardless of the initial VHDL description. This is hardly possible, even for the simple function used in this example.

7.4.3 Tree priority encoder

A priority encoder is a circuit that returns the codes for the highest-priority request. We have discussed it in Chapters 4 and 5 and used different VHDL constructs to describe this circuit. The conditional signal assignment and if statements are natural to describe this function, and they specify the same priority routing network. The shape of the priority routing network is a single cascading chain, somewhat similar to the layout of cascading reduced-xor design in the previous example. Since the critical path is formed along this chain, performance suffers when the number of inputs increases. In the reduced-xor circuit, we can convert the cascading chain into a tree by rearranging the order of xor operations. This is also possible for the priority encoder, although the rearrangement is more involved. This example shows how to create an alternative treelike structure. We use a 16-to-4 priority encoder to demonstrate the scheme.

The VHDL description for the cascading design is straightforward, as shown in Listing 7.24.

Listing 7.24 Cascading description of a priority encoder

```
library ieee;
use ieee.std_logic_1164.all;
entity prio_encoder is
```

```
     port (
 5       r: in std_logic_vector(15 downto 0);
         code: out std_logic_vector(3 downto 0);
         active: out std_logic
     );
   end prio_encoder;
10
   architecture cascade_arch of prio_encoder is
   begin
     code <= "1111"  when r(15)='1'  else
             "1110"  when r(14)='1'  else
15           "1101"  when r(13)='1'  else
             "1100"  when r(12)='1'  else
             "1011"  when r(11)='1'  else
             "1010"  when r(10)='1'  else
             "1001"  when r(9)='1'   else
20           "1000"  when r(8)='1'   else
             "0111"  when r(7)='1'   else
             "0110"  when r(6)='1'   else
             "0101"  when r(5)='1'   else
             "0100"  when r(4)='1'   else
25           "0011"  when r(3)='1'   else
             "0010"  when r(2)='1'   else
             "0001"  when r(1)='1'   else
             "0000";
     active <= r(15) or r(14) or r(13) or r(12) or
30               r(11) or r(10) or r(9) or r(8) or
               r(7) or r(6) or r(5) or r(4) or
               r(3) or r(2) or r(1) or r(0);
   end cascade_arch;
```

The diagram of the code segment is shown in Figure 7.9, which consists of a chain of 15 2-to-1 multiplexers.

To develop a tree design, we start with smaller priority encoders and then rearrange them to the desired layout. Design in this example uses a 4-to-2 priority encoder. The function table and block diagram of a 4-to-2 decoder are shown in Figure 7.10(a). The block diagram of the 16-to-4 tree priority encoder is shown in Figure 7.10(b).

The basic skeleton consists of a two-level tree. The 16 input requests are divided into four groups and fed to four 4-to-2 priority encoders in the first level. Each 4-to-2 priority encoder performs two functions. First, they generate active signals, act1, act2, act3 and act4, to indicate whether a request occurs in a particular group. Each active signal can be interpreted as the request signal of that particular group. Second, due to the clever arrangement of input connection, their output codes, code_g3, code_g2, code_g1 and code_g0, form the two LSBs of the final 4-bit code. For example, if the highest-priority request is r(9), its code is "1001". The r(9) signal is connected to the second 4-to-2 priority encoder in the first level and its output, code_g2, is "01", which is the two LSBs of "1001".

There is only one 4-to-2 priority encoder in the second level. Its inputs are the four "group request" signals from the first level. The output, code_msb, is the code of the group with the highest-priority request, which forms the two MSBs of the 4-bit code signal. We also need a 4-to-1 multiplexer in the second level. The code_msb signal is used to select and route the 2 LSBs from the proper group to final output.

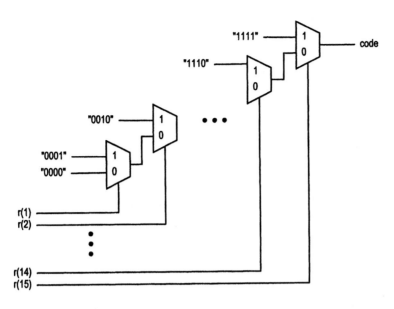

Figure 7.9   Cascading priority encoder.

Since a 4-to-2 priority encoder is used repeatedly, we use component instantiation in the code. The 4-to-2 priority encoder is coded as a regular cascading design, as shown in Listing 7.25.

Listing 7.25   4-to-2 priority encoder

```
library ieee;
use ieee.std_logic_1164.all;
entity prio42 is
    port(
s        r4: in std_logic_vector(3 downto 0);
         code2: out std_logic_vector(1 downto 0);
         act42: out std_logic
    );
end prio42;
10
  architecture cascade_arch of prio42 is
  begin
      code2 <= "11"  when r4(3)='1'  else
               "10"  when r4(2)='1'  else
15             "01"  when r4(1)='1'  else
               "00";
      act42 <= r4(3) or r4(2) or r4(1) or r4(0);
  end cascade_arch;
```

The VHDL code for the tree design is shown in Listing 7.26, which basically follows the diagram of Figure 7.10(b). The code uses VHDL component instantiation, which is briefly reviewed in Section 2.2.2 and discussed in Chapter 13.

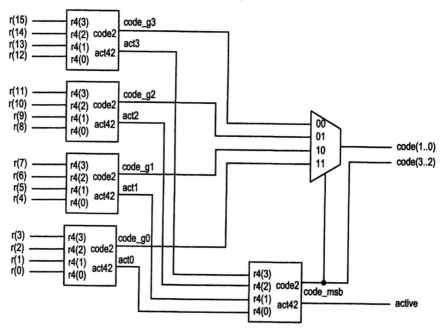

r4	code2	act42
1---	11	1
01--	10	1
001-	01	1
0001	00	1
0000	00	0

(a) 4-to-2 priority encoder

(b) 16-to-4 priority encoder using 4-to-2 priority encoders

Figure 7.10   Tree priority encoder.

Listing 7.26   16-to-4 priority encoder

```vhdl
architecture tree_arch of prio_encoder is
    component prio42 is
        port(
            r4: in std_logic_vector(3 downto 0);
 5          code2: out std_logic_vector(1 downto 0);
            act42: out std_logic
        );
    end component;
    signal code_g3, code_g2, code_g1, code_g0, code_msb:
10      std_logic_vector(1 downto 0);
    signal tmp: std_logic_vector(3 downto 0);
    signal act3, act2, act1, act0: std_logic;
begin
    -- four 1st-stage 4-to-2 priority encoders
15  unit_level_0_0: prio42
        port map(r4=>r(3 downto 0), code2=>code_g0,
                act42=>act0);
    unit_level_0_1: prio42
        port map(r4=>r(7 downto 4), code2=>code_g1,
20              act42=>act1);
    unit_level_0_2: prio42
        port map(r4=>r(11 downto 8), code2=>code_g2,
                act42=>act2);
    unit_level_0_3: prio42
25      port map(r4=>r(15 downto 12), code2=>code_g3,
                act42=>act3);
    -- 2nd stage 4-to-2 priority encoder
    tmp <= act3 & act2 & act1 & act0;
    unit_level_2: prio42
30      port map(r4=>tmp, code2=>code(3 downto 2),
                act42=>active);
    -- 2 MSBs of code
    code(3 downto 2) <= code_msb;
    -- 2 LSBs of code
35  with code_msb select
        code(1 downto 0) <= code_g3 when "11",
                            code_g2 when "10",
                            code_g1 when "01",
                            code_g0 when others;
40 end tree_arch;
```

Now let us analyze the critical path of two designs. The critical path of the first cascading design consists of fifteen 2-to-1 multiplexers. The critical path of the tree design consists of two 4-to-2 priority encoders plus one 4-to-1 multiplexer. Since the 4-to-2 priority encoder uses the regular cascading design, it is constructed by three 2-to-1 multiplexers. Thus, the critical path of the tree design consists of six 2-to-1 multiplexers and one 4-to-1 multiplexer. It is much shorter than that of the cascading design.

Although software can perform a certain degree of optimization during synthesis, the optimization tends to be local and a good initial description can make a significant impact on the final implementation. This is especially true as the number of input requests increases. We can further refine the design by utilizing a tree of 2-to-1 priority encoders inside the

4-to-2 priority encoder. The 16-to-4 priority encoder now becomes a tree consisting of four levels of 2-to-1 priority encoders.

A major drawback of the tree design is the code complexity. Revising the code for different input sizes is very involved. An alternative scalable design is discussed in Chapter 15.

7.4.4 Barrel shifter revisited

We discussed the design of a barrel shifter in Section 7.3.5. This design suffers several problems, and an alternative is developed in this section. We first examine the design of an 8-bit rotate-right circuit and then extend it to the complete three-function circuit.

In Section 7.3.5, the rotating circuit is translated directly from the function table and coded by a selected signal assignment statement. The code is repeated in Listing 7.27.

Listing 7.27 Single-level rotate-right circuit

```
    library ieee;
    use ieee.std_logic_1164.all;
    entity rotate_right is
       port(
5          a: in std_logic_vector(7 downto 0);
           amt: in std_logic_vector(2 downto 0);
           y: out std_logic_vector(7 downto 0)
       );
    end rotate_right;
10
    architecture direct_arch of rotate_right is
    begin
       with amt select
          y<= a                                  when "000",
15           a(0) & a(7 downto 1)                when "001",
             a(1 downto 0) & a(7 downto 2) when "010",
             a(2 downto 0) & a(7 downto 3) when "011",
             a(3 downto 0) & a(7 downto 4) when "100",
             a(4 downto 0) & a(7 downto 5) when "101",
20           a(5 downto 0) & a(7 downto 6) when "110",
             a(6 downto 0) & a(7) when others; -- 111
    end direct_arch;
```

This code implies an 8-bit 8-to-1 multiplexer circuit. In actual implementation the 8-bit multiplexer is composed of eight 1-bit 8-to-1 multiplexers, as shown in Figure 7.11.

Although the conceptual diagram seems to be all right, this approach suffers some subtle problems. First, a wide multiplexer cannot be effectively mapped to certain device technologies. Second, since an input data bit is routed to all multiplexers, the connection wires grow on the order of $O(n^2)$. The wiring area becomes congested as the number of inputs grows. Finally, the basic layout of this circuit is a single narrow strip, as in Figure 7.11. This makes placing and routing more difficult.

An alternative design is to do the rotating in levels, as shown in Figure 7.12(a). In each level, a bit of the amt signal indicates whether the input is passed directly to the output or rotated by a fixed amount. Bits 0, 1 and 2 of the amt signal control the routing in levels 0, 1 and 2 respectively. The amounts are different in each level, which are the 2^0, 2^1 and 2^2 positions. After passing three levels, the total number of positions rotated is the summation

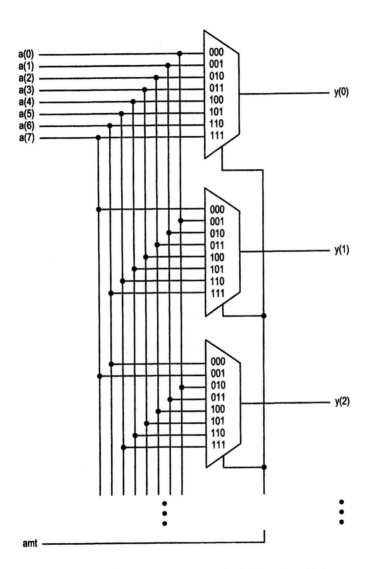

Figure 7.11   Barrel shifter using a single level of 8-to-1 multiplexers.

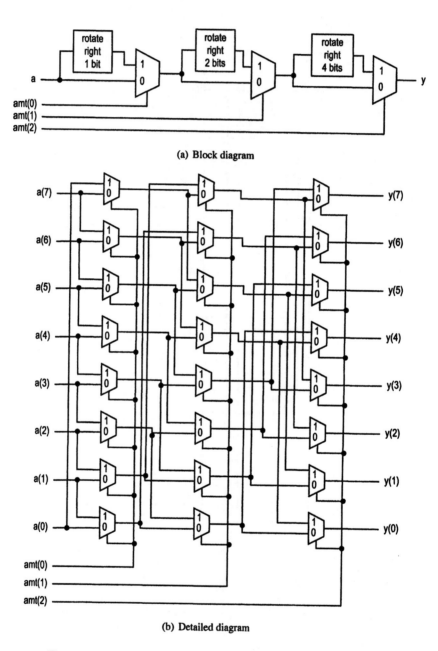

(a) Block diagram

(b) Detailed diagram

Figure 7.12   Barrel shifter using three levels of 2-to-1 multiplexers.

of positions rotated in each level, which is $amt(2)*2^2 + amt(1)*2^1 + amt(0)*2^0$. The VHDL code of this revised design is shown in Listing 7.28.

Listing 7.28   Multilevel rotate-right circuit

```
architecture multi_level_arch of rotate_right is
   signal le0_out, le1_out, le2_out:
      std_logic_vector(7 downto 0);
begin
5  -- level 0, shift 0 or 1 bit
   le0_out <= a(0) & a(7 downto 1) when amt(0)='1' else
               a;
   -- level 1, shift 0 or 2 bits
   le1_out <=
10     le0_out(1 downto 0) & le0_out(7 downto 2)
          when amt(1)='1' else
       le0_out;
   -- level 2, shift 0 or 4 bits
   le2_out <=
15     le1_out(3 downto 0) & le1_out(7 downto 4)
          when amt(2)='1' else
       le1_out;
   -- output
   y <= le2_out;
20 end multi_level_arch;
```

A more detailed diagram of this design is shown in Figure 7.12(b). Note that rotating a fixed amount involves only signal routing and requires no physical components.

Comparing the two designs is more subtle. The first design needs eight 8-to-1 multiplexers and its critical path is the same as the critical path of an 8-to-1 multiplexer. The multilevel design needs eight 2-to-1 multiplexers at each level, and thus a total of twenty-four 2-to-1 multiplexers. Its critical path consists of three levels of 2-to-1 multiplexers. The implementation of these multiplexers is technology dependent and there is no clear-cut answer on circuit size and propagation delay. The additional wiring area and delay of the first design further complicates the comparison. However, when the input becomes large, the wiring and routing will become more problematic in the first design. The regular interconnection pattern of the multilevel design can scale better and thus should be the preferred choice. The VHDL code of multilevel design is also easier to scale. The amount of revision is on the order of $O(\log_2 n)$ rather than $O(n)$, as in the first design.

To extend the rotate-right circuit to incorporate the additional logic shift-right and arithmetic shift-right functions, we can apply the preprocessing idea from Section 7.3.5. Since there are three levels in the new design, preprocessing has to be performed at each level. The revised VHDL code is given in Listing 7.29.

Listing 7.29   Multilevel description of a three-function barrel shifter

```
architecture multi_level_arch of shift3mode is
   signal le0_out, le1_out, le2_out:
      std_logic_vector(7 downto 0);
   signal le0_sin: std_logic;
5  signal le1_sin: std_logic_vector(1 downto 0);
   signal le2_sin: std_logic_vector(3 downto 0);
begin
   -- level 0, shift 0 or 1 bit
```

```
       with lar select
10       le0_sin <= '0'      when "00",
                     a(7) when "01",
                     a(0) when others;
       le0_out <= le0_sin & a(7 downto 1) when amt(0)='1' else
                     a;
15   — level 1, shift 0 or 2 bits
       with lar select
         le1_sin <=
           "00"                      when "00",
           (others => le0_out(7)) when "01",
20         le0_out(1 downto 0)      when others;
       le1_out <= le1_sin & le0_out(7 downto 2)
                     when amt(1)='1' else
                 le0_out;
       — level 2, shift 0 or 4 bits
25   with lar select
         le2_sin <=
           "0000"                    when "00",
           (others => le1_out(7)) when "01",
           le1_out(3 downto 0)      when others;
30   le2_out <= le2_sin & le1_out(7 downto 4)
                     when amt(2)='1' else
                 le1_out;
       — output
       y <= le2_out;
35 end multi_level_arch ;
```

The preprocessing utilizes three 3-to-1 multiplexers, whose widths are 1, 2 and 4 bits respectively, and their overall complexity is similar to the 8-bit 3-to-1 multiplexer used in Section 7.3.5.

7.5 GENERAL CIRCUITS

The examples of previous sections are focused on specific aspects of design and VHDL coding. Several general design examples are presented in this section.

7.5.1 Gray code incrementor

The *Gray code* is a special kind of code in that only a single bit changes between any two successive code words. It minimizes the number of transitions when a signal switches between successive words. A 4-bit Gray code and its corresponding binary code are shown in Table 7.1. A Gray code incrementor is a circuit that generates the next word in Gray code. The function table of a 4-bit Gray code incrementor is shown in Table 7.2. A straightforward design is simply to translate this table into a selected signal assignment statement, as in Listing 7.30.

Listing 7.30   Initial description of a Gray code incrementor

```
library ieee;
use ieee.std_logic_1164.all;
use ieee.numeric_std.all;
```

Table 7.1   4-bit Gray code

Binary code $b_3b_2b_1b_0$	Gray code $g_3g_2g_1g_0$
0000	0000
0001	0001
0010	0011
0011	0010
0100	0110
0101	0111
0110	0101
0111	0100
1000	1100
1001	1101
1010	1111
1011	1110
1100	1010
1101	1011
1110	1001
1111	1000

Table 7.2   Function table of a 4-bit Gray code incrementor

Gray code	Incremented Gray code
0000	0001
0001	0011
0011	0010
0010	0110
0110	0111
0111	0101
0101	0100
0100	1100
1100	1101
1101	1111
1111	1110
1110	1010
1010	1011
1011	1001
1001	1000
1000	0000

```
     entity g_inc is
5      port(
           g: in std_logic_vector(3 downto 0);
           g1: out std_logic_vector(3 downto 0)
       );
     end g_inc ;
10
     architecture table_arch of g_inc is
     begin
        with g select
           g1 <= "0001" when "0000",
15               "0011" when "0001",
                 "0010" when "0011",
                 "0110" when "0010",
                 "0111" when "0110",
                 "0101" when "0111",
20               "0100" when "0101",
                 "1100" when "0100",
                 "1101" when "1100",
                 "1111" when "1101",
                 "1110" when "1111",
25               "1010" when "1110",
                 "1011" when "1010",
                 "1001" when "1011",
                 "1000" when "1001",
                 "0000" when others; — "1000"
30 end table_arch;
```

Although the VHDL code is simple, it is not scalable because the needed revision is on the order of $O(2^n)$. Unfortunately, there is no easy algorithm to derive the next Gray code word directly. Since an algorithm exists for conversion between Gray code and binary code, one possible approach is to derive it indirectly by using a binary incrementor. This design includes three stages:

1. Convert a Gray code word to the corresponding binary word.
2. Increment the binary word.
3. Convert the result back to the Gray code word.

The binary-to-Gray conversion algorithm is based on the following observation: the ith bit (i.e., g_i) of the Gray code word is '1' if the ith bit and $(i+1)$th bit (i.e., b_i and b_{i+1}) of the corresponding binary word are different. This observation can be translated into a logic equation:

$$g_i = b_i \oplus b_{i+1}$$

We can verify this equation by using the 4-bit code of Table 7.1:

$$g_3 = b_3 \oplus 0 = b_3$$
$$g_2 = b_2 \oplus b_3$$
$$g_1 = b_1 \oplus b_2$$
$$g_0 = b_0 \oplus b_1$$

The equation for Gray-to-binary conversion can be obtained by manipulating the previous equation:

$$b_i = g_i \oplus b_{i+1}$$

We can also expand b_{i+1} on the right-hand side recursively. For example, a 4-bit code can be expressed as

$$b_3 = g_3 \oplus 0 = g_3$$
$$b_2 = g_2 \oplus b_3 = g_2 \oplus g_3$$
$$b_1 = g_1 \oplus b_2 = g_1 \oplus g_2 \oplus g_3$$
$$b_0 = g_0 \oplus b_1 = g_0 \oplus g_1 \oplus g_2 \oplus g_3$$

Once we know the conversion algorithm, we can derive the VHDL code. Note that the equations for the Gray-to-binary conversion are very similar to the reduced-xor-vector function discussed in Section 7.4.2. The VHDL code of the new design is shown in Listing 7.31. We use the compact vector form, similar to that in the shared_compact_arch architecture of Listing 7.21, for Gray-to-binary and binary-to-Gray code conversions.

Listing 7.31   Compact description of a Gray code incrementor

```
architecture compact_arch of g_inc is
   constant WIDTH: integer := 4;
   signal b, b1: std_logic_vector(WIDTH-1 downto 0);
begin
   -- Gray to binary
   b <= g xor ('0' & b(WIDTH-1 downto 1));
   -- binary increment
   b1 <= std_logic_vector((unsigned(b)) + 1);
   -- binary to Gray
   g1<= b1 xor ('0' & b1(WIDTH-1 downto 1));
end compact_arch;
```

The new code is independent of the input size and the revision is on the order of $O(1)$. Since each part can easily be identified, this design allows us to utilize the alternative implementation for the adder and Gray-to-binary circuit. If performance is an issue, we can replace them with faster but larger circuits.

7.5.2   Programmable priority encoder

In a regular priority encoder, the order of priority for each request is fixed. For example, the order of eight requests, $r(7), \ldots, r(0)$, is normally $r(7), r(6), \ldots, r(1)$ and $r(0)$. Some applications need to dynamically change the priority of a request to give fair access to each request. In this subsection, we consider a programmable 8-to-3 priority encoder in which the priority can be assigned in a wrapped-around fashion. In addition to the eight regular request signals, the circuit also has a 3-bit control signal, c, which specifies the request that has the highest priority. For example, if c is "011", $r(3)$ has the highest priority and the order of the requests is $r(3), r(2), r(1), r(0), r(7), \ldots, r(4)$. A brute-force design is to utilize eight regular priority encoders and one 8-to-1 multiplexer. Each priority encoder has a fixed request order, and the multiplexer passes the desired code to the output. While this design is straightforward, it is not very efficient.

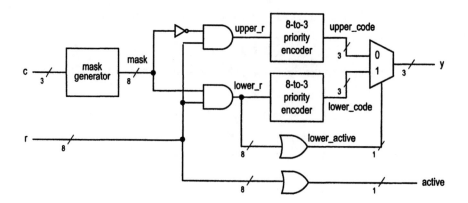

Figure 7.13   Block diagram of a programmable priority encoder.

A better design is shown in Figure 7.13. This design first uses the c signal to generate two 8-bit masks, which are used to clear the upper and lower parts of the requests. For example, if c is "011", the lower mask is "00000111" and the upper mask is "11111000", the inverse of the lower mask. We apply the two masks to the original requests and obtain two new masked requests, in which the lower and upper parts are cleared. For example, if a request is "11011011", the two masked requests will be "11011000" and "00000011". If the active signal is asserted in the lower group, it means that there is a request from that group and its code will be routed to the final output. Otherwise, the code from the upper priority encoder will be routed to the output. If we continue the previous example, the codes from the upper and lower priority encoders are "111" and "001", and since there is a request from the lower group, "001" will be routed to the final output. The VHDL code describing this design is shown in Listing 7.32.

Listing 7.32   Programmable priority encoder

```
library ieee;
use ieee.std_logic_1164.all;
entity fair_prio_encoder is
    port(
5       r: in std_logic_vector(7 downto 0);
        c: in std_logic_vector(2 downto 0);
        code: out std_logic_vector(2 downto 0);
        active: out std_logic
    );
10 end fair_prio_encoder;

    architecture arch of fair_prio_encoder is
        signal mask, lower_r, upper_r:
            std_logic_vector(7 downto 0);
15      signal lower_code, upper_code:
            std_logic_vector(2 downto 0);
        signal lower_active: std_logic;
    begin
        with c select
20          mask <= "00000001" when "000",
                    "00000011" when "001",
```

```
                         "00000111"  when  "010",
                         "00001111"  when  "011",
                         "00011111"  when  "100",
25                       "00111111"  when  "101",
                         "01111111"  when  "110",
                         "11111111"  when  others;
        lower_r <= r and mask;
        upper_r <= r and (not mask);
30      lower_code <=  "111"   when lower_r(7)='1'  else
                       "110"   when lower_r(6)='1'  else
                       "101"   when lower_r(5)='1'  else
                       "100"   when lower_r(4)='1'  else
                       "011"   when lower_r(3)='1'  else
35                     "010"   when lower_r(2)='1'  else
                       "001"   when lower_r(1)='1'  else
                       "000";
        upper_code <=  "111"   when upper_r(7)='1'  else
                       "110"   when upper_r(6)='1'  else
40                     "101"   when upper_r(5)='1'  else
                       "100"   when upper_r(4)='1'  else
                       "011"   when upper_r(3)='1'  else
                       "010"   when upper_r(2)='1'  else
                       "001"   when upper_r(1)='1'  else
45                     "000";
        lower_active <= lower_r(7) or lower_r(6) or lower_r(5) or
                        lower_r(4) or lower_r(3) or lower_r(2) or
                        lower_r(1) or lower_r(0);
        code <= lower_code when lower_active='1' else
50              upper_code;
        active <= r(7) or r(6) or r(5) or r(4) or
                  r(3) or r(2) or r(1) or r(0);
    end arch;
```

The VHDL code is much more efficient than the first design.

7.5.3 Signed addition with status

The definition of the VHDL addition operator is very simple. It takes two operands and returns the summation. In a complex digital system, such as a processor, adders frequently need additional status signals and carry signals. Status signals show various conditions of an addition operation, including zero, sign and overflow. Zero status indicates whether the result is zero, sign status indicates whether the result is a positive or negative number, and overflow status indicates whether overflow occurs during operation. Carry signals pass information between successive additions. For example, if we want to construct a 64-bit adder by using 8-bit adders, we have to utilize the carry signals to convey the relevant carry information. Carry signals include the carry-in signal, which is an input that comes from the previous stage, and the carry-out signal, which is an output signal to be passed to the next stage. We consider the addition of two signed integers in this subsection.

The derivation of status signals is trickier than it first appears because of the overflow condition. Overflow affects the determination of sign and zero status and thus must be determined first. Our derivation of overflow condition is based on the following observations:

- If the two operands have different signs, overflow can never occur since the addition of a positive number and a negative number will always decrease the magnitude.
- If the two operands and the result have the same sign, overflow does not occur since the result is still within the range.
- If the two operands have the same sign but the result has a different sign, overflow occurs. The sign change indicates that the result goes beyond the positive or negative boundary and thus is beyond the range. We can verify this by checking the binary wheel of Figure 7.6.

Let the sign bits of two operands and summation be s_a, s_b and s_s respectively. We can translate our observation into the following logic expression:

$$overflow = (s_a \cdot s_b \cdot s_s') + (s_a' \cdot s_b' \cdot s_s)$$

Once we know the overflow condition, we can determine the zero condition and the sign. Because of the potential of overflow, the addition result may not be 0 even if the summation output is 0. For example, if we add two 4-bit inputs, "1000" and "1000", the summation output is "0000" because of overflow. Thus, the zero condition should be asserted only if the summation output is 0 and there is no overflow.

From our observation on overflow, it is clear that the sign bit of the summation output is not necessarily the sign of a real addition result. In the example above, the sign bit of summation "0000" is '0' while the addition result should be negative. Thus, the sign bit of the addition result is the same as the sign bit of the summation output only if no overflow occurs. It should be inverted otherwise.

Carry signals can be handled by using two extra bits in the internal signals. One bit will be appended to the left to incorporate the carry-out signal. The other bit will be appended to the right to inject the carry-in signal, as explained in Section 7.3.1. The complete VHDL code is shown in Listing 7.33.

Listing 7.33 Signed addition with status

```
library ieee;
use ieee.std_logic_1164.all;
use ieee.numeric_std.all;
entity adder_status is
5    port(
        a,b: in std_logic_vector(7 downto 0);
        cin: in std_logic;
        sum: out  std_logic_vector(7 downto 0);
        cout, zero, overflow, sign: out std_logic
10   );
end adder_status;

architecture arch of adder_status is
    signal a_ext, b_ext, sum_ext: signed(9 downto 0);
15   signal ovf: std_logic;
    alias sign_a: std_logic is a_ext(8);
    alias sign_b: std_logic is b_ext(8);
    alias sign_s: std_logic is sum_ext(8);
begin
20   a_ext <= signed('0' & a & '1');
    b_ext <= signed('0' & b & cin);
    sum_ext <= a_ext + b_ext;
```

				a_3	a_2	a_1	a_0	multiplicand
$\times$				b_3	b_2	b_1	b_0	multiplier
				$a_3 b_0$	$a_2 b_0$	$a_1 b_0$	$a_0 b_0$	
			$a_3 b_1$	$a_2 b_1$	$a_1 b_1$	$a_0 b_1$		
		$a_3 b_2$	$a_2 b_2$	$a_1 b_2$	$a_0 b_2$			
$+$	$a_3 b_3$	$a_2 b_3$	$a_1 b_3$	$a_0 b_3$				
y_7	y_6	y_5	y_4	y_3	y_2	y_1	y_0	product

Figure 7.14   Multiplication as a summation of $a_i b_j$ terms.

```
     ovf <= (sign_a and sign_b and (not sign_s)) or
            ((not sign_a) and (not sign_b) and sign_s);
25   cout <= sum_ext(9);
     sign <= sum_ext(8) when ovf='0' else
            not sum_ext(8);
     zero <= '1' when (sum_ext(8 downto 1)=0 and ovf='0') else
            '0';
30   overflow <= ovf;
     sum <= std_logic_vector(sum_ext(8 downto 1));
   end arch;
```

7.5.4 Combinational adder-based multiplier

A multiplier is a fairly complicated circuit. The synthesis of the VHDL multiplication operator depends on the individual software and the underlying target technology, and cannot always be done automatically. In this example, we study a simple, portable, though not optimal, combinational adder-based multiplier.

The multiplier is based on the algorithm we learned in elementary school. The multiplication of two 4-bit numbers is illustrated in Figure 7.14, which are aligned in a specific two-dimensional pattern. This algorithm includes three tasks:

1. Multiply the digits of the multiplier (b_3, b_2, b_1 and b_0 of Figure 7.14) by the multiplicand (A of Figure 7.14) one at a time to obtain $b_3 * A$, $b_2 * A$, $b_1 * A$ and $b_0 * A$. Since b_i is a binary digit, it can only be 0 or 1, and thus $b_i * A$ can only be 0 or A. The $b_i * A$ operation becomes bitwise and operation of b_i and the digits of A; that is,

$$b_i * A = (a_3 \cdot b_i, \ a_2 \cdot b_i, \ a_1 \cdot b_i, \ a_0 \cdot b_i)$$

2. Shift $b_i * A$ to left i positions.
3. Add the shifted $b_i * A$ terms to obtain the final product.

The VHDL code of an 8-bit multiplier based on this algorithm is shown in Listing 7.34. We first construct an 8-bit vector, $b_i b_i b_i b_i b_i b_i b_i b_i$, for each b_i to facilitate the bitwise and operation. The vector is used to generate shifted $b_i * A$ terms. Note that padding 0's are inserted around $b_i * A$ to form a 16-bit signal. The shifted $b_i * A$ terms are then summated by seven adders, which are arranged as a tree to increase performance, to obtain the final result.

Listing 7.34 Initial description of an adder-based multiplier

```
library ieee;
use ieee.std_logic_1164.all;
use ieee.numeric_std.all;
entity mult8 is
    port(
        a, b: in std_logic_vector(7 downto 0);
        y: out std_logic_vector(15 downto 0)
    );
end mult8;

architecture comb1_arch of mult8 is
    constant WIDTH: integer:=8;
    signal au, bv0, bv1, bv2, bv3, bv4, bv5, bv6, bv7:
        unsigned(WIDTH-1 downto 0);
    signal p0,p1,p2,p3,p4,p5,p6,p7,prod:
        unsigned(2*WIDTH-1 downto 0);
begin
    au <= unsigned(a);
    bv0 <= (others=>b(0));
    bv1 <= (others=>b(1));
    bv2 <= (others=>b(2));
    bv3 <= (others=>b(3));
    bv4 <= (others=>b(4));
    bv5 <= (others=>b(5));
    bv6 <= (others=>b(6));
    bv7 <= (others=>b(7));
    p0 <="00000000" & (bv0 and au);
    p1 <="0000000" & (bv1 and au) & "0";
    p2 <="000000" & (bv2 and au) & "00";
    p3 <="00000" & (bv3 and au) & "000";
    p4 <="0000" & (bv4 and au) & "0000";
    p5 <="000" & (bv5 and au) & "00000";
    p6 <="00" & (bv6 and au) & "000000";
    p7 <="0" & (bv7 and au) & "0000000";
    prod <= ((p0+p1)+(p2+p3))+((p4+p5)+(p6+p7));
    y <= std_logic_vector(prod);
end comb1_arch;
```

Adders are the major components of this design. For a circuit with an n-bit multiplicand and an n-bit multiplier, the product has $2n$ bits. The shifted b_i*A has to be extended to $2n$ bits, and thus the design needs $n-1$ $2n$-bit adders. The code can easily be expanded for a larger multiplier, and the needed revision is on the order of $O(n)$.

One way to reduce the size of this circuit is to add shifted b_i*A terms in sequence. This reduces the width of the adder to $n+1$ bits. Operation of the new design is illustrated in Figure 7.15.

We first obtain b_0*A and form the first partial product pp0. To accommodate the carry-out of future addition, one extra bit is appended to the left of b_0*A. Note that the LSB of prod (i.e., prod(0)) is the same as the LSB of pp0 (i.e., pp0(0)), and the pp0(0) bit has no effect on the remaining addition operations. We need only add the upper bits of the pp0 to b_1*A to form the next partial sum, pp1. Note that prod(1) is same as pp1(0), and pp1(0) has no effect on the remaining additions. We can repeat the process to obtain other

$$
\begin{array}{rrrrl}
 & a_3 & a_2 & a_1 & a_0 & \text{multiplicand} \\
\times & b_3 & b_2 & b_1 & b_0 & \text{multiplier}
\end{array}
$$

				a_3b_0	a_2b_0	a_1b_0	a_0b_0	
			$pp0_4$	$pp0_3$	$pp0_2$	$pp0_1$	$pp0_0$	partial product $pp0$
		$+$	a_3b_1	a_2b_1	a_1b_1	a_0b_1		
		$pp1_4$	$pp1_3$	$pp1_2$	$pp1_1$	$pp1_0$		partial product $pp1$
	$+$	a_3b_2	a_2b_2	a_1b_2	a_0b_2			
	$pp2_4$	$pp2_3$	$pp2_2$	$pp2_1$	$pp2_0$			partial product $pp2$
$+$	a_3b_3	a_2b_3	a_1b_3	a_0b_3				
$pp3_4$	$pp3_3$	$pp3_2$	$pp3_1$	$pp3_0$				partial product $pp3$

$pp3_4$	$pp3_3$	$pp3_2$	$pp3_1$	$pp3_0$	$pp2_0$	$pp1_0$	$pp0_0$	product $prod$

Figure 7.15 Multiplication as successive summation.

partial sums in sequence. This design still needs $n - 1$ adders, but the width of the adders is decreased from $2n$ to $n + 1$, about one half of the original size. The VHDL code of an 8-bit multiplier based on this algorithm is shown in Listing 7.35.

Listing 7.35 More efficient description of an adder-based multiplier

```vhdl
architecture comb2_arch of mult8 is
    constant WIDTH: integer:=8;
    signal au,bv0,bv1,bv2,bv3,bv4,bv5,bv6,bv7:
        unsigned(WIDTH-1 downto 0);
    signal pp0,pp1,pp2,pp3,pp4,pp5,pp6,pp7:
        unsigned(WIDTH downto 0);
    signal prod: unsigned(2*WIDTH-1 downto 0);
begin
    au <= unsigned(a);
    bv0 <= (others=>b(0));
    bv1 <= (others=>b(1));
    bv2 <= (others=>b(2));
    bv3 <= (others=>b(3));
    bv4 <= (others=>b(4));
    bv5 <= (others=>b(5));
    bv6 <= (others=>b(6));
    bv7 <= (others=>b(7));
    pp0 <= "0" & (bv0 and au);
    pp1 <= ("0" & pp0(WIDTH downto 1)) + ("0" & (bv1 and au));
    pp2 <= ("0" & pp1(WIDTH downto 1)) + ("0" & (bv2 and au));
    pp3 <= ("0" & pp2(WIDTH downto 1)) + ("0" & (bv3 and au));
    pp4 <= ("0" & pp3(WIDTH downto 1)) + ("0" & (bv4 and au));
    pp5 <= ("0" & pp4(WIDTH downto 1)) + ("0" & (bv5 and au));
    pp6 <= ("0" & pp5(WIDTH downto 1)) + ("0" & (bv6 and au));
    pp7 <= ("0" & pp6(WIDTH downto 1)) + ("0" & (bv7 and au));
    prod <= pp7 & pp6(0) & pp5(0) & pp4(0) & pp3(0) &
            pp2(0) & pp1(0) & pp0(0);
```

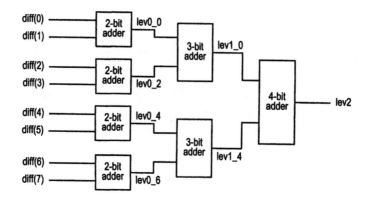

Figure 7.16   Block diagram of a population counter.

```
   y <= std_logic_vector(prod);
end comb2_arch;
```

7.5.5 Hamming distance circuit

A Hamming distance of two words is the number of bit positions in which the two words differ. For example, the Hamming distance of two 8-bit words "00010011" and "10010010" is 2 since the bits at position 0 (LSB) and position 7 (MSB) are different. The Hamming distance is used in some error correction and data compression applications. This example considers a circuit that calculates the Hamming distance of two 8-bit inputs.

Our design has two basic steps. The first step determines the bits that are different and marks them as '1'. The second step counts the number of 1's in the word, a function known as a *population counter*. For example, consider the inputs "00010011" and "10010010". The first step returns "10000001" since the bits at positions 0 and 7 are different, and the second step returns 2 since there are two 1's in the word.

We can implement the first step by using a simple bitwise xor operation. Recall that the 1-bit xor function returns '1' only if the input is "01" or "10". It can be interpreted that the function returns '1' if two inputs are different. Thus, after applying a bitwise xor operation, we can mark all the bits that are different.

Design of the population counter is more difficult. Our first design is shown in Figure 7.16. It counts the number of 1's by stages. In the first level of the circuit, we divide the 8 bits into four pairs and add the 1's in each pair. Four 2-bit adders are needed to perform the operation. In the second level, we pair the results and add them again. Two 3-bit adders are needed. The process is repeated one more time in the third level to obtain the final result. The VHDL code is shown in Listing 7.36.

Listing 7.36   Initial description of a Hamming distance circuit

```
library ieee;
use ieee.std_logic_1164.all;
use ieee.numeric_std.all;
entity hamming is
5    port(
        a,b: in std_logic_vector(7 downto 0);
```

```
        y: out std_logic_vector(3 downto 0)
      );
   end hamming;

   architecture effi_arch of hamming is
      signal diff: unsigned(7 downto 0);
      signal lev0_0, lev0_2, lev0_4, lev0_6:
         unsigned(1 downto 0);
      signal lev1_0, lev1_4: unsigned(2 downto 0);
      signal lev2: unsigned(3 downto 0);
   begin
      diff <= unsigned(a xor b);
      lev0_0 <= ('0' & diff(0)) + ('0' & diff(1));
      lev0_2 <= ('0' & diff(2)) + ('0' & diff(3));
      lev0_4 <= ('0' & diff(4)) + ('0' & diff(5));
      lev0_6 <= ('0' & diff(6)) + ('0' & diff(7));
      lev1_0 <= ('0' & lev0_0) + ('0' & lev0_2);
      lev1_4 <= ('0' & lev0_4) + ('0' & lev0_6);
      lev2 <= ('0' & lev1_0) + ('0' & lev1_4);
      y <= std_logic_vector(lev2);
   end effi_arch;
```

Although this population counter design is fairly efficient, the code is somewhat tedious. An alternative design is to use a clever shifting and masking scheme to rearrange the input and utilize a fixed-size 8-bit adder at each level. Assume that the 8-bit input to the population counter is $d_7 d_6 d_5 d_4 d_3 d_2 d_1 d_0$. The algorithm is summarized below.

- **Level 0.** We first split and rearrange the original input into two words, $0d_6 0d_4 0d_2 0d_0$ and $0d_7 0d_5 0d_3 0d_1$, and then add them by an 8-bit adder. Assume that the result is $e_7 e_6 e_5 e_4 e_3 e_2 e_1 e_0$. Because of the locations of the 0's, this adder performs essentially four 2-bit additions, and $e_1 e_0$, $e_3 e_2$, $e_5 e_4$ and $e_7 e_6$ are $d_0 + d_1$, $d_2 + d_3$, $d_4 + d_5$ and $d_6 + d_7$ respectively.

- **Level 1.** We perform splitting and addition operations similar to those at level 0. However, the input now is split into $00e_5 e_4 00e_1 e_0$ and $00e_7 e_6 00e_3 e_2$. The result should be in the form of $f_7 f_6 f_5 f_4 f_3 f_2 f_1 f_0$. Note that f_7 and f_3 should be 0. The adder actually performs two 3-bit additions, and $f_2 f_1 f_0$ and $f_6 f_5 f_4$ are $e_1 e_0 + e_3 e_2$ and $e_5 e_4 + e_7 e_6$ respectively.

- **Level 2.** We repeat the same operation except that the input is split into $0000 f_3 f_2 f_1 f_0$ and $0000 f_7 f_6 f_5 f_4$. The result should be in the form of $0000 g_3 g_2 g_1 g_0$.

The VHDL code based on this design is shown in Listing 7.37. The splitting and rearrangement of the input can be done by masking and shifting. For example, the mask in level 0 is mask0, "01010101". After performing bitwise and operation of $d_7 d_6 d_5 d_4 d_3 d_2 d_1 d_0$ and mask0, we obtain the first input word, $0d_6 0d_4 0d_2 0d_0$. We can obtain the second input word in a similar way after first shifting $d_7 d_6 d_5 d_4 d_3 d_2 d_1 d_0$ to the right one position. The operations are similar at levels 1 and 2 but have different masking patterns and amounts of shifting.

Listing 7.37 Compact description of a Hamming distance circuit

```
architecture compact_arch of hamming is
   signal diff, lev0, lev1, lev2: unsigned(7 downto 0);
   constant MASK0: unsigned(7 downto 0) := "01010101";
   constant MASK1: unsigned(7 downto 0) := "00110011";
```

```
 5    constant MASK2: unsigned(7 downto 0) := "00001111";

    begin
       diff <= unsigned(a xor b);
       lev0 <= (diff and MASK0) +
10               (('0'& diff(7 downto 1)) and MASK0);
       lev1 <= (lev0 and MASK1) +
                (("00" & lev0(7 downto 2)) and MASK1);
       lev2 <= (lev1 and MASK2) +
                (("0000" & lev1(7 downto 4)) and MASK2);
15    y <= std_logic_vector(lev2(3 downto 0));
    end compact_arch;
```

This design requires more adder bits than the first version. However, its code is more compact and the needed revision is on the order of $O(\log_2 n)$.

7.6 SYNTHESIS GUIDELINES

- Operators can be shared in mutually exclusive branches by proper routing of the input operands and/or result. It is more beneficial for complex operators.

- Many operations have certain common functionality. The hardware resource can be shared by these operations.

- RT-level code can outline the general layout of the circuit. A tree- or rectangle-shaped description can help the synthesis process and placement and routing process to derive a more efficient circuit.

7.7 BIBLIOGRAPHIC NOTES

Developing efficient design and VHDL codes requires the insight and in-depth knowledge of the problem at hand. The digital systems texts, *Digital Design Principles and Practices* by J. F. Wakerly and *Contemporary Logic Design* by R. H. Katz, provide detailed discussion on the construction of many commonly used parts, such as decoders, encoders, comparators and adders. Bibliography in Chapter 15 provides more references on the design and algorithms of multiplier and arithmetic functions.

Problems

7.1 Consider an arithmetic circuit that can perform four operations: a+b, a−b, a+1 and a−1, where a and b are 16-bit unsigned numbers and the desired operation is specified by a 2-bit control signal, ctrl.

   (a) Design the circuit using two adders, one incrementor and one decrementor. Derive the VHDL code.
   (b) Design the circuit using only one adder. Derive the VHDL code.
   (c) Synthesize the two designs with an ASIC device. Compare the areas and performances.
   (d) Synthesize the two designs with an FPGA device. Compare the areas and performances.

7.2   Design a circuit that converts an 8-bit signed input to 8-bit *sign-magnitude* output (where the MSB is the sign bit and the remaining 7 bits are magnitude). Use a minimal number of relational and arithmetic operators in your design. Draw the top-level diagram and derive the VHDL code.

7.3   Extend the dual-mode comparator of Section 7.3.2 to include sign-magnitude mode. Use only one 7-bit comparator in your design. Derive the VHDL code.

7.4   Consider a 16-bit shifting circuit that can perform rotating right or rotating left. Use selected signal assignment statements similar to that in Section 7.3.5 to implement the shifting function.

(a) Design the circuit using one rotate-right circuit, one rotate-left circuit and one 2-to-1 multiplexer to select the desired result. Derive the VHDL code.

(b) Design the circuit using one rotate-right circuit with a pre- and post-processing reversing circuit. The reversing circuit either passes the original input or reverses the input bit-wise (e.g., if a 4-bit input $a_3 a_2 a_1 a_0$ is used, the reversed output becomes $a_0 a_1 a_2 a_3$). Derive the VHDL code.

(c) Draw block diagrams of the two designs and analyze and compare their size and performance.

(d) Synthesize the two designs with an ASIC device. Compare the areas and performances.

(e) Synthesize the two designs with an FPGA device. Compare the areas and performances.

7.5   Consider a reduced-xor-vector function with 16 inputs. Design the circuit using a parallel-prefix structure similar to that of Figure 7.8(b) and derive the VHDL code.

7.6   We can further refine the tree priority encoder in Section 7.4.3 by using 2-to-1 priority encoders.

(a) Design a tree-structured 16-to-4 priority encoder using 2-to-1 priority encoders. The design should have four levels. Draw the block diagram and derive the VHDL code accordingly.

(b) Synthesize the new design and the two designs in Section 7.4.3 with an ASIC device. Compare the areas and performances.

(c) Synthesize the new design and the two designs in Section 7.4.3 with an FPGA device. Compare the areas and performances.

7.7   A leading zero counting circuit counts the number of consecutive 0's of an input. Consider a circuit with a 16-bit input.

(a) Design the circuit using one conditional signal assignment statement and derive the VHDL code.

(b) Derive a smaller 4-bit leading-zero counting circuit first. Design a 16-bit treelike leading-zero counting circuit using 4-bit counting circuits. Derive the VHDL code.

(c) Synthesize the two designs with an ASIC device. Compare the areas and performances.

(d) Synthesize the two designs with an FPGA device. Compare the areas and performances.

7.8   Design a 16-bit rotate-left shifting circuit using the multilevel structure discussed in Section 7.4.4.

7.9   Repeat Problem 7.4, but design the shifting circuit using the multilevel structure discussed in Section 7.4.4. Compare the area and performance with those in Problem 7.4.

7.10   We define the distance from the Gray code word a to the Gray code word b as the number of transitions from code word a to code word b. For example, consider the 4-bit Gray code words "0101" and "1111" as a and b. The distance from a to b is 4 since four transitions are needed (i.e., "0101" $\Rightarrow$ "0100" $\Rightarrow$ "1100" $\Rightarrow$ "1101" $\Rightarrow$ "1111"). Design a circuit to calculate the distance of two 4-bit Gray code words and derive the VHDL code.

7.11   Although the code is compact, the synthesize result of the `compact_arch` architecture of the Gray code incrementor may not be more efficient than the `table_arch` architecture.

(a) Synthesize the two designs with an ASIC device. Compare the areas and performances.

(b) Extend the two designs for 8-bit Gray code. The `table_arch` architecture now has 2^8 entries. You may need to write a program (using C, Java etc.) to generate the VHDL code. Synthesize the two 8-bit designs with an ASIC device. Compare the areas and performances.

(c) Synthesize the two 8-bit designs with an FPGA device. Compare the areas and performances.

(d) If you have enough hardware resources, repeat parts (b) and (c) by gradually increasing the design to 10-, 12-, 14- and 16-bit inputs.

7.12   Design a priority encoder that returns the codes of the highest and second-highest priority requests. The input is an 8-bit `req` signal and the outputs are `code1`, `code2`, `valid1` and `valid2`, which are the 3-bit codes and 1-bit valid signals of the highest and second-highest priority requests respectively.

7.13   Many instrument panels use binary-coded-decimal (BCD) format, in which 10 decimal digits are coded by using 4 bits. During an addition operation, if the sum of a digit exceeds 9, 10 will be subtracted from the current digit and a carry is generated for the next digit. Design a 3-digit BCD adder which has two 12-bit inputs, representing two 3-digit BCD numbers, and an output, which is a 4-digit (16-bit) BCD number. Draw the top-level diagram and derive the VHDL code accordingly.

7.14   In an analog amplifier, the output voltage becomes saturated (i.e., reaching the most positive voltage, $+V_{cc}$, or the most negative voltage, $-V_{cc}$) when the output exceeds the maximal range. In some digital signal processing applications, we wish to design an 8-bit signed saturation adder that mimics the behavior of an analog amplifier; i.e., if the addition result overflows, the result becomes the most positive or the most negative numbers. Draw the top-level diagram and derive the VHDL code accordingly.

7.15   The two multipliers in Section 7.5.4 utilize seven 16-bit adders and seven 8-bit adders respectively.

(a) Determine the critical path for both designs.

(b) In an area-optimized adder, the propagation is proportional to the number of bits in the adder (i.e., on the order of $O(n)$). Assume that both designs utilize this kind of adder. Compare the propagation delays for the two designs.

7.16   Revise the designs in Section 7.5.4 to accommodate inputs in signed integer format. Derive the VHDL code. (*Hint*: An 8-bit 2's-complement number, $a_7 a_6 a_5 a_4 a_3 a_2 a_1 a_0$, has a value of $-a_7*2^7 + a_6*2^6 + a_5*2^5 + \cdots + a_0*2^0$.)

7.17 Design an 8-bit combinational divider based on a long-division algorithm, the one you learned in elementary school. The inputs are 8-bit dividend and divisor in unsigned format, and the outputs are 8-bit quotient and remainder. Derive the VHDL code. (*Hint*: Division can be done by a sequence of "comparing and subtracting" operations. This operation takes two inputs, a and b, and returns a if $a < b$ and returns $a - b$ otherwise.)

7.18 One way to implement the population counter in Section 7.5.5 is to exhaustively construct a function table and use a single selected signal assignment statement to implement the table.

    (a) Derive the VHDL code for an 8-bit population counter based on function table design. You may need to write a program (using C, Java etc.) to generate the VHDL code.

    (b) Synthesize this design and the other two designs in Section 7.5.5 with an ASIC device. Compare the areas and performances.

    (c) Synthesize this design and the other two designs in Section 7.5.5 with an FPGA device. Compare the areas and performances.

    (d) Repeat parts (a) to (c) for 10- and 12-bit inputs.

CHAPTER 8

SEQUENTIAL CIRCUIT DESIGN: PRINCIPLE

A sequential circuit is a circuit that has an internal state, or memory. A *synchronous* sequential circuit, in which all memory elements are controlled by a global synchronizing signal, greatly simplifies the design process and is the most important design methodology. Our focus is on this type of circuit. In this chapter and the next chapter, we examine the VHDL description of basic memory elements and study the design of sequential circuits with a "regular structure." Chapters 10, 11 and 12 discuss the design of sequential circuits with a "random structure" (finite state machine) and circuits based on register transfer methodology.

8.1 OVERVIEW OF SEQUENTIAL CIRCUITS

8.1.1 Sequential versus combinational circuits

A combinational circuit, by definition, is a circuit whose output, after the initial transient period, is a function of current input. It has no internal state and therefore is "memoryless" about the past events (or past inputs). A sequential circuit, on the other hand, has an *internal state*, or *memory*. Its output is a function of current input as well as the internal state. The internal state essentially "memorizes" the effect of the past input values. The output thus is affected by current input value as well as past input values (or the entire sequence of input values). That is why we call a circuit with internal state a *sequential circuit*.

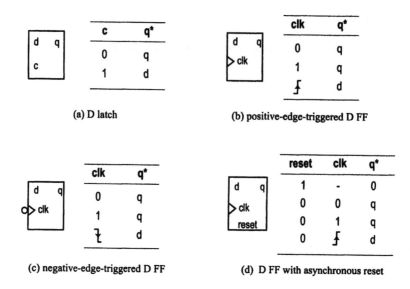

(a) D latch

(b) positive-edge-triggered D FF

(c) negative-edge-triggered D FF

(d) D FF with asynchronous reset

Figure 8.1   D latch and D FF.

8.1.2 Basic memory elements

We can add memory to a circuit in two ways. One way is to add closed feedback loops in a combinational circuit, in which the memory is implicitly manifested as system states. Because of potential timing hazards and racing, this approach is very involved and not suitable for synthesis.

The other way is to use predesigned memory components. All device libraries have certain memory cells, which are carefully designed and thoroughly analyzed. These elements can be divided into two broad categories: *latch* and *flip-flop* (*FF*). We review the basic characteristics of a D-type latch (or just D latch) and D-type FF (or just D FF).

D latch   The symbol and function table of a D latch are shown in Figure 8.1(a). Note that we use * to represent the next value, and thus q* means the next value of q. The c and d inputs can be considered as a control signal and data input respectively. When c is asserted, input data, d, is passed directly to output, q. When c is deasserted, the output remains the same as the previous value. Since the operation of the D latch depends on the level of the control signal, we say that it is *level sensitive*. A representative timing diagram is shown in the q_latch output of Figure 8.2. Note that input data is actually stored into the latch at the falling edge of the control signal.

Since the latch is "transparent" when c is asserted, it may cause racing if a loop exists in the circuit. For example, the circuit in Figure 8.3 attempts to swap the contents of two latches. Unfortunately, racing occurs when c is asserted. Because of the potential complication of timing, we normally do not use latches in synthesis.

D FF   The symbol and function table of a *positive-edge-triggered* D FF are shown in Figure 8.1(b). D FF has a special control signal known as a *clock signal*, which is labeled clk in the diagram. The D FF is activated only when the clock signal changes from '0' to '1', which is known as the *rising edge* of the clock. At other times, its output remains the same

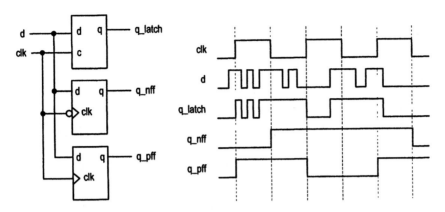

Figure 8.2   Simplified timing diagram of D latch and D FFs.

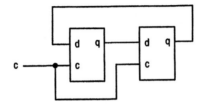

Figure 8.3   Data swapping using D latches.

as its previous value. In other words, at the rising edge of the clock, a D FF takes a sample of input data, stores the value into memory, and passes the value to output. The output, which reflects the stored value, does not change until the next rising edge. Since operation of the D FF depends on the edge of the clock signal, we say that it is *edge sensitive*. A representative timing diagram is shown in the q_pff output of Figure 8.2. Note that the clock signal, clk, is functioning as a sampling signal, which takes a sample of the input data, d, at the rising edge. The clock signal plays a key role in a sequential circuit and we add a small triangle, as in the clk port in Figure 8.1(b), to emphasize use of an edge-triggered FF.

The operation of a *negative-edge-triggered* D FF is similar except that sampling is performed at the falling edge of the clock. Its symbol and function table are shown in Figure 8.1(c). A representative timing diagram is shown in the q_nff output of Figure 8.2.

The sampling property of FFs has several advantages. First, variations and glitches between two rising edges have no effect on the content of the memory. Second, there will be no race condition in a closed feedback loop. If we reconstruct the swapping circuit of Figure 8.3 by replacing the D latches with the D FFs, the D FFs swap their contents at each rising edge of the clock and the circuit functions as expected. The disadvantage of the D FF is its circuit size, which is about twice as large as that of a D latch. Since its benefits far outweigh the size disadvantage, today's sequential circuits normally utilize D FFs as the storage elements.

The timing of a D FF is more involved than that of a combinational component. The timing diagram is shown in Figure 8.4. There are three main timing parameters:

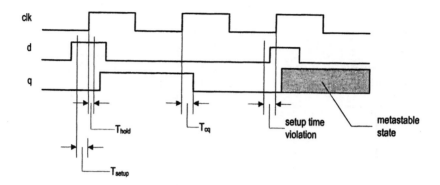

Figure 8.4   Detailed timing diagram of a D FF.

- T_{cq}: *clock-to-q delay*, the propagation delay required for the d input to show up at the q output after the sampling edge of the clock signal.
- T_{setup}: *setup time*, the time interval in which the d signal must be stable *before* the clock edge.
- T_{hold}: *hold time*, the time interval in which the d signal must be stable *after* the clock edge.

T_{cq} corresponds roughly to the propagation delay of a combinational component. T_{setup} and T_{hold}, on the other hand, are *timing constraints*. They specify that the d signal must be stable in a small window around the sampling edge of the clock. If the d signal changes within the setup or hold time window, which is known as *setup time violation* or *hold time violation*, the D FF may enter a *metastable state*, in which the q becomes neither '0' nor '1'. The issue of metastability is discussed in Chapter 16.

8.1.3   Synchronous versus asynchronous circuits

The clock signal of FFs plays a key role in sequential circuit design. According to the arrangement of the clock, we can divide the sequential circuits into the following classes:

- *Globally synchronous circuit* (or simply *synchronous circuit*). A globally synchronous circuit uses FFs as memory elements, and all FFs are controlled (i.e., *synchronized*) by a single global clock signal. Synchronous design is the most important methodology used to design and develop large, complex digital systems. It not only facilitates the synthesis but also simplifies the verification, testing, and prototyping process. Our discussion is focused mainly on this type of circuit.
- *Globally asynchronous locally synchronous circuit*. Sometimes physical constraints, such as the distance between components, prevent the distribution of a single clock signal. In this case, a system may be divided into several smaller subsystems. Since a subsystem is smaller, it can follow the synchronous design principle. Thus, subsystems are synchronous internally. Since each subsystem utilizes its own clock, operation between the subsystems is asynchronous. We need special interface circuits between the subsystems to ensure correct operation. Chapter 16 discusses the design of the interface circuits.
- *Globally asynchronous circuit*. A globally asynchronous circuit does not use a clock signal to coordinate the memory operation. The state of a memory element changes independently. Globally asynchronous circuits can be divided into two categories.

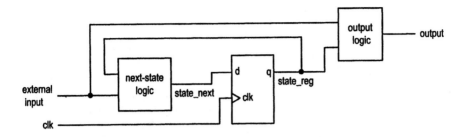

Figure 8.5   Conceptual diagram of a synchronous sequential circuit.

The first category comprises circuits that consist of FFs but do not use the clock in a disciplined way. One example is the *ripple counter*, in which the clock port of an FF is connected to the output of the previous FF. Utilizing FFs in this way is a poor design practice. The second category includes the circuits that contain "clockless" memory components, such as a latch or a combinational circuit with closed feedback loops. This kind of circuit is sometimes simply referred to as an *asynchronous circuit*. The design of asynchronous circuits is very different from that of synchronous circuits and is not recommended for HDL synthesis. The danger is demonstrated by an example in Section 8.3.

8.2 SYNCHRONOUS CIRCUITS

8.2.1 Basic model of a synchronous circuit

The basic diagram of a synchronous circuit is shown in Figure 8.5. The memory element, frequently know as a *state register*, is a collection of D FFs, synchronized by a common global clock signal. The output of the register (i.e., the content stored in the register), the state_reg signal, represents the internal state of the system. The *next-state logic* is a combinational circuit that determines the next state of the system. The *output logic* is another combinational circuit that generates the external output signal. Note that the output depends on the external input signal and the current state of the register. The circuit operates as follows:

- At the rising edge of the clock, the value of the state_next signal (appearing at the d port) is sampled and propagated to the q port, which becomes the new value of the state_reg signal. The value is also stored in FFs and remains unchanged for the rest of the clock period. It represent the *current state* of the system.
- Based on the value of the state_reg signal and external input, the next-state logic computes the value of the state_next signal and the output logic computes the value of external output.
- At the next rising edge of the clock, the new value of the state_next signal is sampled and the state_reg signal is updated. The process then repeats.

To satisfy the timing constraints of the FFs, the clock period must be large enough to accommodate the propagation delay of the next-state logic, the clock-to-q delay of the FFs and the setup time of the FFs. This aspect is discussed in Section 8.6.

There are several advantages of synchronous design. First, it simplifies circuit timing. Satisfying the timing constraints (i.e., avoiding setup time and hold time violation) is one of the most difficult design tasks. When a circuit has hundreds or even thousands of FFs and each FF is driven by an individual clock, the design and analysis will be overwhelming. Since in a synchronous circuit all FFs are driven by the identical clock signal, the sampling of the clock edge occurs simultaneously. We only need to consider the timing constraints of a single memory component. Second, the synchronous model clearly separates the combinational circuits and the memory element. We can easily isolate the combinational part of the system, and design and analyze it as a regular combinational circuit. Third, the synchronous design can easily accommodate the timing hazards. As we discussed in Section 6.5.3, the timing hazards are unavoidable in a large synthesized combinational circuit. In a synchronous circuit, inputs are sampled and stored at the rising edge of the clock. The glitches do not matter as long as they are settled at the time of sampling. Instead of considering all the possible timing scenarios, we only need to focus on worst-case propagation delays of the combinational circuit.

8.2.2 Synchronous circuits and design automation

The synchronous model essentially reduces a complex sequential circuit to a single closed feedback loop and greatly simplifies the design process. We only need to analyze the timing of a simple loop. Once it is done, the memory elements can be isolated and separated from the circuit. The sequential design now becomes a combinational design and we can apply the previous optimization and synthesizing schemes of combinational circuits to construct sequential circuits. Because of this, the synchronous model is the most dominant methodology in today's design environment. Most EDA tools are based on this model.

The benefit of synchronous methodology is not just limited to synthesis. It can facilitate the other tasks of the development process. The impact of synchronous methodology is summarized below.

- *Synthesis.* Since we can separate the memory elements, the system is reduced to a combinational circuit. All optimization algorithms and techniques used in combinational circuit synthesis can be applied accordingly.
- *Timing analysis.* The analysis involves only a single closed feedback loop. It is straightforward once the propagation delay of the combination circuit is known. Thus, the timing analysis of the sequential circuit is essentially reduced to the timing analysis of its combinational part.
- *Cycle-based simulation.* Cycle-based simulation ignores the exact propagation delay but simulates the circuit operation from one clock cycle to another clock cycle. Since we can easily identify the memory elements and their clock, cycle-based simulation can be used for synchronous design.
- *Testing.* One key testing technique is to use scan registers to shift in test patterns and shift out the results. Because the memory elements are isolated, we can easily replace them with scan registers when needed.
- *Design reuse.* The main timing constraint of the synchronous design is embedded in the period of the clock signal (to be discussed in Section 8.6), which depends mainly on the propagation delay of the combination part. As long as the clock period is large enough, the same design can be implemented by different device technologies.
- *Hardware emulation.* Because the same synchronous design can be targeted to different device technologies, it is possible to first construct the design in FPGA technology, run and verify the circuit at a slower clock rate, and then fabricate it in ASIC technology.

8.2.3 Types of synchronous circuits

Based on the "representation and transition patterns" of state, we divide synchronous circuits into three types. These divisions are informal, just for clarity of coding. The three types of sequential circuits are:

- *Regular sequential circuit*. The state representation and state transitions have a simple, *regular* pattern, as in a counter and a shift register. Similarly, the next-state logic can be implemented by regular, structural components, such as an incrementor and shifter.
- *Random sequential circuit*. The state transitions are more complicated and there is no special relation between the states and their binary representations. The next-state logic must be constructed from scratch (i.e., by *random* logic). This kind of circuit is known as a *finite state machine* (*FSM*).
- *Combined sequential circuit*. A combined sequential circuit consists of both a regular sequential circuit and an FSM. The FSM is used to control operation of the regular sequential circuit. This kind of circuit is based on the *register transfer methodology* and is sometimes known as *finite state machine with data path* (*FSMD*).

We discuss the design and description of regular sequential circuits in this chapter and the next chapter, and we cover the FSM and FSMD in Chapters 10, 11 and 12.

8.3 DANGER OF SYNTHESIS THAT USES PRIMITIVE GATES

As we discussed earlier, an asynchronous sequential circuit can be constructed from scratch by adding a feedback loop to the combinational components. Although asynchronous circuits potentially can run faster and consume less power, designing an asynchronous circuit is difficult because of the potential races and oscillations. The design procedure is totally different from the synchronous methodology, and we should avoid using normal EDA software to synthesize asynchronous circuits. Since this book focuses on RT-level synthesis, we do not discuss this topic in detail. The following example illustrates the potential danger of using the normal synthesis procedure to construct an asynchronous circuit.

Consider the D latch discussed in Section 8.1.2. We can easily translate the truth table into VHDL code, as shown in Listing 8.1.

Listing 8.1   D latch from scratch

```
library ieee;
use ieee.std_logic_1164.all;
entity dlatch is
    port(
        c: in std_logic;
        d: in std_logic;
        q: out std_logic
    );
end dlatch;

architecture demo_arch of dlatch is
    signal q_latch: std_logic;
begin
    process(c,d,q_latch)
    begin
```

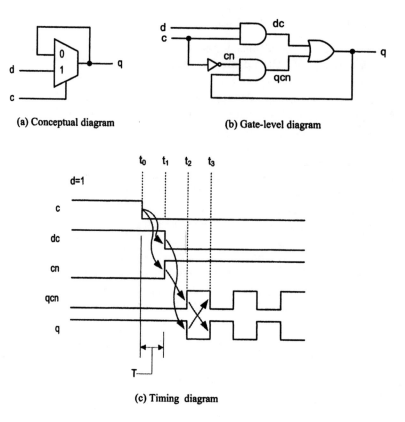

(a) Conceptual diagram   (b) Gate-level diagram

(c) Timing diagram

Figure 8.6   Synthesizing a D latch from scratch.

```
      if (c='1') then
          q_latch <= d;
      else
          q_latch <= q_latch;
20        end if;
      end process;
      q <= q_latch;
   end demo_arch;
```

Synthesis software can normally recognize that this code is for a D latch and should infer
a predesigned D-latch cell from the cell library accordingly. For demonstration purposes,
let us try to use simple gates to synthesize it from scratch. We can derive the conceptual
diagram and expand it to a gate-level diagram following the procedure to synthesize a
combinational circuit, as shown in Figure 8.6(a) and (b).

At first glance, the circuit is just like a combinational circuit except that the output is
looped back as an input. However, there is a serious timing problem for this circuit. Let us
assume that all gates have a propagation delay of T and the wire delays are negligible, and
that c, d and q are '1' initially. Now consider what happens when c changes from '1' to '0'
at time t_0. According to the function table, we expect that q should be latched to the value
of d and thus should remain '1'. Following the circuit diagram, we can derive a detailed
timing diagram, as shown in Figure 8.6(c). The events are summarized below.

- At t_0, c changes to '0'.
- At t_1 (after a delay of T), dc and cn change.
- At t_2 (after a delay of $2T$), qcn changes (due to cn) and q changes (due to dc).
- At t_3 (after a delay of $3T$), q changes (due to qcn) and qcn changes (due to q).

Clearly, the output q continues to oscillate at a period of $2T$ and the circuit is unstable.

Recall that in Section 6.5.4, we discussed delay-sensitive circuit, in which the correctness of circuit function depends on the delays of various components. Asynchronous circuits belong to this category and thus are not suitable for synthesis. If we really wish to implement an asynchronous circuit from scratch, it is better to do it manually using a schematic rather than relying on synthesis.

8.4 INFERENCE OF BASIC MEMORY ELEMENTS

All device libraries have predesigned memory cells. Internally, these cells are designed as asynchronous sequential circuits. They are carefully crafted and thoroughly analyzed and verified. These cells are treated as "leaf units," and no further synthesis or optimization will be performed. The previous section has shown the danger of deriving a memory element from scratch. To avoid this, we must express our intent clearly and precisely in VHDL code so that these predesigned latches or FFs can be inferred. While we should be innovative about the design, it is a good idea to follow the standard VHDL description of latch and FF to avoid any unwanted surprise.

8.4.1 D latch

The function table of a D latch was shown in Figure 8.1(a). The corresponding VHDL code is shown in Listing 8.2. It is the standard description. Synthesis software should infer a predesigned D latch from the device library.

Listing 8.2 D latch

```
library ieee;
use ieee.std_logic_1164.all;
entity dlatch is
    port(
        c: in std_logic;
        d: in std_logic;
        q: out std_logic
    );
end dlatch;

architecture arch of dlatch is
begin
    process(c,d)
    begin
        if (c='1') then
            q <= d;
        end if;
    end process;
end arch;
```

In this code, the value of d is passed to q when c is '1'. Note that there is no else branch in the if statement. According to the VHDL definition, q will keep its previous value when c is not '1' (i.e., c is '0'). This is just what we want for the D latch. Alternatively, we can explicitly include the else branch to express that q has its previous value when c is '0', as in the VHDL code in Listing 8.1. The code is not as compact or clear and is not recommended.

8.4.2 D FF

Positive-edge-triggered D FF   The function table of a positive-edge-triggered D FF was shown in Figure 8.1(b). The corresponding VHDL code is shown in Listing 8.3. This is a standard description and should be recognized by all synthesis software. A predesigned D FF should be inferred accordingly.

Listing 8.3   D FF

```
    library ieee;
    use ieee.std_logic_1164.all;
    entity dff is
        port(
5           clk: in std_logic;
            d: in std_logic;
            q: out std_logic
        );
    end dff;
10
    architecture arch of dff is
    begin
        process(clk)
        begin
15          if (clk'event and clk='1') then
                q <= d;
            end if;
        end process;
    end arch;
```

The key expression to infer the D FF is the Boolean expression

```
clk'event and clk='1'
```

The 'event term is a VHDL attribute returning true when there is a change in signal value (i.e., an *event*). Thus, when clk'event is true, it means that the value of clk has changed. When the clk='1' expression is true, it means that the new value of clk is '1'. When both expressions are true, it indicates that the clk signal changes to '1', which is the rising edge of the clk signal.

The if statement states that at the rising of the clk signal, q gets the value of d. Since there is no else branch, it means that q keeps its previous value otherwise. Thus, the VHDL code accurately describes the function of a D FF. Note that the d signal is not in the sensitivity list. It is reasonable since the output only responds to clk and does nothing when d changes its value.

We can also add an extra condition clk'last_value='0' to the Boolean expression:

```
clk'event and clk='1' and clk'last_value='0'
```

to ensure that the transition is from '0' to '1' rather than from a metavalue to '1'. This may affect simulation but has no impact on synthesis. The above Boolean expression is defined as a function, `rising_edge()`, in the IEEE `std_logic_1164` package. We can rewrite the previous VHDL code as

```
architecture arch of dff is
begin
   process(clk)
   begin
      if rising_edge(clk) then
         q <= d;
      end if;
   end process;
end arch;
```

We can also use wait statement inside the process to infer the D FF:

```
architecture wait_arch of dff is
begin
   process
   begin
      wait until clk'event and clk='1';
      q <= d ;
   end process ;
end wait_arch;
```

However, since the sensitivity list makes the code easier to understand, we do not use this format in this book.

Theoretically, a then branch can be added to the code:

```
if (clk'event and clk='1') then
   q <= d;
else
   q <= '1';
end if;
```

Although it is syntactically correct, it is meaningless for synthesis purpose.

Negative-edge-triggered D FF A negative-edge-triggered D FF is similar to a positive-edge-triggered D FF except that the input data is sampled at the falling edge of the clock. To specify the falling edge, we must revise the Boolean expression of the if statement:

```
if (clk'event and clk='0') then
```

We can also use the Boolean expression

```
clk'event and clk='0' and clk'last_value='1'
```

to ensure the '1' to '0' transition or use the shorthand function, `falling_edge()`, defined in the IEEE `std_logic_1164` package.

D FF with asynchronous reset A D FF may contain an asynchronous reset signal that clears the D FF to '0'. The symbol and function table are shown in Figure 8.1(d). Note that the reset operation does not depend on the level or edge of the clock signal. Actually, we can consider that it has a higher priority than the clock-controlled operation. The VHDL code is shown in Listing 8.4.

Listing 8.4   D FF with asynchronous reset

```
   library ieee;
   use ieee.std_logic_1164.all;
   entity dffr is
      port(
5        clk: in std_logic;
         reset: in std_logic;
         d: in std_logic;
         q: out std_logic
      );
10 end dffr;

   architecture arch of dffr is
   begin
      process(clk,reset)
15    begin
         if (reset='1') then
            q <='0';
         elsif (clk'event and clk='1') then
            q <= d;
20       end if;
      end process;
   end arch;
```

Both the reset and clk signals are in the sensitivity list since either can invoke the process. When the process is invoked, it first checks the reset signal. If it is '1', the D FF is cleared to '0'. Otherwise, the process continues checking the rising-edge condition, as in a regular D FF. Note that there is no else branch.

Since the reset operation is independent of the clock, it cannot be synthesized from a regular D FF. A D FF with asynchronous reset is another leaf unit. The synthesis software recognizes this format and should infer the desired D FF cell from the device library.

Asynchronous reset, as its name implies, is not synchronized by the clock signal and thus should not be used in normal synchronous operation. The major use of a reset signal is to clear the memory elements and set the system to an initial state. Once the system enters the initial state, it starts to operate synchronously and will never use the reset signal again. In many digital systems, a short reset pulse is generated when the power is turned on.

Some D FFs may also have an asynchronous preset signal that sets the D FF to '1'. The VHDL code is shown in Listing 8.5.

Listing 8.5   D FF with asynchronous reset and preset

```
   library ieee;
   use ieee.std_logic_1164.all;
   entity dffrp is
      port(
5        clk: in std_logic;
         reset, preset: in std_logic;
         d: in std_logic;
         q: out std_logic
      );
10 end dffrp;

   architecture arch of dffrp is
```

```
        begin
            process(clk,reset,preset)
15          begin
                if (reset='1') then
                    q <='0';
                elsif (preset='1') then
                    q <= '1';
20              elsif (clk'event and clk='1') then
                    q <= d;
                end if;
            end process;
        end arch;
```

Since the asynchronous signal is normally used for system initialization, a single preset or reset signal should be adequate most of the time.

8.4.3 Register

A register is a collection of a D FFs that is driven by the same clock and reset signals. The VHDL code of an 8-bit register is shown in Listing 8.6.

Listing 8.6   Register

```
    library ieee;
    use ieee.std_logic_1164.all;
    entity reg8 is
        port(
5           clk: in std_logic;
            reset: in std_logic;
            d: in std_logic_vector(7 downto 0);
            q: out std_logic_vector(7 downto 0)
        );
10  end reg8;

    architecture arch of reg8 is
    begin
        process(clk,reset)
15      begin
            if (reset='1') then
                q <=(others=>'0');
            elsif (clk'event and clk='1') then
                q <= d;
20          end if;
        end process;
    end arch;
```

The code is similar to D FF except that the d input and the q output are now 8 bits wide. We use the symbol of D FF for the register. The size of the register can be derived by checking the bus width marks of the input and output connections.

8.4.4 RAM

Random access memory (RAM) can be considered as a collection of latches with special interface circuits. It is used to provide massive storage. While technically it is possible to synthesize a RAM from scratch by assembling D-latch cells and control circuits, the result is bulky and inefficient. Utilizing the device library's predesigned RAM module, whose memory cells are crafted and optimized at the transistor level, is a much better alternative. Although the basic structure of RAMs is similar, their sizes, speeds, interfaces, and timing characteristics vary widely, and thus it is not possible to derive a portable, device-independent VHDL code to infer the desired RAM module. We normally need to use explicit component instantiation statement for this task.

8.5 SIMPLE DESIGN EXAMPLES

The most effective way to derive a sequential circuit is to follow the block diagram in Figure 8.5. We first identify and separate the memory elements and then derive the next-state logic and output logic. After separating the memory elements, we are essentially designing the combinational circuits, and all the schemes we learned earlier can be applied accordingly. A clear separation between memory elements and combinational circuits is essential for the synthesis of large, complex design and is helpful for the verification and testing processes. Our VHDL code description follows this principle and we always use an isolated VHDL segment to describe the memory elements.

Since identifying and separating the memory elements is the key in deriving a sequential circuit, we utilize the following coding practice to emphasize the existence of the memory elements:

- Use an individual VHDL code segment to infer memory elements. The segment should be the standard description of a D FF or register.
- Use the suffix _reg to represent the output of a D FF or a register.
- Use the suffix _next to indicate the next value (the d input) of a D FF or a register.

We examine a few simple, representative sequential circuits in this section and study more sophisticated examples in Chapter 9.

This coding practice may make the code appear to be somewhat cumbersome, especially for a simple circuit. However, its long-term benefits far outweigh the inconvenience. The alternative coding style, which mixes the memory elements and combinational circuit in one VHDL segment, is discussed briefly in Section 8.7.

8.5.1 Other types of FFs

There are other types of FFs, such as D FF with an enable signal, JK FF and T FF. They were popular when a digital system was constructed by SSI components because they may reduce the number of IC chips on a printed circuit board. Since all these FFs can be synthesized by a D FF, they are not used today. The following subsections show how to construct them from a D FF.

D FF with enable   Consider a D FF with an additional enable signal. The function table is shown in Figure 8.7(a). Note that the enable signal, en, has an effect only at the rising edge of the clock. This means that the signal is synchronized to the clock. At the rising edge of the clock, the FF samples both en and d. If en is '0', which means that the FF is not enabled, FF keeps its previous value. On the other hand, if en is '1', the FF is enabled and functions as a regular D FF. The VHDL code is shown in Listing 8.7.

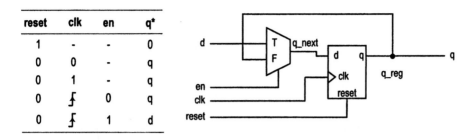

reset	clk	en	q*
1	-	-	0
0	0	-	q
0	1	-	q
0	ʃ	0	q
0	ʃ	1	d

(a) Function table   (b) Conceptual diagram

Figure 8.7   D FF with an enable signal.

Listing 8.7   D FF with an enable signal

```vhdl
library ieee;
use ieee.std_logic_1164.all;
entity dff_en is
    port(
        clk: in std_logic;
        reset: in std_logic;
        en: in std_logic;
        d: in std_logic;
        q: out std_logic
    );
end dff_en;

architecture two_seg_arch of dff_en is
    signal q_reg: std_logic;
    signal q_next: std_logic;
begin
    -- D FF
    process(clk,reset)
    begin
        if (reset='1') then
            q_reg <= '0';
        elsif (clk'event and clk='1') then
            q_reg <= q_next;
        end if;
    end process;
    -- next-state logic
    q_next <= d when en ='1' else
                q_reg;
    -- output logic
    q <= q_reg;
end two_seg_arch;
```

The VHDL code follows the basic sequential block diagram and is divided into three segments: a memory element, next-state logic and output logic. The memory element is a regular D FF. The next-state logic is implemented by a conditional signal assignment statement. The q_next signal can be either d or the original content of the FF, q_reg,

reset	clk	t	q*
1	-	-	0
0	0	-	q
0	1	-	q
0	∫	0	q
0	∫	1	q'

(a) Function table

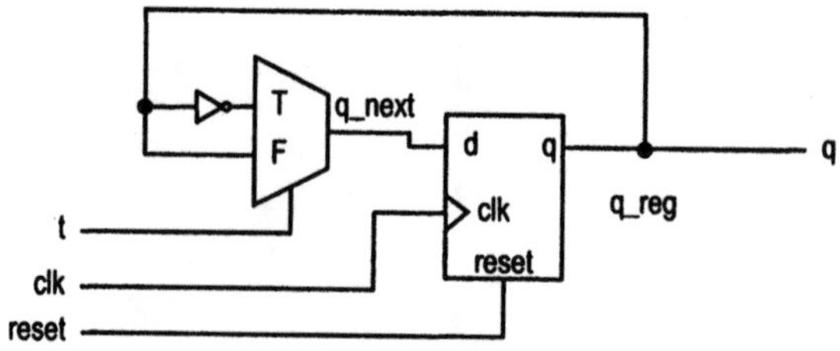

(b) Conceptual diagram

Figure 8.8 T FF.

depending on the value of en. At the rising edge of the clock, q_next will be sampled and stored into the memory element. The output logic is simply a wire that connects the output of the register to the q port.

The conceptual diagram is shown in Figure 8.7(b). To obtain the diagram, we first separate and derive the memory element, and then derive the combinational circuit using the procedure described in Chapter 4.

T FF A T FF has a control signal, t, which specifies whether the FF to invert (i.e., *toggle*) its content. The function table of a T FF is shown in Figure 8.8(a). Note that the t signal is sampled at the rising edge of the clock. The VHDL code is shown in Listing 8.8, and the conceptual diagram is shown in Figure 8.8(b).

Listing 8.8   T FF

```
     library ieee;
     use ieee.std_logic_1164.all;
     entity tff is
        port(
5          clk: in std_logic;
           reset: in std_logic;
           t: in std_logic;
           q: out std_logic
        );
10   end tff;

     architecture two_seg_arch of tff is
        signal q_reg: std_logic;
        signal q_next: std_logic;
15   begin
        -- D FF
        process(clk,reset)
        begin
           if (reset='1') then
20             q_reg <= '0';
           elsif (clk'event and clk='1') then
              q_reg <= q_next;
           end if;
        end process;
```

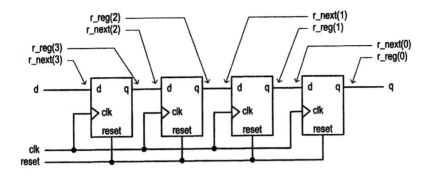

Figure 8.9   4-bit free-running shift-right register.

```
25      -- next-state logic
        q_next <= q_reg when t='0' else
                  not(q_reg);
        -- output logic
        q <= q_reg;
30  end two_seg_arch;
```

8.5.2 Shift register

A shift register shifts the content of the register left or right 1 bit in each clock cycle. One major application of a shifter register is to send parallel data through a serial line. In the transmitting end, a data word is first loaded to register in parallel and is then shifted out 1 bit at a time. In the receiving end, the data word is shifted in 1 bit at a time and reassembled.

Free-running shift-right register   A free-running shift register performs the shifting operation continuously. It has no other control signals. A 4-bit free-running shift-right register is shown in Figure 8.9. We can rearrange the FFs and align them vertically, as in Figure 8.10(a). After grouping the four FFs together and treating them as a single memory block, we transform the circuit into the basic sequential circuit block diagram in Figure 8.10(b). The VHDL code can be derived according to the block diagram, as in Listing 8.9.

Listing 8.9   Free-running shift-right register

```
library ieee;
use ieee.std_logic_1164.all;
entity shift_right_register is
   port(
5      clk, reset: in std_logic;
       d: in std_logic;
       q: out std_logic
   );
end shift_right_register;
10
   architecture two_seg_arch of shift_right_register is
      signal r_reg: std_logic_vector(3 downto 0);
      signal r_next: std_logic_vector(3 downto 0);
```

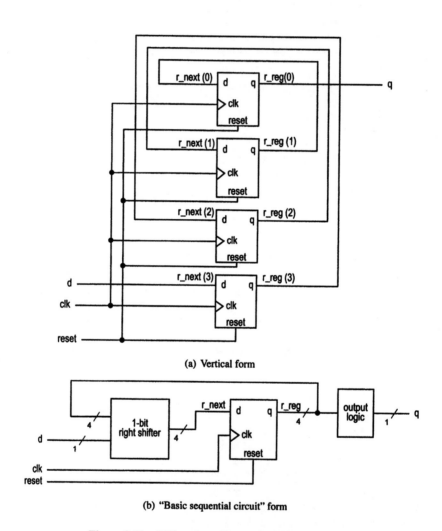

(a) Vertical form

(b) "Basic sequential circuit" form

Figure 8.10   Shift register diagram in different forms.

```
     begin
15      —— register
     process(clk,reset)
     begin
        if (reset='1') then
           r_reg <= (others=>'0');
20      elsif (clk'event and clk='1') then
           r_reg <= r_next;
        end if;
     end process;
        —— next-state logic (shift right 1 bit)
25   r_next <= d & r_reg(3 downto 1);
        —— output
     q <= r_reg(0);
  end two_seg_arch;
```

The VHDL code follows the basic sequential circuit block diagram, and the key is the code for the next-state logic. The statement

```
     r_next <= d & r_reg(3 downto 1);
```

indicates that the original register content is shifted to the right 1 bit and a new bit, d, is inserted to the left. The memory element part of the code is the standard description of a 4-bit register.

Universal shift register   A universal shift register can load a parallel data word and perform shifting in either direction. There are four operations: load, shift right, shift left and pause. A control signal, ctrl, specifies the desired operation. The VHDL code is shown in Listing 8.10. Note that the d(0) input and the q(3) output are used as serial-in and serial-out for the shift-left operation, and the d(3) input and the q(0) output are used as serial-in and serial-out for the shift-right operation. The block diagram is shown in Figure 8.11.

Listing 8.10   Universal shift register

```
  library ieee;
  use ieee.std_logic_1164.all;
  entity shift_register is
     port(
5       clk, reset: in std_logic;
        ctrl: in std_logic_vector(1 downto 0);
        d: in std_logic_vector(3 downto 0);
        q: out std_logic_vector(3 downto 0)
     );
10 end shift_register;

  architecture two_seg_arch of shift_register is
     signal r_reg: std_logic_vector(3 downto 0);
     signal r_next: std_logic_vector(3 downto 0);
15 begin
        —— register
     process(clk,reset)
     begin
        if (reset='1') then
```

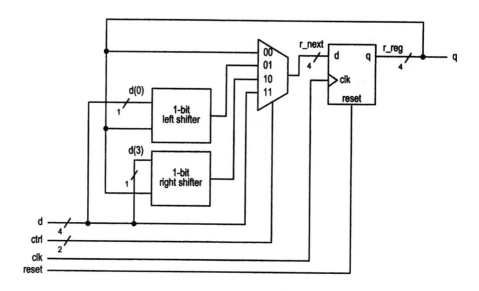

Figure 8.11   4-bit universal register.

```
20              r_reg <= (others=>'0');
          elsif (clk'event and clk='1') then
              r_reg <= r_next;
          end if;
       end process;
25     -- next-state logic
       with ctrl select
          r_next <=
             r_reg                              when "00", --pause
             r_reg(2 downto 0) & d(0)  when "01", --shift left;
30           d(3) & r_reg(3 downto 1)  when "10", --shift right;
             d                              when others; -- load
       -- output logic
       q <= r_reg;
    end two_seg_arch;
```

8.5.3 Arbitrary-sequence counter

A sequential counter circulates a predefined sequence of states. The next-state logic determines the patterns in the sequence. For example, if we need a counter to cycle through the sequence of "000", "011", "110", "101" and "111", we can construct a combinational circuit with a function table that specifies the desired patterns, as in Table 8.1.

The VHDL code is shown in Listing 8.11. Again, the code follows the basic block diagram of Figure 8.5. A conditional signal assignment statement is used to implement the function table.

Listing 8.11   Arbitrary-sequence counter

```
library ieee;
use ieee.std_logic_1164.all;
```

Table 8.1   Patterns of an arbitrary-sequence counter

Input pattern	Next pattern
000	011
011	110
110	101
101	111
111	000

```vhdl
entity arbi_seq_counter4 is
    port(
5       clk, reset: in std_logic;
        q: out std_logic_vector(2 downto 0)
    );
end arbi_seq_counter4;

10 architecture two_seg_arch of arbi_seq_counter4 is
       signal r_reg: std_logic_vector(2 downto 0);
       signal r_next: std_logic_vector(2 downto 0);
   begin
       -- register
15     process(clk,reset)
       begin
           if (reset='1') then
               r_reg <= (others=>'0');
           elsif (clk'event and clk='1') then
20             r_reg <= r_next;
           end if;
       end process;
       -- next-state logic
       r_next <= "011" when r_reg="000" else
25                "110" when r_reg="011" else
                 "101" when r_reg="110" else
                 "111" when r_reg="101" else
                 "000";   -- r_reg ="111"
       -- output logic
30     q <= r_reg;
   end two_seg_arch;
```

8.5.4 Binary counter

A binary counter circulates through a sequence that resembles the unsigned binary number. For example, a 3-bit binary counter cycles through "000", "001", "010", "011", "100", "101", "110" and "111", and then repeats.

Free-running binary counter   An n-bit binary counter has a register with n FFs, and its output is interpreted as an unsigned integer. A free-running binary counter increments the content of the register every clock cycle, counting from 0 to $2^n - 1$ and then repeating. In addition to the register output, we assume that there is a status signal, max_pulse, which is asserted when the counter is in the all-one state. The VHDL code of a 4-bit binary counter is shown in Listing 8.12.

Listing 8.12 Free-running binary counter

```
  library ieee;
  use ieee.std_logic_1164.all;
  use ieee.numeric_std.all;
  entity binary_counter4_pulse is
5   port(
        clk, reset: in std_logic;
        max_pulse: out std_logic;
        q: out std_logic_vector(3 downto 0)
      );
10 end binary_counter4_pulse;

  architecture two_seg_arch of binary_counter4_pulse is
     signal r_reg: unsigned(3 downto 0);
     signal r_next: unsigned(3 downto 0);
15 begin
     -- register
     process(clk,reset)
     begin
        if (reset='1') then
20         r_reg <= (others=>'0');
        elsif (clk'event and clk='1') then
           r_reg <= r_next;
        end if;
     end process;
25   -- next-state logic (incrementor)
     r_next <= r_reg + 1;
     -- output logic
     q <= std_logic_vector(r_reg);
     max_pulse <= '1' when r_reg="1111" else
30                 '0';
  end two_seg_arch;
```

The next-state logic consists of an incrementor, which calculates the new value for the next state of the register. Note that the definition requests the 4-bit binary counter counts in a wrapped-around fashion; i.e., when the counter reaches the maximal number, "1111", it should return to "0000" and start over again. It seems that we should replace statement

```
  r_next <= r_reg + 1;
```

with

```
  r_next <= (r_reg + 1) mod 16;
```

However, in the IEEE numeric_std package, the definition of + on the unsigned data type is modeled after a hardware adder, which behaves like wrapping around when the addition result exceeds the range. Thus, the original statement is fine. While correct, using the **mod** operator is redundant. It may confuse some synthesis software since the **mod** operator cannot be synthesized. The output logic uses a conditional signal assignment statement to implement the desired pulse. The conceptual diagram is shown in Figure 8.12.

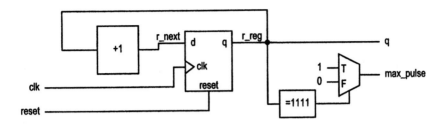

Figure 8.12   Conceptual diagram of a free-running binary counter.

Table 8.2   Function table of a featured binary counter

syn_clr	load	en	q*	Operation
1	–	–	$00\cdots00$	synchronous clear
0	1	–	d	parallel load
0	0	1	q+1	count
0	0	0	q	pause

Featured binary counter   Rather than leaving the counter in the free-running mode, we can exercise more control. The function table in Table 8.2 shows a binary counter with additional features. In the counter, we can synchronously clear the counter to 0, load a specific value, and enable or pause the counting. The VHDL code is shown in Listing 8.13.

Listing 8.13   Featured binary counter

```
library ieee;
use ieee.std_logic_1164.all;
use ieee.numeric_std.all;
entity binary_counter4_feature is
   port(
       clk, reset: in std_logic;
       syn_clr, en, load: in std_logic;
       d: in std_logic_vector(3 downto 0);
       q: out std_logic_vector(3 downto 0)
   );
end binary_counter4_feature;

architecture two_seg_arch of binary_counter4_feature is
   signal r_reg: unsigned(3 downto 0);
   signal r_next: unsigned(3 downto 0);
begin
   -- register
   process(clk,reset)
   begin
       if (reset='1') then
           r_reg <= (others=>'0');
       elsif (clk'event and clk='1') then
           r_reg <= r_next;
       end if;
```

```
25    end process;
      --- next-state logic
      r_next <= (others=>'0') when syn_clr='1' else
                unsigned(d)    when load='1' else
                r_reg + 1      when en ='1' else
30               r_reg;
      --- output logic
      q <= std_logic_vector(r_reg);
   end two_seg_arch;
```

8.5.5 Decade counter

Instead of utilizing all possible 2^n states of an n-bit binary counter, we sometime only want
the counter to circulate through a subset of the states. We define a mod-m counter as a
binary counter whose states circulate from 0 to $m - 1$ and then repeat. Let us consider the
design of a mod-10 counter, also known as a *decade counter*. The counter counts from 0
to 9 and then repeats. We need at least 4 bits ($\lceil \log_2 10 \rceil$) to accommodate the 10 possible
states, and the output is 4 bits wide. The VHDL description is shown in Listing 8.14.

Listing 8.14   Decade counter

```
   library ieee;
   use ieee.std_logic_1164.all;
   use ieee.numeric_std.all;
   entity mod10_counter is
5     port(
         clk, reset: in std_logic;
         q: out std_logic_vector(3 downto 0)
      );
   end mod10_counter;
10
   architecture two_seg_arch of mod10_counter is
      constant TEN: integer := 10;
      signal r_reg: unsigned(3 downto 0);
      signal r_next: unsigned(3 downto 0);
15 begin
      --- register
      process(clk,reset)
      begin
         if (reset='1') then
20          r_reg <= (others=>'0');
         elsif (clk'event and clk='1') then
            r_reg <= r_next;
         end if;
      end process;
25    --- next-state logic
      r_next <= (others=>'0') when r_reg=(TEN-1) else
                r_reg + 1;
      --- output logic
      q <= std_logic_vector(r_reg);
30 end two_seg_arch;
```

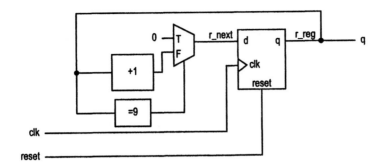

Figure 8.13   Conceptual diagram of a decade counter.

The key to this design is the next-state logic. When the counter reaches 9, as indicated by the condition `r_reg=(TEN-1)`, the next value will be 0. Otherwise, the next value will be incremented by 1. The conceptual diagram is shown in Figure 8.13.

We can rewrite the next-state logic as

```
r_next <= (r_reg + 1) mod 10;
```

Although the code is compact and clean, it cannot be synthesized due to the complexity of the **mod** operator.

8.5.6 Programmable mod-m counter

We can easily modify the code of the previous decade counter to a mod-m counter for any m. However, the counter counts a fixed, predefined sequence. In this example, we design a "programmable" 4-bit mod-m counter, in which the value of m is specified by a 4-bit input signal, m, which is interpreted as an unsigned number. The range of m is from "0010" to "1111", and thus the counter can be programmed as a mod-2, mod-3, ... , or mod-15 counter.

The maximal number in the counting sequence of a mod-m counter is $m - 1$. Thus, when the counter reaches $m - 1$, the next state should be 0. Our first design is based on this observation. The VHDL code is similar to the decade counter except that we need to replace the `r_reg=(TEN-1)` condition of the next-state logic with `r_reg=(unsigned(m)-1)`. The code is shown in Listing 8.15.

Listing 8.15   Initial description of a programmable mod-m counter

```
library ieee;
use ieee.std_logic_1164.all;
use ieee.numeric_std.all;
entity prog_counter is
 5   port(
        clk, reset: in std_logic;
        m: in std_logic_vector(3 downto 0);
        q: out std_logic_vector(3 downto 0)
    );
10 end prog_counter;

   architecture two_seg_clear_arch of prog_counter is
```

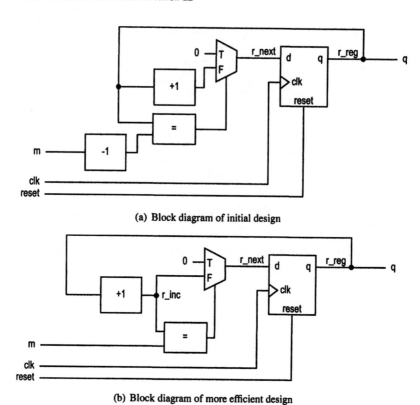

(a) Block diagram of initial design

(b) Block diagram of more efficient design

Figure 8.14 Block diagrams of a programmable mod-m counter.

```
      signal r_reg: unsigned(3 downto 0);
      signal r_next: unsigned(3 downto 0);
15  begin
      -- register
      process(clk,reset)
      begin
        if (reset='1') then
20        r_reg <= (others=>'0');
        elsif (clk'event and clk='1') then
          r_reg <= r_next;
        end if;
      end process;
25    -- next-state logic
      r_next <= (others=>'0') when r_reg=(unsigned(m)-1) else
                r_reg + 1;
      -- output logic
      q <= std_logic_vector(r_reg);
30  end two_seg_clear_arch;
```

The conceptual diagram of this code is shown in Figure 8.14(a). The next-state logic consists of an incrementor, a decrementor and a comparator. There is an opportunity for sharing. Note that the Boolean expression

```
r_reg=(unsigned(m)-1)
```

can also be written as

```
(r_reg+1)=unsigned(m)
```

Since the r_req+1 operation is needed for incrementing operation, we can use it in comparison and eliminate the decrementor. The revised VHDL code is shown in Listing 8.16.

Listing 8.16 More efficient description of a programmable mod-m counter

```
architecture two_seg_effi_arch of prog_counter is
    signal r_reg: unsigned(3 downto 0);
    signal r_next, r_inc: unsigned(3 downto 0);
begin
5   -- register
    process(clk,reset)
    begin
        if (reset='1') then
            r_reg <= (others=>'0');
10        elsif (clk'event and clk='1') then
            r_reg <= r_next;
        end if;
    end process;
    -- next-state logic
15   r_inc <= r_reg + 1;
    r_next <= (others=>'0') when r_inc=unsigned(m) else
              r_inc;
    -- output logic
    q <= std_logic_vector(r_reg);
20 end two_seg_effi_arch;
```

Note that we employ a separate statement for the shared expression:

```
r_inc <= r_reg + 1;
```

and use the r_inc signal for both comparison and incrementing. The diagram of the revised code is shown in Figure 8.14(b).

8.6 TIMING ANALYSIS OF A SYNCHRONOUS SEQUENTIAL CIRCUIT

The timing of a combinational circuit is characterized primarily by the propagation delay, which is the time interval required to generate a stable output response from an input change. The timing characteristic of a sequential circuit is different because of the constraints imposed by memory elements. The major timing parameter in a sequential circuit is the *maximal clock rate*, which embeds the effect of the propagation delay of the combination circuit, the clock-to-q delay of the register and the setup time constraint of the register. Other timing issues include the condition to avoid hold time violation and I/O-related timing parameters.

8.6.1 Synchronized versus unsynchronized input

Satisfying the setup and hold time constraints is the most crucial task in designing a sequential circuit. One motivation behind synchronous design methodology is to group all FFs together and control them with the same clock signal. Instead of considering the constraints of tens or hundreds of FFs, we can treat them as one memory component and deal with the timing constraint of a *single* register.

The conceptual diagram of Figure 8.5 can be considered as a simplified block diagram for all synchronous sequential circuits. In this diagram, FFs and registers are grouped together as the state register. The input of this register is the state_next signal. It is generated by next-state logic, which is a combinational logic with two inputs, including the external input and the output of the state register, state_reg. To study the timing constraint of the state register, we need to examine the impact of the two inputs of the next-state logic. Our discussion considers the following effects:

- The effect of the state_reg signal.
- The effect of *synchronized* external input.
- The effect of *unsynchronized* external input.

Since the state_reg signal is the output of the state register, it is synchronized by the same clock. A closed feedback loop is formed in the diagram through this signal. The timing analysis of a synchronous sequential circuit focuses mainly on this loop and is discussed in Section 8.6.2 .

A *synchronized* external input means that the generation of the input signal is controlled by the same clock signal, possibly from a subsystem of the same design. The timing analysis is somewhat similar to the closed-loop analysis describe above, and is discussed in Section 8.6.5.

An *unsynchronized* external input means that the input signal is generated from an external source or an independent subsystem. Since the system has no information about the unsynchronized external input, it cannot prevent timing violations. For this kind of input, we must use an additional *synchronization circuit* to synchronize the signal with the system clock. This issue is be discussed in Chapter 16.

8.6.2   Setup time violation and maximal clock rate

In Figure 8.5, the output of the register is processed via next-state logic, whose output becomes the new input to the register. To analyze the timing, we have to study the operation of this closed feedback loop and examine the state_reg and state_next signals. The state_reg signal is the output of the register, and it also serves as the input to the next-state logic. The state_next signal is the input of the register, and it is also the output of the next-state logic.

Maximal clock rate   The timing diagram in Figure 8.15 shows the responses of the state_reg and state_next signals during one clock cycle. At time t_0, the clock changes from '0' to '1'. We assume that the state_next signal has stabilized and doesn't change within the setup and hold time periods. After the clock-to-q delay (i.e., T_{cq}), the register's output, state_reg, becomes available at time t_1, which is $t_0 + T_{cq}$. Since state_reg is the input of the next-state logic, the next-state logic responds accordingly. We define the propagation delays of the fastest and slowest responses as $T_{next(min)}$ and $T_{next(max)}$ respectively. In the timing diagram, the state_next signal changes at t_2, which is $t_1 + T_{next(min)}$, and becomes stabilized at t_3, which is $t_1 + T_{next(max)}$. At time t_5, a new rising clock edge arrives and the current clock cycle ends. The state_next is sampled at t_5 and the process repeats again. t_5 is determined by the period (T_c) of the clock signals, which is $t_0 + T_c$.

Now let us examine the impact of the setup time constraint. The setup time constraint indicates that the state_next signal must be stabilized at least T_{setup} before the next

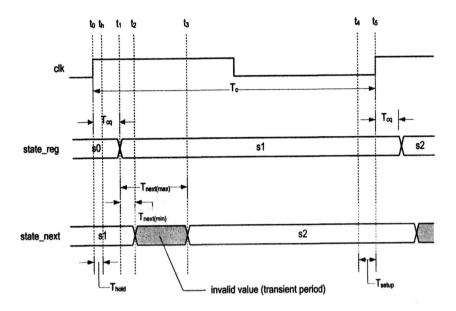

Figure 8.15   Timing analysis of a basic sequential circuit.

sampling edge at t_5. This point is labeled t_4 in the timing diagram. To satisfy the setup time constraint, the state_next signal must be stabilized before t_4. This requirement translates into the condition

$$t_3 < t_4$$

From the timing diagram, we see that

$$t_3 = t_0 + T_{cq} + T_{next(max)}$$

and

$$t_4 = t_5 - T_{setup} = t_0 + T_c - T_{setup}$$

We can rewrite the inequality equation as

$$t_0 + T_{cq} + T_{next(max)} < t_0 + T_c - T_{setup}$$

which is simplified to

$$T_{cq} + T_{next(max)} + T_{setup} < T_c$$

This shows the role of the clock period on a sequential circuit. To avoid setup time violation, the minimal clock period must be

$$T_{c(min)} = T_{cq} + T_{next(max)} + T_{setup}$$

The clock period is the main parameter to characterize the timing and performance of a sequential circuit. We commonly use the maximal clock rate or frequency, the reciprocal of the minimal period, to describe the performance of a sequential circuit, as in a 500-MHz counter or 2-GHz processor.

Clock rate examples   For a given technology, the T_{cq} and T_{setup} of a D FF are obtained from the data sheet. We can determine the maximal clock rate of a sequential circuit once the propagation delay of the next-state logic is known. This information can only be determined after synthesis and placement and routing. However, we can calculate and estimate the rate of some simple examples.

Assume that we use the technology discussed in Section 6.2.6, and T_{cq} and T_{setup} of its D FF cell are 1 and 0.5 ns respectively. The delay information of combinational components can be obtained from Table 6.2. Let us first consider the free-running shift register of Section 8.5.2. The next-state logic of the shift register only involves the routing of the input and output signals. If we assume that the wiring delay is negligible, its propagation delay is 0. The minimal clock period and maximal clock rate become

$$T_{c(min)} = T_{cq} + T_{setup} = 1.5 \text{ ns}$$

$$f_{max} = \frac{1}{T_{cq} + T_{setup}} = \frac{1}{1.5 \text{ ns}} \approx 666.7 \text{ MHz}$$

Clearly, this is the maximal clock rate that can be achieved with this particular technology.

The second example is an 8-bit free-running binary counter, similar to the 4-bit version of Section 8.5.4. The next-state logic of this circuit is the incrementor, as shown in Figure 8.12. If we choose the incrementor that is optimized for area, the clock rate for this 8-bit binary counter is

$$f_{max} = \frac{1}{T_{cq} + T_{8\_bit\_inc(area)} + T_{setup}} = \frac{1}{1 \text{ ns} + 2.4 \text{ ns} + 0.5 \text{ ns}} \approx 256.4 \text{ MHz}$$

If we increase the size of the counter, a wider incrementor must be utilized, and the propagation delay of the incrementor is increased accordingly. The clock rate of a 16-bit binary counter is reduced to

$$f_{max} = \frac{1}{T_{cq} + T_{16\_bit\_inc(area)} + T_{setup}} = \frac{1}{1 \text{ ns} + 5.5 \text{ ns} + 0.5 \text{ ns}} \approx 142.9 \text{ MHz}$$

and the clock rate of a 32-bit counter is reduced to

$$f_{max} = \frac{1}{T_{cq} + T_{32\_bit\_inc(area)} + T_{setup}} = \frac{1}{1 \text{ ns} + 11.6 \text{ ns} + 0.5 \text{ ns}} \approx 76.3 \text{ MHz}$$

To increase the performance of a binary counter, we must reduce the value of $T_{cq} + T_{next(max)} + T_{setup}$. Since T_{cq} and T_{setup} are determined by the intrinsic characteristics of FFs, they cannot be altered unless we switch to a different device technology. The only way to increase performance is to reduce the propagation delay of the incrementor. If we replace the incrementors that are optimized for delay, the clock rates of the 8-, 16- and 32-bit binary counters are increased to

$$f_{max} = \frac{1}{T_{cq} + T_{8\_bit\_inc(delay)} + T_{setup}} = \frac{1}{1 \text{ ns} + 1.5 \text{ ns} + 0.5 \text{ ns}} \approx 333.3 \text{ MHz}$$

$$f_{max} = \frac{1}{T_{cq} + T_{16\_bit\_inc(delay)} + T_{setup}} = \frac{1}{1 \text{ ns} + 3.3 \text{ ns} + 0.5 \text{ ns}} \approx 208.3 \text{ MHz}$$

and

$$f_{max} = \frac{1}{T_{cq} + T_{32\_bit\_inc(delay)} + T_{setup}} = \frac{1}{1 \text{ ns} + 7.5 \text{ ns} + 0.5 \text{ ns}} \approx 111.1 \text{ MHz}$$

respectively.

8.6.3 Hold time violation

The impact of the hold time constraint is somewhat different from the setup time constraint. Hold time, T_{hold}, is the time period that the input signal must be stabilized after the sampling edge. In the timing diagram of Figure 8.15, it means that the state_next must be stable between t_0 and t_h, which is $t_0 + T_{hold}$. Note that the earliest time that state_next changes is at time t_2. To satisfy the hold time constraint, we must ensure that

$$t_h < t_2$$

From the timing diagram, we see that

$$t_2 = t_0 + T_{cq} + T_{next(min)}$$

and

$$t_h = t_0 + T_{hold}$$

The inequality becomes

$$t_0 + T_{hold} < t_0 + T_{cq} + T_{next(min)}$$

which is simplified to:

$$T_{hold} < T_{cq} + T_{next(min)}$$

$T_{next(min)}$ depends on the complexity of next-state logic. In some applications, such as the shift register, the output of one FF is connected to the input of another FF, and the propagation delay of the next-state logic is the wire delay, which can be close to 0. Thus, in the worst-case scenario, the inequality becomes

$$T_{hold} < T_{cq}$$

Note that both parameters are the intrinsic timing parameters of the FF, and the inequality has nothing to do with the next-state logic. Manufacturers usually guarantee that their devices satisfy this condition. Thus, we need not worry about the hold time constraint unless the clock edge cannot arrive at all FFs at the same time. We discuss this issue in Chapter 16.

8.6.4 Output-related timing considerations

The closed feedback diagram in Figure 8.5 is the core of a sequential system. In addition, there are also external inputs and outputs. Let us first consider the output part of the circuit. The output signal of a sequential circuit can be divided into the *Moore-typed output* (or just *Moore output*) and *Mealy-typed output* (or just *Mealy output*). For Moore output, the output signal is a function of *system state* (i.e., the output of the register) only. On the other hand, for Mealy output, the output signal is a function of *system state and the external input*. The two types of output can coexist, as shown in Figure 8.16. The main timing parameter for both types of outputs is T_{co}, the time required to obtain a valid output signal after the rising edge of the clock. The value of T_{co} is the summation of T_{cq} and T_{output} (the propagation delay of the output logic); that is,

$$T_{co} = T_{cq} + T_{output}$$

For Mealy output, there exists a path in which the input can affect the output directly. The propagation delay from input to output is simply the combinational propagation delay of output logic.

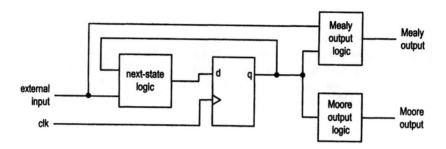

Figure 8.16   Output circuits of a sequential circuit.

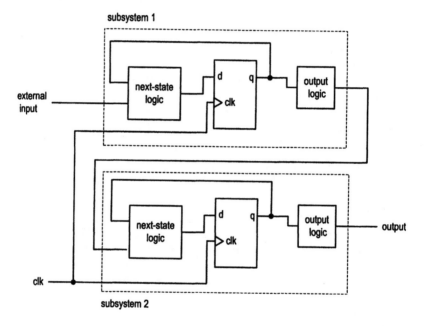

Figure 8.17   Input timing of two synchronous subsystems.

8.6.5 Input-related timing considerations

In a large design, a system may contain several synchronous subsystems. Thus, it is possible that an input comes from a subsystem that is controlled and synchronized by the same clock. The block diagram of this situation is shown in Figure 8.17. Note that the two subsystems are controlled by the same clock and thus are synchronous. At the rising edge of the clock, the register of subsystem 1 samples a new input value. After $T_{co(system1)}$, its new output, which is the input for the next-state logic of subsystem 2, becomes available. At this point the timing analysis is identical to that in Section 8.6.2. To avoid setup time violation, the timing of the two circuits must satisfy the following condition:

$$T_{co(system1)} + T_{next(max)} + T_{setup} < T_c$$

Note that $T_{next(max)}$, the propagation delay of next-state logic, is somewhat different from the calculation used in Section 8.6.2. The $T_{next(max)}$ here is the propagation delay from the external input to state_next, whereas $T_{next(max)}$ used in earlier minimal clock period calculation in Section 8.6.2 is the propagation delay from the internal register output (i.e., state_reg) to state_next. To be more accurate, we should separate the two constraints. The constraint for the closed loop is

$$T_{cq} + T_{next(max\ of\ state\_reg-to-state\_next)} + T_{setup} < T_{c1}$$

and the constraint for the external input is

$$T_{co(system1)} + T_{next(max\ of\ ext\_input-to-state\_next)} + T_{setup} < T_{c2}$$

We usually determine the clock period based on the calculation of T_{c1}. If T_{c2} turns out to be greater than T_{c1}, we normally redesign the I/O buffer rather than slowing down the clock rate of the entire system. For example, we can employ an extra input buffer for the external input of subsystem 2. Although this approach delays the external input by one clock cycle, it reduces the $T_{co(system1)}$ to T_{cq} in the second constraint.

8.7 ALTERNATIVE ONE-SEGMENT CODING STYLE

So far, all VHDL coding follows the basic block diagram of Figure 8.5 and separates the memory elements from the rest of the logic. Alternatively, we can describe the memory elements and the next-state logic in a single process segment. For a simple circuit, this style appears to be more compact. However, it becomes involved and error-prone for more complex circuits. In this section, we use some earlier examples to illustrate the one-segment VHDL description and the problems associated with this style.

8.7.1 Examples of one-segment code

D FF with enable   Consider the D FF with an enable signal in Listing 8.7. It can be rewritten in one-segment style, as in Listing 8.17.

Listing 8.17   One-segment description of a D FF with enable

```
architecture one_seg_arch of dff_en is
begin
    process(clk,reset)
    begin
        if (reset='1') then
            q <='0';
        elsif (clk'event and clk='1') then
            if (en='1') then
                q <= d;
            end if;
        end if;
    end process;
end one_seg_arch;
```

The code is similar to a regular D FF except that there is an if statement inside the elsif branch:

```
if (en='1') then
    q <= d;
end if;
```

The interpretation the code is that at the rising edge of clk, if en is '1', q gets the value of the d input. Note that there is no else branch in the previous statement. It implies that if en is not '1', q will keep its previous value, which should be the value of the register's output. Thus, the code correctly describes the function of the en signal. In the actual implementation, "keep its previous value" is achieved by sampling the FF's output and again stores the value back to the FF. This point is elaborated in the next example.

T FF   Consider the T FF in Listing 8.8. It can be rewritten in one-segment style, as in Listing 8.18.

Listing 8.18   One-segment description of a T FF

```
    architecture one_seg_arch of tff is
       signal q_reg: std_logic;
    begin
       process(clk, reset)
 5     begin
          if reset='1' then
             q_reg <= '0';
          elsif (clk'event and clk='1') then .
             if (t='1') then
10               q_reg <= not q_reg;
             end if;
          end if;
       end process;
       q <= q_reg;
15  end one_seg_arch;
```

We use an internal signal, q_reg, to represent the content and the output of an FF. The statement

```
    q_reg <= not q_reg;
```

may appear strange at first glance. So let us examine it in more detail. The q_reg signal on the right-hand side represents the output value of the FF, and the **not** q_reg expression forms the new value of q_reg. This value has no effect on the FF until the process is activated and the clk'event **and** clk='1' condition is true, which specified the occurrence of the rising edge of the clk signal. At this point the value is assigned to q_reg (actually, stored into the FF named q_reg). Thus, the code correctly describes the desired function. Note that if this statement is an isolated concurrent signal assignment statement, a closed combinational feedback loop is formed, in which the output and input of an inverter are tied together.

As in the previous example, the inner if statement has no else branch, and thus q_reg will keep its previous value if the t='1' condition is false. In actual implementation, "keep its previous value" is achieved by sampling the FF's output and storing the value back to the FF. Thus, the more descriptive if statement can be written as

```
    if (t='1') then
       q_reg <= not(q_reg);
    else
       q_reg <= q_reg;
    end if;
```

Featured binary counter   Consider the featured binary counter in Listing 8.13. We can convert it into one-segment code, as in Listing 8.19.

Listing 8.19   One-segment description of a featured binary counter

```vhdl
architecture one_seg_arch of binary_counter4_feature is
   signal r_reg: unsigned(3 downto 0);
begin
   -- register & next-state logic
   process(clk,reset)
   begin
      if (reset='1') then
         r_reg <= (others=>'0');
      elsif (clk'event and clk='1') then
         if syn_clr='1' then
            r_reg <= (others=>'0');
         elsif load='1' then
            r_reg <= unsigned(d);
         elsif en ='1' then
            r_reg <= r_reg + 1;
         end if;
      end if;
   end process;
   -- output logic
   q <= std_logic_vector(r_reg);
end one_seg_arch;
```

The key to this code is the incrementing part, which is done using the statement

```vhdl
r_reg <= r_reg + 1;
```

The interpretation of r_reg in this statement is similar to that in T FF except that the **not** operation is replaced by incrementing.

Free-running binary counter   Consider the 4-bit free-running binary counter in Listing 8.12. The first attempt to convert it to a single-segment style is shown in Listing 8.20.

Listing 8.20   Incorrect one-segment description of a free-running binary counter

```vhdl
architecture not_work_one_seg_glitch_arch
                     of binary_counter4_pulse is
   signal r_reg: unsigned(3 downto 0);
begin
   process(clk,reset)
   begin
      if (reset='1') then
         r_reg <= (others=>'0');
      elsif (clk'event and clk='1') then
         r_reg <= r_reg + 1;
         if r_reg="1111" then
            max_pulse <= '1';
         else
            max_pulse <= '0';
         end if;
      end if;
   end process;
```

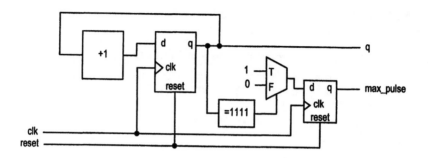

Figure 8.18 Free-running binary counter with an unintended output buffer.

```
q <= std_logic_vector(r_reg);
end not_work_one_seg_glitch_arch;
```

The output logic does not function as we expected. Because the statement

```
if r_reg="1111" then
   max_pulse <= '1';
else
   max_pulse <= '0';
end if;
```

is inside the `clk'event and clk='1'` branch, a 1-bit register is inferred for the `max_pulse` signal. The register works as a buffer and delays the output by one clock cycle, and thus the `max_pulse` signal will be asserted when `r_reg="0000"`. The block diagram of this code is shown in Figure 8.18.

To correct the problem, we have to move the output logic outside the process, as in Listing 8.21.

Listing 8.21 Correct one-segment description of a free-running binary counter

```
architecture work_one_seg_glitch_arch
                    of binary_counter4_pulse is
   signal r_reg: unsigned(3 downto 0);
begin
5     process(clk,reset)
      begin
         if (reset='1') then
            r_reg <= (others=>'0');
         elsif (clk'event and clk='1') then
10          r_reg <= r_reg + 1;
         end if;
      end process;
      q <= std_logic_vector(r_reg);
      max_pulse <= '1' when r_reg="1111" else
15                    '0';
   end work_one_seg_glitch_arch;
```

Programmable counter   Consider the programmable mod-m counter in Listing 8.16. The first attempt to reconstruct the `two_seg_effi_arch` architecture in one-segment coding style is shown in Listing 8.22.

Listing 8.22   Incorrect one-segment description of a programmable counter

```
architecture not_work_one_arch of prog_counter is
    signal r_reg: unsigned(3 downto 0);
begin
    process(clk,reset)
s   begin
        if reset='1' then
            r_reg <= (others=>'0');
        elsif (clk'event and clk='1') then
            r_reg <= r_reg+1;
10          if (r_reg=unsigned(m)) then
                r_reg<= (others=>'0');
            end if;
        end if;
    end process;
15  q <= std_logic_vector(r_reg);
end not_work_one_arch;
```

The code does not work as specified. Recall that a signal will not be updated until the end of the process. Thus, r_reg is updated to r_reg+1 in the end. When the comparison r_reg=unsigned(m) is performed, the old value of r_reg is used. Because the correct r_reg value is late for one clock, the counter counts one extra value. The code actually specified a mod-$(m + 1)$ counter instead.

To correct the problem, we must move the incrementing operation outside the process so that it can be performed concurrently with the process. The modified VHDL code is shown in Listing 8.23.

Listing 8.23   Correct one-segment description of a programmable counter

```
architecture work_one_arch of prog_counter is
    signal r_reg: unsigned(3 downto 0);
    signal r_inc: unsigned(3 downto 0);
begin
s   process(clk,reset)
    begin
        if reset='1' then
            r_reg <= (others=>'0');
        elsif (clk'event and clk='1') then
10          if (r_inc=unsigned(m)) then
                r_reg <= (others=>'0');
            else
                r_reg <= r_inc;
            end if;
15      end if;
    end process;
    r_inc <= r_reg + 1;
    q <= std_logic_vector(r_reg);
end work_one_arch;
```

8.7.2 Summary

When we combine the memory elements and next-state logic in the same process, it is much harder to "visualize" the circuit and to map the VHDL statements into hardware components. This style may make code more compact for a few simple circuits, as in the first three examples. However, when a slightly more involved feature is needed, as the max_pulse output or the incrementor sharing of the last two examples, the one-segment style makes the code difficult to understand and error-prone. Although we can correct the problems, the resulting code contains extra statements and is far worse than the codes in Section 8.5. Furthermore, since the combinational logic and memory elements are mixed in the same process, it is more difficult to perform optimization and to fine-tune the combinational circuit. In summary, although the two-segment code may occasionally appear cumbersome, its benefits far outweigh the inconvenience, and we generally use this style in this book.

8.8 USE OF VARIABLES IN SEQUENTIAL CIRCUIT DESCRIPTION

We have learned how to infer an FF or a register from a signal. It is done by using the clk'event **and** clk='1' condition to indicate the rising edge of the clock signal. Any signal assigned under this condition is required to keep its previous value, and thus an FF or a register is inferred accordingly.

A variable can also be assigned under the clk'event **and** clk='1' condition, but its implication is different because a variable is local to the process and its value is not needed outside the process. If a variable is *assigned a value before it is used*, it will get a value every time when the process is invoked and there is no need to keep its previous value. Thus, no memory element is inferred. On the other hand, if a variable is *used before it is assigned a value*, it will use the value from the previous process execution. The variable has to memorize the value between the process invocations, and thus an FF or a register will be inferred.

Since using a variable to infer memory is more error-prone, we generally prefer to use a signal for this task. The major use of variables is to obtain an intermediate value inside the clk'event **and** clk='1' branch without introducing an unintended register. This can best be explained by an example. Let us consider a simple circuit that performs an operation a **and** b and stores the result into an FF at the rising edge of the clock. We use three outputs to illustrate the effect of different coding attempts. The VHDL code is shown in Listing 8.24.

Listing 8.24   Using a variable to infer an FF

```
library ieee;
use ieee.std_logic_1164.all;
entity varaible_ff_demo is
    port(
        a,b,clk: in std_logic;
        q1,q2,q3: out std_logic
    );
end varaible_ff_demo;

architecture arch of varaible_ff_demo is
    signal tmp_sig1: std_logic;
begin
```

```
        — attempt 1
        process (clk)
15      begin
            if (clk'event and clk='1') then
                tmp_sig1 <= a and b;
                q1 <= tmp_sig1;
            end if;
20      end process;
        — attempt 2
        process (clk)
            variable   tmp_var2: std_logic;
        begin
25          if (clk'event and clk='1') then
                tmp_var2 := a and b;
                q2 <= tmp_var2;
            end if;
        end process;
30      — attempt 3
        process (clk)
            variable   tmp_var3: std_logic;
        begin
            if (clk'event and clk='1') then
35              q3 <= tmp_var3;
                tmp_var3 := a and b;
            end if;
        end process;
    end arch;
```

In the first attempt, we try to use the tmp_sig1 signal for the temporary result. However, since the tmp_sig1 signal is inside the clk'event and clk='1' branch, an unintended D FF is inferred. The two statements

```
tmp_sig1 <= a and b;
q1 <= tmp_sig1;
```

are interpreted as follows. At the rising edge of the clk signal, the value of a and b will be sampled and stored into an FF named tmp_sig1, and the old value (not current value of a and b) from the tmp_sig1 signal will be stored into an FF named q1. The diagram is shown in Figure 8.19(a).

The value of a and b is delayed by the unintended buffer, and thus this description fails to meet the specification. Since both statements are signal assignment statements, we will obtain the same result if we switch the order of the two statements.

The second attempt uses a variable, tmp_var2, for the temporary result and the statements become

```
tmp_var2 := a and b;
q2 <= tmp_var2;
```

Note that the tmp_var2 variable is first assigned a value and then used in the next statement. Thus, no memory element is inferred and the circuit meets the specification. The diagram is shown in Figure 8.19(b).

The third attempt uses a variable, tmp_var3, for the temporary result. It is similar to the second process except that the order of the two statements is reversed:

```
q3 <= tmp_var3;
```

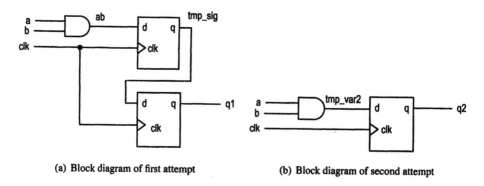

(a) Block diagram of first attempt   (b) Block diagram of second attempt

Figure 8.19   Register inference with a variable.

```
tmp_var3 := a and b;
```

In this code, the `tmp_var3` variable is first used before it is assigned a value. According to the VHDL definition, the value of `tmp_var3` from the previous process invocation will be used. An FF will be inferred to store the previous value. Thus, the circuit described by the third attempt is the same as that of the first attempt, which contains an unwanted buffer.

We can use a variable to overcome the problem of the one-segment programmable mod-m counter in Listing 8.22. The revised code is shown in Listing 8.25.

Listing 8.25   Variable description of a programmable counter

```
architecture variable_arch of prog_counter is
    signal r_reg: unsigned(3 downto 0);
begin
    process(clk,reset)
s       variable q_tmp: unsigned(3 downto 0);
    begin
        if reset='1' then
            r_reg <= (others=>'0');
        elsif (clk'event and clk='1') then
10          q_tmp := r_reg + 1;
            if (q_tmp=unsigned(m)) then
                r_reg <= (others=>'0');
            else
                r_reg <= q_tmp;
15          end if;
        end if;
    end process;
    q <= std_logic_vector(r_reg);
end variable_arch;
```

Instead of using the `r_reg` signal, we create a variable, `q_tmp`, to store the intermediate result of the incrementing operation. Unlike the signal assignment, the variable assignment takes effect immediately, and thus the code functions as intended.

8.9 SYNTHESIS OF SEQUENTIAL CIRCUITS

In Chapter 6, we examined the synthesis procedure for a combinational circuit. The synthesis of a sequential circuit is identical to this procedure but has two extra steps:

1. Identify and separate the memory elements from the circuit.
2. Select the proper leaf cells from the device library to realize the memory elements.
3. Synthesize the remaining combinational circuit.

If we follow the recommended coding style, the memory elements are specified in individual VHDL segments and thus can be easily inferred and properly instantiated by synthesis software. Once this is done, the remaining process is identical to the synthesis of a combinational circuit.

While synthesizing a combinational circuit, we can include a timing constraint to specify the desired maximal propagation delay, and the synthesis software will try to obtain a circuit to meet this constraint. For a sequential circuit, we can specify the desired maximal clock rate. In a synchronous design, this constraint can easily be translated into the maximal propagation delay of the combinational next-state logic, as indicated by the minimal clock period equation. Thus, all the optimization schemes used in combinational circuits can also be applied to sequential circuit synthesis.

In summary, when we design and code a sequential circuit in a disciplined way, synthesizing it is just like the synthesis of a combinational circuit. We can apply the analysis and optimization schemes developed for combinational circuits to sequential circuit design.

8.10 SYNTHESIS GUIDELINES

- Strictly follow the synchronous design methodology; i.e., all registers in a system should be synchronized by a common global clock signal.

- Isolate the memory components from the VHDL description and code them in a separate segment. One-segment coding style is not advisable.

- The memory components should be coded clearly so that a predesigned cell can be inferred from the device library.

- Avoid synthesizing a memory component from scratch.

- Asynchronous reset, if used, should be only for system initialization. It should not be used to clear the registers during regular operation.

- Unless there is a compelling reason, a variable should not be used to infer a memory component.

8.11 BIBLIOGRAPHIC NOTES

Design and analysis of intermediate-sized synchronous sequential circuits are covered by standard digital systems texts, such as *Digital Design Principles and Practices* by J. F. Wakerly and *Contemporary Logic Design* by R. H. Katz. The former also has a section on the derivation and analysis of asynchronous sequential circuits.

Problems

8.1 Repeat the timing analysis of Section 8.3 for the circuit shown in Figure 8.6 with the following assumptions and examine the q output.

- The propagation delay of the inverter is T and the propagation delays of and and or gates are $2T$.
- The propagation delay of the inverter is $2T$ and the propagation delays of and and or gates are T.

8.2 The SR latch is defined in the left table below. Some device library does not have an SR-latch cell. Instead of synthesizing it from scratch using combinational gates, we want to do this by using a D latch. Derive the VHDL code for this design. The code should contain a standard VHDL description to infer a D latch and a combinational segment that maps the s and r signals to the d and c ports of the D latch to achieve the desired function.

s	r	q*
0	0	q
0	1	0
1	0	1
1	1	not allowed

SR latch

j	k	clk	q*
-	-	0	q
-	-	1	q
0	0	∫	q
0	1	∫	0
1	0	∫	1
1	1	∫	q'

JK FF

8.3 A JK FF is defined as in the right table above. Use a D FF and a combinational circuit to design the circuit. Derive the VHDL code and draw the conceptual diagram for this circuit.

8.4 If we replace the D FFs of the free-running shift register of Section 8.5.2 with D latches and connect the external clock signal to the c ports of all D latches, discuss what will happen to the circuit.

8.5 Expand the design of the universal shift register of Section 8.5.2 to include rotate-right and rotate-left operations. To accommodate the revision, the ctrl signal has to be extended to 3 bits. Derive the VHDL code for this circuit.

8.6 Consider an 8-bit free-running up–down binary counter. It has a control signal, up. The counter counts up when the up signal is '1' and counts down otherwise. Derive the VHDL code for this circuit and draw the conceptual top-level diagram.

8.7 Consider a 4-bit counter that counts from 3 ("0011") to 12 ("1100") and then wraps around. If the counter enters an unused state (such as "0000") because of noise, it will restart from "0011" at the next rising edge of the clock. Derive the VHDL code for this circuit and draw the conceptual top-level diagram.

8.8 Redesign the arbitrary counter of Section 8.5.3 using a mod-5 counter and special output decoding logic. Derive the VHDL code for this design.

8.9   Design a programmable frequency divider. In addition to clock and reset, it has a control signal, c, which is a 4-bit signal interpreted as an unsigned number. The circuit has an output signal, pulse, whose frequency is controlled by c. If the clock frequency is f and the value of c is m, the frequency of the pulse signal will be $\frac{f}{2m}$. For example, if c is "0101", the frequency of the pulse signal be $\frac{f}{25}$. Derive the VHDL code for this circuit.

8.10   Assume that we have a 1-MHz clock signal. Design a circuit that generates a 1-Hz output pulse with a 50% duty cycle (i.e., 50% of '1' and 50% of '0'). Derive the VHDL code for this circuit.

8.11   Consider the block diagram of the decade counter in Figure 8.13. Let T_{cq} and T_{setup} of the D FF be 1 and 0.5 ns, and the propagation delays of the incrementor, comparator and multiplexer be 5, 3 and 0.75 ns respectively. Assume that no further optimization will be performed during synthesis. Determine the maximal clock rate.

8.12   Consider the two block diagrams of the programmable mod-m counter in Figure 8.14. Assume that no further optimization will be performed during synthesis. Use the timing information in Problem 8.11 to determine the maximal clock rates of the two configurations.

CHAPTER 9

SEQUENTIAL CIRCUIT DESIGN: PRACTICE

After learning the basic model and coding style, we explore more sophisticated regular sequential circuits in this chapter. The design examples show the implementation of a variety of counters, the use of registers as fast, temporary storage, and the construction of a "pipeline" to increase the throughput of certain combinational circuits.

9.1 POOR DESIGN PRACTICES AND THEIR REMEDIES

Synchronous design is the most important design methodology for developing a large, complex, reliable digital system. In the past, some poor, non-synchronous design practices were used. Those techniques failed to follow the synchronous principle and should be avoided in RT-level design. Before continuing with more examples, we examine those practices and their remedies. The most common problems are:

- Misuse of the asynchronous reset.
- Misuse of the gated clock.
- Misuse of the derived clock.

Some of those practices were used when a system was realized by SSI and MSI devices and the silicon real estate and printed circuit board were a premium. Designers tended to cut corners to save a few chips. These legacy practices are no longer applicable in today's design environment and should be avoided. The following subsections show how to remedy these poor non-synchronous design practices.

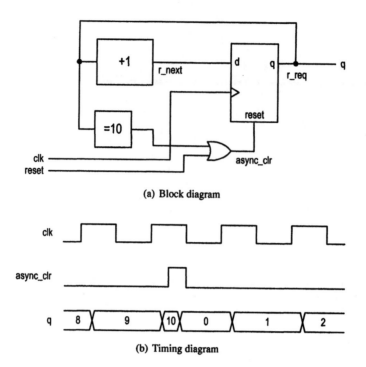

(a) Block diagram

(b) Timing diagram

Figure 9.1   Decade counter using asynchronous reset.

In few special situations, such as the interface to an external system and low power design, the use of multiple clocks and asynchrony may be unavoidable. This kind of design cannot easily be incorporated into the regular synthesis and testing flow. It should be treated differently and separated from the regular sequential system development. We discuss the asynchronous aspect in Chapter 16.

9.1.1   Misuse of asynchronous signals

In a synchronous design, we utilize only asynchronous reset or preset signals of FFs for system initialization. These signals should not be used in regular operation. A decade (mod-10) counter based on asynchronous reset is shown in Figure 9.1(a). The idea behind the design is to clear the counter to "0000" immediately after the counter reaches "1010". The timing diagram is shown in Figure 9.1(b). If we want, we can write VHDL code for this design, as in Listing 9.1.

Listing 9.1   Decade counter using an asynchronous reset signal

```
library ieee;
use ieee.std_logic_1164.all;
use ieee.numeric_std.all;
entity mod10_counter is
  port(
      clk, reset: in std_logic;
      q: out std_logic_vector(3 downto 0)
  );
end mod10_counter;
```

```
10  architecture poor_async_arch of mod10_counter is
       signal r_reg: unsigned(3 downto 0);
       signal r_next: unsigned(3 downto 0);
       signal async_clr: std_logic;
15  begin
       -- register
       process(clk,async_clr)
       begin
          if (async_clr='1') then
20            r_reg <= (others=>'0');
          elsif (clk'event and clk='1') then
             r_reg <= r_next;
          end if;
       end process;
25     -- asynchronous clear
       async_clr <= '1' when (reset='1' or r_reg="1010") else
                    '0';
       -- next state logic
       r_next <= r_reg + 1;
30     -- output logic
       q <= std_logic_vector(r_reg);
    end poor_async_arch;
```

There are several problems with this design. First, the transition from state "1001" (9) to "0000" (0) is noisy, as shown in Figure 9.1(b). In that clock period, the counter first changes from "1001" (9) to "1010" (10) and then clears to "0000" (0) after the propagation delay of the comparator and reset. Second, this design is not very reliable. A combinational circuit is needed to generate the clear signal, and glitches may exist. Since the signal is connected to the asynchronous reset of the register, the register will be cleared to "0000" whenever a glitch occurs. Finally, because the asynchronous reset is used in normal operation, we cannot apply the timing analysis technique of Section 8.6. It is very difficult to determine the maximal operation clock rate for this design.

The remedy for this design is to load "0000" in a synchronous fashion. We can use a multiplexer to route "0000" or the incremented result to the input of the register. The code was discussed in Section 8.5.5 and is listed in Listing 9.2 for comparison. In terms of the circuit complexity, the synchronous design requires an additional 4-bit 2-to-1 multiplexer.

Listing 9.2 Decade counter using a synchronous clear signal

```
   architecture two_seg_arch of mod10_counter is
      signal r_reg: unsigned(3 downto 0);
      signal r_next: unsigned(3 downto 0);
   begin
5     -- register
      process(clk,reset)
      begin
         if (reset='1') then
            r_reg <= (others=>'0');
10       elsif (clk'event and clk='1') then
            r_reg <= r_next;
         end if;
      end process;
```

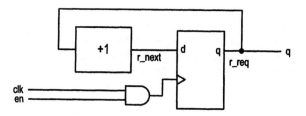

Figure 9.2   Disabling FF with a gated clock.

```
       — next—state logic
15     r_next <= (others=>'0') when r_reg=9 else
               r_reg + 1;
       — output logic
       q <= std_logic_vector(r_reg);
   end two_seg_arch;
```

9.1.2 Misuse of gated clocks

Correct operation of a synchronous circuit relies on an accurate clock signal. Since the clock signal needs to drive hundreds or even thousands of FFs, it uses a special distribution network and its treatment is very different from that of a regular signal. We should not manipulate the clock signal in RT-level design.

One bad RT-level design practice is to use a gated clock to suspend system operation, as shown in Figure 9.2. The intention of the design is to pause the counter operation by disabling the clock signal. The design suffers from several problems. First, since the enable signal, en, changes independent of the clock signal, the output pulse can be very narrow and cause the counter to malfunction. Second, if the en signal is not glitch-free, the glitches will be passed through the and cell and be treated as clock edges by the counter. Finally, since the and cell is included in the clock path, it may interfere with the construction and analysis of the clock distribution network.

The remedy for this design is to use a synchronous enable signal for the register, as discussed in Section 8.5.1. We essentially route the register output as a possible input. If the en signal is low, the same value is sampled and stored back to the register and the counter appears to be "paused." The VHDL codes for the original and revised designs are shown in Listings 9.3 and 9.4. In terms of the circuit complexity, the synchronous design requires an additional 2-to-1 multiplexer.

Listing 9.3   Binary counter with a gated clock

```
   library ieee;
   use ieee.std_logic_1164.all;
   use ieee.numeric_std.all;
   entity binary_counter is
5    port(
         clk, reset: in std_logic;
         en: in std_logic;
         q: out std_logic_vector(3 downto 0)
     );
10 end binary_counter;
```

```
    architecture gated_clk_arch of binary_counter is
       signal r_reg: unsigned(3 downto 0);
       signal r_next: unsigned(3 downto 0);
15     signal gated_clk: std_logic;
    begin
       -- register
       process(gated_clk,reset)
       begin
20        if (reset='1') then
             r_reg <= (others=>'0');
          elsif (gated_clk'event and gated_clk='1') then
             r_reg <= r_next;
          end if;
25     end process;
       -- gated clock
       gated_clk <= clk and en;
       -- next-state logic
       r_next <= r_reg + 1;
30     -- output logic
       q <= std_logic_vector(r_reg);
    end gated_clk_arch;
```

Listing 9.4 Binary counter with a synchronous enable signal

```
    architecture two_seg_arch of binary_counter is
       signal r_reg: unsigned(3 downto 0);
       signal r_next: unsigned(3 downto 0);
    begin
5      -- register
       process(clk,reset)
       begin
          if (reset='1') then
             r_reg <= (others=>'0');
10        elsif (clk'event and clk='1') then
             r_reg <= r_next;
          end if;
       end process;
       -- next-state logic
15     r_next <= r_reg + 1 when en='1' else
                 r_reg;
       -- output logic
       q <= std_logic_vector(r_reg);
    end two_seg_arch;
```

Power consumption is one important design criterion in today's digital system. A commonly used technique is to gate the clock to reduce the unnecessary transistor switching activities. However, this practice should not be done in RT-level code. The system should be developed and coded as a normal sequential circuit. After synthesis and verification, we can apply special power optimization software to replace the enable logic with a gated clock systematically.

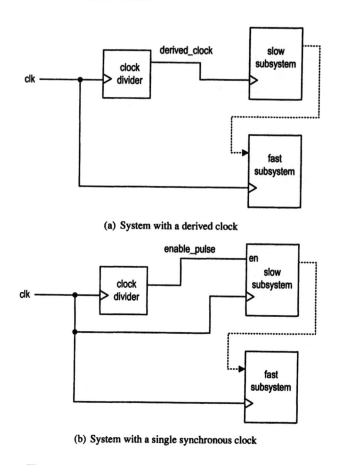

(a) System with a derived clock

(b) System with a single synchronous clock

Figure 9.3   System composed of fast and slow subsystems.

9.1.3   Misuse of derived clocks

A large digital system may consist of subsystems that operate in different paces. For example, a system may contain a fast processor and a relatively slow I/O subsystem. One way to accommodate the slow operation is to use a clock divider (i.e., a counter) to derive a slow clock for the subsystem. The block diagram of this approach is shown in Figure 9.3(a). There are several problems with this approach. The most serious one is that the system is no longer synchronous. If the two subsystems interact, as shown by the dotted line in Figure 9.3(a), the timing analysis becomes very involved. The simple timing model of Section 8.6 can no longer be applied and we must consider two clocks that have different frequencies and phases. Another problem is the placement and routing of the multiple clock signals. Since a clock signal needs a special driver and distribution network, adding derivative clock signals makes this process more difficult. A better alternative is to add a synchronous enable signal to the slow subsystem and drive the subsystem with the same clock signal. Instead of generating a derivative clock signal, the clock divider generates a low-rate single-clock enable pulse. This scheme is shown in Figure 9.3(b).

Let us consider a simple example. Assume that the system clock is 1 MHz and we want a timer that counts in minutes and seconds. The first design is shown in Figure 9.4(a). It first utilizes a mod-1000000 counter to generate a 1-Hz squared wave, which is used as a

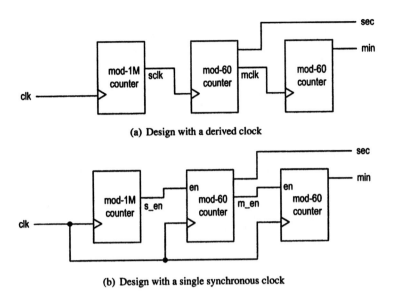

(a) Design with a derived clock

(b) Design with a single synchronous clock

Figure 9.4   Second and minute counter.

1-Hz clock to drive the second counter. The second counter is a mod-60 counter, which in turn generates a $\frac{1}{60}$-Hz signal to drive the clock of the minute counter. The VHDL code is shown in Listing 9.5. It consists of a mod-1000000 counter and two mod-60 counters. The output logic of the mod-1000000 counter and one mod-60 counter utilizes comparators to generate 50% duty-cycle pulses, which are used as the clocks in successive stages.

Listing 9.5   Second and minute counter with derived clocks

```
library ieee;
use ieee.std_logic_1164.all;
use ieee.numeric_std.all;
entity timer is
5    port(
         clk, reset: in std_logic;
         sec,min: out std_logic_vector(5 downto 0)
     );
end timer;

10
architecture multi_clock_arch of timer is
     signal r_reg: unsigned(19 downto 0);
     signal r_next: unsigned(19 downto 0);
     signal s_reg, m_reg: unsigned(5 downto 0);
15   signal s_next, m_next: unsigned(5 downto 0);
     signal sclk, mclk: std_logic;
begin
     -- register
     process(clk,reset)
20   begin
         if (reset='1') then
             r_reg <= (others=>'0');
         elsif (clk'event and clk='1') then
```

```
                  r_reg <= r_next;
25         end if;
      end process;
      -- next-state logic
      r_next <= (others=>'0') when r_reg=999999 else
                r_reg + 1;
30    -- output logic
      sclk <= '0' when r_reg < 500000 else
              '1';
      -- second divider
      process(sclk,reset)
35    begin
         if (reset='1') then
            s_reg <= (others=>'0');
         elsif (sclk'event and sclk='1') then
            s_reg <= s_next;
40       end if;
      end process;
      -- next-state logic
      s_next <= (others=>'0') when s_reg=59 else
                s_reg + 1;
45    -- output logic
      mclk <= '0' when s_reg < 30 else
              '1';
      sec <= std_logic_vector(s_reg);
      -- minute divider
50    process(mclk,reset)
      begin
         if (reset='1') then
            m_reg <= (others=>'0');
         elsif (mclk'event and mclk='1') then
55          m_reg <= m_next;
         end if;
      end process;
      -- next-state logic
      m_next <= (others=>'0') when m_reg=59 else
60             m_reg + 1;
      -- output logic
      min <= std_logic_vector(m_reg);
   end multi_clock_arch;
```

To convert the design to a synchronous circuit, we need to make two revisions. First, we add a synchronous enable signal for the mod-60 counter. The enable signal functions as the en signal discussed in examples in Section 8.5.1. When it is deasserted, the counter will pause and remain in the same state. Second, we have to replace the 50% duty cycle clock pulse with a one-clock-period enable pulse, which can be obtained by decoding a specific value of the counter. The revised diagram is shown in Figure 9.4(b), and the VHDL code is shown in Listing 9.6.

Listing 9.6 Second and minute counter with enable pulses

```
architecture single_clock_arch of timer is
   signal r_reg: unsigned(19 downto 0);
   signal r_next: unsigned(19 downto 0);
```

```
     signal s_reg, m_reg: unsigned(5 downto 0);
5    signal s_next, m_next: unsigned(5 downto 0);
     signal s_en, m_en: std_logic;
  begin
     -- register
     process(clk,reset)
10   begin
        if (reset='1') then
           r_reg <= (others=>'0');
           s_reg <= (others=>'0');
           m_reg <= (others=>'0');
15      elsif (clk'event and clk='1') then
           r_reg <= r_next;
           s_reg <= s_next;
           m_reg <= m_next;
        end if;
20   end process;
     -- next-state logic/output logic for mod-1000000 counter
     r_next <= (others=>'0') when r_reg=999999 else
                 r_reg + 1;
     s_en <= '1' when r_reg = 500000 else
25              '0';
     -- next state logic/output logic for second divider
     s_next <= (others=>'0') when (s_reg=59 and s_en='1') else
                 s_reg + 1      when s_en='1' else
                 s_reg;
30   m_en <= '1' when s_reg=30 and s_en='1' else
                 '0';
     -- next-state logic for minute divider
     m_next <= (others=>'0') when (m_reg=59 and m_en='1') else
                 m_reg + 1      when m_en='1' else
35               m_reg;
     -- output logic
     sec <= std_logic_vector(s_reg);
     min <= std_logic_vector(m_reg);
  end single_clock_arch;
```

9.2 COUNTERS

A counter can be considered as a circuit that circulates its internal state through a set of patterns. The patterns dictate the complexity of the next-state logic and the performance of the counter. Some applications require patterns with specific characteristics. We studied several counters in Sections 8.5. These counters are variations of the binary counter, which follows the basic binary counting sequence. This section introduces several other types of commonly used counters.

9.2.1 Gray counter

An n-bit Gray counter also circulates through all 2^n states. Its counting sequence follows the Gray code sequence, in which only one bit is changed between successive code words.

The design and VHDL description are similar to those of a binary counter except that we need to replace the binary incrementor with the Gray code incrementor of Section 7.5.1. The VHDL code of a 4-bit Gray counter is shown in Listing 9.7.

Listing 9.7   Gray counter

```
library ieee;
use ieee.std_logic_1164.all;
use ieee.numeric_std.all;
entity gray_counter4 is
5    port(
        clk, reset: in std_logic;
        q: out std_logic_vector(3 downto 0)
     );
  end gray_counter4;
10
  architecture arch of gray_counter4 is
     constant WIDTH: natural := 4;
     signal g_reg: unsigned(WIDTH-1 downto 0);
     signal g_next, b, b1: unsigned(WIDTH-1 downto 0);
15 begin
     -- register
     process(clk,reset)
     begin
        if (reset='1') then
20          g_reg <= (others=>'0');
        elsif (clk'event and clk='1') then
           g_reg <= g_next;
        end if;
     end process;
25   -- next-state logic
     -- Gray to binary
     b <= g_reg xor ('0' & b(WIDTH-1 downto 1));
     -- binary increment
     b1 <= b+1;
30   -- binary to Gray
     g_next <= b1 xor ('0' & b1(WIDTH-1 downto 1));
     -- output logic
     q <= std_logic_vector(g_reg);
  end arch;
```

9.2.2   Ring counter

A ring counter is constructed by connecting the serial-out port to the serial-in port of a shift register. The basic sketch of a 4-bit shift-to-right ring counter and its timing diagram are shown in Figure 9.5. After the "0001" pattern is loaded, the counter circulates through "1000", "0100", "0010" and "0001" states, and then repeats.

There are two methods of implementing a ring counter. The first method is to load the initial pattern during system initialization. Consider a 4-bit ring counter. We can set the counter to "0001" when the reset signal is asserted. After initialization, the reset signal is deasserted and the counter enters normal synchronous operation and circulates through the patterns. The VHDL code of a 4-bit ring counter is shown in Listing 9.8.

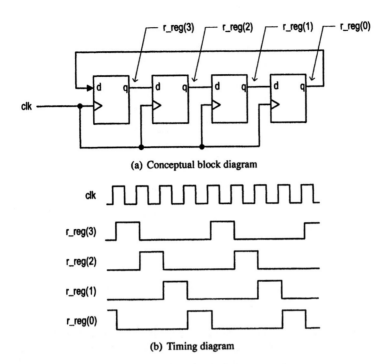

(a) Conceptual block diagram

(b) Timing diagram

Figure 9.5 Sketch of a 4-bit ring counter.

Listing 9.8 Ring counter using asynchronous initialization

```
library ieee;
use ieee.std_logic_1164.all;
entity ring_counter is
    port(
5       clk, reset: in std_logic;
        q: out std_logic_vector(3 downto 0)
    );
end ring_counter;

10 architecture reset_arch of ring_counter is
    constant WIDTH: natural := 4;
    signal r_reg: std_logic_vector(WIDTH-1 downto 0);
    signal r_next: std_logic_vector(WIDTH-1 downto 0);
begin
15  -- register
    process(clk,reset)
    begin
        if (reset='1') then
            r_reg <= (0=>'1', others=>'0');
20      elsif (clk'event and clk='1') then
            r_reg <= r_next;
        end if;
    end process;
    -- next-state logic
25  r_next <= r_reg(0) & r_reg(WIDTH-1 downto 1);
```

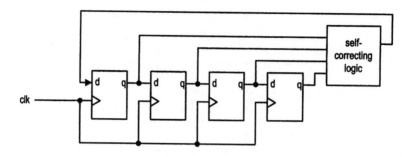

Figure 9.6 Block diagram of a self-correcting ring counter.

```
                        — output logic
                     q <= r_reg;
                  end reset_arch;
```

Note that the q_reg is initialized with "0001" by the statement

```
r_reg <= (0=>'1', others=>'0');
```

The alternative method is to utilize a self-correcting logic to feed the serial-in port with the correct pattern. The block diagram of a 4-bit self-correcting ring counter is shown in Figure 9.6. The design is based on the observation that a '1' can only be shifted into the shift register if the current three MSBs of the register are "000". If any of the three MSBs is not '0', the self-correcting logic generates a '0' and shifts it into the register. This process continues until the three MSBs become "000" and a '1' is shifted in afterward. Note that this scheme works even when the register contains an invalid pattern initially. For example, if the initial value of the register is "1101", the logic will gradually shift in 0's and return to the normal circulating sequence. Because of this property, the circuit is known as *self-correcting*.

The VHDL code for this design is shown in Listing 9.9. Note that no special input pattern is needed during system initialization, and the all-zero pattern is used in the code.

Listing 9.9   Ring counter using self-correcting logic

```
architecture self_correct_arch of ring_counter is
    constant WIDTH: natural := 4;
    signal r_reg, r_next: std_logic_vector(WIDTH-1 downto 0);
    signal s_in: std_logic;
5 begin
    — register
    process(clk,reset)
    begin
        if (reset='1') then
10          r_reg <= (others=>'0');
        elsif (clk'event and clk='1') then
            r_reg <= r_next;
        end if;
    end process;
15  — next-state logic
    s_in <= '1' when r_reg(WIDTH-1 downto 1)="000" else
            '0';
```

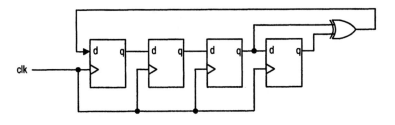

Figure 9.7 Block diagram of a 4-bit LFSR.

```
      r_next <= s_in & r_reg(WIDTH-1 downto 1);
      -- output logic
 20   q <= r_reg;
    end self_correct_arch;
```

In a ring counter, an n-bit register can generate only n states, which is much smaller than the possible 2^n states of a binary counter. Despite its inefficiency, a ring counter offers several benefits. First, each bit of a ring counter is in the 1-out-of-n format. It requires no decoding logic and is glitch-free. Second, the output of a ring counter is out of phase, and the n output bits of an n-bit ring counter form a set of n-phase signals. For example, in the timing diagram of the 4-bit ring counter, each bit is activated for one-fourth of the period and only one bit is activated at a particular phase. Finally, the ring counter is extremely fast. For the reset_arch architecture, the next-state logic consists only of connection wires. If we assume that the wiring delay is negligible, the maximal clock rate becomes $\frac{1}{T_{cq}+T_{setup}}$, which is the fastest clock rate that can be achieved by a sequential circuit for a given technology.

9.2.3 LFSR (linear feedback shift register)

The *linear feedback shift register* (*LFSR*) is a shift register that utilizes a special feedback circuit to generate the serial input value. The feedback circuit is essentially the next-state logic. It performs xor operation on certain bits of the register and forces the register to cycle through a set of unique states. In a properly designed n-bit LFSR, we can use a few xor gates to force the register to circulate through $2^n - 1$ states. The diagram of a 4-bit LFSR is shown in Figure 9.7. The two LSB signals of the register are xored to generate a new value, which is fed back to the serial-in port of the shift register. Assume that the initial state of register is "1000". The circuit will circulate through the 15 (i.e., $2^4 - 1$) states as follows: "1000", "0100", "0010", "1001", "1100", "0110", "1011", "0101", "1010", "1101", "1110", "1111", "0111", "0011", "0001".

Note that the "0000" state is not included and constitutes the only missing state. If the LFSR enters this state accidentally, it will be stuck in this state.

The construction of LFSRs is based on the theoretical study of finite fields. The term *linear* comes from the fact that the general feedback equation of an LFSR is described by an expression of the *and* and *xor* operators, which form a linear system in algebra. The theoretical study shows some interesting properties of LFSRs:

- An n-bit LFSR can cycle through up to $2^n - 1$ states. The all-zero state is excluded from the sequence.
- A feedback circuit to generate maximal number of states exists for any n.

Table 9.1 Feedback expression of LFSR

Register size	Feedback expression
2	$q_1 \oplus q_0$
3	$q_1 \oplus q_0$
4	$q_1 \oplus q_0$
5	$q_2 \oplus q_0$
6	$q_1 \oplus q_0$
7	$q_3 \oplus q_0$
8	$q_4 \oplus q_3 \oplus q_2 \oplus q_0$
16	$q_5 \oplus q_4 \oplus q_3 \oplus q_0$
32	$q_{22} \oplus q_2 \oplus q_1 \oplus q_0$
64	$q_4 \oplus q_3 \oplus q_1 \oplus q_0$
128	$q_{29} \oplus q_{17} \oplus q_2 \oplus q_0$

- The sequence generated by the feedback circuit is *pseudorandom*, which means that the sequence exhibits a certain statistical property and appears to be random.

The feedback circuit depends on the number of bits of the LFSR and is determined on an ad hoc basis. Despite its irregular pattern, the feedback expressions are very simple, involving either one or three xor operators most of the time. Table 9.1 lists the feedback expressions for register sizes between 2 and 8 as well as several larger values. We assume that the output of the n-bit shift register is $q_{n-1}, q_{n-2}, \ldots, q_1, q_0$. The result of the feedback expression is to be connected to the serial-in port of the shift register (i.e., the input of the $(n-1)$th FF).

Once we know the feedback expression, the coding of LFSR is straightforward. The VHDL code for a 4-bit LFSR is shown in Listing 9.10. Note that the LFSR cannot be initialized with the all-zero pattern. In pseudo number generation, the initial value of the sequence is known as a *seed*. We use a constant to define the initial value and load it into the LFSR during system initialization.

Listing 9.10 LFSR

```
library ieee;
use ieee.std_logic_1164.all;
entity lfsr4 is
   port(
s      clk, reset: in std_logic;
       q: out std_logic_vector(3 downto 0)
   );
end lfsr4;

10 architecture no_zero_arch of lfsr4 is
   signal r_reg, r_next: std_logic_vector(3 downto 0);
   signal fb: std_logic;
   constant SEED: std_logic_vector(3 downto 0):="0001";
begin
15   -- register
   process(clk,reset)
   begin
       if (reset='1') then
```

```
              r_reg <= SEED;
20         elsif (clk'event and clk='1') then
              r_reg <= r_next;
          end if;
      end process;
      -- next-state logic
25    fb <= r_reg(1) xor r_reg(0);
      r_next <= fb & r_reg(3 downto 1);
      -- output logic
      q <= r_reg;
   end no_zero_arch;
```

The unique properties of LFSR make them useful in a variety of applications. The first type of application utilizes its pseudorandomness property to scramble and descramble data, as in testing, encryption and modulation. The second type takes advantages of its simple combinational feedback circuit. For example, we can use just three xor gates in a 32-bit LFSR to cycle through $2^{32} - 1$ states. By comparison, we need a fairly large 32-bit incrementor to cycle through 2^{32} states in a binary counter. By the component information of Table 6.2, the gate counts for three xor gates and a 32-bit incrementor are 9 and 113 respectively, and their propagation delays are 0.8 and 11.6 ns respectively. Thus, an LFSR can replace other counters for applications in which the order of the counting states is not important. This is a clever design technique. For example, we can use three xor gates to implement a 128-bit LFSR, and it takes about 10^{12} years for a 100-GHz system to circulate all the possible $2^{128} - 1$ states.

The all-zero state is excluded in a pure LFSR. It is possible to use an additional circuit to insert the all-zero state into the counting sequence so that an n-bit LFSR can circulate through all 2^n states. This scheme is based on the following observation. In any LFSR, a '1' will be shifted in after the "00$\cdots$01" state since the all-zero state is not possible. In other words, the feedback value will be '1' when the $n - 1$ MSBs are 0's and the state following "00$\cdots$01" will always be "10$\cdots$00". The revised design will insert the all-zero state, "00$\cdots$00", between the "00$\cdots$01" and "10$\cdots$00" states. Let the output of an n-bit shift register be $q_{n-1}, q_{n-2}, \ldots, q_1, q_0$ and the original feedback signal of the LFSR be f_b. The modified feedback value f_{zero} has the expression

$$f_{zero} = f_b \oplus (q'_{n-1} \cdot q'_{n-2} \cdots q'_2 \cdot q'_1)$$

This expression can be analyzed as follows:

- The expression $q'_{n-1} \cdot q'_{n-2} \cdots q'_2 \cdot q'_1$ indicates the condition that the $n - 1$ MSBs are 0's. This condition can only be true when the LFSR is in "00$\cdots$01" or "00$\cdots$00" state.
- If the condition above is false, the value of f_{zero} is f_b since $f_b \oplus 0 = f_b$. This implies that the circuit will shift in a regular feedback value and follow the original sequence except for the "00$\cdots$01" or "00$\cdots$00" states.
- If the current state of the register is "00$\cdots$01", the value of f_b should be '1' and the expression $f_b \oplus (q'_{n-1} \cdot q'_{n-2} \cdots q'_2 \cdot q'_1)$ becomes $1 \oplus 1$. Thus, a '0' will be shifted into the register at the next rising edge of the clock and the next state will be "00$\cdots$00".
- If the current state of the register is "00$\cdots$00", the value of f_b should be '0' and the expression $f_b \oplus (q'_{n-1} \cdot q'_{n-2} \cdots q'_2 \cdot q'_1)$ becomes $0 \oplus 1$. Thus, a '1' will be shifted into the register at the next rising edge of the clock and the next state will be "10$\cdots$00".
- Once the shift register reaches "10$\cdots$00", it returns to the regular LFSR sequence.

The analysis clearly shows that the modified feedback circuit can insert the all-zero state between the "00 $\cdots$ 01" and "10 $\cdots$ 00" states. Technically, the and operation of the revised feedback expression destroys the "linearity," and thus the circuit is no longer a linear feedback shift register. The modified design is sometimes known as a *Bruijn counter*.

Once understanding the form and operation of the modified feedback expression, we can easily incorporate it into the VHDL code. The revised code is shown in Listing 9.11. We use the statement

```
zero <= '1' when r_reg(3 downto 1)="000" else
        '0';
```

to obtain the result of the $q'_{n-1} \cdot q'_{n-2} \cdots q'_2 \cdot q'_1$ expression. Note that the all-zero state can be loaded into the register during system initialization.

Listing 9.11   LFSR with the all-zero state

```
architecture with_zero_arch of lfsr4 is
   signal r_reg, r_next: std_logic_vector(3 downto 0);
   signal fb, zero, fzero: std_logic;
   constant seed: std_logic_vector(3 downto 0):="0000";
 5 begin
   -- register
   process(clk,reset)
   begin
      if (reset='1') then
10       r_reg <= seed;
      elsif (clk'event and clk='1') then
         r_reg <= r_next;
      end if;
   end process;
15 -- next-state logic
   fb <= r_reg(1) xor r_reg(0);
   zero <= '1' when r_reg(3 downto 1)="000" else
           '0';
   fzero <= zero xor fb;
20 r_next <= fzero & r_reg(3 downto 1);
   -- output logic
   q <= r_reg;
end with_zero_arch;
```

9.2.4 Decimal counter

A decimal counter circulates the patterns in binary-coded decimal (BCD) format. The BCD code uses 4 bits to represent a decimal number. For example, the BCD code for the three-digit decimal number 139 is "0001 0011 1001". The decimal counter follows the decimal counting sequence and the number following 139 is 140, which is represented as "0001 0100 0000".

One possible way to construct a decimal counter is to design a BCD incrementor for the next-state logic, just like a regular incrementor in a binary counter. Because of the cumbersome implementation of the BCD incrementor, this method is not efficient. A better alternative is to divide the counter into stages of decade counters and use special enable logic to control the increment of the individual decade counters.

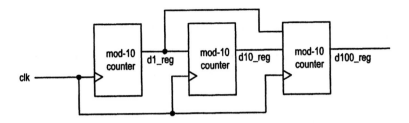

Figure 9.8 Block diagram of a BCD counter.

Consider a 3-digit (12-bit) decimal counter that counts from 000 to 999 and then repeats. It can be implemented by cascading three special decade counters, and the sketch is shown in Figure 9.8. The leftmost decade counter represents the least significant decimal digit. It is a regular mod-10 counter that counts from 0 to 9 (i.e., from "0000" to "1001") and repeats. The middle decade counter is a mod-10 counter with a special enable circuit. It increments only when the least significant decimal digit reaches 9. The rightmost decade counter represents the most significant decimal digit, and it increments only when the two least significant decimal digits are equal to 99. Note that when the counter reaches 999, it will return to 000 at the next rising edge of the clock. The VHDL codes are shown in Listings 9.12 and 9.13. The former uses three conditional signal assignment statements and the latter uses a nested if statement to check whether the counter reaches --9, -99 or 999.

Listing 9.12 Three-digit decimal counter using conditional concurrent statements

```
library ieee;
use ieee.std_logic_1164.all;
use ieee.numeric_std.all;
entity decimal_counter is
5    port(
         clk, reset: in std_logic;
         d1, d10, d100: out std_logic_vector(3 downto 0)
    );
end decimal_counter;
10
architecture concurrent_arch of decimal_counter is
    signal d1_reg, d10_reg, d100_reg: unsigned(3 downto 0);
    signal d1_next, d10_next, d100_next: unsigned(3 downto 0);
begin
15  -- register
    process(clk,reset)
    begin
        if (reset='1') then
            d1_reg <= (others=>'0');
20          d10_reg <= (others=>'0');
            d100_reg <= (others=>'0');
        elsif (clk'event and clk='1') then
            d1_reg <= d1_next;
            d10_reg <= d10_next;
25          d100_reg <= d100_next;
        end if;
    end process;
```

```
     -- next-state logic
     d1_next <= "0000" when d1_reg=9 else
30             d1_reg + 1;
     d10_next <= "0000" when (d1_reg=9 and d10_reg=9) else
                d10_reg + 1 when d1_reg=9 else
                d10_reg;
     d100_next <=
35     "0000" when (d1_reg=9 and d10_reg=9 and d100_reg=9) else
       d100_reg + 1 when (d1_reg=9 and d10_reg=9) else
       d100_reg;
     -- output
     d1 <= std_logic_vector(d1_reg);
40   d10 <= std_logic_vector(d10_reg);
     d100 <= std_logic_vector(d100_reg);
  end concurrent_arch;
```

Listing 9.13 Three-digit decimal counter using a nested if statement

---

```
architecture if_arch of decimal_counter is
   signal d1_reg, d10_reg, d100_reg: unsigned(3 downto 0);
   signal d1_next, d10_next, d100_next: unsigned(3 downto 0);
begin
5    -- register
   process(clk,reset)
   begin
      if (reset='1') then
         d1_reg <= (others=>'0');
10        d10_reg <= (others=>'0');
         d100_reg <= (others=>'0');
      elsif (clk'event and clk='1') then
         d1_reg <= d1_next;
         d10_reg <= d10_next;
15        d100_reg <= d100_next;
      end if;
   end process;
   -- next-state logic
   process(d1_reg,d10_reg,d100_reg)
20   begin
      d10_next <= d10_reg;
      d100_next <= d100_reg;
      if d1_reg/=9 then
         d1_next <= d1_reg + 1;
25    else -- reach --9
         d1_next <= "0000";
         if d10_reg/=9 then
            d10_next <= d10_reg + 1;
         else -- reach -99
30           d10_next <= "0000";
            if d100_reg/=9 then
               d100_next <= d100_reg + 1;
            else -- reach 999
               d100_next <= "0000";
35           end if;
```

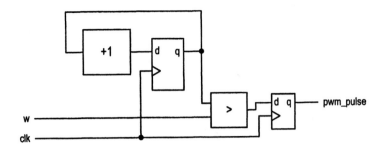

Figure 9.9 Block diagram of a PWM circuit.

```
        end if;
      end if;
    end process;
    -- output
40  d1 <= std_logic_vector(d1_reg);
    d10 <= std_logic_vector(d10_reg);
    d100 <= std_logic_vector(d100_reg);
  end if_arch;
```

9.2.5 Pulse width modulation circuit

Instead of using the counting patterns directly, some applications generate output signals based on the state of the counter. One example is a pulse width modulation (PWM) circuit. In a square wave, the *duty cycle* is defined as the percentage of time that the signal is asserted as '1' in a period. For example, the duty cycle of a symmetric square wave is 50% since the signal is asserted half of the period. A PWM circuit generates an output pulse with an adjustable duty cycle. It is frequently used to control the on–off time of an external system.

Consider a PWM circuit whose duty cycle can be adjusted in increments of $\frac{1}{16}$, i.e., the duty cycle can be $\frac{1}{16}, \frac{2}{16}, \frac{3}{16}, \ldots, \frac{15}{16}, \frac{16}{16}$. A 4-bit control signal, w, which is interpreted as an unsigned integer, specifies the desired duty cycle. The duty cycle will be $\frac{16}{16}$ when w is "0000", and will be $\frac{w}{16}$ otherwise. This circuit can be implemented by a mod-16 counter with a special output circuit, and the conceptual diagram is shown in Figure 9.9.

The mod-16 counter circulates through 16 patterns. An output circuit compares the current pattern with the w signal and asserts the output pulse when the counter's value is smaller than w. The output pulse's period is 16 times the clock period, and $\frac{w}{16}$ of the period is asserted. The VHDL code is shown in Listing 9.14. Note that an additional Boolean expression, w="0000", is included to accommodate the special condition. We also add an output buffer to remove any potential glitch.

Listing 9.14 PWM circuit

```
  library ieee;
  use ieee.std_logic_1164.all;
  use ieee.numeric_std.all;
  entity pwm is
5   port(
        clk, reset: in std_logic;
        w: in std_logic_vector(3 downto 0);
```

```
            pwm_pulse: out std_logic
        );
10  end pwm;

    architecture two_seg_arch of pwm is
        signal r_reg: unsigned(3 downto 0);
        signal r_next: unsigned(3 downto 0);
15      signal buf_reg: std_logic;
        signal buf_next: std_logic;
    begin
        -- register & output buffer
        process(clk,reset)
20      begin
            if (reset='1') then
                r_reg <= (others=>'0');
                buf_reg <= '0';
            elsif (clk'event and clk='1') then
25              r_reg <= r_next;
                buf_reg <= buf_next;
            end if;
        end process;
        -- next-state logic
30      r_next <= r_reg + 1;
        -- output logic
        buf_next <=
            '1' when (r_reg<unsigned(w)) or (w="0000") else
            '0';
35      pwm_pulse <= buf_reg;
    end two_seg_arch;
```

9.3 REGISTERS AS TEMPORARY STORAGE

Instead of being dedicated to a specific circuit, such as a counter, registers can also be used as general-purpose storage or buffer to store data. Since the circuit size of a D FF is several times larger than that of a RAM cell, using registers as massive storage is not cost-effective. They are normally used to construct small, fast temporal storage in a large digital system. This section examines various storage structures, including a register file, register-based first-in-first-out buffer and register-based look-up table.

9.3.1 Register file

A register file consists of a set of registers. To reduce the amount of wiring and I/O signals, the register file provides only one write port and few read ports, which are shared by all registers for data access. Each register is assigned a binary address as an identifier, and an external system uses the address to specify which register is to be involved in the operation. The storage and retrieval operations are known as the *write* and *read* operations respectively. A processor normally.includes a register file as fast temporary storage.

In this subsection, we illustrate the design and coding of a register file with four 16-bit registers and three I/O ports, which include one write port and two read ports. The data signals are labeled w_data, r_data0 and r_data1, and the port addresses are labeled

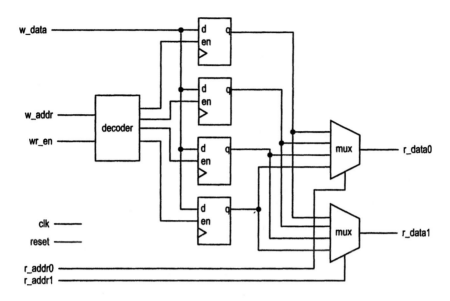

Figure 9.10   Block diagram of a register file.

w_addr, r_addr0 and r_addr1. There is also a control signal, wr_en, which is the write enable signal to indicate whether a write operation can be performed.

The conceptual diagram is shown in Figure 9.10. The design consists of three major parts: registers with enable signals, a write decoding circuit, and read multiplexing circuits. There are four 16-bit registers, each register with an individual enable signal, en. The en signal is synchronous and indicates whether the input data can be stored into the register. Its function is identical to the FF example in Section 8.5.1.

The write decoding circuit examines the wr_en signal and decodes the write port address. If the wr_en signal is asserted, the decoding circuit functions as a regular 2-to-2^2 binary decoder that asserts one of the four en signals of the corresponding register. The w_data signal will be sampled and stored into the corresponding register at the rising edge of the clock.

The read multiplexing circuit consists of two 4-to-1 multiplexers. It utilizes r_addr0 and r_addr1 as the selection signals to route the desired register outputs to the read ports.

Note that the registers are structured as a two-dimensional 4-by-16 array of D FFs and would best be represented by a two-dimensional data type. There is no predefined two-dimensional data type in the IEEE std_logic_1164 package, and thus we must create a user-defined data type. One way to do it is to create a user-defined "array of arrays" data type. Assume that the number of rows and columns of an array are ROW and COL respectively. The data type and signal declaration can be written as

```
type aoa_type is array (ROW-1 downto 0) of
       std_logic_vector(COL-1 downto 0);
signal s: aoa_type;
```

We can use s[i] to access the ith row of the array and use s[i][j] to access the jth bit of the ith row of the array.

Once understanding the basic block diagram and data type, we can derive the VHDL code accordingly. The VHDL code is shown in Listing 9.15. The register and corresponding enabling circuit are described by two processes. The decoding circuit is described in another process. If wr_en is not asserted, en will be "0000" and no register will be updated. Otherwise, one bit of the en signal will asserted according to the value of the w_addr signal. The read ports are described as two multiplexers.

Listing 9.15 2^2-by-16 register file

```
   library ieee;
   use ieee.std_logic_1164.all;
   entity reg_file is
      port(
5        clk, reset: in std_logic;
         wr_en: in std_logic;
         w_addr: in std_logic_vector(1 downto 0);
         w_data: in std_logic_vector(15 downto 0);
         r_addr0, r_addr1: in std_logic_vector(1 downto 0);
10       r_data0, r_data1: out std_logic_vector(15 downto 0)
      );
   end reg_file;

   architecture no_loop_arch of reg_file is
15    constant W: natural:=2; -- number of bits in address
      constant B: natural:=16; -- number of bits in data
      type reg_file_type is array (2**W-1 downto 0) of
           std_logic_vector(B-1 downto 0);
      signal array_reg: reg_file_type;
20    signal array_next: reg_file_type;
      signal en: std_logic_vector(2**W-1 downto 0);
   begin
      -- register
      process(clk,reset)
25    begin
         if (reset='1') then
            array_reg(3) <= (others=>'0');
            array_reg(2) <= (others=>'0');
            array_reg(1) <= (others=>'0');
30          array_reg(0) <= (others=>'0');
         elsif (clk'event and clk='1') then
            array_reg(3) <=  array_next(3);
            array_reg(2) <=  array_next(2);
            array_reg(1) <=  array_next(1);
35          array_reg(0) <=  array_next(0);
         end if;
      end process;
      -- enable logic for register
      process(array_reg,en,w_data)
40    begin
         array_next(3) <= array_reg(3);
         array_next(2) <= array_reg(2);
         array_next(1) <= array_reg(1);
         array_next(0) <= array_reg(0);
45       if en(3)='1' then
            array_next(3) <= w_data;
         end if;
```

```
            if en(2)='1' then
                array_next(2) <= w_data;
50          end if;
            if en(1)='1' then
                array_next(1) <= w_data;
            end if;
            if en(0)='1' then
55              array_next(0) <= w_data;
            end if;
        end process;
        -- decoding for write address
        process(wr_en,w_addr)
60      begin
            if (wr_en='0') then
                en <= (others=>'0');
            else
                case w_addr is
65                  when "00" =>    en <= "0001";
                    when "01" =>    en <= "0010";
                    when "10" =>    en <= "0100";
                    when others => en <= "1000";
                end case;
70          end if;
        end process;
        -- read multiplexing
        with r_addr0 select
            r_data0 <=  array_reg(0) when "00",
75                      array_reg(1) when "01",
                        array_reg(2) when "10",
                        array_reg(3) when others;
        with r_addr1 select
            r_data1 <=  array_reg(0) when "00",
80                      array_reg(1) when "01",
                        array_reg(2) when "10",
                        array_reg(3) when others;
    end no_loop_arch;
```

Although the description is straightforward, the code is not very compact. The code will be cumbersome and lengthy for a larger register file. A more effective description and the proper use of two-dimensional data types are discussed in Chapter 15.

9.3.2 Register-based synchronous FIFO buffer

A first-in-first-out (FIFO) buffer acts as "elastic" storage between two subsystems. The conceptual diagram is shown in Figure 9.11. One subsystem stores (i.e., writes) data into the buffer, and the other subsystem retrieves (i.e., reads) data from the buffer and removes it from the buffer. The order of data retrieval is same as the order of data being stored, and thus the buffer is known as a *first-in-first-out buffer*. If two subsystems are synchronous (i.e., driven by the same clock), we need only one clock for the FIFO buffer and it is known as a *synchronous* FIFO buffer.

The most common way to construct a FIFO buffer is to add a simple control circuit to a generic memory array, such as a register file or RAM. We can arrange the generic memory

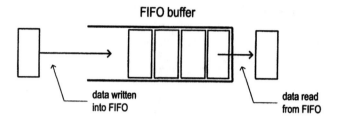

Figure 9.11   Conceptual diagram of a FIFO buffer.

array as a circular queue and use two pointers to mark the beginning and end of the FIFO buffer. The conceptual sketch is shown in Figure 9.12(c). The first pointer, known as the *write pointer* (labeled wr ptr), points to the first empty slot in front of the buffer. During a write operation, the data is stored in this designed slot, and the write pointer advances to the next slot (i.e., incremented by 1). The second pointer, known as the *read pointer* (labeled rd ptr), points to the end of the buffer. During a read operation, the data is retrieved and the read pointer advances one slot, effectively releasing the slot for future write operations.

Figure 9.12 shows a sequence of write and read operations and the corresponding growth and shrinking of the buffer. Initially, both the read and write pointers point to the 0 address, as in Figure 9.12(a). Since the buffer is empty, no read operation is allowed at this time. After a write operation, the write pointer increments and the buffer contains one item in the 0 address, as in Figure 9.12(b). After a few more write operations, the write pointer continues to increase and the buffer expands accordingly, as in Figure 9.12(c). A read operation is performed afterward. The read pointer advances in the same direction, and the previous slot is released, as in Figure 9.12(d). After several more write operations, the buffer is full, as in Figure 9.12(f), and no write operation is allowed. Several read operations are then performed, and the buffer eventually shrinks to 0, as in Figure 9.12(g), (h) and (i).

The block diagram of a register-based FIFO is shown in Figure 9.13. It consists of a register file and a control circuit, which generates proper read and write pointer values and status signals. Note that the FIFO buffer doesn't have any explicit external address signal. Instead, it utilizes two control signals, wr and re, for write and read operations. At the rising edge of the clock, if the wr signal is asserted and the buffer is not full, the corresponding input data will be sampled and stored into the buffer. The output data from the FIFO is always available. The re signal might better be interpreted as a "remove" signal. If it is asserted at the rising edge and the buffer is not empty, the FIFO's read pointer advances one position and makes the current slot available. After the internal delays of the incrementing and routing, new output data is available in FIFO's output port.

During FIFO operation, an overflow occurs when the external system attempts to write new data when the FIFO is full, and an underflow occurs when the external system attempts to read (i.e., remove) a slot when the FIFO is empty. To ensure correct operation, a FIFO buffer must include the full and empty status signals for the two special conditions. In a properly designed system, the external systems should check the status signals before attempting to access the FIFO.

The major components of a FIFO control circuit are two counters, whose outputs function as write and read pointers respectively. During regular operation, the write counter advances one position when the wr signal is asserted at the rising edge of the clock, and the read counter advances one position when the re signal is asserted. We normally prefer to add

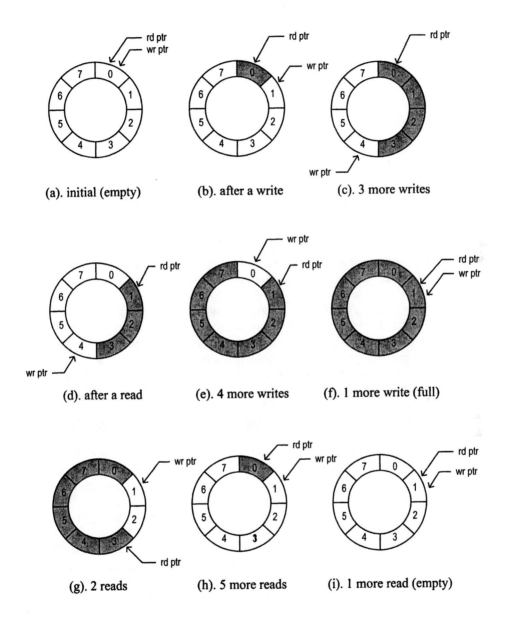

Figure 9.12 Circular-queue implementation of a FIFO buffer.

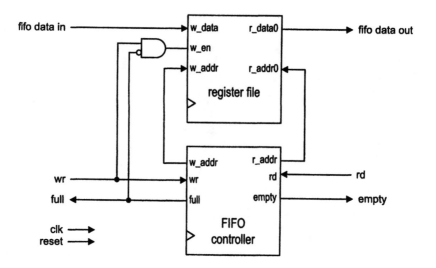

Figure 9.13   Block diagram of a register-based FIFO buffer.

some safety precautions to ensure that data will not be written into a full buffer or removed from an empty buffer. Under these conditions, the counters will retain the previous values.

The difficult part of the control circuit is the handling of two special conditions in which the FIFO buffer is empty or full. When the FIFO buffer is empty, the read pointer is the same as the write pointer, as shown in Figure 9.12(a) and (i). Unfortunately, this is also the case when the FIFO buffer is full, as shown in Figure 9.12(f). Thus, we cannot just use read and write pointers to determine full and empty conditions. There are several schemes to generate the status signals, and all of them involve additional circuitry and FFs. We examine two schemes in this subsection.

FIFO control circuit with augmented binary counters   The first method is to use the binary counters for the read and write pointers and increase their sizes by 1 bit. We can determine the full or empty condition by comparing the MSBs of the two pointers. This scheme can be best explained and observed by an example. Consider a FIFO with 3-bit address (i.e., 2^3 words). Two 4-bit counters will be used for the read and write pointers. The counters and the status of a sequence of operations are shown in Table 9.2. The three LSBs of the read and write pointers are used as addresses to access the register file and wrap around after eight increments. They are equal when the FIFO is empty or full. The MSBs of the read and write pointers can be used to distinguish the two conditions. The two bits are the same when the FIFO is empty. After eight write operations, the MSB of the write pointers flips and becomes the opposite of the MSB of the read pointer. The opposite values in MSBs indicate that the FIFO is full. After eight read operations, the MSB of the read pointer flips and becomes identical to the MSB of the write pointer, which indicates that the FIFO is empty again. A more detailed block diagram of this scheme is shown in Figure 9.14.

The VHDL code of a 4-word FIFO controller is shown in Listing 9.16. A constant, N, is used inside the architecture body to indicate the number of address bits. Note that the w_ptr_reg and r_ptr_reg signals, which are the write and read pointers, are increased to N + 1 bits.

Table 9.2 Representative sequence of FIFO operations

Write pointer	Read pointer	Operation	Status
0 000	0 000	initialization	empty
0 111	0 000	after 7 writes	
1 000	0 000	after 1 write	full
1 000	0 100	after 4 reads	
1 100	0 100	after 4 writes	full
1 100	1 011	after 7 reads	
1 100	1 100	after 1 read	empty
0 011	1 100	after 7 writes	
0 100	1 100	after 1 write	full
0 100	0 100	after 8 reads	empty

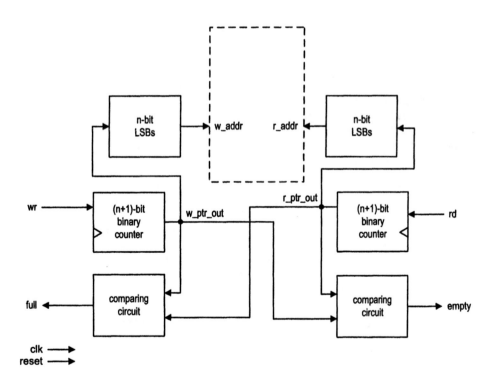

Figure 9.14 Detailed block diagram of an augmented-binary-counter FIFO control circuit.

Listing 9.16   FIFO control circuit with augmented binary counters

```vhdl
library ieee;
use ieee.std_logic_1164.all;
use ieee.numeric_std.all;
entity fifo_sync_ctrl4 is
    port(
        clk, reset: in std_logic;
        wr, rd: in std_logic;
        full, empty: out std_logic;
        w_addr, r_addr: out std_logic_vector(1 downto 0)
    );
end fifo_sync_ctrl4;

architecture enlarged_bin_arch of fifo_sync_ctrl4 is
    constant N: natural:=2;
    signal w_ptr_reg, w_ptr_next: unsigned(N downto 0);
    signal r_ptr_reg, r_ptr_next: unsigned(N downto 0);
    signal full_flag, empty_flag: std_logic;
begin
    -- register
    process(clk,reset)
    begin
        if (reset='1') then
            w_ptr_reg <= (others=>'0');
            r_ptr_reg <= (others=>'0');
        elsif (clk'event and clk='1') then
            w_ptr_reg <= w_ptr_next;
            r_ptr_reg <= r_ptr_next;
        end if;
    end process;
    -- write pointer next-state logic
    w_ptr_next <=
        w_ptr_reg + 1 when wr='1' and full_flag='0' else
        w_ptr_reg;
    full_flag <=
        '1' when r_ptr_reg(N) /=w_ptr_reg(N) and
                 r_ptr_reg(N-1 downto 0)=w_ptr_reg(N-1 downto 0)
            else
        '0';
    -- write port output
    w_addr <= std_logic_vector(w_ptr_reg(N-1 downto 0));
    full <= full_flag;
    -- read pointer next-state logic
    r_ptr_next <=
        r_ptr_reg + 1 when rd='1' and empty_flag='0' else
        r_ptr_reg;
    empty_flag <= '1' when r_ptr_reg=w_ptr_reg else
                     '0';
    -- read port output
    r_addr <= std_logic_vector(r_ptr_reg(N-1 downto 0));
    empty <= empty_flag;
end enlarged_bin_arch;
```

To complete the FIFO buffer, we combine the control circuit and the register file, as shown in Figure 9.13. This can be done by merging the previous register file VHDL code with the FIFO controller code. A more systematic approach is to use component instantiation, which is discussed in Chapter 13.

FIFO control circuit with status FFs   An alternative design of a FIFO control circuit is to keep track of the state of the empty and full conditions and to use this information, combined with the wr and rd signals, to determine the new conditions. This scheme does not require augmented counters but needs two extra FFs to record the empty and full statuses. During system initialization, the full status FF is set to '0' and the empty status FF is set to '1'. After initialization, the wr and rd signals are examined at the rising edge of the clock, and the pointers and the FFs are modified according to the following rules:

- wr and rd are "00": Since no operation is specified, pointers and FFs remain in the previous state.
- wr and rd are "11": Write and read operations are performed simultaneously. Since the net size of the buffer remains the same, the empty and full conditions will not change. Both pointers advance one position.
- wr and rd are "10": This indicates that only a write operation is performed. We must first make sure that the buffer is not full. If that is the case, the write pointer advances one position and the empty status FF should be deasserted. The advancement may make the buffer full. This condition happens if the *next value* of the write pointer is equal to the current value of the read pointer (i.e., the write pointer catches up to the read pointer). If this condition is true, the full status FF will be set to '1' accordingly.
- wr and rd are "01": This indicates that only a read operation is performed. We must first make sure that the buffer is not empty. If that is the case, the read pointer advances one position and the full status FF should be deasserted. The advancement may make the buffer empty. This condition happens if the *next value* of the read pointer is equal to the current value of the write pointer (i.e., the read pointer catches up to the write pointer). If this condition is true, the empty status FF will be set to '1' accordingly.

The VHDL code for this scheme is shown in Listing 9.17. In this code, we combine the next-state logic of the pointers and FFs into a single process and use a case statement to implement the desired operations under various wr and rd combinations.

Listing 9.17   FIFO controller with status FFs

```
architecture lookahead_bin_arch of fifo_sync_ctrl4 is
   constant N: natural:=2;
   signal w_ptr_reg, w_ptr_next: unsigned(N-1 downto 0);
   signal w_ptr_succ: unsigned(N-1 downto 0);
   signal r_ptr_reg, r_ptr_next: unsigned(N-1 downto 0);
   signal r_ptr_succ: unsigned(N-1 downto 0);
   signal full_reg, empty_reg: std_logic;
   signal full_next, empty_next: std_logic;
   signal wr_op: std_logic_vector(1 downto 0);
begin
   -- register
   process(clk,reset)
   begin
      if (reset='1') then
         w_ptr_reg <= (others=>'0');
```

```vhdl
                  r_ptr_reg <= (others=>'0');
              elsif (clk'event and clk='1') then
                  w_ptr_reg <= w_ptr_next;
                  r_ptr_reg <= r_ptr_next;
20            end if;
       end process;
       -- status FF
       process(clk,reset)
       begin
25            if (reset='1') then
                  full_reg <= '0';
                  empty_reg <= '1';
              elsif (clk'event and clk='1') then
                  full_reg <= full_next;
30                empty_reg <= empty_next;
              end if;
       end process;
       -- successive value for the write and read pointers
       w_ptr_succ <= w_ptr_reg + 1;
35     r_ptr_succ <= r_ptr_reg + 1;
       -- next-state logic
       wr_op <= wr & rd;
       process(w_ptr_reg,w_ptr_succ,r_ptr_reg,r_ptr_succ,
                 wr_op,empty_reg,full_reg)
40     begin
           w_ptr_next <= w_ptr_reg;
           r_ptr_next <= r_ptr_reg;
           full_next <= full_reg;
           empty_next <= empty_reg;
45         case wr_op is
              when "00" => -- no op
              when "10" => -- write
                  if (full_reg /= '1') then -- not full
                      w_ptr_next <= w_ptr_succ;
50                    empty_next <= '0';
                      if (w_ptr_succ=r_ptr_reg) then
                          full_next <='1';
                      end if;
                  end if;
55            when "01" => -- read
                  if (empty_reg /= '1') then -- not empty
                      r_ptr_next <= r_ptr_succ;
                      full_next <= '0';
                      if (r_ptr_succ=w_ptr_reg) then
60                        empty_next <='1';
                      end if;
                  end if;
              when others => -- write/read;
                  w_ptr_next <= w_ptr_succ;
65                r_ptr_next <= r_ptr_succ;
          end case;
       end process;
       -- write port output
```

```
     w_addr <= std_logic_vector(w_ptr_reg);
70   full <= full_reg;
     r_addr <= std_logic_vector(r_ptr_reg);
     empty <= empty_reg;
end lookahead_bin_arch;
```

FIFO control circuit with a non-binary counter   For the previous two FIFO control circuit implementations, the two incrementors used in the binary counters consume the most hardware resources. If we examine operation of the read and write pointers closely, there is no need to access the register in binary sequence. Any order of access is fine as long as the two pointers circulate through the identical sequence. If we can derive a circuit to generate the status signals, other types of counters can be used for the pointers.

In the first scheme, we enlarge the binary counter and use the extra MSB to determine the status. This approach is based on the special property of the binary counting sequence and cannot easily be modified for other types of counters.

In the second scheme, the status signal relies on the successive value of the counter, and thus this scheme can be applied to any type of counter. Because of its simple next-state logic, LFSR is the best choice. It replaces the incrementor of a binary counter with a few xor cells and can significantly improve circuit size and performance, especially for a large FIFO address space.

Modifying the VHDL code is straightforward. Let us consider a FIFO controller with a 4-bit address. In the original code, the following two statements generate the successive values:

```
w_ptr_succ <= w_ptr_reg + 1;
r_ptr_succ <= r_ptr_reg + 1;
```

They can be replaced by the next-state logic of a 4-bit LFSR:

```
w_ptr_succ <=
    (w_ptr_reg(1) xor w_ptr_reg(0)) & w_ptr_reg(3 downto 1);
r_ptr_succ <=
    (r_ptr_reg(1) xor r_ptr_reg(0)) & r_ptr_reg(3 downto 1);
```

We must also revise the asynchronous reset portion of the code to initialize the counters for a non-zero value.

Recall that an n-bit LFSR circulates through only $2^n - 1$ states, and thus the size of the FIFO buffer is reduced by one accordingly. For a large n, the impact of the reduction is very small. We can also use a Bruijn counter if the entire 2^n address space is required.

9.3.3   Register-based content addressable memory

In a register file, each register in the file is assigned a unique address. When using a register to store a data item, we associate the item with the address, and access (i.e., read or write) this item via the address. An alternative way to identify a data item is to associate each item with a unique "key" and use this key to access the data item. This organization is known as *content addressable memory (CAM)*. A CAM is used in applications that require high-speed search, such as cache memory management and network routing.

The operation of a CAM can best be explained by a simple example. Consider a network router that examines the 16-bit destination field of an incoming packet and routes it to one of the eight output ports. A 4-word CAM stores information regarding the most frequently

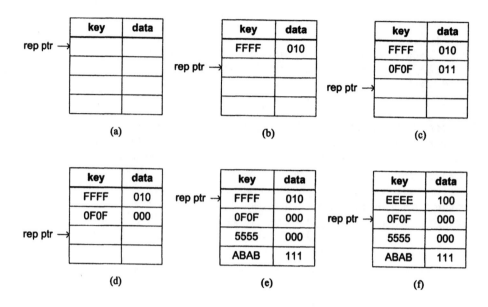

Figure 9.15 Operation of a conceptual CAM.

routed packets. The conceptual sketch is shown in Figure 9.15(a). The CAM includes four words, and each word is composed of a key–data pair. The key of the CAM is the 16-bit destination field, and the data is the 3-bit output port number. Since there are 2^{16} possible combinations for the key, which is far greater than the 4-word capacity of the CAM, we may need to remove an old key–data pair to make room for the new incoming pair. A replacement pointer, labeled `rep ptr` in the diagram, indicates the location of the word to be removed.

Let us first examine a sequence of write operations:

1. Write ($FFFF_{16}$, 010_2), which means an item with a key of $FFFF_{16}$ and data of 010_2. Since the CAM is empty and no key exists, the item is stored into the CAM, as shown in Figure 9.15(b). The replacement pointer advances accordingly.
2. Write ($0F0F_{16}$, 011_2). Since no existing key matches the new input key, the item is stored into the CAM, as shown in Figure 9.15(c).
3. Write ($0F0F_{16}$, 000_2). The input key matches an existing key in the CAM. The corresponding data of the key is replaced by the new data 000_2, as shown in Figure 9.15(d).
4. Write two new items. The CAM is now full, as shown in Figure 9.15(e). We assume that the replacement pointer moves in a round-robin fashion and thus returns to the first location of the table.
5. Write ($EEEE_{16}$, 100_2). Since the CAM is full now, a word must be removed to make place for the new item. The content of the first location is discarded for the new item, as shown in Figure 9.15(f).

To perform a read operation, we present the key as the input, and the data associated with the key will be routed to the output. For example, if the key is $EEEE_{16}$, the output of the CAM becomes 100_2. Since there is a chance that the input key does not match any stored key, a CAM usually contains an additional output signal, `hit`, to indicate whether there is a match (i.e., a *hit*).

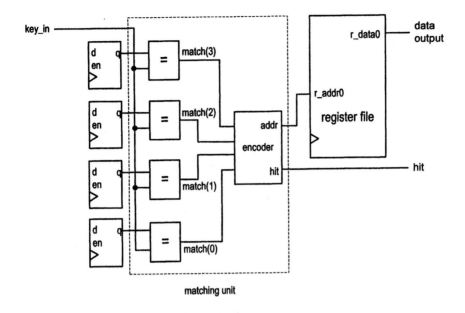

Figure 9.16   Matching circuit of a 4-word CAM.

Similar to SRAM, many technology libraries contain predesigned CAM modules that are constructed and optimized at the transistor level. Although the density of these modules is very high, accessing a CAM cell requires more time than is required by an FF. To improve performance, we sometimes want to use FFs to implement a small, fast synchronous CAM in the critical part of the system.

The major difference between a register file and a CAM is that the CAM uses a key, instead of an address, to identify a data item. One way to construct a CAM is to separate the storage into two register arrays, one for the data and one for the key. In our discussion, we call them a data file and a key file respectively. The data file is organized as a register file and uses an address to access its data. The key file contains a matching circuit, which compares the input key with the content of the key array and generates the corresponding address of the matched key. The address is then used to access the data stored in the data file.

The implementation of a CAM is fairly involved and we start with the read operation. The most unique component is the key file's matching circuit. The block diagram of the matching circuit of a 4-entry CAM and the relevant circuits is shown in Figure 9.16. The output of each register of the key array is compared with the current value of the input key (i.e., the key_in signal) and a 1-bit matching signal (i.e., the match(i) signal) is generated accordingly. Since each stored key is unique, at most one can be matched. We use the hit signal to indicate whether a match occurs (i.e., whether the input key is a "hit" or a "miss"). If the key_in signal is a hit, one bit of the 4-bit match signal is asserted. A 2^2-to-2 binary encoder generates the binary code of the matched location. The code is then used as the read port address of the data file, and the corresponding data item is routed to the output. Note that the output is not valid if the hit signal is not asserted.

The function of the key file is somewhat like a "reversed read operation" of a register file. In a register file, we present the address as an input and obtain the content of the

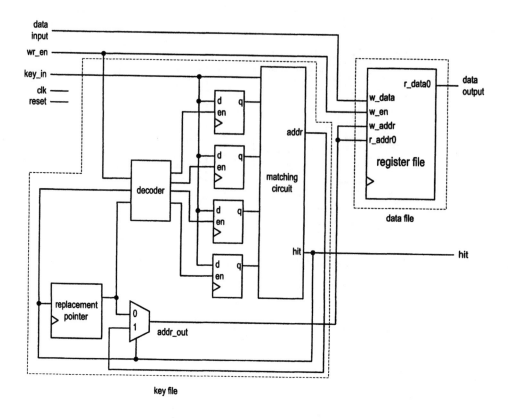

Figure 9.17 Block diagram of a 4-word register-based CAM.

corresponding register. On the other hand, we present a key (which is like the content) as the input to a key file and obtain the address where the key is stored.

The write operation is more complicated because of the possible miss condition. During a write operation, we present a key–data pair. If the key is a hit, the matching circuit will generate the write address of the data file, and the input data will be stored into the corresponding location. If the key is a miss, several tasks must be performed:

- Find an available register in the key array.
- Store the input key into this register.
- Store the data into the corresponding address in the data file.

The block diagram of the complete 4-word CAM is shown in Figure 9.17. The write operation of the data file is controlled by the external wr_en signal and the address is specified by the addr_out signal. The addr_out signal has two possible sources, one from the matching circuit and one from the replacement pointer. The first address is used if the input key is a hit. The *replacement pointer* is a circuit that keeps track of the available register location in the key array. The circuit updates the value when a miss occurs during the write operation. Its output value is used if the input key leads to a miss.

The writing operation of the key array is controlled by a decoding circuit similar to that of a register file. A key register can be written only if the wr_en signal is asserted and a miss occurs. If this is the case, the input key will be loaded into a register in the key array with the address specified by the replacement pointer.

The capacity of a register-based CAM is normally small, and the CAM is used to keep the "most frequently used" key–data pairs. When it is full and a miss occurs, a stored pair must be discarded to make place for a new pair. The *replacement policy* determines how to select the pair, and this policy is implemented by the replacement pointer circuit. One simple policy is the FIFO policy, which can be implemented by a binary counter. Initially, the CAM is empty and the counter is zero. The counter increments as the key–data pairs are stored into the CAM. The CAM is full when the counter reaches its maximal value. When a new pair comes and a miss occurs, the counter returns to 0 and wraps around. This corresponds to overwriting (i.e., discarding) the pair with the oldest key and storing the new pair in its location, achieving the desired FIFO policy.

A register file normally has one write port and several read ports. In theory, the same configuration can be achieved in a CAM by presenting several keys in parallel and using several matching circuits to generate multiple addresses. However, this is not common because of the complexity of the comparison circuit, and we normally use one input key signal, as in Figure 9.17. The read operation will be performed if the wr_en signal is not asserted.

The VHDL code of the data file is similar to the register file discussed in Section 9.3.1. We can even use a regular register file by connecting the addr_out signal to the w_addr and r_addr0 signals of the register file, as in Figure 9.17. The VHDL code of the key file of a 4-word, 16-bit CAM is shown in Listing 9.18. It follows the block diagram of Figure 9.17. A 2-bit binary counter is used to implement the FIFO replacement policy.

Listing 9.18   Key file of a 4-word CAM

```
library ieee;
use ieee.std_logic_1164.all;
use ieee.numeric_std.all;
entity key_file is
    port(
        clk, reset: in std_logic;
        wr_en: in std_logic;
        key_in: in std_logic_vector(15 downto 0);
        hit: out std_logic;
        addr_out: out std_logic_vector(1 downto 0)
    );
end key_file;

architecture no_loop_arch of key_file is
    constant WORD: natural:=2;
    constant BIT: natural:=16;
    type reg_file_type is array (2**WORD-1 downto 0) of
        std_logic_vector(BIT-1 downto 0);
    signal array_reg: reg_file_type;
    signal array_next: reg_file_type;
    signal en: std_logic_vector(2**WORD-1 downto 0);
    signal match: std_logic_vector(2**WORD-1 downto 0);
    signal rep_reg, rep_next: unsigned(WORD-1 downto 0);
    signal addr_match: std_logic_vector(WORD-1 downto 0);
    signal wr_key, hit_flag: std_logic;
begin
    -- register
    process(clk, reset)
```

```
       begin
30        if (reset='1') then
             array_reg(3) <= (others=>'0');
             array_reg(2) <= (others=>'0');
             array_reg(1) <= (others=>'0');
             array_reg(0) <= (others=>'0');
35        elsif (clk'event and clk='1') then
             array_reg(3) <=  array_next(3);
             array_reg(2) <=  array_next(2);
             array_reg(1) <=  array_next(1);
             array_reg(0) <=  array_next(0);
40        end if;
       end process;
       -- enable logic for register
       process(array_reg,en,key_in)
       begin
45        array_next(3) <= array_reg(3);
          array_next(2) <= array_reg(2);
          array_next(1) <= array_reg(1);
          array_next(0) <= array_reg(0);
          if en(3)='1' then
50           array_next(3) <= key_in;
          end if;
          if en(2)='1' then
             array_next(2) <= key_in;
          end if;
55        if en(1)='1' then
             array_next(1) <= key_in;
          end if;
          if en(0)='1' then
             array_next(0) <= key_in;
60        end if;
       end process;

       -- decoding for write address
       wr_key <= '1' when (wr_en='1' and hit_flag='0') else
65              '0';
       process(wr_key,rep_reg)
       begin
          if (wr_key='0') then
             en <= (others=>'0');
70        else
             case rep_reg  is
                when "00" =>   en <= "0001";
                when "01" =>   en <= "0010";
                when "10" =>   en <= "0100";
75              when others => en <= "1000";
             end case;
          end if;
       end process;

80     -- replacement pointer
       process(clk,reset)
```

```
      begin
        if (reset='1') then
           rep_reg <= (others=>'0');
 85     elsif (clk'event and clk='1') then
           rep_reg <= rep_next;
        end if;
      end process;
      rep_next <= rep_reg + 1 when wr_key='1' else
 90                 rep_reg;

      -- key comparison
      process(array_reg,key_in)
      begin
 95     match <= (others=>'0');
        if array_reg(3)=key_in then
           match(3) <= '1';
        end if;
        if array_reg(2)=key_in then
100        match(2) <= '1';
        end if;
        if array_reg(1)=key_in then
           match(1) <= '1';
        end if;
105     if array_reg(0)=key_in then
           match(0) <= '1';
        end if;
      end process;
      -- encoding
110   with match select
        addr_match <=
           "00" when "0001",
           "01" when "0010",
           "10" when "0100",
115        "11" when others;
      -- hit
      hit_flag <= '1' when match /="0000" else '0';
      --output
      hit <= hit_flag;
120   addr_out <= addr_match when (hit_flag='1') else
                   std_logic_vector(rep_reg);
  end no_loop_arch;
```

As in the register file and FIFO buffer, the code will be cumbersome for a larger CAM. A more systematic approach is discussed in Chapter 15.

9.4 PIPELINED DESIGN

Pipeline is an important technique to increase the performance of a system. The basic idea is to overlap the processing of several tasks so that more tasks can be completed in the same amount of time. If a combinational circuit can be divided into stages, we can insert buffers (i.e., registers) at proper places and convert the circuit into a pipelined design. This section introduces the concept of pipeline and shows how to add pipeline to the combinational multiplier discussed in Chapter 8.

9.4.1 Delay versus throughput

Before we study the pipelined circuit, it will be helpful to first understand the difference between delay and throughput, two criteria used to examine the performance of a system. *Delay* is the time required to complete one task, and *throughput* is the number of tasks that can be completed per unit time. The two are related but are not identical.

To illustrate the concept, let us use the bank ATM machine transaction as an example. Assume that a bank branch originally has only one ATM machine and it takes 3 minutes to complete a transaction. The delay to complete one transaction is 3 minutes and the maximal throughput is 20 transactions per hour. If the bank wishes to increase the performance of its ATM system, there are two options. The first option is to use a newer, faster ATM machine. For example, the bank can install a machine that takes only 1.5 minutes to complete a transaction. The second option is to add another machine so that there are two ATM machines running in parallel. For the first option, the delay becomes 1.5 minutes and the maximal throughput increases to 40 transactions per hour. For the second option, the transaction delay experienced by a user is still 3 minutes and thus remains the same. However, since there are two ATM machines, the system's maximal throughput is doubled to 40 transactions per hour. In summary, the first option reduces the delay in an individual transaction and increases the throughput at the same time, whereas the second option can only improve the throughput.

Adding pipeline to a combinational circuit is somewhat like the second option and can only increase a system's throughput. It will not reduce the delay in an individual task. Actually, because of the overhead introduced by the registers and non-ideal stage division, the delay will be worse than in the non-pipelined design.

9.4.2 Overview on pipelined design

Pipelined laundry　The pipelining technique can be applied to a task that is processed in stages. To illustrate the concept, let us consider the process of doing laundry. Assume that we do a load of laundry in three stages, which are washing, drying and folding, and that each stage takes 20 minutes. For non-pipelined processing, a new load cannot start until the previous load is completed. The time line for processing four loads of laundry is shown in Figure 9.18(a). It takes 240 minutes (i.e., $4*3*20$ minutes) to complete the four loads. In terms of the performance criteria, the delay of processing one load is 60 minutes and the throughput is $\frac{1}{60}$ load per minute (i.e., four loads in 240 minutes).

If we examine the process carefully, there is room for improvement. After 20 minutes, the washing stage is done and the washer is idle. We can start a new load at this point rather than waiting for completion of the entire laundry process. Since each stage takes the same amount of time, there will be no contention in subsequent stages. The time line of the pipelined version of four loads is shown in Figure 9.18(b). It takes 120 minutes to complete the four loads. In terms of performance, the delay in processing one load remains 60 minutes. However, the throughput is increased to $\frac{2}{60}$ load per minute (i.e., 4 loads in 120 minutes). If we process k loads, it will take $40+20k$ minutes. The throughput becomes $\frac{k}{40+20k}$ load per minute. If k is large, the throughput approaches $\frac{1}{20}$ load per minute, which is three times better than that of the non-pipelined process.

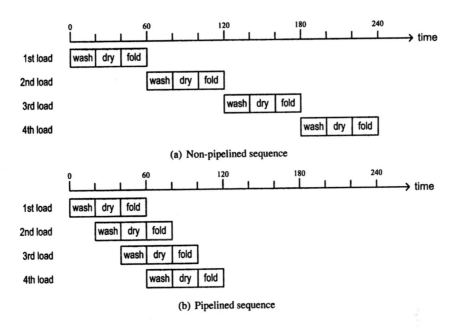

(a) Non-pipelined sequence

(b) Pipelined sequence

Figure 9.18   Timing diagrams of pipelined and non-pipelined laundry sequences.

This example shows an ideal pipelined situation, in which a task can be divided into stages of identical delays. If washing, drying and folding take 15, 25 and 20 minutes respectively, we must accommodate the stage with the longest delay or a conflict will occur. For example, when washing is done at the first stage, we have to wait for 10 minutes before putting a new load into the dryer. Under this scenario, we can only start a new load every 25 minutes at best. The delay to complete one load is increased from 60 minutes to 75 minutes (i.e., 3*25 minutes) now. The throughput for k loads becomes $\frac{k}{50+25k}$ load per minute and approaches $\frac{1}{25}$ load per minute when k is large. Note that while the pipelined processing helps improving the throughput, it actually increases the processing delay for a single load.

Pipelined combinational circuit   The same pipeline concept can be applied to combinational circuits. We can divide a combinational circuit in stages so that the processing of different tasks can be overlapped, as in the laundry example. To ensure that the signals in each stage flow in the correct order and to prevent any potential race, we must add a register between successive stages, as shown in the four-stage pipelined combinational circuit of Figure 9.19. An output buffer is also included in the last stage. A register functions as a "flood gate" that regulates signal flow. It ensures that the signals can be passed to the next stage only at predetermined points. The clock period of the registers should be large enough to accommodate the slowest stage. At a faster stage, output will be blocked by the register even when the processing has been completed earlier. The output data at each stage will be sampled and stored into registers at the rising edge of the clock. These data will be used as input for the next stage and remain unchanged in the remaining part of the clock period. At the end of the clock period, the new output data are ready. They will be sampled and passed to the next stage (via the register) at the next rising edge of the clock.

The effectiveness of the pipelined circuit is judged by two performance criteria, the delay and the throughput. Consider the previous four-stage pipelined combinational circuit. Assume that the original propagation delays of the four stages are T_1, T_2, T_3 and T_4

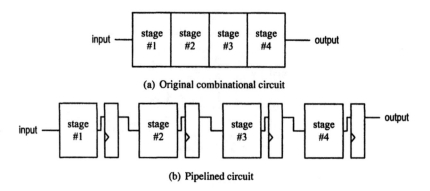

(a) Original combinational circuit

(b) Pipelined circuit

Figure 9.19   Construction of a four-stage pipelined circuit.

respectively. Let T_{max} be the longest propagation delay among the four stages:

$$T_{max} = \max(T_1, T_2, T_3, T_4)$$

The clock period needs to accommodate the longest delay plus the overhead introduced by the buffer register in each stage, which includes the setup time (T_{setup}) and the clock-to-q delay (T_{cq}) of the register. Thus, the minimal clock period, T_c, is

$$T_c = T_{max} + T_{setup} + T_{cq}$$

In the original non-pipelined combinational circuit, the propagation delay in processing one data item is

$$T_{comb} = T_1 + T_2 + T_3 + T_4$$

For the pipelined design, processing a data requires four clock cycles and the propagation delay becomes

$$T_{pipe} = 4T_c = 4T_{max} + 4(T_{setup} + T_{cq})$$

This is clearly worse than the propagation delay of the original circuit.

The second performance criterion is the throughput. Since there is no overlapping when the data is processed, the maximal throughput of the non-pipelined design is $\frac{1}{T_{comb}}$. The throughput of the pipelined design can be derived by calculating the time required to complete k consecutive data items. When the process starts, the pipeline is empty. It takes $3T_c$ to fill the first three stages, and the pipeline does not generate an output in the interval. After this, the pipeline is full and the circuit can produce one output in each clock cycle. Thus, it requires $3T_c + kT_c$ to process k data items. The throughput is $\frac{k}{3T_c + kT_c}$ and approaches $\frac{1}{T_c}$ as k becomes very large.

In an ideal pipelined scenario, the propagation delay of each stage is identical (which implies that $T_{max} = \frac{T_{comb}}{4}$), and the register overhead (i.e., $T_{setup} + T_{cq}$) is comparably small and can be ignored. T_{pipe} can be simplified as

$$T_{pipe} = 4T_c \approx 4T_{max} = T_{comb}$$

The throughput becomes

$$\frac{1}{T_c} \approx \frac{1}{T_{max}} = \frac{4}{T_{comb}}$$

This implies that the pipeline imposes no penalty on the delay but increases the throughput by a factor of 4.

The discussion of four-stage pipelined design can be generalized to an N-stage pipeline. In the ideal scenario, the delay to process one data item remains unchanged and the throughput can be increased N-fold. This suggests that it is desirable to have more stages in the pipeline. However, when N becomes large, the propagation delay of each stage becomes smaller. Since $T_{setup} + T_{cq}$ of the register remains the same, its impact becomes more significant and can no longer be ignored. Thus, extremely large N has less effect and may even degrade the performance. In reality, it is also difficult, if not impossible, to keep dividing the original combinational circuit into smaller and smaller stages.

When discussing the throughput of a pipelined system, we have to be aware of the condition to obtain maximal throughput. The assumption is that the external data are fed into the pipeline at a rate of $\frac{1}{T_c}$ so that the pipeline is filled all the time. If the external input data cannot be issued fast enough, there will be slack (a "bubble") inside the pipeline, and the throughput will be decreased accordingly. If the external data is issued only sporadically, the pipelined design will not improve the performance at all.

9.4.3 Adding pipeline to a combinational circuit

Although we can add pipeline to any combinational circuit by inserting registers into the intermediate stages, the pipelined version may not provide better performance. The previous analysis shows that the good candidate circuits for effective pipeline design should include the following characteristics:

- There is enough input data to feed the pipelined circuit.
- The throughput is a main performance criterion.
- The combinational circuit can be divided into stages with similar propagation delays.
- The propagation delay of a stage is much larger than the setup time and the clock-to-q delay of the register.

If a circuit is suitable for the pipelined design, we can convert the original circuit and derive the VHDL code by the following procedure:

1. Derive the block diagram of the original combinational circuit and arrange the circuit as a cascading chain.
2. Identify the major components and estimate the relative propagation delays of these components.
3. Divide the chain into stages of similar propagation delays.
4. Identify the signals that cross the boundary of the chain.
5. Insert registers for these signals in the boundary.

This procedure is illustrated by the examples in the following subsections.

Simple pipelined adder-based multiplier   The adder-based multiplier discussed in Section 7.5.4 uses multiple adders to sum the bit products in stages and thus is a natural match for a pipelined design. Our design is based on the scheme used in the comb1_arch architecture of Listing 7.34. To reduce the clutter in the block diagram, we use a 5-bit multiplier to demonstrate the design process. The design approach can easily be extended to an 8-bit or larger multiplier.

The two major components are the adder and bit-product generation circuit. To facilitate the pipeline design process, we can arrange these components in cascade. The rearranged block diagram is shown in Figure 9.20(a). The circuit to generate the bit product is labeled BP in the diagram. Since the bit-product generation circuit involves only bitwise and operation

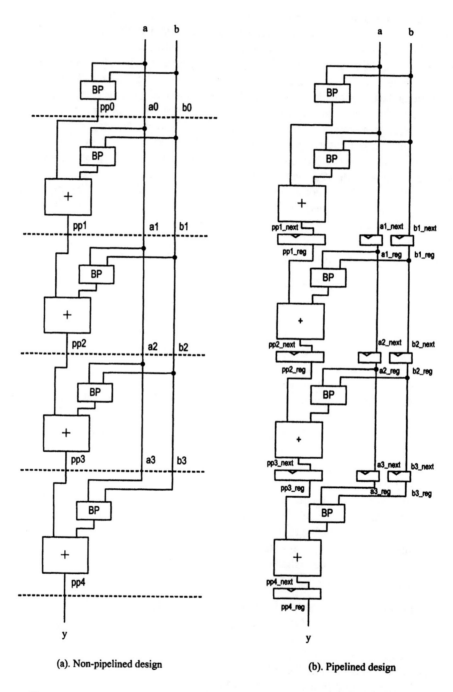

(a). Non-pipelined design   (b). Pipelined design

Figure 9.20   Block diagrams of non-pipelined and four-stage pipelined multipliers.

and the padding of 0's, its propagation delay is small. We combine it with the adder to form a stage. The division of the the circuit is shown in Figure 9.20(a), in which the boundary of each stage is shown by a dashed line. To help the coding, a signal is given a unique name in each stage. For example, the a signal is renamed a0, a1, a2 and a3 in the zeroth, first, second and third stages of the pipeline respectively. Since no addition is needed to generate the first partial product (i.e., the pp0 signal), the zeroth and first stages can be merged into a single stage later.

For a signal crossing the stage boundary line, a register is needed between the two stages. There are two types of registers. The first type of register is used to accommodate the computation flow and to store the intermediate results, which are the partial products, pp1, pp2, pp3 and pp4, in the diagram. The second type of register preserves the information needed for each stage, which are a1, a2, a3, b1, b2 and b3. The function of these registers is less obvious. In this pipeline, the processing at each stage depends on the partial product result from the preceding stage, as well as the values of the a and b signals. Note that four multiplications are performed concurrently in the pipeline, each with its own values for the a and b signals. As the partial product calculation progresses through the pipeline, these values must go with the calculation in each stage. The second type of register essentially carries the original values along the pipeline so that a correct copy of the input data is available in each stage. The completed pipelined multiplier with proper registers is shown in Figure 9.20(b).

Following the diagram, we can derive the VHDL code accordingly. The code of the rearranged combinational multiplier is shown in Listing 9.19. The creation of the new signal names is only for later use and should have no effect on synthesis.

Listing 9.19 Non-pipelined multiplier in cascading stages

```
library ieee;
use ieee.std_logic_1164.all;
use ieee.numeric_std.all;
entity mult5 is
   port(
        clk, reset: in std_logic;
        a, b: in std_logic_vector(4 downto 0);
        y: out std_logic_vector(9 downto 0)
   );
end mult5;

architecture comb_arch of mult5 is
   constant WIDTH: integer:=5;
   signal a0, a1, a2, a3: std_logic_vector(WIDTH-1 downto 0);
   signal b0, b1, b2, b3: std_logic_vector(WIDTH-1 downto 0);
   signal bv0, bv1, bv2, bv3, bv4:
        std_logic_vector(WIDTH-1 downto 0);
   signal bp0, bp1, bp2, bp3, bp4:
        unsigned(2*WIDTH-1 downto 0);
   signal pp0, pp1, pp2, pp3, pp4:
        unsigned(2*WIDTH-1 downto 0);
begin
   -- stage 0
   bv0 <= (others=>b(0));
   bp0 <=unsigned("00000" & (bv0 and a));
   pp0 <= bp0;
```

```
       a0 <= a;
       b0 <= b;
       -- stage 1
30     bv1 <= (others=>b0(1));
       bp1 <=unsigned("0000" & (bv1 and a0) & "0");
       pp1 <= pp0 + bp1;
       a1 <= a0;
       b1 <= b0;
35     -- stage 2
       bv2 <= (others=>b1(2));
       bp2 <=unsigned("000" & (bv2 and a1) & "00");
       pp2 <= pp1 + bp2;
       a2 <= a1;
40     b2 <= b1;
       -- stage 3
       bv3 <= (others=>b2(3));
       bp3 <=unsigned("00" & (bv3 and a2) & "000");
       pp3 <= pp2 + bp3;
45     a3 <= a2;
       b3 <= b2;
       -- stage 4
       bv4 <= (others=>b3(4));
       bp4 <=unsigned("0" & (bv4 and a3) & "0000");
50     pp4 <= pp3 + bp4;
       -- output
       y <= std_logic_vector(pp4);
    end comb_arch;
```

When converting the circuit into a pipelined version, we first add the specifications for the registers and then reconnect the input and output of each stage to the registers. Instead of using the output from the preceding stage, each stage of pipeline circuit obtains its input from the boundary register. Similarly, the output of each stage is now connected to the input of the register rather than feeding directly to the next stage. For example, the pp2 signal of the non-pipelined circuit is generated in the second stage and is then used in the third stage:

```
       -- stage 2
    pp2 <= pp1 + bp2;
       -- stage 3
    pp3 <= pp2 + bp3;
```

In the pipelined design, the signal should be stored in a register, and the code becomes

```
       -- register
       if (reset ='1') then
          pp2_reg <= (others=>'0');
       elsif (clk'event and clk='1') then
          pp2_reg <= pp2_next;
       end if;
          . . .
       -- stage 2
    pp2_next <= pp1_reg + bp2;
       -- stage 3
    pp3_next <= pp2_reg + bp3;
```

The complete VHDL code of the four-stage pipelined circuit is shown in Listing 9.20.

Listing 9.20   Four-stage pipelined multiplier

```vhdl
architecture four_stage_pipe_arch of mult5 is
   constant WIDTH: integer :=5;
   signal a1_reg, a2_reg, a3_reg:
      std_logic_vector(WIDTH-1 downto 0);
   signal a0, a1_next, a2_next, a3_next:
      std_logic_vector(WIDTH-1 downto 0);
   signal b1_reg, b2_reg, b3_reg:
      std_logic_vector(WIDTH-1 downto 0);
   signal b0, b1_next, b2_next, b3_next:
      std_logic_vector(WIDTH-1 downto 0);
   signal bv0, bv1, bv2, bv3, bv4:
      std_logic_vector(WIDTH-1 downto 0);
   signal bp0, bp1, bp2, bp3, bp4:
      unsigned(2*WIDTH-1 downto 0);
   signal pp1_reg, pp2_reg, pp3_reg, pp4_reg:
      unsigned(2*WIDTH-1 downto 0);
   signal pp0, pp1_next, pp2_next, pp3_next, pp4_next:
      unsigned(2*WIDTH-1 downto 0);
begin
   -- pipeline registers (buffers)
   process(clk,reset)
   begin
      if (reset ='1') then
         pp1_reg <= (others=>'0');
         pp2_reg <= (others=>'0');
         pp3_reg <= (others=>'0');
         pp4_reg <= (others=>'0');
         a1_reg <= (others=>'0');
         a2_reg <= (others=>'0');
         a3_reg <= (others=>'0');
         b1_reg <= (others=>'0');
         b2_reg <= (others=>'0');
         b3_reg <= (others=>'0');
      elsif (clk'event and clk='1') then
         pp1_reg <= pp1_next;
         pp2_reg <= pp2_next;
         pp3_reg <= pp3_next;
         pp4_reg <= pp4_next;
         a1_reg <= a1_next;
         a2_reg <= a2_next;
         a3_reg <= a3_next;
         b1_reg <= b1_next;
         b2_reg <= b2_next;
         b3_reg <= b3_next;
      end if;
   end process;

   -- merged stage 0 & 1 for pipeline
   bv0 <= (others=>b(0));
   bp0 <=unsigned("00000" & (bv0 and a));
   pp0 <= bp0;
   a0 <= a;
```

```
     b0  <= b;
     --

55   bv1  <= (others=>b0(1));
     bp1  <=unsigned("0000" & (bv1 and a0) & "0");
     pp1_next  <= pp0 + bp1;
     a1_next  <= a0;
     b1_next  <= b0;
60   -- stage 2
     bv2  <= (others=>b1_reg(2));
     bp2  <=unsigned("000" & (bv2 and a1_reg) & "00");
     pp2_next  <= pp1_reg + bp2;
     a2_next  <= a1_reg;
65   b2_next  <= b1_reg;
     -- stage 3
     bv3  <= (others=>b2_reg(3));
     bp3  <=unsigned("00" & (bv3 and a2_reg) & "000");
     pp3_next  <= pp2_reg + bp3;
70   a3_next  <= a2_reg;
     b3_next  <= b2_reg;
     -- stage 4
     bv4  <= (others=>b3_reg(4));
     bp4  <=unsigned("0" & (bv4 and a3_reg) & "0000");
75   pp4_next  <= pp3_reg + bp4;
     -- output
     y <= std_logic_vector(pp4_reg);
   end four_stage_pipe_arch;
```

We can adjust the number of stages by adding or removing buffer registers. For example, we can reduce the number of pipeline stages by removing the registers in the first and third stages and create a two-stage pipelined multiplier. The revised VHDL code is shown in Listing 9.21.

Listing 9.21   Two-stage pipelined multiplier

```
   architecture two_stage_pipe_arch of mult5 is
     constant WIDTH: integer:=5;
     signal a2_reg: std_logic_vector(WIDTH-1 downto 0);
     signal a0, a1, a2_next, a3:
5       std_logic_vector(WIDTH-1 downto 0);
     signal b2_reg: std_logic_vector(WIDTH-1 downto 0);
     signal b0, b1, b2_next, b3:
        std_logic_vector(WIDTH-1 downto 0);
     signal bv0, bv1, bv2, bv3, bv4:
10      std_logic_vector(WIDTH-1 downto 0);
     signal bp0, bp1, bp2, bp3, bp4:
        unsigned(2*WIDTH-1 downto 0);
     signal pp2_reg, pp4_reg: unsigned(2*WIDTH-1 downto 0);
     signal pp0, pp1, pp2_next, pp3, pp4_next:
15      unsigned(2*WIDTH-1 downto 0);
   begin
     -- pipeline registers (buffers)
     process(clk,reset)
     begin
20       if (reset ='1') then
```

```
                pp2_reg <= (others=>'0');
                pp4_reg <= (others=>'0');
                a2_reg <= (others=>'0');
                b2_reg <= (others=>'0');
25          elsif (clk'event and clk='1') then
                pp2_reg <= pp2_next;
                pp4_reg <= pp4_next;
                a2_reg <= a2_next;
                b2_reg <= b2_next;
30          end if;
        end process;

        -- stage 0
        bv0 <= (others=>b(0));
35      bp0 <=unsigned("00000" & (bv0 and a));
        pp0 <= bp0;
        a0 <= a;
        b0 <= b;
        -- stage 1
40      bv1 <= (others=>b0(1));
        bp1 <=unsigned("0000" & (bv1 and a0) & "0");
        pp1 <= pp0 + bp1;
        a1 <= a0;
        b1 <= b0;
45      -- stage 2 ( with  buffer )
        bv2 <= (others=>b1(2));
        bp2 <=unsigned("000" & (bv2 and a1) & "00");
        pp2_next <= pp1 + bp2;
        a2_next <= a1;
50      b2_next <= b1;
        -- stage 3
        bv3 <= (others=>b2_reg(3));
        bp3 <=unsigned("00" & (bv3 and a2_reg) & "000");
        pp3 <= pp2_reg + bp3;
55      a3 <= a2_reg;
        b3 <= b2_reg;
        -- stage 4 ( with  buffer )
        bv4 <= (others=>b3(4));
        bp4 <=unsigned("0" & (bv4 and a3) & "0000");
60      pp4_next <= pp3 + bp4;
        -- output
        y <= std_logic_vector(pp4_reg);
    end two_stage_pipe_arch;
```

More efficient pipelined adder-based multiplier   We can make some improvements
of the initial pipelined design. First, as discussed in Section 7.5.4, we can use a smaller
$(n + 1)$-bit adder to replace the $2n$-bit adder in an n-bit multiplier. The same technique can
be applied to the pipelined version. Second, we can reduce the size of the partial-product
register. This is based on the observation that the valid LSBs of the partial products grow
incrementally in each stage, from $n + 1$ bits to $2n$ bits. There is no need to use a $2n$-bit
register to carry the invalid MSBs in every stage. In the previous example, we can use a
5-bit register for the initial partial product (i.e., the pp0 signal), and increase the size by 1 in

each stage. Finally, we can reduce the size of the registers that hold the b signal. Note that only the b_i bit of b is needed to obtain the bit product at the ith stage. Once the calculation is done, the b_i bit can be discarded. Instead of using n-bit registers to carry b, we can drop one LSB bit in each stage and reduce the register size decrementally. In the previous example, we can drop the register bits for b_0 and b_1 in the first stage, the bits for b_2 in the second stage and so on. The VHDL code of the revised design is shown in Listing 9.22.

Listing 9.22 More efficient four-stage pipelined multiplier

```
    architecture effi_4_stage_pipe_arch of mult5 is
      signal a1_reg, a2_reg, a3_reg:
        std_logic_vector(4 downto 0);
      signal a0, a1_next, a2_next, a3_next:
5       std_logic_vector(4 downto 0);
      signal b0: std_logic_vector(4 downto 1);
      signal b1_next, b1_reg: std_logic_vector(4 downto 2);
      signal b2_next, b2_reg: std_logic_vector(4 downto 3);
      signal b3_next, b3_reg: std_logic_vector(4 downto 4);
10     signal bv0, bv1, bv2, bv3, bv4:
        std_logic_vector(4 downto 0);
      signal bp0, bp1, bp2, bp3, bp4: unsigned(5 downto 0);
      signal pp0: unsigned(5 downto 0);
      signal pp1_next, pp1_reg: unsigned(6 downto 0);
15     signal pp2_next, pp2_reg: unsigned(7 downto 0);
      signal pp3_next, pp3_reg: unsigned(8 downto 0);
      signal pp4_next, pp4_reg: unsigned(9 downto 0);
    begin
      -- pipeline registers (buffers)
20     process(clk,reset)
      begin
        if (reset ='1') then
          pp1_reg <= (others=>'0');
          pp2_reg <= (others=>'0');
25         pp3_reg <= (others=>'0');
          pp4_reg <= (others=>'0');
          a1_reg <= (others=>'0');
          a2_reg <= (others=>'0');
          a3_reg <= (others=>'0');
30         b1_reg <= (others=>'0');
          b2_reg <= (others=>'0');
          b3_reg <= (others=>'0');
        elsif (clk'event and clk='1') then
          pp1_reg <= pp1_next;
35         pp2_reg <= pp2_next;
          pp3_reg <= pp3_next;
          pp4_reg <= pp4_next;
          a1_reg <= a1_next;
          a2_reg <= a2_next;
40         a3_reg <= a3_next;
          b1_reg <= b1_next;
          b2_reg <= b2_next;
          b3_reg <= b3_next;
        end if;
45   end process;
```

```
        —— merged stage 0 & 1 for pipeline
        bv0 <= (others=>b(0));
        bp0 <=unsigned('0' & (bv0 and a));
50      pp0 <= bp0;
        a0 <= a;
        b0 <= b(4 downto 1);
        ——
        bv1 <= (others=>b0(1));
55      bp1 <=unsigned('0' & (bv1 and a0));
        pp1_next(6 downto 1) <= ('0' & pp0(5 downto 1)) + bp1;
        pp1_next(0) <= pp0(0);
        a1_next <= a0;
        b1_next <= b0(4 downto 2);
60      —— stage 2
        bv2 <= (others=>b1_reg(2));
        bp2 <=unsigned('0' & (bv2 and a1_reg));
        pp2_next(7 downto 2) <= ('0' & pp1_reg(6 downto 2)) + bp2;
        pp2_next(1 downto 0) <= pp1_reg(1 downto 0);
65      a2_next <= a1_reg;
        b2_next <= b1_reg(4 downto 3);
        —— stage 3
        bv3 <= (others=>b2_reg(3));
        bp3 <=unsigned('0' & (bv3 and a2_reg));
70      pp3_next(8 downto 3) <= ('0' & pp2_reg(7 downto 3)) + bp3;
        pp3_next(2 downto 0) <= pp2_reg(2 downto 0);
        a3_next <= a2_reg;
        b3_next(4) <= b2_reg(4);
        —— stage 4
75      bv4 <= (others=>b3_reg(4));
        bp4 <=unsigned('0' & (bv4 and a3_reg));
        pp4_next(9 downto 4) <= ('0' & pp3_reg(8 downto 4)) + bp4;
        pp4_next(3 downto 0) <= pp3_reg(3 downto 0);
        —— output
80      y <= std_logic_vector(pp4_reg);
     end effi_4_stage_pipe_arch;
```

Tree-shaped pipelined multiplier   Discussion in Section 7.5.4 shows that we can re-arrange a cascading network to reduce the propagation delay. In an n-bit combinational multiplier, the critical path consists of $n - 1$ adders in a cascading network. The critical path can be reduced to $\lceil \log_2 n \rceil$ adders when a tree-shaped network is used. The same scheme can be applied to the pipelined multiplier. The 5-bit tree-shaped combinational circuit is shown in Figure 9.21(a). The five bit products are first evaluated in parallel and then fed into the tree-shaped network. The pipelined version is shown in Figure 9.21(b). It is divided into three stages and the required registers are shown as dark bars. Note that one bit product has to be carried through two stages. The VHDL code is given in Listing 9.23.

Listing 9.23   Tree-shaped three-stage pipelined multiplier

```
architecture tree_pipe_arch of mult5 is
   constant WIDTH: integer:=5;
   signal bv0, bv1, bv2, bv3, bv4:
      std_logic_vector(WIDTH-1 downto 0);
```

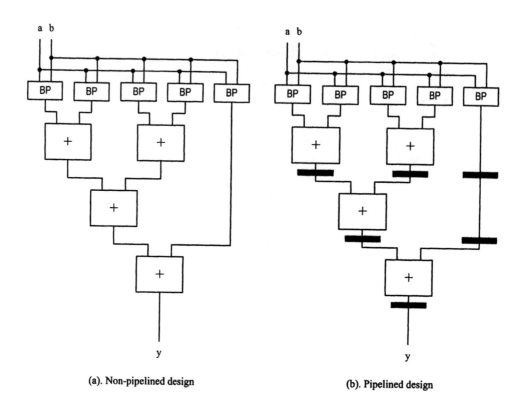

(a). Non-pipelined design

(b). Pipelined design

Figure 9.21   Block diagrams of tree-shaped non-pipelined and pipelined multipliers.

```
 5      signal bp0, bp1, bp2, bp3, bp4:
            unsigned(2*WIDTH-1 downto 0);
        signal bp4_s1_reg, bp4_s2_reg:
            unsigned(2*WIDTH-1 downto 0);
        signal bp4_s1_next, bp4_s2_next:
10          unsigned(2*WIDTH-1 downto 0);
        signal pp01_reg, pp23_reg, pp0123_reg, pp01234_reg:
            unsigned(2*WIDTH-1 downto 0);
        signal pp01_next, pp23_next, pp0123_next, pp01234_next:
            unsigned(2*WIDTH-1 downto 0);
15 begin
        -- pipeline registers (buffers)
        process(clk,reset)
        begin
            if (reset ='1') then
20              pp01_reg   <= (others=>'0');
                pp23_reg   <= (others=>'0');
                pp0123_reg <= (others=>'0');
                pp01234_reg <= (others=>'0');
                bp4_s1_reg <= (others=>'0');
25              bp4_s2_reg <= (others=>'0');
```

```
          elsif (clk'event and clk='1') then
              pp01_reg <= pp01_next;
              pp23_reg <= pp23_next;
              pp0123_reg <= pp0123_next;
30            pp01234_reg <= pp01234_next;
              bp4_s1_reg <= bp4_s1_next;
              bp4_s2_reg <= bp4_s2_next;
          end if;
      end process;
35
      -- stage 1
      -- bit product
      bv0 <= (others=>b(0));
      bp0 <=unsigned("00000" & (bv0 and a));
40    bv1 <= (others=>b(1));
      bp1 <=unsigned("0000" & (bv1 and a) & "0");
      bv2 <= (others=>b(2));
      bp2 <=unsigned("000" & (bv2 and a) & "00");
      bv3 <= (others=>b(3));
45    bp3 <=unsigned("00" & (bv3 and a) & "000");
      bv4 <= (others=>b(4));
      bp4 <=unsigned("0" & (bv4 and a) & "0000");
      -- adder
      pp01_next <= bp0 + bp1;
50    pp23_next <= bp2 + bp3;
      bp4_s1_next <= bp4;
      -- stage 2
      pp0123_next <= pp01_reg + pp23_reg;
      bp4_s2_next <= bp4_s1_reg;
55    -- stage 3
      pp01234_next <= pp0123_reg + bp4_s2_reg;
      -- output
      y <= std_logic_vector(pp01234_reg);
  end tree_pipe_arch;
```

In terms of performance, the delay in the tree-shaped multiplier is smaller since it has only three pipelined stages. The improvement will become more significant for a larger multiplier. On the other hand, the throughput of the two pipelined designs is similar because they have a similar clock rate. Both can generate a new multiplication result in each clock cycle.

Although the division of the adder-based multiplier appears to be reasonable, it is not optimal. Examining the circuit in "finer granularity" can shed light about the data dependency on the internal structure and lead to a more efficient partition. This issue is discussed in Section 15.4.2.

9.4.4 Synthesis of pipelined circuits and retiming

The major step of adding pipeline to a combinational circuit is to divide the circuit into adequate stages. To achieve this goal, we must know the propagation delays of the relevant components. However, since the components will be transformed, merged and optimized during synthesis and wiring delays will be introduced during placement and routing, this information cannot easily be determined at the RT level.

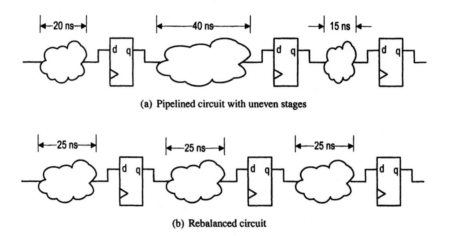

(a) Pipelined circuit with uneven stages

(b) Rebalanced circuit

Figure 9.22 Example of circuit retiming.

Except for a highly regular structure, such as the previous adder-based multiplier example, partitioning a circuit into proper stages is difficult. We may need to synthesize major components and even some subsystems in advance to obtain rough estimations of the propagation delays, and then use this information to guide the division.

More sophisticated synthesis software can automate this task to some degree. It is known as *retiming*. For example, consider the three-stage pipelined circuit shown in Figure 9.22(a). The combinational circuits are shown as clouds with their propagation delays. The division of the original combinational circuit is not optimal and thus creates three uneven stages. In regular synthesis software, optimization can be done only for a combinational circuit, and thus the three combinational circuits of Figure 9.22(a) are processed independently. On the other hand, synthesis software with retiming capability can examine the overall circuit and move combinational circuits crossing the register boundaries. A rebalanced implementation is shown in Figure 9.22(b). This tool is especially useful if the combinational circuits are random and do not have an easily recognizable structure.

9.5 SYNTHESIS GUIDELINES

- Asynchronous reset, if used, should be only for system initialization. It should not be used to clear the registers during regular operation.

- Do not manipulate or gate the clock signal. Most desired operations can be achieved by using a register with an enable signal.

- LFSR is an effective way to construct a counter. It can be used when the counting patterns are not important.

- Throughput and delay are two performance criteria. Adding pipeline to a combinational circuit can increase the throughput but not reduce the delay.

- The main task of adding pipeline to a combinational circuit is to divide the circuit into balanced stages. Software with retiming capability can aid in this task.

9.6 BIBLIOGRAPHIC NOTES

Although the implementation of an LFSR is simple, it has lots of interesting properties and a wide range of applications. The text, *Built In Test for VLSI: Pseudorandom Techniques* by Paul H. Bardell et al., has an in-depth discussion on the fundamentals and implementation of LFSRs. The application note of Xilinx, *Efficient Shift Registers, LFSR Counters, and Long Pseudorandom Sequence Generators*, includes a table that lists LFSR feedback expressions from 2 to 160 bits.

A pipeline is a commonly used technique to increase the performance of a processor. Its design is more involved because of the external data dependency. The text, *Computer Organization and Design: The Hardware/Software Interface, 3rd edition*, by David A. Patterson and John L. Hennessy, provides comprehensive coverage of this topic.

Problems

9.1   Consider the decade counter shown in Figure 9.1. Let T_{inc}, T_{comp} and T_{or} be the propagation delays of the incrementor, comparator and or cell, and T_{setup}, T_{cq} and T_{rq} be the setup time, clock-to-q delay and reset-to-q delay of the register. Determine the maximal clock rate of this counter.

9.2   Consider the following asynchronous counter constructed with T FFs:

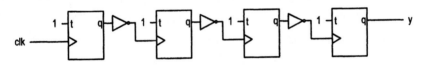

(a) Draw the waveform for the clock and the output of four FFs.
(b) Describe the operation of this counter.
(c) Design a synchronous counter that performs the same task and derive the VHDL code accordingly.

9.3   For the 4-bit ring counter discussed in Section 9.2.2, the output of the 4 FFs appears to be out of phase. Let $T_{cq(0)}$ and $T_{cq(1)}$ be the clock-to-q delays when the q output of an FF becomes '0' and '1' respectively. Note that $T_{cq(0)}$ and $T_{cq(1)}$ may not always be equal. Perform a detailed timing analysis to determine whether a ring counter can produce true non-overlapping four-phase signals.

9.4   Design a 4-bit self-correction synchronous counter that circulates a single '0' (i.e., circulates the "1110" pattern).

9.5   Revise the design of the 4-bit LFSR in Section 9.2.3 to include the "0000" pattern but exclude the "1111" pattern.

9.6   Let the propagation delay of an xor cell be 4 ns, the propagation delay of an n-bit incrementor be $6n$ ns, and the setup time and clock-to-q delay of the register be 2 and 3 ns respectively.
(a) Determine the maximal operation rates of a 4-bit LFSR and a binary counter.
(b) Determine the maximal operation rates of an 8-bit LFSR and a binary counter.
(c) Determine the maximal operation rates of a 16-bit LFSR and a binary counter.
(d) Determine the maximal operation rates of a 64-bit LFSR and a binary counter.

9.7 Use the timing data from Problem 9.6. In addition, let the propagation delay of an invertor, 2-input and cell and 2-input or cell be 2 ns.

(a) Use these cells to implement the additional comparison circuit needed in the 4-bit Brujin counter in Section 9.2.3.

(b) Determine the maximal operation rates of the 4-bit Brujin counter.

(c) Repeats parts (a) and (b) for an 8-bit Brujin counter.

9.8 An alternative way to design a BCD counter is to use a BCD adder.

(a) Design a 3-digit BCD incrementor that adds 1 to a 3-digit 12-bit BCD operand, and derive the VHDL code.

(b) Use this circuit to implement a 3-digit BCD counter and derive the VHDL code.

(c) Compare the circuit complexity between this design and the counter discussed in Section 9.2.4.

9.9 For the PWM circuit in Section 9.2.5, can we replace the binary counter with a Brujin counter? Explain.

9.10 The PWM circuit can control the duty cycle, but its frequency is fixed. If the original frequency of the clock signal is f_{clk}, the frequency of the PWM circuit in Section 9.2.5 is $\frac{f_{clk}}{2^4}$. We can extend the PWM circuit to a programmable pulse generator by adding additional control signal to specify the desired frequency. Let k be a 4-bit signal that is interpreted as an unsigned divisor. The frequency of the new output pulse will be $\frac{f_{clk}}{k*2^4}$ if k is not 0 and will be $\frac{f_{clk}}{2^4}$ if k is 0. Design this circuit and derive the VHDL code.

9.11 A *stack* is a buffer in which the data is stored and retrieved in *first-in-last-out* fashion. In a synchronous stack, it should consist of the following I/O signals:

- w_data and r_data: data to be written (also known as *pushed*) into and read (also known as *popped*) from the stack.
- push and pop: control signals to enable the push or pop operation.
- full and empty: status signals.
- clk and reset: the clock and reset signals.

We can use a register file to construct this circuit. by following the design approach of the FIFO buffer.

(a) Draw a top-level diagram similar to that in Figure 9.13.

(b) Consider a stack of four words. Derive the VHDL code of the control circuit.

9.12 In the CAM of Section 9.3.3, a binary encoding circuit is included in the key file circuit to generate the address. This address is then decoded by the decoding circuit in the register file. We can eliminate both encoding and decoding circuits to make the design more efficient. Derive the VHDL code for the revised register file and the key file.

9.13 We can add a "mask" input to the CAM so that only a portion of the key will be used for search. For example, if the key is 16 bits and the mask input is "0000000011111111", only the eight LSBs of the key will be used for search. If a search finds multiple matches, the address with the smallest value will be used. Revise the key file of Section 9.3.3 to include this feature and derive the VHDL code.

9.14 Consider a combinational circuit that requires 128 ns to process input data and assume that it can always be divided into smaller parts of equal propagation delays. Let T_{cq} and T_{setup} of the register be 1 and 3 ns respectively. Determine the throughput and delay

(a) of the original circuit.

(b) if the circuit is converted into a 2-stage pipeline.

(c) if the circuit is converted into a 4-stage pipeline.

(d) if the circuit is converted into an 8-stage pipeline.

(e) if the circuit is converted into a 16-stage pipeline.

(f) if the circuit is converted into a 32-stage pipeline.

9.15   Convert the reduced-xor circuit in Section 7.4.1 into a four-stage pipelined circuit.

CHAPTER 10

FINITE STATE MACHINE: PRINCIPLE AND PRACTICE

A finite state machine (FSM) is a sequential circuit with "random" next-state logic. Unlike the regular sequential circuit discussed in Chapters 8 and 9, the state transitions and event sequence of an FSM do not exhibit a simple pattern. Although the basic block diagram of an FSM is similar to that of a regular sequential circuit, its design procedure is different. The derivation of an FSM starts with a more abstract model, such as a state diagram or an algorithm state machine (ASM) chart. Both show the interactions and transitions between the internal states in graphical formats. In this chapter, we study the representation, timing and implementation issues of an FSM as well as derivation of the VHDL code. Our emphasis is on the application of an FSM as the control circuit for a large, complex system, and our discussion focuses on the issues related to this aspect. As in previous chapters, our discussion is limited to the synchronous FSM, in which the state register is controlled by a single global clock.

10.1 OVERVIEW OF FSMS

As its name indicates, a *finite state machine (FSM)* is a circuit with internal states. Unlike the regular sequential circuits discussed in Chapters 8 and 9, state transition of an FSM is more complicated and the sequence exhibits no simple, regular pattern, as in a counter or shift register. The next-state logic has to be constructed from scratch and is sometimes known as "random" logic.

Formally, an FSM is specified by five entities: symbolic states, input signals, output signals, next-state function and output function. A state specifies a unique internal condition

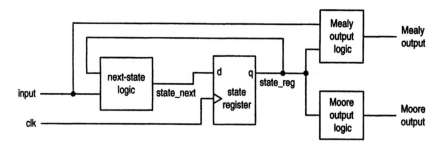

Figure 10.1   Block diagram of an FSM.

of a system. As time progresses, the FSM transits from one state to another. The new state is determined by the next-state function, which is a function of the current state and input signals. In a *synchronous FSM*, the transition is controlled by a clock signal and can occur only at the triggering edge of the clock. As we discussed in Section 8.2, our study strictly follows the synchronous design methodology, and thus coverage is limited to the synchronous FSM.

The output function specifies the value of the output signals. If it is a *function of the state* only, the output is known as a *Moore output*. On the other hand, if it is a *function of the state and input signals*, the output is known as a *Mealy output*. An FSM is called a *Moore machine* or *Mealy machine* if it contains only Moore outputs or Mealy outputs respectively. A complex FSM normally has both types of outputs. The differences and implications of the two types of outputs are discussed in Section 10.4.

The block diagram of an FSM is shown in Figure 10.1. It is similar to the block diagram of a regular sequential circuit. The state register is the memory element that stores the state of the FSM. It is synchronized by a global clock. The next-state logic implements the next-state function, whose input is the current state and input signals. The output logic implements the output function. This diagram includes both Moore output logic, whose input is the current state, and Mealy output logic, whose input is the current state and input signals. The main application of an FSM is to realize operations that are performed in a sequence of steps. A large digital system usually involves complex tasks or algorithms, which can be expressed as a sequence of actions based on system status and external commands. An FSM can function as the control circuit (known as the *control path*) that coordinates and governs the operations of other units (known as the *data path*) of the system. Our coverage of FSM focuses on this aspect. The actual construction of such systems is discussed in the next two chapters. FSMs can also be used in many simple tasks, such as detecting a unique pattern from an input data stream or generating a specific sequence of output values.

10.2   FSM REPRESENTATION

The design of an FSM normally starts with an abstract, graphic description, such as a state diagram or an ASM chart. Both descriptions utilize symbolic state notations, show the transition among the states and indicate the output values under various conditions. A state diagram or an ASM chart can capture all the needed information (i.e., state, input, output, next-state function, and output function) in a single graph.

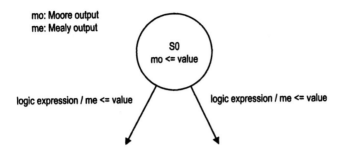

Figure 10.2 Notation for a state.

10.2.1 State diagram

A state diagram consists of nodes, which are drawn as circles (also known as *bubbles*), and one-direction transition arcs. The notation for nodes and arcs is shown in Figure 10.2. A *node* represents a unique state of the FSM and it has a unique symbolic name. An *arc* represents a transition from one state to another and is labeled with the condition that will cause the transition. The condition is expressed as a logic expression composed of input signals. An arc will be taken when the corresponding logic expression is evaluated to be logic '1'.

The output values are also specified on the state diagram. The Moore output is a function of state and thus is naturally placed inside the state bubble. On the other hand, the Mealy output depends on both state and input and thus is placed under the condition expression of the transition arcs. To reduce the clutter, we list only the output signals that are activated or asserted. An output signal will assume the default, unasserted value (*not* don't-care) if it is not listed inside the state bubble or under the logic expression of an arc. We use the following notation for an asserted output value:

```
signal_name <= asserted value;
```

In general, an asserted signal will be logic '1' unless specified otherwise.

The state diagram can best be explained by an example. Figure 10.3 shows the state diagram of a hypothetical memory controller FSM. The controller is between a processor and a memory chip, interpreting commands from the processor and then generating a control sequence accordingly. The commands, *mem*, *rw* and *burst*, from the processor constitute the input signals of the FSM. The *mem* signal is asserted to high when a memory access is required. The *rw* signal indicates the type of memory access, and its value can be either '1' or '0', for memory read and memory write respectively. The *burst* signal is for a special mode of a memory read operation. If it is asserted, four consecutive read operations will be performed. The memory chip has two control signals, *oe* (for output enable) and *we* (for write enable), which need to be asserted during the memory read and memory write respectively. The two output signals of the FSM, *oe* and *we*, are connected to the memory chip's control signals. For comparison purpose, we also add an artificial Mealy output signal, *we_me*, to the state diagram.

Initially, the FSM is in the idle state, waiting for the *mem* command from the processor. Once *mem* is asserted, the FSM examines the value of *rw* and moves to either the read1 state or the write state. These input conditions can be formalized to logic expressions, as shown in the transition arcs from the idle state:

- *mem′*: represents that no memory operation is required.

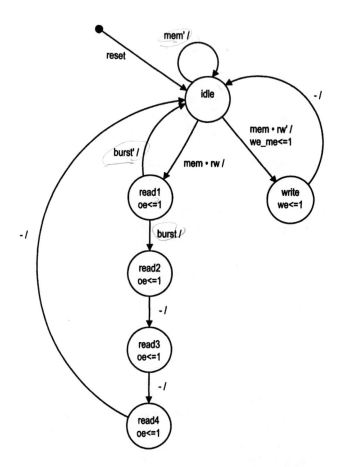

Figure 10.3   State diagram of a memory controller FSM.

- $mem \cdot rw$: represents that a memory read operation is required.
- $mem \cdot rw'$: represents that a memory write operation is required.

The results of these logic expressions are checked at the rising edge of the clock. If the mem' expression is true (i.e., mem is '0'), the FSM stays in the idle state. If the $mem \cdot rw$ expression is true (i.e., both mem and rw are '1'), the FSM moves to the read1 state. Once it is there, the oe signal is activated, as indicated in the state bubble. On the other hand, if the $mem \cdot rw'$ expression is true (i.e., mem is '1' and rw is '0'), the FSM moves to the write state and activates the we signal.

After the FSM reaches the read1 state, the $burst$ signal is examined at the next rising edge of the clock. If it is '1', the FSM will go through read2, read3 and read4 states in the next three clock cycles and then return to the idle state. Otherwise, the FSM returns to the idle state. We use the notation "−" to represent the "always true" condition. After the FSM reaches the write state, it will return to the idle state at the next rising edge of the clock.

The we_me signal is asserted only when the FSM is in the idle state and the $mem \cdot rw'$ expression is true. It will be deactivated when the FSM moves away from the idle state (i.e., to the write state). It is a Mealy output since its value depends on the state and the input signals (i.e., mem and rw).

In practice, we usually want to force an FSM into an initial state during system initialization. It is frequently done by an asynchronous reset signal, similar to the asynchronous reset signal used in a register of a regular sequential circuit. Sometimes a solid dot is used to indicate this transition, as shown in Figure 10.3. This transition is only for system initialization and has no effect on normal FSM operation.

10.2.2 ASM chart

An *algorithmic state machine (ASM)* chart is an alternative method for representing an FSM. Although an ASM chart contains the same amount of information as a state diagram, it is more descriptive. We can use an ASM chart to specify the complex sequencing of events involving commands (input) and actions (output), which is the hallmark of complex algorithms. An ASM chart representation can easily be transformed to VHDL code. It can also be extended to describe FSMD (FSM with a data path), which is discussed in the next two chapters.

An ASM chart is constructed of a network of ASM blocks. An *ASM block* consists of one state box and an optional network of decision boxes and conditional output boxes. A typical ASM block is shown in Figure 10.4. The *state box*, as its name indicates, represents a state in an FSM. It is identified by a symbolic state name on the top left corner of the state box. The action or output listed inside the box describes the desired output signal values when the FSM enters this state. Since the outputs rely on the state only, they correspond to the Moore outputs of the FSM. To reduce the clutter, we list only signals that are activated or asserted. An output signal will assume the default, unasserted value if it is not listed inside the box. We use the same notation for an asserted output signal:

```
signal_name <= asserted value;
```

Again, we assume that an asserted signal will be logic '1' unless specified otherwise.

A *decision box* tests an input condition to determine the exit path of the current ASM block. It contains a Boolean expression composed of input signals and plays a similar role to the logic expression in the transition arc of a state diagram. Because of the flexibility of the Boolean expression, it can describe more complex conditions, such as

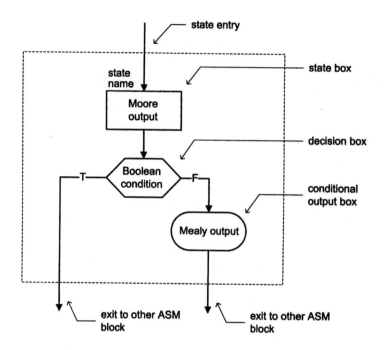

Figure 10.4   ASM block.

(a > b) **and** (c /= 1). Depending on the value of the Boolean expression, the FSM can follow either the *true path* or the *false path*, which are labeled as T or F in the exit paths of the decision box. If necessary, we can cascade multiple decision boxes inside an ASM block to describe a complex condition.

A *conditional output box* also lists asserted output signals. However, it can only be placed after an exit path of a decision box. It implies that these output signals can be asserted only if the condition of the previous decision box is met. Since the condition is composed of a Boolean expression of input signals, these output signals' values depend on the current state and input signals, and thus they are Mealy outputs. Again, to reduce clutter, we place a conditional output box in an ASM block only when the corresponding output signal is asserted. The output signal assumes the default, unasserted value when there is no conditional output box.

Since an ASM chart is another way of representing an FSM, an ASM chart can be converted to a state diagram and vice versa. An ASM block corresponds to a state and its transition arcs of a state diagram. The key for the conversion is the transformation between the logic expressions of the transition arcs in a state diagram and the decision boxes in an ASM chart.

The conversion can best be explained by examining several examples. The first example is shown in Figure 10.5. It is an FSM with no branching arches. The state diagram and the ASM chart are almost identical.

The second example is shown in Figure 10.6. The FSM has two transition arcs from the s0 state and has a Mealy output, y. The logic expressions a and a' of the transition arches are translated into a decision box with Boolean expression a = 1. Note that the two states are transformed into two ASM blocks. The decision and conditional output boxes are not new states, just actions associated with the ASM block s0.

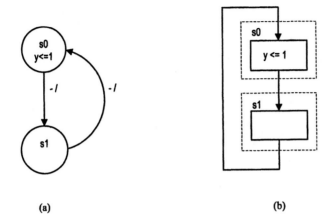

Figure 10.5   Example 1 of state diagram and ASM chart conversion.

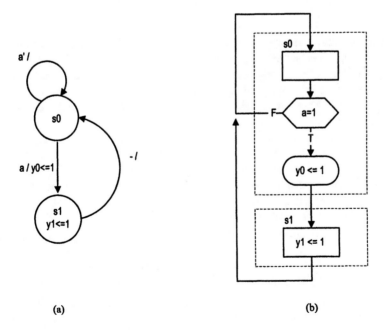

Figure 10.6   Example 2 of state diagram and ASM chart conversion.

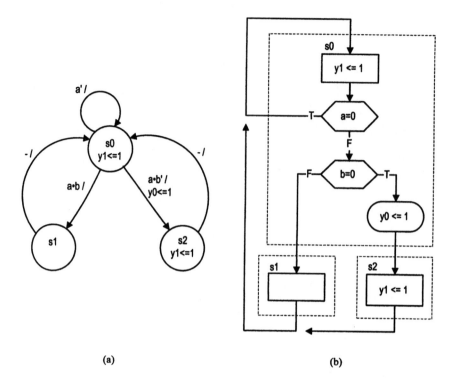

(a)   (b)

Figure 10.7 Example 3 of state diagram and ASM chart conversion.

The third example is shown in Figure 10.7. The transitions from the s0 state are more involved. We can translate the logic expressions a' and $a \cdot b'$ directly into two decision boxes of conditions a = 0 and (a = 1) **and** (b = 0). However, closer examination shows that the second decision box is on the false path of the first decision box, which implies that a is '1'. Thus, we can eliminate the a = 1 condition from the second decision box and make the decision simpler and more descriptive.

The fourth example is shown in Figure 10.8. The output of the FSM is more complex and depends on various input conditions. The state diagram needs multiple logic expressions in the transition arc to express various input conditions. The ASM chart can accommodate the situation and is more descriptive. Finally, the ASM chart of the previous memory controller FSM, whose state diagram is shown in Figure 10.3, is shown in Figure 10.9.

Since an ASM chart is used to model an FSM, two rules apply:

1. For a given input combination, there is one unique exit path from the current ASM block.

2. The exit path of an ASM block must always lead to a state box. The state box can be the state box of the current ASM block or a state box of another ASM block.

Several common errors are shown in Figure 10.10. The ASM chart of Figure 10.10(a) violates the first rule. There are two exit paths if a and b are both '1', and there is no exit path if a and b are both '0'. The ASM chart of Figure 10.10(b) also violates the first rule since there is no exit path when the condition of the decision box is false. The ASM chart of Figure 10.10(c) violates the second rule because the exit path of the bottom ASM block does not enter the top ASM block via the state box. The second rule essentially states that

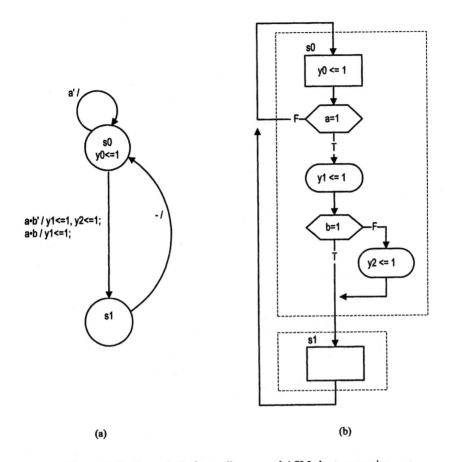

(a)   (b)

Figure 10.8   Example 4 of state diagram and ASM chart conversion.

the decision boxes and conditional output boxes are associated with a single ASM block and they cannot be shared by other ASM blocks.

An ASM chart and a state diagram contain the same information. Because of the use of decision boxes and flowchart-like graphs, an ASM chart can accommodate the complex conditions involved in state transitions and Mealy outputs, as shown in the third and fourth examples. On the other hand, an ASM chart may be cumbersome for an FSM with simple, straightforward state transitions, and a state diagram is preferred. We use mostly state diagrams in this chapter, but use mainly extended ASM charts while discussing the RT methodology in Chapters 11 and 12.

10.3   TIMING AND PERFORMANCE OF AN FSM

10.3.1   Operation of a synchronous FSM

While a state diagram or an ASM chart shows all the states and transitions, it does not provide information about *when* a transition takes place. In a synchronous FSM, the state transition is controlled by the rising edge of the system clock. Mealy output and Moore output are not directly related to the clock but are responding to input or state change. However, since

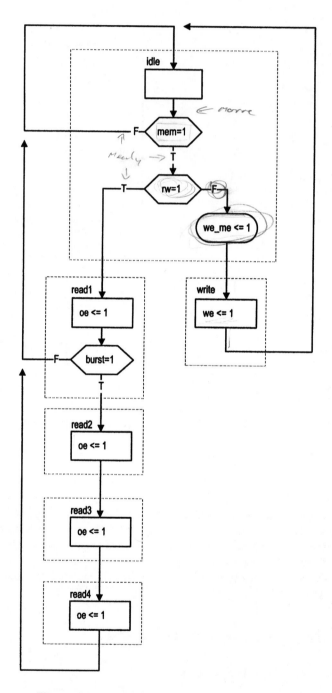

Figure 10.9   ASM chart of a memory controller FSM.

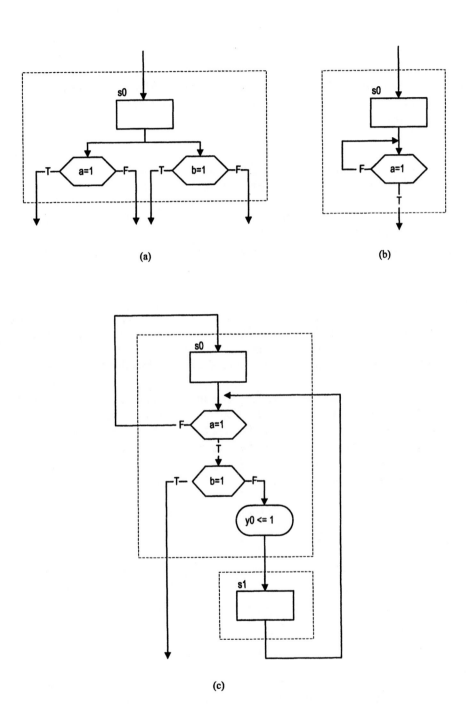

Figure 10.10 Common errors in ASM chart construction.

a Moore output depends only on the state, its transition is indirectly synchronized by the clock.

The timing of a synchronous FSM can best be explained by examining the operation of an ASM block. In an ASM chart, each ASM block represents a state of the FSM. Instead of moving "continuously" from one block to another block, as in a traditional flowchart, the transitions between ASM blocks can occur only at the rising edge of the clock. The operation of an ASM block transition can be interpreted as follows:

1. At the rising edge of the clock, the FSM enters a new state (and thus a new ASM block).
2. During the clock period, the FSM performs several operations. It activates the Moore output signals asserted in this state. It evaluates various Boolean expressions of the decision boxes and activates the Mealy output signals accordingly.
3. At the next rising edge of the clock (which is the end of the current clock period), the results of Boolean expressions are examined simultaneously, an exit path is determined, and the FSM enters the designated new ASM block.

A state and its transitions in a state diagram are interpreted in the same manner.

10.3.2 Performance of an FSM

When an FSM is synthesized, the physical components introduce propagation delays. Since the block diagram of an FSM is almost identical to that of a regular sequential circuit, the timing analysis of an FSM is similar to that of a regular sequential circuit, as discussed in Section 8.6. The main timing parameters associated with the block diagram of Figure 10.1 are:

- $T_{cq}, T_{setup}, T_{hold}$: the clock-to-q delay, setup time and hold time of the state register.
- $T_{next(max)}$: the maximal propagation delay of the next-state logic.
- $T_{output(mo)}$: the propagation delay of output logic for the Moore output.
- $T_{output(me)}$: the propagation delay of output logic for the Mealy output.

As in a regular sequential circuit, the performance of an FSM is characterized by the maximal clock rate (or minimal clock period). The minimal clock period is

$$T_c = T_{cq} + T_{next(max)} + T_{setup}$$

and the maximal clock rate is

$$f = \frac{1}{T_{cq} + T_{next(max)} + T_{setup}}$$

Since an FSM is frequently used as the controller, the response of the output signal is also important. A Moore output is characterized by the clock-to-output delay, which is

$$T_{co(mo)} = T_{cq} + T_{output(mo)}$$

A Mealy output may respond to the change of a state or an input signal. The former is characterized by the clock-to-output delay, similar to the Moore output:

$$T_{co(me)} = T_{cq} + T_{output(me)}$$

The latter is just the propagation delay of Mealy output logic, which is $T_{output(me)}$.

10.3.3   Representative timing diagram

The timing diagram helps us to better understand the operation of an FSM and generation of the output signals. It is especially critical when an FSM is used as a control circuit. One tricky part regarding the FSM timing concerns the rising edge of the clock. In an ideal FSM, there is no propagation delay, and thus the state and output signal change at the edge. If the state or output is fed to other synchronous components, which take a sample at the rising edge, it is difficult to determine what the value is. In reality, this will not happen since there is always a clock-to-q delay from the state register. To avoid confusion, this delay should always be included in the timing diagram.

A detailed, representative timing diagram of a state transition is shown in Figure 10.11. It is based on the FSM shown in Figure 10.6. We assume that the next state of the FSM (the state_next signal) is s0 initially. At t_1, the rising edge of the clock, the state register samples the state_next signal. After T_{cq} (at t_2), the state register stores the value and reflects the value in its output, the state_reg signal. This means that the FSM moves to the s0 state. At t_3, the a input changes from '0' to '1'. According to the ASM chart, the condition of the decision box is met and the true branch is taken. In terms of the circuit, the change of the a signal activates both the next-state logic and the Mealy output logic. After the delay of T_{next} (at t_4), the state_next signal changes to s1. Similarly, the Mealy output, y0, changes to '1' after $T_{output(me)}$ (at t_5). At t_6, the a signal switches back to '0'. The state_next and y0 signals respond accordingly. Note that the change of the state_next signal has no effect on the state register (i.e., the state of the FSM). At t_7, the a signal changes to '1' again, and thus the state_next and y0 signals become s1 and '1' after the delays. At t_8, the current period ends and a new rising edge occurs. The state register samples the state_next signal and stores the s1 value into the register. After T_{cq} (at t_9), the register obtains its new value and the FSM moves to the s1 state. The change in the state_reg signal triggers the next-state logic, Mealy output logic and Moore output logic. After the T_{next} delay (at t_{10}), the next-state logic generates a new value of s0. We assume that $T_{output(mo)}$ and $T_{output(me)}$ are similar. After this delay (at t_{11}), the Mealy output, y0, is deactivated, and the Moore output, y1, is activated. The y1 signal remains asserted for the entire clock cycle. At t_{12}, a new clock edge arrives, the state_reg signal changes to s0 after the T_{cq} delay (at t_{13}), and the FSM returns to the s0 state. The y1 signal is deactivated after the $T_{output(mo)}$ delay (at t_{14}).

The timing diagram illustrates the major difference between an ASM chart and a regular flowchart. In an ASM chart, the state transition (or ASM block transition) occurs only at the rising edge of the clock signal. Within the clock period, the Boolean condition and the next state may change but have no effect on the system state. The new state is determined solely by the values sampled at the rising edge of the clock.

10.4   MOORE MACHINE VERSUS MEALY MACHINE

As we discussed in Section 10.1, an FSM can be classified into a Moore machine or a Mealy machine. In theoretical computer science, a Moore machine and a Mealy machine are considered to have similar computation capability (both can recognize "regular expressions"), although a Mealy machine normally accomplishes the same task with fewer states. When the FSM is used as a control circuit, the control signals generated by a Moore machine and a Mealy machine have different timing characteristics. Understanding the subtle timing difference is critical for the correctness and efficiency of a control circuit. We use a simple

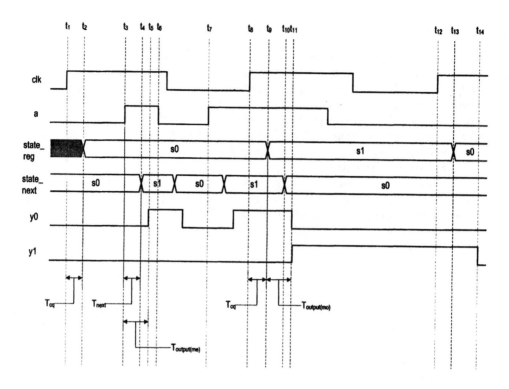

Figure 10.11 FSM timing diagram.

edge detection circuit to illustrate the difference between a Mealy machine and a Moore machine.

10.4.1 Edge detection circuit

We assume that a synchronous system is connected to a slowly varying input signal, strobe, which can be asserted to '1' for a long time (much greater than the clock period of the FSM). An edge detection circuit is used to detect the rising edge of the strobe signal. It generates a "short" pulse when the strobe signal changes from '0' to '1'. The width of the output pulse is about the same or less than a clock period of the FSM. Since the intention is to show the difference between a Mealy machine and a Moore machine, we are deliberately vague about the specification of the width and timing of the output pulse.

The basic design idea is to construct an FSM that has a zero state and a one state, which represent that the input has been '0' or '1' for a long period of time respectively. The FSM has a single input signal, strobe, and a single output signal. The output will be asserted "momentarily" when the FSM transits from the zero state to the one state.

We first consider a design based on a Moore machine. The state diagram is shown in Figure 10.12(a). There are three states. In addition to the zero and one states, the FSM also has an edge state. When strobe becomes '1' in the zero state, it implies that strobe changes from '0' to '1'. The FSM moves to the edge state, in which the output signal, p1, is asserted. In normal operation, strobe should continue to be '1' and the FSM moves to the one state at the next rising edge of the clock and stays there until strobe returns to '0'.

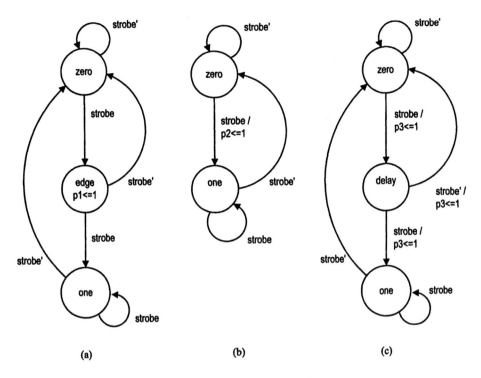

Figure 10.12   Edge detector state diagrams.

If strobe is really short and changes to '0' in the edge state, the FSM will return to the zero state. A representative timing diagram is shown in the top portion of Figure 10.13.

The second design is based on a Mealy machine. The state diagram is shown in Figure 10.12(b). It consists of only the zero and one states. When strobe changes from '0' to '1' in the zero state, the FSM moves to the one state. From the state diagram, it seems that the output signal, p2, is asserted when the FSM transit from the zero state to the one state. Actually, p2 is asserted in the zero state whenever strobe is '1'. When the FSM moves to the one state, p2 will be deasserted. The timing diagram is shown in the middle portion of Figure 10.13.

For demonstration purposes, we also include a version that combines both types of outputs. The third design inserts a delay state into the Mealy machine–based design and prolongs the output pulse for one extra clock cycle. The state diagram is shown in Figure 10.12(c). In this design, the FSM will assert the output, p3, in the zero state, as in the second design. However, the FSM moves to the delay state afterward and forces p3 to be asserted for another clock cycle by placing the assertion on both transition edges of the delay state. Note that since p3 is asserted in the delay state under all transition arcs, it implies that p3 will be asserted in the delay state regardless of the input condition. The behavior of the FSM in the delay state is similar to the edge state of the Moore machine–based design, and we can also move the output assertion, p3<=1, into the bubble of the delay state. The timing diagram is shown in the bottom portion of Figure 10.13.

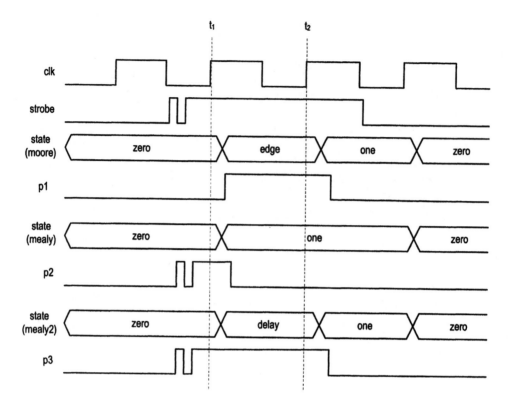

Figure 10.13   Edge detector timing diagram.

10.4.2 Comparison of Moore output and Mealy output

All three edge detector designs can generate a "short" pulse when the input changes from '0' to '1', but there are subtle differences. Understanding the differences is the key to deriving a correct and efficient FSM and an FSM-based control circuit.

There are three major differences between the Moore machine and Mealy machine–based designs. First, a Mealy machine normally requires fewer states to perform the same task. This is due to the fact that its output is a function of states and external inputs, and thus several possible output values can be specified in one state. For example, in the zero state of the second design, p2 can be either '0' or '1', depending on the value of strobe. Thus, the Mealy machine–based design requires only two states whereas the Moore machine–based design requires three states.

Second, a Mealy machine can generate a faster response. Since a Mealy output is a function of input, it changes whenever the input meets the designated condition. For example, in Mealy machine–based design, if the FSM is in the zero state, p2 is asserted immediately after strobe changes from '0' to '1', as shown in the timing diagram. On the other hand, a Moore machine reacts indirectly to input changes. The Moore machine–based design also senses the changes of strobe in the zero state. However, it has to wait until the next state (i.e., the edge state) to respond. The change causes the FSM to move to the edge state. At the next rising edge of the clock, the FSM moves to this state and p1 responds accordingly, as shown in the timing diagram. In a synchronous system, the distinction between a Mealy output and a Moore output normally means a delay of one clock cycle. Recall that the input signal of a synchronous system is sampled only at the rising edge of the clock. Let us assume that the output of the edge detection circuit is used by another synchronous system. Consider the first transition edge of strobe in Figure 10.13. The p2 signal can be sampled at t_1. However, the p1 signal is not available at that time because of the clock-to-q delay and output logic delay. Its value can be sampled only by the next rising edge at t_2.

The third difference involves the control of the width and timing of the output signal. In a Mealy machine, the width of an output signal is determined by the input signal. The output signal is activated when the input signal meets the designated condition and is normally deactivated when the FSM enters a new state. Thus, its width varies with input and can be very narrow. Also, a Mealy machine is susceptible to glitches in the input signal and passes these undesired disturbances to the output. This is shown in the p2 signal of Figure 10.13. On the other hand, the output of a Moore machine is synchronized with the clock edge and its width is about the same as a clock period. It is not susceptible to glitches from the input signal. Although the output logic can still introduce glitches, this can be overcome by clever output buffering schemes, which are discussed in Section 10.7.

As mentioned earlier, our focus on FSM is primarily on its application as a control circuit. From this perspective, selection between a Mealy machine and a Moore machine depends on the need of control signals. We can divide control signals into two categories: edge-sensitive and level-sensitive. An *edge-sensitive* control signal is used as input for a sequential circuit synchronized by the same clock. A simple example is the enable signal of a counter. Since the signal is sampled only at the rising edge of the clock, the width of the signal and the existence of glitches do not matter as long as it is stable during the setup and hold times of the clock edge. Both the Mealy and the Moore machines can generate output signals that meet this requirement. However, a Mealy machine is preferred since it uses fewer states and responds one clock faster than does a Moore machine. Note that the p3 signal generated by the modified Mealy machine will be active for two clock edges and is actually incorrect for an edge-sensitive control signal.

A *level-sensitive* control signal means that a signal has to be asserted for a certain amount of time. When asserted, it has to be stable and free of glitches. A good example is the write enable signal of an SRAM chip. A Moore machine is preferred since it can accurately control the activation time of its output, and can shield the control signal from input glitches. Because of the potential glitches, the p3 signal is again not desirable.

10.5 VHDL DESCRIPTION OF AN FSM

The block diagram of an FSM shown in Figure 10.1 is similar to that of the regular sequential circuit shown in Figure 8.5. Thus, derivation of VHDL code for an FSM is similar to derivation for a regular sequential circuit. We first identify and separate the memory elements and then derive the next-state logic and output logic. There are two differences in the derivation. The first is that symbolic states are used in an FSM description. To capture this kind of representation, we utilize VHDL's *enumeration data type* for the state registers. The second difference is in the derivation of the next-state logic. Instead of using a regular combinational circuit, such as an incrementor or shifter, we have to construct the code according to a state diagram or ASM chart.

We use the previous memory controller FSM to show the derivation procedure in the following subsections.

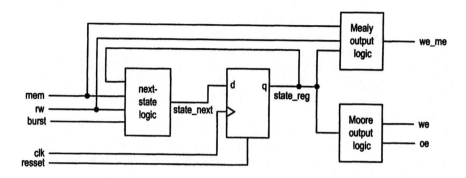

Figure 10.14   Block diagram of a memory controller FSM.

10.5.1 Multi-segment coding style

The first method is to derive the VHDL code according to the blocks of a block diagram, and we call it the multi-segment coding style. The block diagram of the previous memory controller is shown in Figure 10.14. There are four blocks and we use a VHDL code segment for each block. The complete VHDL code is shown in Listing 10.1.

Listing 10.1   Multi-segment memory controller FSM

```
    library ieee;
    use ieee.std_logic_1164.all;
    entity mem_ctrl is
      port(
5       clk, reset: in std_logic;
        mem, rw, burst: in std_logic;
        oe, we, we_me: out std_logic
      );
    end mem_ctrl ;
10
    architecture mult_seg_arch of mem_ctrl is
      type mc_state_type is
        (idle, read1, read2, read3, read4, write);
      signal state_reg, state_next: mc_state_type;
15  begin
      -- state register
      process(clk,reset)
      begin
        if (reset='1') then
20        state_reg <= idle;
        elsif (clk'event and clk='1') then
          state_reg <= state_next;
        end if;
      end process;
25    -- next-state logic
      process(state_reg,mem,rw,burst)
      begin
        case state_reg is
          when idle =>
30          if mem='1' then
```

```
                 if rw='1' then
                     state_next <= read1;
                 else
                     state_next <= write;
35               end if;
             else
                 state_next <= idle;
             end if;
         when write =>
40           state_next <= idle;
         when read1 =>
             if (burst='1') then
                 state_next <= read2;
             else
45               state_next <= idle;
             end if;
         when read2 =>
             state_next <= read3;
         when read3 =>
50           state_next <= read4;
         when read4 =>
             state_next <= idle;
     end case;
 end process;
55 -- Moore output logic
 process(state_reg)
 begin
     we <= '0'; -- default value
     oe <= '0';  -- default value
60   case state_reg is
         when idle =>
         when write =>
             we <= '1';
         when read1 =>
65           oe <= '1';
         when read2 =>
             oe <= '1';
         when read3 =>
             oe <= '1';
70       when read4 =>
             oe <= '1';
     end case;
 end process;
 -- Mealy output logic
75 process(state_reg,mem,rw)
 begin
     we_me <= '0'; -- default value
     case state_reg is
         when idle =>
80           if (mem='1') and (rw='0') then
                 we_me <= '1';
             end if;
         when write =>
```

```
                 when read1 =>
85               when read2 =>
                 when read3 =>
                 when read4 =>
           end case;
        end process;
90 end mult_seg_arch;
```

Inside the architecture declaration, we use the VHDL's enumeration data type. The data type is declared as

```
type mc_state_type is (idle,read1,read2,read3,read4,write);
```

The syntax of the enumeration data type statement is very simple:

```
type type_name is (list_of_all_possible_values);
```

It simply enumerates all possible values in a list. In this particular example, we list all the symbolic state names. The next statement then uses this newly defined type as the data type for the state register's input and output:

```
signal state_reg, state_next: mc_state_type;
```

The architecture body is divided into four code segments. The first segment is for the state register. Its code is like that of a regular register except that a user-defined data type is used for the signal. We use an asynchronous reset signal for initialization. The state register is cleared to the idle state when the reset signal is asserted.

The second code segment is for the next-state logic and is the key part of the FSM description. It is patterned after the ASM chart of Figure 10.9. We use a case statement with state_reg as the selection expression. The state_reg signal is the output of the state register and represents the current state of the FSM. Based on its value and input signal, the next state, denoted by the state_next signal, can be determined. As shown in the previous segment, the next state will be stored into the state register and becomes the new state at the rising edge of the clock. The state_next signal can be derived directly from the ASM block. For a simple ASM block, such as the read2 block, there is only one exit path and the state_next signal is very straightforward:

```
state_next <= idle;
```

For a block with multiple exit paths, we can use if statements to code the decision boxes. The Boolean condition inside a decision box can be directly translated to the Boolean expression of the if statement, and the two exit paths can be expressed as the then branch and the else branch of the if statement. Thus, we can follow the decision boxes and derive the VHDL code for the state_next signal accordingly. For example, in the idle block, the cascade decision boxes can be translated into a nested if statement:

```
if mem='1' then
   if rw='1' then
      state_next <= read1;
   else
      state_next <= write;
   end if;
else
   state_next <= idle;
end if;
```

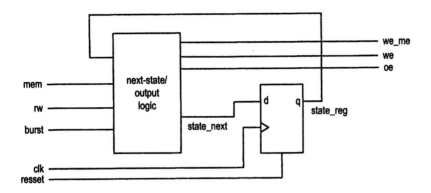

Figure 10.15   Block diagram of a two-segment memory controller FSM.

Note that the ASM has three possible exit paths from the `idle` block, and thus the `state_next` signal has three possible values.

The third code segment is the Moore output logic. Again, we use a case statement with `state_reg` as the selection expression. Note that since the Moore output is a function of state only, no input signal is in the sensitive list. Our code follows the ASM chart. Two sequential signal assignment statements are used to represent the default output value:

```
we <= '0';
oe <= '0';
```

If an output signal is asserted inside a state box, we put a signal assignment statement in the corresponding choice in the VHDL code to overwrite the default value.

The fourth code segment is the Mealy output logic. Note that some input signal is now in the sensitive list. Again, following the ASM chart, we use a case statement with `state_reg` as the selection expression and use an if statement for the decision box. The Mealy output, the `we_me` signal, will be assigned to the designated value according to the input condition.

We intentionally use the case statement to demonstrate the relationship between the code and the ASM chart. It may become somewhat cumbersome. The segment can also be written in a more compact but ad hoc way. For example, the Mealy output logic segment can be rewritten as

```
we_me <= '1' when ((state_reg=idle) and (mem='1') and
                    (rw='0')) else
         '0';
```

10.5.2 Two-segment coding style

The two-segment coding style divides an FSM into a state register segment and a combinational circuit segment, which integrates the next-state logic, Moore output logic and Mealy output logic. In VHDL code, we need to merge the three segments and move the `state_next`, `oe`, `we` and `we_me` signals into a single process. The block diagram is shown in Figure 10.15. The architecture body of this revised code is shown in Listing 10.2.

Listing 10.2   Two-segment memory controller FSM

```
architecture two_seg_arch of mem_ctrl is
    type mc_state_type is
```

```vhdl
              (idle, read1, read2, read3, read4, write);
       signal state_reg, state_next: mc_state_type;
  begin
       -- state register
       process(clk,reset)
       begin
           if (reset='1') then
               state_reg <= idle;
           elsif (clk'event and clk='1') then
               state_reg <= state_next;
           end if;
       end process;
       -- next-state logic and output logic
       process(state_reg,mem,rw,burst)
       begin
           oe <= '0';      -- default values
           we <= '0';
           we_me <= '0';
           case state_reg is
               when idle =>
                   if mem='1' then
                       if rw='1' then
                           state_next <= read1;
                       else
                           state_next <= write;
                           we_me <= '1';
                       end if;
                   else
                       state_next <= idle;
                   end if;
               when write =>
                   state_next <= idle;
                   we <= '1';
               when read1 =>
                   if (burst='1') then
                       state_next <= read2;
                   else
                       state_next <= idle;
                   end if;
                   oe <= '1';
               when read2 =>
                   state_next <= read3;
                   oe <= '1';
               when read3 =>
                   state_next <= read4;
                   oe <= '1';
               when read4 =>
                   state_next <= idle;
                   oe <= '1';
           end case;
       end process;
  end two_seg_arch;
```

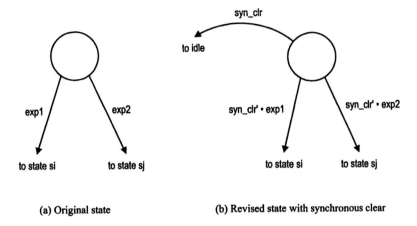

(a) Original state

(b) Revised state with synchronous clear

Figure 10.16   Adding synchronous clear to a state diagram.

10.5.3   Synchronous FSM initialization

An alternative for the asynchronous initialization is to use a synchronous clear signal. To achieve this goal, we have to add an additional transition arc for every state. The logic expression of this arc corresponds to the assertion of the synchronous clear signal and is given preference over other conditions. Assume that the syn_clr signal is added to the FSM for this purpose and an FSM will be forced to the idle state when the syn_clr signal is asserted. The required revision for a state is shown in Figure 10.16.

Although revising a state diagram or an ASM chart introduces a significant amount of clutter, this can be done easily in VHDL. We just add an extra if statement to check the syn_clr signal in the next-state logic segment. If the condition syn_clr='1' is true, the idle value will be assigned to the state_next signal. Otherwise, the FSM takes the else branch and performs the normal transition. The needed revisions for the memory controller FSM example are shown below.

```
entity mem_ctrl is
port (
    syn_clr: in std_logic;    -- new input
    . . .

architecture mult_seg_arch of mem_ctrl is
. . .
begin
. . .
    -- next-state logic
    process(state_reg,mem,rw,burst,syn_clr)
    begin
        if (syn_clr='1') then    -- synchronous clear
            state_next <= idle;
        else    -- original state_next values
            case state_reg is
                when idle =>
```

```
        end case;
      end if;
  end process;
  . . .
```

10.5.4 One-segment coding style and its problem

We may be tempted to make the code more compact and describe the FSM in a single segment, as shown in Listing 10.3.

Listing 10.3   One-segment memory controller FSM

```
  architecture one_seg_wrong_arch of mem_ctrl is
    type mc_state_type is
        (idle, read1, read2, read3, read4, write);
    signal state_reg: mc_state_type;
5 begin
    process(clk,reset)
    begin
      if (reset='1') then
          state_reg <= idle;
10    elsif (clk'event and clk='1') then
          oe <= '0';       -- default values
          we <= '0';
          we_me <= '0';
          case state_reg is
15          when idle =>
              if mem='1' then
                  if rw='1' then
                      state_reg <= read1;
                  else
20                    state_reg <= write;
                      we_me <= '1';
                  end if;
              else
                  state_reg <= idle;
25            end if;
            when write =>
              state_reg <= idle;
              we <= '1';
            when read1 =>
30            if (burst='1') then
                  state_reg <= read2;
              else
                  state_reg <= idle;
              end if;
35            oe <= '1';
            when read2 =>
              state_reg <= read3;
              oe <= '1';
            when read3 =>
40            state_reg <= read4;
              oe <= '1';
```

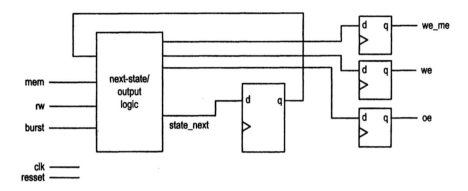

Figure 10.17   FSM with unwanted output buffers.

```
          when read4 =>
             state_reg <= idle;
             oe <= '1';
45        end case;
       end if;
     end process;
 end one_seg_wrong_arch;
```

Unfortunately, this code suffers the same problem as that of the similar regular sequential circuit code discussed in Section 8.7. Recall that a left-hand-side signal within the clk'event and clk='1' branch infers a register. While this is the desired effect for the state_reg signal, three unwanted registers are inferred for the oe, we and we_me signals, as shown in Figure 10.17 (for clarity, the connection lines for the clk and reset signals are not shown). These signals are delayed by one clock cycle and the code does not meet the specification described by the ASM chart. Although we can fix the problem by using a separate process for the output logic, the resulting code is less clear. We generally refrain from this style of coding.

10.5.5   Synthesis and optimization of FSM

After dividing a sequential circuit into a register and a combinational circuit, we can apply RT-level optimization techniques for the combinational circuit. However, these techniques are mainly for regular combinational circuits. The next-state logic and output logic of the FSMs are normally random in nature since the code includes primarily case and if statements and does not involve complex operators. These circuits are implemented by gate-level components, and there is very little optimization that we can do when writing RT-level VHDL code. Utilizing two-segment coding provides some degree of sharing since the Boolean expressions inside the decision boxes are used by both next-state logic and output logic.

Theoretically, there is a technique to identify the "equivalent states" of an FSM. We can merge these states into one state and thus reduce the number of states of the FSM. However, in a properly designed FSM, the chance of finding a set of equivalent states is very slim, and this technique is not always applied in the design and synthesis process.

There is one other unique opportunity to reduce the complexity of the combinational circuit of the FSM: assigning proper binary representations for the symbolic states. This issue is discussed in the next section.

The multi- and two-segment coding approach of previous subsections is very general and we can use the two VHDL listings as templates. The key to developing good VHDL code for an FSM is the derivation of an efficient and correct state diagram or ASM chart. Once it is completed, obtaining VHDL code becomes more or less a mechanical procedure. Some design entry software can accept a graphical state diagram and convert it to VHDL code automatically.

10.6   STATE ASSIGNMENT

Our discussion of FSM so far utilizes only symbolic states. During synthesis, each symbolic state has to be mapped to a unique binary representation so that the FSM can be realized by physical hardware. *State assignment* is the process of mapping symbolic values to binary representations.

10.6.1   Overview of state assignment

For a synchronous FSM, the circuit is not delay sensitive and is immune to hazards. As long as the clock period is large enough, the synthesized circuit will function properly for any state assignment. However, physical implementation of next-state logic and output logic is different for each assignment. A good assignment can reduce the circuit size and decrease the propagation delays, which in turn, increases the clock rate of the FSM.

An FSM with n symbolic states requires a state register of at least $\lceil \log_2 n \rceil$ bits to encode all possible symbolic values. We sometimes utilize more bits for other purposes. There are several commonly used state assignment schemes:

- *Binary* (or *sequential*) *assignment*: assigns states according to a binary sequence. This scheme uses a minimal number of bits and needs only a $\lceil \log_2 n \rceil$-bit register.
- *Gray code assignment*: assigns states according to a Gray code sequence. This scheme also uses a minimal number of bits. Because only one bit changes between the successive code words in the sequence, we may reduce the complexity of next-state logic if assigning successive code words to neighboring states.
- *One-hot assignment*: assigns one bit for each state, and thus only a single bit is '1' (or "hot") at a time. For an FSM with n states, this scheme needs an n-bit register.
- *Almost one-hot assignment*: is similar to the one-hot assignment except that the all-zero representation ("0 $\cdots$ 0") is also included. The all-zero state is frequently used as the initial state since it can easily be reached by asserting the asynchronous reset signal of D FFs. This scheme needs an $(n-1)$-bit register for n states.

Although one-hot and almost one-hot assignments need more register bits, empirical data from various studies show that these assignments may reduce the circuit size of next-state logic and output logic. Table 10.1 illustrates these schemes used for the previous memory controller FSM.

Obtaining the optimal assignment is very difficult. For example, if we choose the one-hot scheme for an FSM with n states, there are $n!$ (which is worse than 2^n) possible assignments. It is not practical to obtain the optimal assignment by examining all possible combinations. However, there exists special software that utilizes heuristic algorithms that can obtain a good, suboptimal assignment.

Table 10.1   State assignment example

	Binary assignment	Gray code assignment	One-hot assignment	Almost one-hot assignment
idle	000	000	000001	00000
read1	001	001	000010	00001
read2	010	011	000100	00010
read3	011	010	001000	00100
read4	100	110	010000	01000
write	101	111	100000	10000

10.6.2   State assignment in VHDL

In some situations, we may want to specify the state assignment for an FSM manually. This can be done implicitly or explicitly. In *implicit state assignment*, we keep the original enumeration data type but pass the desired assignment by other mechanisms. The VHDL standard does not define any rule for mapping the values of an enumeration data type to a set of binary representations. It is performed during synthesis. One way to pass the desired statement assignment to software is to use a VHDL feature, known as a *user attribute*, to set a "directive" to guide operation of the software. A user attribute has no effect on the semantics of VHDL code and is recognized only by the software that defines it. The IEEE 1076.6 RTL synthesis standard defines an attribute named enum_encoding for encoding the values of an enumeration data type. This attribute can be used for state assignment. For example, if we wish to assign the binary representations "0000", "0100", "1000", "1001", "1010" and "1011" to the idle, write, read1, read2, read3 and read4 states of the memory controller FSM, we can add the following VHDL segment to the original code:

```
type mc_state_type is (idle,write,read1,read2,read3,read4);
attribute enum_encoding: string;
attribute enum_encoding of mc_state_type:
          type is "0000 0100 1000 1001 1010 1011";
```

This user attribute is very common and should be accepted by most synthesis software.

Synthesis software normally provides several simple state assignment schemes similar to the ones discussed in the previous subsection. If we don't utilize a user attribute, we can specify the desired scheme as a parameter while invoking the software. If nothing is specified, the software will perform the state assignment automatically. It normally selects between binary assignment and one-hot assignment, depending on the characteristics of the targeting device technology. We can also use specialized FSM optimization software to obtain a good, suboptimal assignment.

We can *explicitly* specify the desired state assignment by replacing the symbolic values with the actual binary representations, and use the std_logic_vector data type for this purpose. To demonstrate this scheme, we incorporate the previous state assignment into the memory controller FSM. The revised multi-segment VHDL code is shown in Listing 10.4.

Listing 10.4   Explicit user-defined state assignment

```
architecture state_assign_arch of mem_ctrl is
   constant idle:  std_logic_vector(3 downto 0):="0000";
   constant write: std_logic_vector(3 downto 0):="0100";
   constant read1: std_logic_vector(3 downto 0):="1000";
```

```vhdl
 5      constant read2: std_logic_vector(3 downto 0):="1001";
        constant read3: std_logic_vector(3 downto 0):="1010";
        constant read4: std_logic_vector(3 downto 0):="1011";
        signal state_reg,state_next: std_logic_vector(3 downto 0);
    begin
10     -- state register
        process(clk,reset)
        begin
           if (reset='1') then
              state_reg <= idle;
15         elsif (clk'event and clk='1') then
              state_reg <= state_next;
           end if;
        end process;
        -- next-state logic
20      process(state_reg,mem,rw,burst)
        begin
           case state_reg is
              when idle =>
                 if mem='1' then
25                  if rw='1' then
                       state_next <= read1;
                    else
                       state_next <= write;
                    end if;
30               else
                    state_next <= idle;
                 end if;
              when write =>
                 state_next <= idle;
35            when read1 =>
                 if (burst='1') then
                    state_next <= read2;
                 else
                    state_next <= idle;
40               end if;
              when read2 =>
                 state_next <= read3;
              when read3 =>
                 state_next <= read4;
45            when read4 =>
                 state_next <= idle;
              when others =>
                 state_next <= idle;
           end case;
50      end process;
        -- Moore output logic
        process(state_reg)
        begin
           we <= '0'; -- default value
55         oe <= '0'; -- default value
           case state_reg is
              when idle =>
```

```
         when write =>
             we <= '1';
60       when read1 =>
             oe <= '1';
         when read2 =>
             oe <= '1';
         when read3 =>
65           oe <= '1';
         when read4 =>
             oe <= '1';
         when others =>
       end case;
70    end process;
      -- Mealy output logic
      we_me <= '1' when ((state_reg=idle) and (mem='1') and
                         (rw='0')) else
             '0';
75 end state_assign_arch;
```

In this code, we use std_logic_vector(3 downto 0) as the state register's data type. Six constants are declared to represent the six symbolic state names. Because of the choice of the constant names, the appearance of the code is very similar to that of the original code. However, the name here is just an alias of a binary representation, but the name in the original code is a value of the enumeration data type. One difference in the next-state logic code segment is an extra when clause:

```
when others =>
    state_next <= idle;
```

This revision is necessary since the selection expression of the case statement, state_reg, now is with the std_logic_vector(3 downto 0) data type, and thus has 9^4 possible combinations. The **when others** clause is used to cover all the unused combinations. This mean that when the FSM reaches an unused binary representation (e.g., "1111"), it will return to the idle state in the next clock cycle. We can also use

```
when others =>
    state_next <= "----";
```

if the software accepts the don't-care expression. A **when others** clause is also added for the Moore output code segment.

The explicit state assignment allows us to have more control over the FSM but makes the code more difficult to maintain and prevents the use of FSM optimization software. Unless there is a special need, using an enumeration data type for state representation is preferred.

10.6.3 Handling the unused states

When we map the symbolic states of an FSM to binary representations, there frequently exist unused binary representations (or states). For example, there are six states in the memory controller FSM. If the binary assignment is used, a 3-bit (i.e., $\lceil \log_2 6 \rceil$) register is needed. Since there are 2^3 possible combinations from 3 bits, two binary states are not used in the mapping. If one-hot state assignment is used, there are 58 (i.e., $2^6 - 6$) unused states.

During the normal operation, the FSM will not reach these states; however, it may accidentally enter an unused state due to noise or an external disturbance. One question is what we should do if the FSM reaches an unused state.

In certain applications, we can simply ignore the situation. It is because we assume that the error will never happen, or, if it happens, the system can never recover. In the latter case, there is nothing we can do with the error.

On the other hand, some applications can resume from a short period of anomaly and continue to run. In this case we have to design an FSM that can recover from the unused states. It is known as a *fault-tolerant* or *safe FSM*. For an FSM coded with an explicit state assignment, incorporating this feature is straightforward. We just specify the desired action in the **when others** clause of the case statement. For example, the state_assign_arch architecture in Listing 10.4 is a safe FSM. The code specifies that the FSM returns to the idle state if it enters an unused state:

```
when others =>
    state_next <= idle;
```

If desired, we can revise the code to add an error state for special error handling:

```
when others =>
    state_next <= error;
```

There is no easy way to specify a safe FSM if the enumeration data type is used. Since all possible values of the enumeration data type are used in the case statement of the next-state logic, there is no unused state in VHDL code. The unused states emerge only later during synthesis, and thus they cannot be handled in VHDL code. Some software accepts an artificially added **when others** clause for the unused states. However, by VHDL definition, this clause is redundant and may not be interpreted consistently by different synthesis software.

10.7 MOORE OUTPUT BUFFERING

We can add a buffer by inserting a register or a D FF to any output signal. The purpose of an output buffer is to remove glitches and minimize the clock-to-output delay (T_{co}). The disadvantage of this approach is that the output signal is delayed by one clock cycle.

Since the output of an FSM is frequently used for control purposes, we sometimes need a fast, glitch-free signal. We can apply the regular output buffering scheme to a Mealy or Moore output signal. The buffered signal, of course, is delayed by one clock cycle. For a Moore output, it is possible to obtain a buffered signal without the delay penalty. The following subsections discuss how to design an FSM to achieve this goal.

10.7.1 Buffering by clever state assignment

In a typical Moore machine, we need combinational output logic to implement the output function, as shown in Figure 10.1. Since the Moore output is not a function of input signals, it is shielded from the glitches of the input signals. However, the state transition and output logic may still introduce glitches to the output signals. There are two sources of glitches. The first is the possible simultaneous multiple-bit transitions of the state register, as from the "111" state to the "000" state. Even the register bits are controlled by the same clock, the clock-to-q delay of each D FF may be slightly different, and thus a glitch may show up in the output signal. The second source is the possible hazards inside the output logic.

Table 10.2 State assignment for the memory controller FSM output buffering

	q_3q_2 (oe)(we)	q_1q_0	$q_3q_2q_1q_0$
idle	00	00	0000
read1	10	00	1000
read2	10	01	1001
read3	10	10	1010
read4	10	11	1011
write	01	00	0100

Recall that the clock-to-output delay (T_{co}) is the sum of the clock-to-q delay (T_{cq}) of the register and the propagation delay of the output logic. The existence of the output logic clearly increases the clock-to-output delay.

One way to reduce the effect of the output logic is to eliminate it completely by clever state assignment. In this approach, we first allocate a register bit to each Moore output signal and specify its value according to the output function. Again, let us consider the memory controller FSM. We can assign two register bits according to the output values.of the oe and we signals, as shown in the first column of Table 10.2. Since some states may have the same output patterns, such as the read1, read2, read3 and read4 states of the memory controller, we need to add additional register bits to ensure that each state is mapped to a unique binary representation. In this example, we need at least two extra bits to distinguish the four read states, as shown in the second column of Table 10.2. We then can complete the state assignment by filling the necessary values for the idle and write states, as shown in the third column of Table 10.2. In this state assignment, the value of oe is identical to the value of state_reg(3), and the value of we is identical to the value of state_reg(2). In other words, the output function can be realized by connecting the output signals to the two register bits, and the output logic is reduced to wires. This implementation removes the sources of glitches and reduces T_{co} to T_{cq}.

This design requires manual state assignment and access to individual register bits. Only the explicit state assignment can satisfy the requirement. The state_assign_arch architecture in Listing 10.4 actually uses the state assignment from Table 10.2. We can replace the Moore output logic code segment by connecting the output signals directly to the register's output:

```
—— Moore output logic
oe <= state_reg(3);
we <= state_reg(2);
```

Because the state register is also used as an output buffer, this approach potentially uses fewer register bits for certain output patterns. The disadvantage of this method is the manual manipulation of the state assignment. It becomes tedious as the number of states or output signals grows larger. Furthermore, the assignment has to be modified whenever the number of output signals is changed, the number of states is changed, or the output function is modified. This makes the code error-prone and difficult to maintain.

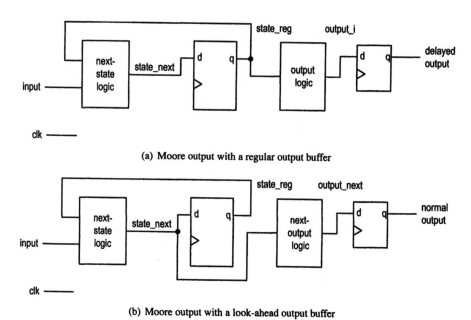

(a) Moore output with a regular output buffer

(b) Moore output with a look-ahead output buffer

Figure 10.18   Block diagrams of output buffering schemes.

10.7.2 Look-ahead output circuit for Moore output

A more systematic approach to Moore output buffering is to use a *look-ahead output circuit*. The basic idea is to buffer the *next output value* to cancel the one-clock delay introduced by the output buffer. In most systems, we don't know a signal's next or future value. However, in an FSM, the next value of the state register is generated by next-state logic and is always available.

This scheme can best be explained by examining the basic FSM block diagram. The block diagram of an FSM with a regular output buffering structure is shown in Figure 10.18(a). The output signals, of course, are delayed by one clock cycle. To cancel the effect of the delay, we can feed the output buffer with the next output value. After being delayed by one clock cycle, the next output value becomes the current output value, which is the desired output. Obtaining the next output is very straightforward. Recall that the current output is a function of the current state, which is the output of the state register, labeled state_reg in the diagram. The next output should be a function of the next state, which is the output of next-state logic, labeled state_next in the diagram. To obtain the next output, we need only disconnect the input of the output logic from the state_reg signal and reconnect it to the state_next signal, as shown in Figure 10.18(b).

Once understanding the block diagrams of Figure 10.18, we can develop the VHDL code accordingly. Again, we use the memory controller FSM as an example. The we_me output will be ignored since it is irrelevant to the Moore output buffering. Note that the state register and next-state logic are the same as in the original block diagram, and only the Moore output logic part is modified. For comparison purposes, we show the VHDL codes for both diagrams. The codes are based on the mutli_seg_arch architecture of Section 10.5.1. The code of the memory controller FSM with a regular output buffer is shown in Listing 10.5.

Listing 10.5   FSM with a regular output buffer

```
architecture plain_buffer_arch of mem_ctrl is
    type mc_state_type is
        (idle, read1, read2, read3, read4, write);
    signal state_reg, state_next: mc_state_type;
5   signal oe_i, we_i, oe_buf_reg, we_buf_reg: std_logic;
begin
    -- state register
    process(clk,reset)
    begin
10      if (reset='1') then
            state_reg <= idle;
        elsif (clk'event and clk='1') then
            state_reg <= state_next;
        end if;
15  end process;
    -- output buffer
    process(clk,reset)
    begin
        if (reset='1') then
20          oe_buf_reg <= '0';
            we_buf_reg <= '0';
        elsif (clk'event and clk='1') then
            oe_buf_reg <= oe_i;
            we_buf_reg <= we_i;
25      end if;
    end process;
    -- next-state logic
    process(state_reg,mem,rw,burst)
    begin
30      case state_reg is
            when idle =>
                if mem='1' then
                    if rw='1' then
                        state_next <= read1;
35                  else
                        state_next <= write;
                    end if;
                else
                    state_next <= idle;
40              end if;
            when write =>
                state_next <= idle;
            when read1 =>
                if (burst='1') then
45                  state_next <= read2;
                else
                    state_next <= idle;
                end if;
            when read2 =>
50              state_next <= read3;
            when read3 =>
                state_next <= read4;
```

```
                when read4 =>
                    state_next <= idle;
55          end case;
        end process;
        -- Moore output logic
        process(state_reg)
        begin
60          we_i <= '0'; -- default value
            oe_i <= '0';  -- default value
            case state_reg is
                when idle =>
                when write =>
65                  we_i <= '1';
                when read1 =>
                    oe_i <= '1';
                when read2 =>
                    oe_i <= '1';
70              when read3 =>
                    oe_i <= '1';
                when read4 =>
                    oe_i <= '1';
            end case;
75      end process;
        -- output
        we <= we_buf_reg;
        oe <= oe_buf_reg;
    end plain_buffer_arch;
```

In this code, we rename the original output signals of the output logic with a post-fix "_i" (for intermediate output signals). These signals are then connected to the output buffers.

To obtain the VHDL code for the look-ahead output buffer, we change the input of the output logic. This can be done by substituting the state_reg signal with the state_next signal in the case statement and the sensitivity list of the process. To make the code more descriptive, we use the post-fix "_next" for the next output signals. The modified code is shown in Listing 10.6.

Listing 10.6 FSM with a look-ahead output buffer

```
architecture look_ahead_buffer_arch of mem_ctrl is
    type mc_state_type is
        (idle, read1, read2, read3, read4, write);
    signal state_reg, state_next: mc_state_type;
5   signal oe_next,we_next,oe_buf_reg,we_buf_reg: std_logic;
    begin
        -- state register
        process(clk,reset)
        begin
10          if (reset='1') then
                state_reg <= idle;
            elsif (clk'event and clk='1') then
                state_reg <= state_next;
            end if;
15      end process;
        -- output buffer
```

```vhdl
      process(clk,reset)
      begin
         if (reset='1') then
20          oe_buf_reg <= '0';
            we_buf_reg <= '0';
         elsif (clk'event and clk='1') then
            oe_buf_reg <= oe_next;
            we_buf_reg <= we_next;
25       end if;
      end process;
      -- next-state logic
      process(state_reg,mem,rw,burst)
      begin
30       case state_reg is
            when idle =>
               if mem='1' then
                  if rw='1' then
                     state_next <= read1;
35                else
                     state_next <= write;
                  end if;
               else
                  state_next <= idle;
40             end if;
            when write =>
               state_next <= idle;
            when read1 =>
               if (burst='1') then
45                state_next <= read2;
               else
                  state_next <= idle;
               end if;
            when read2 =>
50             state_next <= read3;
            when read3 =>
               state_next <= read4;
            when read4 =>
               state_next <= idle;
55       end case;
      end process;
      -- look-ahead output logic
      process(state_next)
      begin
60       we_next <= '0'; -- default value
         oe_next <= '0'; -- default value
         case state_next is
            when idle =>
            when write =>
65             we_next <= '1';
            when read1 =>
               oe_next <= '1';
            when read2 =>
               oe_next <= '1';
```

```
70              when read3 =>
                    oe_next <= '1';
                when read4 =>
                    oe_next <= '1';
            end case;
75    end process;
      -- output
      we <= we_buf_reg;
      oe <= oe_buf_reg;
   end look_ahead_buffer_arch;
```

The look-ahead buffer is a very effective scheme for buffering Moore output. It provides a glitch-free output signal and reduces T_{co} to T_{cq}. Furthermore, this scheme has no effect on the next-state logic or state assignment and needs only minimal modification over the original code.

10.8 FSM DESIGN EXAMPLES

Our focus on the FSM is to use it as the control circuit in large systems. Such systems involve a data path that is composed of regular sequential circuits, and are discussed in Chapters 11 and 12. This section shows several simple stand-alone FSM applications.

10.8.1 Edge detection circuit

The VHDL code for the Moore machine–based edge detection design of Section 10.4.1 is shown in Listing 10.7. The code is based on the state diagram of Figure 10.12(a) and is done in multi-segment style.

Listing 10.7   Edge detector with regular Moore output

```
   library ieee;
   use ieee.std_logic_1164.all;
   entity edge_detector1 is
      port(
5          clk, reset: in std_logic;
           strobe: in std_logic;
           p1: out std_logic
       );
   end edge_detector1;
10
   architecture moore_arch of edge_detector1 is
      type state_type is (zero, edge, one);
      signal state_reg, state_next: state_type;
   begin
15    -- state register
      process(clk,reset)
      begin
          if (reset='1') then
              state_reg <= zero;
20        elsif (clk'event and clk='1') then
              state_reg <= state_next;
          end if;
```

Table 10.3   State assignment for edge detector output buffering

State	state_reg(1) (p1)	state_reg(0)
zero	0	0
edge	1	0
one	0	1

```vhdl
     end process;
     -- next-state logic
25   process(state_reg,strobe)
     begin
        case state_reg is
           when zero=>
              if strobe= '1' then
30               state_next <= edge;
              else
                 state_next <= zero;
              end if;
           when edge =>
35            if strobe= '1' then
                 state_next <= one;
              else
                 state_next <= zero;
              end if;
40         when one =>
              if strobe= '1' then
                 state_next <= one;
              else
                 state_next <= zero;
45            end if;
        end case;
     end process;
     -- Moore output logic
     p1 <= '1' when state_reg=edge else
50           '0';
  end moore_arch;
```

Assume that we want the output signal to be glitch-free. We can do it by using the clever state assignment or look-ahead output buffer scheme. One possible state assignment is shown in Table 10.3, and the VHDL code is shown in Listing 10.8.

Listing 10.8   Edge detector with clever state assignment

```vhdl
architecture clever_assign_buf_arch of edge_detector1 is
   constant zero: std_logic_vector(1 downto 0):= "00";
   constant edge: std_logic_vector(1 downto 0):= "10";
   constant one: std_logic_vector(1 downto 0) := "01";
5  signal state_reg,state_next: std_logic_vector(1 downto 0);
begin
   -- state register
   process(clk,reset)
```

```
     begin
10      if (reset='1') then
            state_reg <= zero;
        elsif (clk'event and clk='1') then
            state_reg <= state_next;
        end if;
15   end process;
     -- next-state logic
     process(state_reg, strobe)
     begin
        case state_reg is
20         when zero=>
               if strobe= '1' then
                   state_next <= edge;
               else
                   state_next <= zero;
25             end if;
           when edge =>
               if strobe= '1' then
                   state_next <= one;
               else
30                 state_next <= zero;
               end if;
           when others =>
               if strobe= '1' then
                   state_next <= one;
35             else
                   state_next <= zero;
               end if;
        end case;
     end process;
40   -- Moore output logic
     p1 <= state_reg(1);
  end clever_assign_buf_arch;
```

The VHDL code for the look-ahead output circuit scheme is given in Listing 10.9.

Listing 10.9 Edge detector with a look-ahead output buffer

```
architecture look_ahead_arch of edge_detector1 is
   type state_type is (zero, edge, one);
   signal state_reg, state_next: state_type;
   signal p1_reg, p1_next: std_logic;
5 begin
   -- state register
   process(clk, reset)
   begin
      if (reset='1') then
10       state_reg <= zero;
      elsif (clk'event and clk='1') then
         state_reg <= state_next;
      end if;
   end process;
15 -- output buffer
```

```
        process(clk,reset)
        begin
            if (reset='1') then
                p1_reg <= '0';
20          elsif (clk'event and clk='1') then
                p1_reg <= p1_next;
            end if;
        end process;
        -- next-state logic
25      process(state_reg,strobe)
        begin
            case state_reg is
                when zero=>
                    if strobe= '1' then
30                      state_next <= edge;
                    else
                        state_next <= zero;
                    end if;
                when edge =>
35                  if strobe= '1' then
                        state_next <= one;
                    else
                        state_next <= zero;
                    end if;
40              when one =>
                    if strobe= '1' then
                        state_next <= one;
                    else
                        state_next <= zero;
45                  end if;
            end case;
        end process;
        -- look-ahead output logic
        p1_next <= '1' when state_next=edge else
50                      '0';
        -- output
        p1 <= p1_reg;
    end look_ahead_arch;
```

Note that in this particular example the clever statement assignment scheme can be implemented by using 2 bits (i.e., two D FFs) but the look-ahead output circuit scheme needs at least three D FFs (2 bits for the state register and 1 bit for the output buffer).

The VHDL code for the Mealy output–based design is shown in Listing 10.10. The code is based on the state diagram of Figure 10.12(b).

Listing 10.10 Edge detector with Mealy output

```
library ieee;
use ieee.std_logic_1164.all;
entity edge_detector2 is
    port(
5       clk, reset: in std_logic;
        strobe: in std_logic;
        p2: out std_logic
```

```
        );
     end edge_detector2;
10
     architecture mealy_arch of edge_detector2 is
        type state_type is (zero, one);
        signal state_reg, state_next: state_type;
     begin
15      -- state register
        process(clk,reset)
        begin
           if (reset='1') then
              state_reg <= zero;
20         elsif (clk'event and clk='1') then
              state_reg <= state_next;
           end if;
        end process;
        -- next-state logic
25      process(state_reg,strobe)
        begin
           case state_reg is
              when zero=>
                 if strobe= '1' then
30                  state_next <= one;
                 else
                    state_next <= zero;
                 end if;
              when one =>
35               if strobe= '1' then
                    state_next <= one;
                 else
                    state_next <= zero;
                 end if;
40         end case;
        end process;
        -- Mealy output logic
        p2 <= '1' when (state_reg=zero) and (strobe='1') else
              '0';
45 end mealy_arch;
```

An alternative to deriving an edge detector is to treat it as a regular sequential circuit and design it in an ad hoc manner. One possible implementation is shown in Figure 10.19. The D FF in this circuit delays the strobe signal for one clock cycle and its output is the "previous value" of the strobe signal. The output of the and cell is asserted when the previous value of the strobe signal is '0' and the current value of the strobe signal is '1', which implies a positive transition edge of the strobe signal. The output signal is like a Mealy output since its value depends on the register's state and input signal. The VHDL code is shown in Listing 10.11. The entity declaration is identical to the Mealy machine–based edge detector in Listing 10.10.

Listing 10.11 Edge detector using direct implementation

```
architecture direct_arch of edge_detector2 is
   signal delay_reg: std_logic;
begin
```

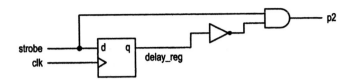

Figure 10.19   Direct implementation of an edge detector.

```
      — delay register
  5   process(clk,reset)
      begin
          if (reset='1') then
              delay_reg <= '0';
          elsif (clk'event and clk='1') then
  10              delay_reg <= strobe;
          end if;
      end process;
      — decoding logic
      p2 <= (not delay_reg) and strobe;
  15 end direct_arch;
```

Although the code is compact for this particular case, this ad hoc approach can only be applied to simple designs. For example, if the requirement specifies a glitch-free Moore output, it is very difficult to derive the circuit this way. Actually, we can easily verify that this ad hoc design is actually Mealy machine–based design with binary state assignment (i.e., 0 to the zero state and 1 to the one state).

10.8.2 Arbiter

In a large system, some resources are shared by many subsystems. For example, several processors may share the same block of memory, and many peripheral devices may be connected to the same bus. An arbiter is a circuit that resolves any conflict and coordinates the access to the shared resource. This example considers an arbiter with two subsystems, as shown in Figure 10.20. The subsystems communicate with the arbiter by a pair of request and grant signals, which are labeled as $r(1)$ and $g(1)$ for subsystem 1, and as $r(0)$ and $g(0)$ for subsystem 0. When a subsystem needs the resources, it activates the request signal. The arbiter monitors use of the resources and the requests, and grants access to a subsystem by activating the corresponding grant signal. Once its grant signal is activated, a subsystem has permission to access the resources. After the task has been completed, the subsystem releases the resources and deactivates the request signal. Since an arbiter's decision is based partially on the events that occurred earlier (i.e., previous request and grant status), it needs internal states to record what happened in the past. An FSM can meet this requirement.

One critical issue in designing an arbiter is the handling of simultaneous requests. Our first design gives priority to subsystem 1. The state diagram of the FSM is shown in Figure 10.21(a). It consists of three states, waitr, grant1 and grant0. The waitr state indicates that the resources is available and the arbiter is waiting for a request. The grant1 and grant0 states indicate that the resource is granted to subsystem 1 and subsystem 0 respectively. Initially, the arbiter is in the waitr state. If the $r(1)$ input (the request from subsystem 1) is activated at the rising edge of the clock, it grants the resources to subsystem 1 by moving to the grant1 state. The $g(1)$ signal is asserted in this state to

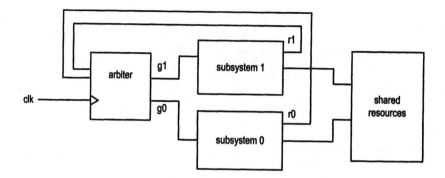

Figure 10.20   Block diagram of an arbiter.

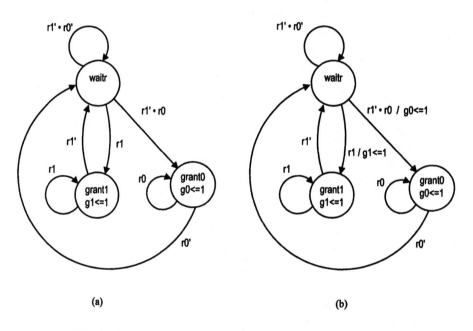

Figure 10.21   State diagrams of a fixed-priority two-request arbiter.

inform subsystem 1 of the availability of the resources. After subsystem 1 completes its usage, it signals the release of the resources by deactivating the r(1) signal. The arbiter returns to the waitr state accordingly.

In the waitr state, if r(1) is not activated and r(0) is activated at the rising edge, the arbiter grants the resources to subsystem 0 by moving to the grant0 state and activates the g(0) signal. Subsystem 0 can then have the resources until it releases them. The VHDL code for this design is shown in Listing 10.12.

Listing 10.12   Arbiter with fixed priority

```
library ieee;
use ieee.std_logic_1164.all;
use ieee.numeric_std.all;
entity arbiter2 is
```

```vhdl
  5    port(
          clk: in std_logic;
          reset: in std_logic;
          r: in std_logic_vector(1 downto 0);
          g: out std_logic_vector(1 downto 0)
 10    );
     end arbiter2;

     architecture fixed_prio_arch of arbiter2 is
        type mc_state_type is (waitr, grant1, grant0);
 15     signal state_reg, state_next: mc_state_type;
     begin
        -- state register
        process(clk,reset)
        begin
 20        if (reset='1') then
               state_reg <= waitr;
           elsif (clk'event and clk='1') then
               state_reg <= state_next;
           end if;
 25     end process;
        -- next-state and output logic
        process(state_reg,r)
        begin
           g <= "00";     -- default values
 30        case state_reg is
              when waitr =>
                 if r(1)='1' then
                    state_next <= grant1;
                 elsif r(0)='1' then
 35                 state_next <= grant0;
                 else
                    state_next <= waitr;
                 end if;
              when grant1 =>
 40              if (r(1)='1') then
                    state_next <= grant1;
                 else
                    state_next <= waitr;
                 end if;
 45              g(1) <= '1';
              when grant0 =>
                 if (r(0)='1') then
                    state_next <= grant0;
                 else
 50                 state_next <= waitr;
                 end if;
                 g(0) <= '1';
           end case;
        end process;
 55 end fixed_prio_arch;
```

If the subsystems are synchronized by the same clock, we can make g(1) and g(0) be Mealy output. The revised state diagram is shown in Figure 10.21(b). This allows the subsystems to obtain the resources one clock cycle earlier. In VHDL code, we modify the code under the `waitr` segment of the case statement to reflect the change. The revised portion becomes

```
when waitr =>
    if r(1)='1' then
        state_next <= grant1;
        g(1) <= '1';    -- newly added line
    elsif r(0)='1' then
        state_next <= grant0;
        g(0) <= '1';    -- newly added line
    else
        state_next <= waitr;
    end if;
```

The resource allocation of the previous design gives priority to subsystem 1. The preferential treatment may cause a problem if subsystem 1 requests the resources continuously. We can revise the state diagram to enforce a fairer arbitration policy. The new policy keeps track of which subsystem had the resources last time and gives preference to the other subsystem if the two request signals are activated simultaneously. The new design has to distinguish two kinds of wait conditions. The first condition is that the resources were last used by subsystem 1 so preference should be given to subsystem 0. The other condition is the reverse of the first. To accommodate the two conditions, we split the original `waitr` state into the `waitr1` and `waitr0` states, in which subsystem 1 and subsystem 0 will be given preferential treatment respectively. The revised state diagram is shown in Figure 10.22. Note that FSM moves from the `grant0` state to the `waitr1` state after subsystem 0 deactivates the request signal, and moves from the `grant1` state to the `waitr0` state after subsystem 1 deactivates the request signal. The revised VHDL code is shown in Listing 10.13.

Listing 10.13 Arbiter with alternating priority

```
architecture rotated_prio_arch of arbiter2 is
    type mc_state_type is (waitr1, waitr0, grant1, grant0);
    signal state_reg, state_next: mc_state_type;
begin
5   -- state register
    process(clk, reset)
    begin
        if (reset='1') then
            state_reg <= waitr1;
10      elsif (clk'event and clk='1') then
            state_reg <= state_next;
        end if;
    end process;
    -- next-state and output logic
15  process(state_reg, r)
    begin
        g <= "00";    -- default values
        case state_reg is
            when waitr1 =>
20              if r(1)='1' then
```

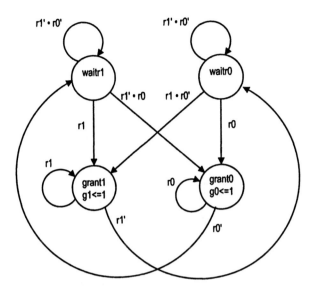

Figure 10.22   State diagram of a fair two-request arbiter.

```
            state_next <= grant1;
        elsif r(0)='1' then
            state_next <= grant0;
        else
25          state_next <= waitr1;
        end if;
    when waitr0 =>
        if r(0)='1' then
            state_next <= grant0;
30      elsif r(1)='1' then
            state_next <= grant1;
        else
            state_next <= waitr0;
        end if;
35  when grant1 =>
        if (r(1)='1') then
            state_next <= grant1;
        else
            state_next <= waitr0;
40      end if;
        g(1) <= '1';
    when grant0 =>
        if (r(0)='1') then
            state_next <= grant0;
45      else
            state_next <= waitr1;
        end if;
        g(0) <= '1';
    end case;
50  end process;
```

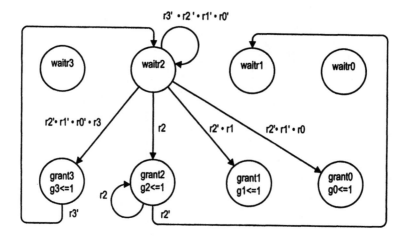

Figure 10.23 Partial state diagram of a four-request arbiter.

end `rotated_prio_arch`;

We can apply the same idea and expand the arbiter to handle more than two requests. The partial state diagram of an arbiter with four requests is shown in Figure 10.23. It assigns priority in round-robin fashion (i.e., subsystem 3, subsystem 2, subsystem 1, subsystem 0, then wrapping around), and the subsystem that obtains the resources will be assigned to the lowest priority next.

10.8.3 DRAM strobe generation circuit

Because of the large number of memory cells, the address signals of a dynamic RAM (DRAM) device are split into two parts, known as row and column. They are sent to the DRAM's address line in a time-multiplexed manner. Two control signals, `ras_n` (row address strobe) and `cas_n` (column address strobe), are strobe signals used to store the address into the DRAM's internal latches. The post-fix "_n" indicates active-low output, the convention used in most memory chips. The simplified timing diagram of a DRAM read cycle is shown in Figure 10.24(a). It is characterized by the following parameters:

- T_{ras}: `ras` access time, the time required to obtain output data after `ras_n` is asserted (i.e., `ras_n` goes to '0').
- T_{cas}: `cas` access time, the time required to obtain output data after `cas_n` is asserted (i.e., `cas_n` goes to '0').
- T_{pr}: precharge time, the time to recharge the DRAM cell to restore the original value (since the cell's content is destroyed by the read operation).
- T_{rc}: read cycle, the minimum elapsed time between two read operations.

The operation of a conventional DRAM device is asynchronous and the device does not have a clock signal. The strobe signals have to be asserted in proper sequence and last long enough to provide the necessary time for decoding, multiplexing and memory cell recharging.

A memory controller is the interface between a DRAM device and a synchronous system. One function of the memory controller is to generate proper strobe signals. This example shows how to use an FSM to accomplish this task. A real memory controller should also

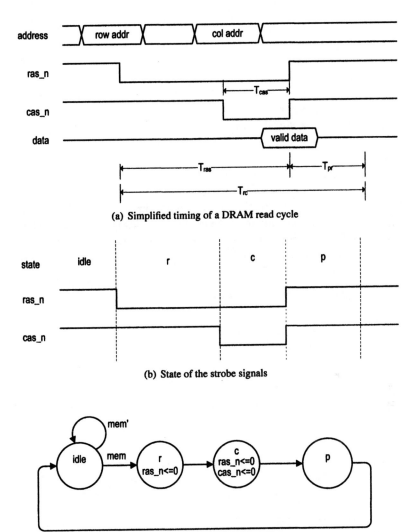

(a) Simplified timing of a DRAM read cycle

(b) State of the strobe signals

(c) State diagram of slow strobe generation

Figure 10.24   Read strobe generation FSM.

contain register and buffer to store address and data and should have extra control signals to coordinate the address bus and data bus operation. A complete memory controller example is discussed in Section 12.3.

Suppose that a DRAM has a read cycle of 120 ns, and T_{ras}, T_{cas} and T_{pr} are 85, 20 and 35 ns respectively. We want to design an FSM that generates the strobe signals, ras_n and cas_n, after the input command signal mem is asserted. The timing diagram of Figure 10.24(a) shows that the ras_n and cas_n signals have to be asserted and deasserted following a specific sequence:

- The ras_n signal is asserted first for at least 65 ns. The output pattern of the FSM is "01" in this interval.
- The cas_n signal is then asserted first for at least 20 ns. The output pattern of the FSM is "00" in this interval.
- The ras_n and cas_n signals are de-asserted first for at least 35 ns. The output pattern of the FSM is "11" in this interval.

Our first design uses a state for a pattern in the sequence and divides a read cycle into three states, namely the r, c and p states, as shown in Figure 10.24(b). The state diagram is shown in Figure 10.24(c). An extra idle state is added to accommodate the no-operation condition. We use a Moore machine since it has better control over the width of the intervals and can be modified to generate glitch-free output. In this design, each pattern lasts for one clock cycle. To satisfy the timing requirement for the three intervals, the clock period has to be at least 65 ns, and it takes 195 ns (i.e., 3*65 ns) to complete a read operation. The VHDL code is shown in Listing 10.14.

Listing 10.14   Slow DRAM read strobe generation FSM with regular output

```
library ieee;
use ieee.std_logic_1164.all;
entity dram_strobe is
    port(
        clk, reset: in std_logic;
        mem: in std_logic;
        cas_n, ras_n: out std_logic
    );
end dram_strobe;

architecture fsm_slow_clk_arch of dram_strobe is
    type fsm_state_type is (idle, r, c, p);
    signal state_reg, state_next: fsm_state_type;
begin
    -- state register
    process(clk,reset)
    begin
        if (reset='1') then
            state_reg <= idle;
        elsif (clk'event and clk='1') then
            state_reg <= state_next;
        end if;
    end process;
    -- next-state logic
    process(state_reg,mem)
    begin
        case state_reg is
```

```
              when idle =>
                 if mem='1' then
30                   state_next <= r;
                 else
                     state_next <= idle;
                 end if;
              when r =>
35               state_next <=c;
              when c =>
                 state_next <=p;
              when p =>
                 state_next <=idle;
40       end case;
     end process;
     -- output logic
     process(state_reg)
     begin
45       ras_n <= '1';
         cas_n <= '1';
         case state_reg is
              when idle =>
              when r =>
50               ras_n <= '0';
              when c =>
                 ras_n <= '0';
                 cas_n <= '0';
              when p =>            .
55       end case;
     end process;
  end fsm_slow_clk_arch;
```

Since the strobe signals are level-sensitive, we have to ensure that these signals are glitch-free. We can revise the previous code to add the look-ahead output buffer, as shown in Listing 10.15.

Listing 10.15   Slow DRAM read strobe generation FSM with a look-ahead output buffer

```
  architecture fsm_slow_clk_buf_arch of dram_strobe is
     type fsm_state_type is (idle,r,c,p);
     signal state_reg, state_next: fsm_state_type;
     signal ras_n_reg, cas_n_reg: std_logic;
5    signal ras_n_next, cas_n_next: std_logic;
  begin
     -- state register and output buffer
     process(clk,reset)
     begin
10       if (reset='1') then
              state_reg <= idle;
              ras_n_reg <= '1';
              cas_n_reg <= '1';
         elsif (clk'event and clk='1') then
15            state_reg <= state_next;
              ras_n_reg <= ras_n_next;
              cas_n_reg <= cas_n_next;
```

```
            end if;
        end process;
20      -- next-state
        process(state_reg,mem)
        begin
            case state_reg is
                when idle =>
25                  if mem='1' then
                        state_next <= r;
                    else
                        state_next <= idle;
                    end if;
30              when r =>
                    state_next <=c;
                when c =>
                    state_next <=p;
                when p =>
35                  state_next <=idle;
            end case;
        end process;
        -- look-ahead output logic
        process(state_next)
40      begin
            ras_n_next <= '1';
            cas_n_next <= '1';
            case state_next is
                when idle =>
45              when r =>
                    ras_n_next <= '0';
                when c =>
                    ras_n_next <= '0';
                    cas_n_next <= '0';
50              when p =>
            end case;
        end process;
        --output
        ras_n <= ras_n_reg;
55      cas_n <= cas_n_reg;
    end fsm_slow_clk_buf_arch;
```

To improve the performance of the memory operation, we can use a smaller clock period to accommodate the differences between the three intervals. For example, we can use a clock with a period of 20 ns and use multiple states for each output pattern. The three output patterns need 4 (i.e., $\lceil\frac{65}{20}\rceil$) states, 1 (i.e., $\lceil\frac{20}{20}\rceil$) state and 2 (i.e., $\lceil\frac{35}{20}\rceil$) states respectively. The revised state diagram is shown in Figure 10.25, in which the original r state is split into r1, r2, r3 and r4 states, and the original p state is split into p1 and p2 states. It now takes seven states, which amounts to 140 ns (i.e., 7*20 ns), to complete a read operation. We can further improve the performance by using a 5-ns clock signal (assuming that the next-state logic and register are fast enough to support it). The three output patterns need 13, 4 and 7 states respectively, and a read operation can be done in 120 ns, the fastest operation speed of this DRAM chip. While still simple, the state diagram becomes tedious to draw. RT methodology (to be discussed in Chapters 11 and 12) can combine counters with FSM and

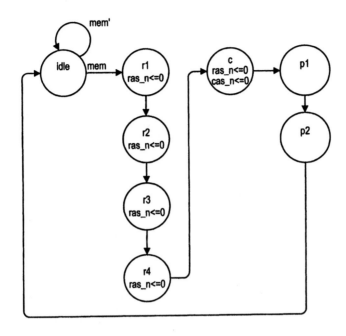

Figure 10.25   State diagram of fast read strobe generation.

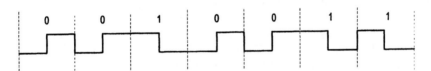

Figure 10.26   Sample waveform of Manchester encoding.

provide a better alternative to implement this type of circuit. In a more realistic scenario, the strobe generation circuit should be part of a large system, and it cannot use an independent clock. The design has to accommodate the clock rate of the main system and adjust the number of states in each pattern accordingly.

10.8.4 Manchester encoding circuit

Manchester code is a coding scheme used to represent a bit in a data stream. A '0' value of a bit is represented as a 0-to-1 transition, in which the lead half is '0' and the remaining half is '1'. Similarly, a '1' value of a bit is represented as a 1-to-0 transition, in which the lead half is '1' and the remaining half is '0'. A sample data stream in Manchester code is shown in Figure 10.26. The Manchester code is frequently used in a serial communication line. Since there is a transition in each bit, the receiving system can use the transitions to recover the clock information.

The Manchester encoder transforms a regular data stream into a Manchester-coded data stream. Because an encoded bit includes a sequence of "01" or "10", two clock cycles are needed. Thus, the maximal data rate is only half of the clock rate. There are two input signals. The d signal is the input data stream, and the v signal indicates whether the d

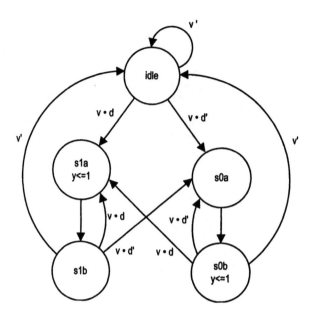

Figure 10.27   State diagram of a Manchester encoder.

signal is valid (i.e., whether there is data to transmit). The d signal should be converted to Manchester code if the v signal is asserted. The output remains '0' otherwise. The state diagram is shown in Figure 10.27. While v is asserted, the FSM starts the encoding process. If d is '0', it travels through the s0a and s0b states. If d is '1', the FSM travels through the s1a and s1b states. Once the FSM reaches the s1b or s0b state, it checks the v signal. If the v signal is still asserted, the FSM skips the idle state and continuously encodes the next input data. The Moore output is used because we have to generate two equal intervals for each bit. The VHDL code is shown in Listing 10.16.

Listing 10.16   Manchester encoder with regular output

```
library ieee;
use ieee.std_logic_1164.all;
entity manchester_encoder is
    port(
5        clk, reset: in std_logic;
        v,d: in std_logic;
        y: out std_logic
    );
end manchester_encoder;
10
    architecture moore_arch of manchester_encoder is
        type state_type is (idle, s0a, s0b, s1a, s1b);
        signal state_reg, state_next: state_type;
    begin
15      -- state register
        process(clk,reset)
        begin
            if (reset='1') then
                state_reg <= idle;
```

```
20        elsif (clk'event and clk='1') then
              state_reg <= state_next;
          end if;
      end process;
      -- next-state logic
25    process(state_reg,v,d)
      begin
          case state_reg is
              when idle=>
                  if v= '0' then
30                    state_next <= idle;
                  else
                      if d= '0' then
                          state_next <= s0a;
                      else
35                        state_next <= s1a;
                      end if;
                  end if;
              when s0a =>
                  state_next <= s0b;
40            when s1a =>
                  state_next <= s1b;
              when s0b =>
                  if v= '0' then
                      state_next <= idle;
45                else
                      if d= '0' then
                          state_next <= s0a;
                      else
                          state_next <= s1a;
50                    end if;
                  end if;
              when s1b =>
                  if v= '0' then
                      state_next <= idle;
55                else
                      if d= '0' then
                          state_next <= s0a;
                      else
                          state_next <= s1a;
60                    end if;
                  end if;
          end case;
      end process;
      -- Moore output logic
65    y <= '1' when state_reg=s1a or state_reg=s0b else
              '0';
  end moore_arch;
```

Because the transition edge of the Manchester code is frequently used by the receiver to recover the clock signal, we should make the output data stream glitch-free. This can be achieved by using the look-ahead output buffer. The revised VHDL code is shown in Listing 10.17.

Listing 10.17   Manchester encoder with a look-ahead output buffer

```
architecture out_buf_arch of manchester_encoder is
    type state_type is (idle, s0a, s0b, s1a, s1b);
    signal state_reg, state_next: state_type;
    signal y_next, y_buf_reg: std_logic;
5 begin
    -- state register and output buffer
    process(clk,reset)
    begin
        if (reset='1') then
10          state_reg <= idle;
            y_buf_reg <= '0';
        elsif (clk'event and clk='1') then
            state_reg <= state_next;
            y_buf_reg <= y_next;
15      end if;
    end process;
    -- next-state logic
    process(state_reg,v,d)
    begin
20      case state_reg is
            when idle=>
                if v='0' then
                    state_next <= idle;
                else
25                  if d= '0' then
                        state_next <= s0a;
                    else
                        state_next <= s1a;
                    end if;
30              end if;
            when s0a =>
                state_next <= s0b;
            when s1a =>
                state_next <= s1b;
35          when s0b =>
                if v='0' then
                    state_next <= idle;
                else
                    if d='0' then
40                      state_next <= s0a;
                    else
                        state_next <= s1a;
                    end if;
                end if;
45          when s1b =>
                if v= '0' then
                    state_next <= idle;
                else
                    if d= '0' then
50                      state_next <= s0a;
                    else
                        state_next <= s1a;
```

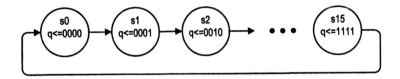

Figure 10.28   State diagram of a free-running mod-16 counter.

```
                  end if ;
               end if ;
55        end case ;
      end process ;
      —— look−ahead output logic
      y_next <= '1' when state_next=s1a or state_next=s0b else
               '0';
60       —— output
      y <= y_buf_reg ;
   end out_buf_arch ;
```

10.8.5   FSM-based binary counter

As discussed in Section 8.2.3, our classification of regular sequential circuits and FSMs (random sequential circuits) is for "design practicality." In theory, all sequential circuits with finite memory can be modeled by FSMs and derived accordingly. This example demonstrates the derivation of an FSM-based binary counter. Let us first consider a free-running 4-bit counter, similar to the one in Section 8.5.4. A 4-bit counter has to traverse 16 (2^4) distinctive states, and thus the FSM should have 16 states. The state diagram is shown in Figure 10.28. Note the regular pattern of transitions.

The FSM can be modified to add more features to this counter and gradually transform it to the featured binary counter of Section 8.5.4. To avoid clutter in the diagram, we use a single generic si state (the ith state of the counter) to illustrate the required modifications. The process is shown in Figure 10.29. We first add the synchronous clear signal, syn_clr, which clears the counter to 0, as in Figure 10.29(b). In the FSM, it corresponds to forcing the FSM to return to the initial state, s0. Note that the logic expressions give priority to the synchronous clear operation. The next step is to add the load operation. This actually involves five input bits, which include the 1-bit control signal, load, and the 4-bit data signal, d. The d signal is the value to be loaded into the counter and it is composed of four individual bits, d3, d2, d1 and d0. The load operation changes the content of the register according to the value of d. In terms of FSM operation, 16 transitions are needed to express the possible 16 next states. The revised diagram is shown in Figure 10.29(c). Finally, we can add the enable signal, en, which can suspend the counting. In terms of FSM operation, it corresponds to staying in the same state. The final diagram is shown in Figure 10.29(d). Note that the logic expressions of the transition arches set the priority of the control signals in the order syn_clr, load and en. Although this design process is theoretically doable, it is very tedious. The diagram will become extremely involved for a larger, say, a 16- or 32-bit, counter. This example shows the distinction between a regular sequential circuit and a random sequential circuit. In Section 12.2, we present a more comprehensive comparison

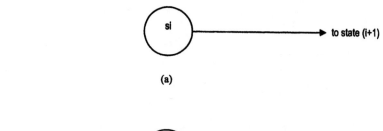

(a)

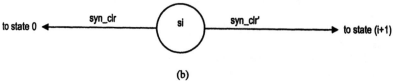

(b)

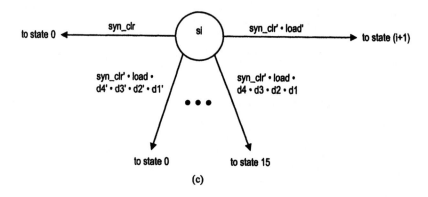

(c)

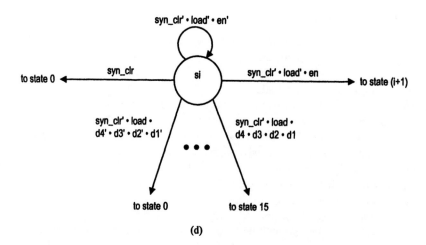

(d)

Figure 10.29   State diagram development of a featured mod-16 counter.

between regular sequential circuits, random sequential circuits and combined sequential circuits, which consist of both regular and random sequential circuits.

10.9 BIBLIOGRAPHIC NOTES

FSM is a standard topic in an introductory digital systems course. Typical digital systems texts, such as *Digital Design Principles and Practices* by J. F. Wakerly and *Contemporary Logic Design* by R. H. Katz, provide comprehensive coverage of the derivation of state diagrams and ASM charts as well as a procedure to realize them manually in hardware. They also show the techniques for state reduction. On the other hand, obtaining optimal state assignment for an FSM is a much more difficult problem. For example, it takes two theoretical texts, *Synthesis of Finite State Machines: Logic Optimization* by T. Villa et al. and *Synthesis of Finite State Machines: Functional Optimization* by T. Kam, to discuss the optimization algorithms.

Problems

10.1 For the "burst" read operation, the memory controller FSM of Section 10.2.1 implicitly specifies that the main system has to activate the rw and mem signals in the first clock cycle and then activate the $burst$ signal in the next clock cycle. We wish to simplify the timing requirement for the main system so that it only needs to issue the command in the first clock cycle (i.e., activates the $burst$ signal at the same time as the rw and mem signals).

(a) Revise the state diagram to achieve this goal.

(b) Convert the state diagram to an ASM chart.

(c) Derive VHDL code according to the ASM chart.

10.2 The memory controller FSM of Section 10.2.1 has to return to the idle state for each memory operation. To achieve better performance, revise the design so that the controller can support "back-to-back" operations; i.e., the FSM can initiate a new memory operation after completing the current operation without first returning to the idle state.

(a) Derive the revised state diagram.

(b) Convert the state diagram to an ASM chart.

(c) Derive VHDL code according to the ASM chart.

10.3 Revise the edge detection circuit of Section 10.4.1 to detect both 0-to-1 and 1-to-0 transitions; i.e., the circuit will generate a short pulse whenever the strobe signal changes state. Use a Moore machine with a minimal number of states to realize this circuit.

(a) Derive the state diagram.

(b) Convert the state diagram to an ASM chart.

(c) Derive VHDL code according to the ASM chart.

10.4 Repeat Problem 10.3, but use a Mealy machine to realize the circuit. The Mealy machine needs only two states.

10.5 In digital communication, a special synchronization pattern, known as a *preamble*, is used to indicate the beginning of a packet. For example, the Ethernet II preamble includes eight repeating octets of "10101010". We wish to design an FSM that generates the "10101010" pattern. The circuit has an input signal, start, and an output, data_out. When start is '1', the "10101010" will be generated in the next eight clock cycles.

(a) Derive the state diagram.

(b) Convert the state diagram to an ASM chart.

(c) Derive VHDL code according to the ASM chart.

(d) Use a clever state assignment to obtain glitch-free output signal. Derive the revised VHDL code.

(e) Use a look-ahead output buffer for the output signal. Derive the revised VHDL code

10.6   Now we wish to design an FSM to detect the "10101010" pattern in the receiving end. The circuit has an input signal, data_in, and an output signal, match. The match signal will be asserted as '1' for one clock period when the input pattern "10101010" is detected.

(a) Derive the state diagram.

(b) Convert the state diagram to an ASM chart.

(c) Derive VHDL code according to the ASM chart.

10.7   Can we apply look-ahead output buffer for Mealy output? Explain.

10.8   The first arbiter of Section 10.8.2 has to return to the waitr state before it can grant the resources to another request. Revise the design so that the arbiter can move from one grant state to another grant state when there is an active request.

(a) Derive the state diagram.

(b) Convert the state diagram to an ASM chart.

(c) Derive VHDL code according to the ASM chart.

10.9   Consider the fair arbiter of Section 10.8.2. Its design is based on the assumption that a subsystem will release the resources voluntarily. An alternative is to use a timeout signal to prevent a subsystem from exhausting the resource. When the timeout signal is asserted, the arbiter will return to a wait state regardless of whether the corresponding request signal is still active.

(a) Derive the revised state diagram.

(b) Convert the state diagram to an ASM chart.

(c) Derive VHDL code according to the ASM chart.

10.10   Redesign the DRAM strobe generation circuit of Section 10.8.3 for a system with a different clock period. Derive the state diagram and determine the required time to complete a read cycle for the following clock periods:

(a) A clock period of 10 ns.

(b) A clock period of 40 ns.

(c) A clock period of 200 ns.

10.11   A Manchester decoder transforms a Manchester-coded data stream back to a regular binary data stream. There are two output signals. The data signal is the recovered data bit, which can be '0' or '1'. The valid signal indicates whether a transition occurs. The valid signal is used to distinguish whether the '0' of the data signal is due to the 0-to-1 transition or inactivity of the data stream.

(a) Derive the state diagram.

(b) Convert the state diagram to an ASM chart.

(c) Derive VHDL code according to the ASM chart.

10.12   Non-return to-zero invert-to ones (NRZI) code is another code used in serial transmission. The output of an NRZI encoder is '0' if the current input value is different from the previous value and is '1' otherwise. Design an NRZI encoder using an FSM and derive the VHDL code accordingly.

10.13   Repeat Problem 10.12, but design an NRZI decoder, which converts a NRZI-coded stream back to a regular binary stream.

10.14   Derive the VHDL code for the FSM-based free-running mod-16 counter of Section 10.8.5.
  (a) Synthesize the code using an ASIC technology. Compare the area and performance (in term of maximal clock rate) of the code in Section 10.8.5.
  (b) Synthesize the code using an FPGA technology. Compare the area and performance of the code in Section 10.8.5.

10.15   Repeat Problem 10.14 for the featured mod-16 counter of Section 10.8.5.

CHAPTER 11

REGISTER TRANSFER METHODOLOGY: PRINCIPLE

To accomplish a complex task, we frequently describe the process by an *algorithm*, which is a sequence of steps or actions. Algorithms are generally implemented by programs written in a traditional programming language (i.e., by software) and executed in a general-purpose computer. However, to obtain better performance and efficiency, it is sometimes beneficial or even necessary to realize an algorithm in custom hardware. The *register transfer methodology (RT methodology)* is a design methodology that describes system operation by a sequence of data transfers and manipulations among the registers. This methodology can support the variables and sequential execution of an algorithm and provide a systematic way to convert an algorithm into hardware.

11.1 INTRODUCTION

11.1.1 Algorithm

An algorithm is a detailed sequence of actions or steps to accomplish a task or to solve a problem. Since the semantics of traditional programming languages is also based on sequential execution, an algorithm can easily be converted into a program using the constructs of these languages. The program is then compiled into the machine instructions and executed in a general-purpose computer. Let us consider a simple task that sums the four elements of an array, divides the sum by 8 and rounds the result to the closest integer. The pseudocode of one possible algorithm is

```
size = 4
sum = 0;
for i in (0 to size-1) do {
    sum = sum + a(i);}
q = sum / 8;
r = sum rem 8;
if (r > 3) {
    q = q + 1;}
outp = q;
```

The algorithm first adds individual elements and stores the result in a variable called sum. It then uses the division (/) and remainder (**rem**) operations to find the quotient and remainder. If the remainder is greater than 3, an extra 1 is added to quotient for rounding. The example demonstrates two basic characteristics of an algorithm:

- *Use of variables.* A variable in an algorithm or pseudocode can be interpreted as a "memory location with a symbolic address" (i.e., the name of the variable). It is used to store an intermediate computation result. For example, in the second statement, 0 is stored into the memory location with a symbolic address of sum. Inside the for loop, a(i) is added with the current content of sum, and then summation is stored back into the same memory location. In the fourth statement, the content of sum is divided by 8, and the result is stored into a memory location with a symbolic address of q.
- *Sequential execution.* The execution of an algorithm is performed sequentially and the order of the steps is important. For example, the summation of the elements must be obtained before the division operation can be performed. Note that the order of execution may rely on certain conditions, as in the for loop and if statements.

In VHDL, the variables and sequential execution are treated as a special case and encapsulated inside a process. Although a description with variables can be synthesized in some cases, the variables are mapped to signals and are not interpreted or realized as "memory locations with symbolic addresses."

11.1.2 Structural data flow implementation

To achieve better performance and efficiency, we frequently want to implement an algorithm in custom hardware. The variable and sequential semantics of algorithm are very different from the concurrent model of hardware. What we have learned so far is to transform "sequential execution" into "structural data flow" by mapping an algorithm into a system of cascading hardware blocks, in which each block represents a statement in the algorithm. For example, we can unroll the loop of the previous algorithm and convert the variables into internal connection signals. Assume that sum is an 8-bit signal. The corresponding VHDL code becomes

```
sum <= 0;
sum0 <= a(0);
sum1 <= sum0 + a(1);
sum2 <= sum1 + a(2);
sum3 <= sum2 + a(3);
q <= "000" & sum3(8 downto 3);
r <= "00000" & sum3(2 downto 0);
outp <= q + 1 when (r > 3) else
        q;
```

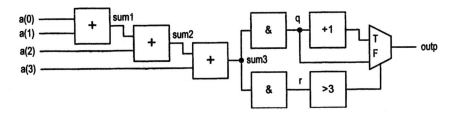

Figure 11.1   Structural data flow implementation.

Note that the sum / 8 and sum rem 8 operations are implemented by concatenation (i.e., &) operations. The corresponding block diagram is shown in Figure 11.1.

Although the circuit can carry out the task, the operation of the hardware is very different from the sequential semantics of the original algorithm. In this construction, the circuit is a pure combinational logic, and the adders and dividers (concatenation operators) execute in parallel. The implementation does not use any concept of variable, and the sequential execution is implicitly embedded in the interconnection of components and the flow of data. To some degree, the synthesis essentially utilizes extra hardware to accelerate the operation. Instead of using a single arithmetic unit of a computer to perform these operations sequentially, the custom hardware utilizes multiple adders and division circuits to calculate the result in parallel.

The structural data flow implementation is not general and can be applied only for simple, trivial algorithms. The following two variations of the previous algorithm illustrate the limitation of this approach. First, let us consider an array with 10 elements. In the pseudocode, this can be done by replacing 4 with 10 in the first statement. This increases the number of loop iterations. We can unroll the loop and derive the structural data flow implementation, which needs nine adders. If the number of elements of the array continues to grow, the number of adders increases accordingly. Clearly, this approach needs excessive hardware resource and is not practical for a larger array. Second, let us assume that the size of the array is not fixed but is specified by an additional input, n. To accomplish this in the algorithm, we only need to substitute n into the first statement and make it size = n. This will be very difficult for structural data flow implementation. Since the hardware cannot expand or shrink dynamically, we have to construct a circuit that can calculate the results for *all* possible values of n and then use a multiplexer to route the desired value to output. The resulting hardware will be extremely complicated, and this approach is not practical in reality.

11.1.3   Register transfer methodology

The previous example shows the limitation and inflexibility of structural data flow implementation. To realize an algorithm in hardware, we need hardware constructs that resemble the variable and sequential execution model. The *register transfer methodology* (*RT methodology*) is aimed for this purpose. The key characteristics of this methodology are:

- Use registers to store the intermediate data and to imitate the variables used in an algorithm.
- Use a custom *data path* to realize all the required register operations.
- Use a custom *control path* to specify the order of the register operations.

We have utilized registers for regular sequential circuits and FSMs in previous chapters. They are usually dedicated to a specific circuit, as in a counter or an FSM. In the RT methodology, the registers are used as general storage that keeps the intermediate computed values, just as the variables of an algorithm. For example, consider a typical statement in pseudocode:

```
a = a + b
```

We can use two registers, a_reg and b_reg, to imitate the a and b variables. When this statement is executed, the content of the a_reg and b_reg registers will be added, and the result will be stored back into the a_reg register at the next rising edge of the clock.

When an algorithm is realized in RT methodology, the necessary data manipulation and data routing are performed by dedicated hardware. For example, an adder is required for the previous statement. The data manipulation circuit, routing network and the registers together are known as the *data path*.

Since an algorithm is described as a sequence of actions, we need a circuit to control *when* and *what* RT operations should take place. The circuit is known as the *control path*. A control path can be realized by an FSM, which can use states to enforce the order of the desired steps and use the decision boxes to imitate the branches and iterations (loops) in an algorithm.

We call this implementation methodology *register transfer methodology* since an algorithm is transformed into a sequence of actions that specifies how the data is manipulated and transferred among registers. A typical RT implementation includes a data path and a control path. We can use an extended FSM to describe the overall system operation. It is known as *FSM with a data path (FSMD)*.

As we mentioned in Section 1.4.3, use of the term *register transfer* is somewhat abused. It sometimes is used rather vaguely to represent a level of abstraction (i.e., the RT level) between the gate and processor levels. In this book, we use the term *RT methodology* for this specific design methodology and the term *RT level* for module-level abstraction.

11.2   OVERVIEW OF FSMD

An FSMD is the key to realizing the RT methodology. This section provides an overview of FSMD, including RT operation, data path, control path and extended ASM chart. The subsequent sections use examples to illustrate the detailed derivation and construction of an FSMD.

11.2.1   Basic RT operation

A basic action in RT methodology is a *register transfer operation*. We use the following notation for an RT operation:

$$r_{dest} \leftarrow f(r_{src1}, r_{src2}, \ldots, r_{srcn})$$

In this notation, the register on the left-hand side (i.e., r_{dest}) is the destination register. The registers on the right-hand side (i.e., r_{src1}, r_{src2} and r_{srcn}) are the source registers and they represent the outputs (i.e., the contents) of these registers. The $f(\cdot)$ function is the operation to be performed. It is an expression composed of source registers and sometimes external inputs. The overall notation means that the new value of r_{dest} is calculated according to $f(r_{src1}, r_{src2}, \ldots, r_{srcn})$, and the result will be stored into r_{dest} at the next rising edge of the clock. Note that the ← notation is not defined in VHDL. It is only used in this book to denote the register transfer operation.

There is no specific restriction on the $f(\cdot)$ function. It can be any expression as long as it can be realized by a combinational circuit. A few representative RT operations are shown below.

- $r \leftarrow 1$: A constant 1 is stored into the r register.
- $r \leftarrow r$: The content of the r register is stored back into itself. The content, of course, remains unchanged.
- $r \leftarrow r \ll 3$: The content of the r register is shifted left three positions and then stored back into itself.
- $r0 \leftarrow r1$: The content of the r1 register is stored (or transferred) into the r0 register.
- $n \leftarrow n - 1$: The content of the n register is decremented by 1 and the result is stored back into itself.
- $y \leftarrow a \oplus b \oplus c \oplus d$: The contents of the a, b, c and d registers are xored and the result is stored into the y register.
- $s \leftarrow a^2 + b^2$: The summation of a squared and b squared is stored into the s register. We can write this expression only if the predesigned combinational multiplier module is available.

The major difference between a variable of an algorithm and a register is that a *system clock* is embedded implicitly in an RT operation. Consider the register operation

$$r_{dest} \leftarrow f(r_{src1}, r_{src2}, \ldots, r_{srcn})$$

Its detailed actions are las follows:

1. At the rising edge of the clock, new data from the source registers is available after the clock-to-q delay of the source registers.
2. The data is computed by a combinational circuit that realizes the $f(\cdot)$ function. We assume that the clock period is long enough to accommodate the propagation delay of the combinational circuit and the setup time of the r_{dest} register. The result is routed to the input of the r_{dest} register.
3. At the next rising edge of the clock, the result will be sampled and stored into the r_{dest} register.

In our discussion of sequential circuits, we use the suffixes _reg and _next for the current output and next input of a register. A more accurate description of an RT operation can be expressed using these suffixes. For example, consider the r1 ← r1 + r2 operation. It actually means

- r1_next <= r1_reg + r2_reg;
- r1_reg <= r1_next at the rising edge of the clock;

Note that the <= notation is used for regular signal assignment.

Realizing an RT operation is straightforward. We basically construct the $f(\cdot)$ function using combinational components and then connect its output to the input of the destination register. Again, consider the r1 ← r1 + r2 operation. It involves an addition in $f(\cdot)$. Its block diagram is shown in Figure 11.2(a) and the corresponding timing diagrams is shown in Figure 11.2(b). Note that the r1 register will not be updated until the next rising edge of the clock.

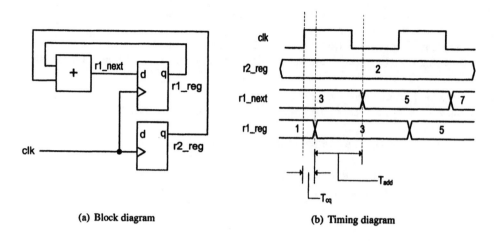

(a) Block diagram   (b) Timing diagram

Figure 11.2   Single RT operation.

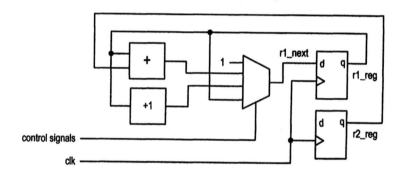

Figure 11.3   Block diagram of a set of RT operations with the same destination register.

11.2.2 Multiple RT operations and data path

An algorithm consists of many steps, and a destination register is not loaded with the same data in these steps. For example, the r1 register may be set to 1 in the initialization step, added with the content of r2 in a summation step, incremented in the two counting steps, and kept unchanged in the final step. Thus, four RT operations use r1 as the destination register:

- r1 ← 1;
- r1 ← r1 + r2;
- r1 ← r1 + 1;
- r1 ← r1;

Because of the multiple possibilities, a multiplexing circuit is needed to route the desired value to the input of the r1 register. The block diagram is shown in Figure 11.3. We can choose the desired RT operation by setting the proper selection signal in the multiplexing circuit.

A design with RT methodology normally involves many registers. We can repeat this procedure for every register. The resulting circuit constitutes the basic, unoptimized data path, which can perform every needed RT operation of an algorithm.

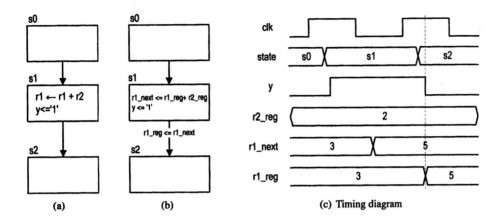

Figure 11.4 RT operation in a segment of an ASMD chart.

11.2.3   FSM as the control path

While a data path realizes all required RT operations in an algorithm, we need a mechanism to specify when and which RT operations should be performed. A control path is used to enforce the order of RT operations and to selectively perform certain RT operations based on the external commands or internal status. A control path can be realized by a custom FSM. An FSM is a natural match for this task for several reasons:

- The state transition of an FSM is performed on a clock-by-clock basis. Since an RT operation is also updated on a clock-by-clock basis, an RT operation can be specified in a state of the FSM.
- An FSM can enforce a specific sequence of actions.
- Upon an examination of input conditions, an FSM can branch to different paths and thus can alter the sequence of actions. This can be used to implement various branch constructs, such as the if and loop statements, in an algorithm.

11.2.4   ASMD chart

Since an RT operation is performed in a state of the FSM, we can extend the FSM to FSMD to indicate the desired RT operation in each state. The state representation and state transition of an FSMD are similar to those of an FSM. However, RT operations, in addition to output signals, are specified in states or transition arcs. We use an extended ASM chart, in which an RT operation can be specified either inside a state box or in a conditional output box, to describe the operation of an FSMD. It is known as an *ASM with a data path chart* (*ASMD chart*).

The construction and operation of an ASMD chart can best be explained by an example. A segment of an ASMD is shown in Figure 11.4(a). An RT operation, $r1 \leftarrow r1 + r2$, is specified in the s1 state. For comparison purposes, we also include a regular activated output, y, in the s1 state. When the FSMD enters the s1 state, the $r1 + r2$ expression is calculated and its result becomes the next value of r1. At the next rising edge of the clock, the FSMD transits from s1 to s2 and r1 is updated with the new value. Note that the r1 register is not updated inside the state box but during the transition between the s1 and s2 states. The new value of r1 is available only when the FSMD reaches the s2 state.

A clumsy, but more accurate, notation is shown in Figure 11.4(b), in which the r1_next signal is calculated in the s1 state, independent of the clock edge, and the r1_reg signal is updated at the transition. Note that the regular signal assignment notation, <=, is used in the diagram.

The timing diagram is shown in Figure 11.4(c). When the FSMD enters the s1 state, the computation of r1 + r2 starts but the output of r1 remains unchanged. Note that the regular output, y, is activated after the clock-to-q delay in the s1 state. At the next rising edge of the clock, the FSMD moves to the s2 state and the new value is sampled and stored into r1. After the clock-to-q delay, the new value is propagated to the output of r1. Note that the y signal is deactivated in the s2 state.

The r1 register samples and stores the input data at every rising edge of the clock. Thus, r1 is updated in the s0 and s2 states as well, even when no operation is needed. Since the system is synchronous, a register cannot be disabled or suspended. Instead, it just keeps its old value by sampling its own output; i.e., performing the r1 ← r1 operation. To reduce the clutter, we don't include this operation in an ASMD chart. If a destination register r is not associated with an RT operation in a state, we assume that it performs the default r ← r operation.

Although the appearances of an ASMD chart and a regular flowchart are somewhat similar, their operations are different. The operation of an FSMD is operated on a clock-by-clock basis. The register in an ASMD chart is not updated until the exit of the current state, and thus an RT operation exhibits some sort of *delayed-store* behavior. The most error-prone part of deriving an ASMD chart is this delayed-store operation. To obtain a correct and efficient ASMD chart, we need to have a clear understanding of the timing of an RT operation and know when a register is updated. Section 11.3.4 provides a comprehensive discussion of this issue.

11.2.5 Basic FSMD block diagram

The conceptual block diagram of an FSMD is shown in Figure 11.5. It is divided into a data path and a control path. The data path can perform all the required RT operations and is composed of three major parts:

- *Data registers*. The registers store the intermediate computation results.
- *Functional units*. The functional units perform the functions specified by RT operations. Typical functional units include an adder, subtractor, incrementor, decrementor and shifter.
- *Routing circuit*. The circuit routes the source registers' outputs to the proper functional units and routes the calculated results from the functional units to proper destination registers. It is normally constructed by customized multiplexers.

A data path normally includes the following input and output signals:

- data input: the external input data, which is to be processed by the FSMD.
- data output: the processed results of the FSMD.
- control signal: input signal used to specify which RT operations should be performed. It is generated by the control path.
- internal status: output signal indicating certain conditions of the data path, such as whether a specific register is 0. This signal is used by the control path to determine the future course of action.

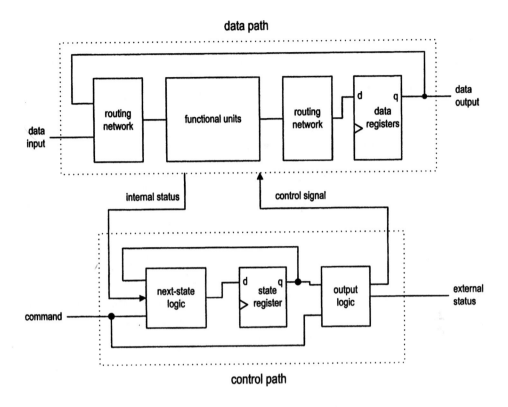

Figure 11.5   Basic block diagram of an FSMD.

The control path is an FSM. As a regular FSM, it contains a state register, next-state logic and output logic. A control path normally includes the following input and output signals:

- command: the external command signal to the FSMD, such as the start of the operation. It is an input to the FSM.
- internal status: signal from the data path, which is also an input to the FSM. The FSM uses it and the external command to determine the next state.
- control signal: output of the FSM used to control data path operation.
- external status: output of the FSM used to indicate the status of the FSMD operation, such as whether the system is busy.

In addition to these signals, the registers of the data path and control path are connected to the same clock signal and to an optional asynchronous reset signal.

Note that the data path resembles a regular sequential circuit, and the control path is an FSM and thus is a random sequential circuit. Therefore, an FSMD can be considered a combined sequential circuit, as discussed in Section 8.2.3. Although the FSMD consists of two types of sequential circuits, both circuits are synchronized by the same clock, and thus the FSMD still follows the same synchronous design methodology.

11.3   FSMD DESIGN OF A REPETITIVE-ADDITION MULTIPLIER

The derivation of an ASMD chart and the construction of an FSMD can best be explained by closely examining several examples. This section illustrates how to convert a simple repetitive-addition multiplication algorithm into an ASMD chart and realize it in hardware. Various alternatives are discussed in subsequent sections.

11.3.1   Converting an algorithm to an ASMD chart

We learned to implement a combinational multiplier in Section 7.5.4. The design utilized multiple adders and is somewhat like the data-flow implementation of Section 11.1.2. An alternative is to use one adder to perform the additions sequentially. Assume that the two operands of the multiplication are a_in and b_in. One simple sequential algorithm is to add a_in repetitively for b_in times. For example, 7*5 can be computed as $7 + 7 + 7 + 7 + 7$. While this method is not efficient, it is simple and we can concentrate on the derivation of the ASMD chart and hardware.

Consider a multiplier with input a_in and b_in, and with output r_out. All three signals are in unsigned integer format. The repetitive-addition algorithm can be formalized in the following pseudocode:

```
if (a_in=0 or b_in=0) then {
   r = 0;}
else{
   a = a_in;
   n = b_in;
   r = 0;
   while (n != 0 ){
      r = r + a;
      n = n - 1;}
}
r_out = r;
```

Note that the ASMD chart does not have a loop construct. Its decision box uses a Boolean condition to choose one of two possible exit paths and thus is somewhat like a combined if and goto statement. To make it closer to an ASMD chart, we convert the while loop using an if statement and two goto statements. The revised pseudocode becomes

```
     if (a_in=0 or b_in=0) then {
        r = 0;}
     else{
        a = a_in;
        n = b_in;
        r = 0;
op:     r = r + a;
        n = n - 1;
        if (n = 0) then{
           goto stop;}
        else{
           goto op;}
     }
stop: r_out = r;
```

To realize this algorithm in hardware, we must first define its input and output signals. The input signals are:

- a_in and b_in: input operands. They are 8-bit signals with the std_logic_vector data type and interpreted as unsigned integers.
- start: command. The multiplier starts operation when the start signal is activated.
- clk: system clock.
- reset: asynchronous reset signal for system initialization.

The output signals are:

- r_out: the product. It is a 16-bit signal with the std_logic_vector data type and interpreted as an unsigned integer.
- ready: external status signal. It is asserted when the multiplication circuit is idle and ready to accept new inputs. It can also be interpreted that the previous operation has been completed and the result is ready.

Note that the start and ready signals are added to accommodate sequential operation. We can imagine that the sequential multiplier is part of a large system. When the main system wants to do a multiplication operation, it first checks the ready signal and then places the two operands on the two data inputs and asserts the start signal. When the start signal is activated, the sequential multiplier takes the two data inputs and begins computation. It activates the ready signal to inform the main system once the computation has been completed.

The ASMD chart is shown in Figure 11.6. It closely follows the pseudo algorithm. It uses n, a and r data registers to imitate the three variables, uses decision boxes to implement the two if statements, and uses RT operations to realize regular sequential statements.

Unlike the pseudocode, in which one statement is executed at a time, the ASMD chart allows some degree of parallelism. When the RT operations are scheduled in the same state, it means that they are performed in the same clock cycle and thus are done in parallel. For example, both $r \leftarrow r + a$ and $n \leftarrow n - 1$ operations are scheduled in the op state. This implies that there are an adder and a decrementor in the physical circuit and that two calculations can be performed simultaneously. In general, we can schedule RT operations in the same state (i.e., the same clock cycle) as long as there is no data dependency, and enough hardware resources are available.

There are four states in the ASMD chart. The idle state indicates that the circuit is currently idle. The ready signal is asserted accordingly. If the start signal is asserted, the FSMD checks whether one of the inputs is zero and branches to the ab0 or load state. In the ab0 state, r is assigned to 0 and the FSMD returns to the idle state. Although not required, we assume that a and n are loaded with a_in and b_in. In the load state, r is initialized to 0, and a and n are loaded with the external input values. The FSMD then enters the loop and iterates through the op (for "operation") state b_in times. In each iteration, it adds the content of a to r and decrements n by 1. The n register is used to keep track of the number of operations. The loop stops when it reaches 0 and the FSMD returns to the idle state. We intentionally use a vague Boolean expression, count_0=1, inside the decision box. This is elaborated in Section 11.3.4. As in an ASM chart, we use a dashed box to represent an ASMD block and to emphasize that all operations inside the block are done in parallel at the same clock cycle.

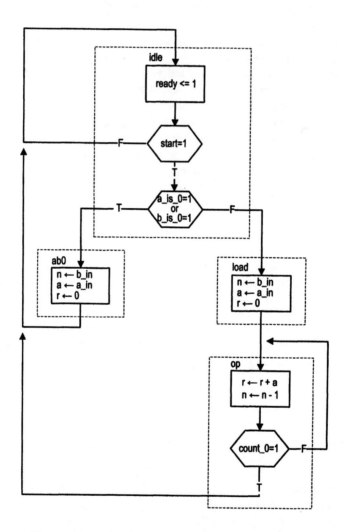

Figure 11.6   ASMD chart of a repetitive-addition multiplier.

11.3.2   Construction of the FSMD

Once the ASMD chart is constructed, more detailed information is available. We can refine the basic sketch of Figure 11.5 and derive a more detailed conceptual block diagram. We first divide the system into a control path and a data path.

The construction of the control path is the same as with the FSM. Recall that the signals inside the decision box constitute the input of the FSM. In the ASMD chart, the Boolean expressions use four signals: start, a_is_0, b_is_0 and count_0. The start signal is the external command, and the other three are internal status signals from the data path. They are asserted when the corresponding conditions are met. The output of the control path includes the external ready status signal and the control signals that specify the RT operations of the data path. In this example, we use the output of the state register as the control signal. The block diagram is shown at the bottom of Figure 11.8.

At first glance, construction of the data path seems to be more involved. However, it can be derived systematically by following simple guidelines. The basic data path can be constructed as follows:

1. List all possible RT operations in the ASMD chart.
2. Group RT operations according to their destination registers.
3. For each group, derive the circuit following the process of Section 11.2.2:
   (a) Construct the destination register.
   (b) Construct the combinational circuits involved in each RT operation.
   (c) Add multiplexing and routing circuits if the destination register is associated with multiple RT operations.
4. Add the necessary circuits to generate the status signals.

The RT operations of the repetitive-addition multiplication ASMD are grouped as follows:

- RT operations with the r register:
  - r ← r (in the idle state)
  - r ← 0 (in the load and ab0 states)
  - r ← r + a (in the op state)
- RT operations with the n register:
  - n ← n (in the idle state)
  - n ← b_in (in the load and ab0 states)
  - n ← n - 1 (in the op state)
- RT operations with the a register:
  - a ← a (in the idle and op states)
  - a ← a_in (in the load and ab0 states)

Note that we must include the default RT operations for the three registers.

Let us consider the circuit associated with the r register. The conceptual diagram is shown in Figure 11.7. It has three possible sources for the input: 0, r and r + a. The routing of the next value is done by an abstract multiplexer, as the one discussed in Section 4.3.2. It uses the output of the state register as the select signal, and its input ports are labeled with the four possible symbolic values. The connection indicates that r_reg is routed to r_next if the state_reg signal is idle, 0 is routed to r_next if the state_reg signal is ab0 or load, and r_reg + a_reg is routed to r_next if the state_reg signal is op.

We can repeat the process for two other registers and use three comparators to implement the three status signals. The complete data path, combined with the control path, is shown

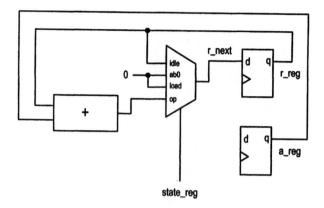

Figure 11.7 Data path associated with the r register.

in Figure 11.8. The clock and reset signals are connected to all registers. To reduce clutter, they are not shown on the diagram. The four major parts of the data path, functional units, routing circuit, data registers and status circuit, are grouped as shaded blocks.

Since this is a simple design, Figure 11.8 is somewhat unnecessarily complicated. For example, the multiplexing circuit for the a_next signal can be replaced by a register with an enable signal. The purpose of the diagram is to illustrate the derivation process. This process is very general and thus can be applied to any properly designed ASMD chart. Since the block diagram will eventually be described by VHDL code and synthesized, the multiplexing circuit will be optimized during logic synthesis.

11.3.3 Multi-segment VHDL description of an FSMD

After understanding the construction of an FSMD, we can derive the VHDL program accordingly. Our first VHDL description follows the detailed block diagram of Figure 11.8. The diagram is divided into seven blocks, which include the state register, next-state logic and output logic of the control path, and the data registers, functional units, routing network and status circuit of the data path. We use a VHDL segment for each block, and the code is shown in Listing 11.1.

Listing 11.1 Multi-segment description of a repetitive-addition multiplier

```
library ieee;
use ieee.std_logic_1164.all;
use ieee.numeric_std.all;
entity seq_mult is
    port(
        clk, reset: in std_logic;
        start: in std_logic;
        a_in, b_in: in std_logic_vector(7 downto 0);
        ready: out std_logic;
        r: out std_logic_vector(15 downto 0)
    );
end seq_mult;

architecture mult_seg_arch of seq_mult is
```

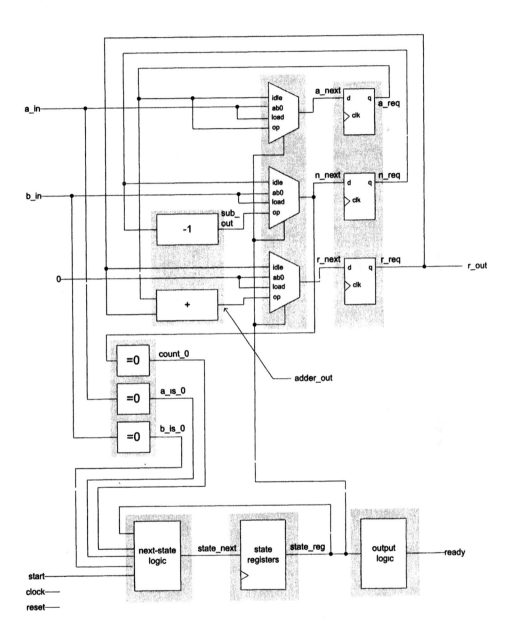

Figure 11.8   Complete block diagram of a repetitive-addition multiplier.

```
15      constant WIDTH: integer:=8;
        type state_type is (idle, ab0, load, op);
        signal state_reg, state_next: state_type;
        signal a_is_0, b_is_0, count_0: std_logic;
        signal a_reg, a_next: unsigned(WIDTH-1 downto 0);
20      signal n_reg, n_next: unsigned(WIDTH-1 downto 0);
        signal r_reg, r_next: unsigned(2*WIDTH-1 downto 0);
        signal adder_out: unsigned(2*WIDTH-1 downto 0);
        signal sub_out: unsigned(WIDTH-1 downto 0);
    begin
25      -- control path: state register
        process(clk,reset)
        begin
           if reset='1' then
              state_reg <= idle;
30         elsif (clk'event and clk='1') then
              state_reg <= state_next;
           end if;
        end process;
        -- control path: next-state/output logic
35      process(state_reg,start,a_is_0,b_is_0,count_0)
        begin
           case state_reg is
              when idle =>
                 if start='1' then
40                  if (a_is_0='1' or b_is_0='1') then
                       state_next <= ab0;
                    else
                       state_next <= load;
                    end if;
45               else
                    state_next <= idle;
                 end if;
              when ab0 =>
                 state_next <= idle;
50            when load =>
                 state_next <= op;
              when op =>
                 if count_0='1' then
                    state_next <= idle;
55               else
                    state_next <= op;
                 end if;
           end case;
        end process;
60      -- control path: output logic
        ready <= '1' when state_reg=idle else '0';
        -- data path: data register
        process(clk,reset)
        begin
65         if reset='1' then
              a_reg <= (others=>'0');
              n_reg <= (others=>'0');
```

```
            r_reg <= (others=>'0');
        elsif (clk'event and clk='1') then
70          a_reg <= a_next;
            n_reg <= n_next;
            r_reg <= r_next;
        end if;
    end process;
75  -- data path: routing multiplexer
    process(state_reg,a_reg,n_reg,r_reg,
            a_in,b_in,adder_out,sub_out)
    begin
        case state_reg is
80          when idle =>
                a_next <= a_reg;
                n_next <= n_reg;
                r_next <= r_reg;
            when ab0 =>
85              a_next <= unsigned(a_in);
                n_next <= unsigned(b_in);
                r_next <= (others=>'0');
            when load =>
                a_next <= unsigned(a_in);
90              n_next <= unsigned(b_in);
                r_next <= (others=>'0');
            when op =>
                a_next <= a_reg;
                n_next <= sub_out;
95              r_next <= adder_out;
        end case;
    end process;
    -- data path: functional units
    adder_out <= ("00000000" & a_reg) + r_reg;
100 sub_out <= n_reg - 1;
    -- data path: status
    a_is_0 <= '1' when a_in="00000000" else '0';
    b_is_0 <= '1' when b_in="00000000" else '0';
    count_0 <= '1' when n_next="00000000" else '0';
105 -- data path: output
    r <= std_logic_vector(r_reg);
end mult_seg_arch;
```

11.3.4 Use of a register value in a decision box

The key to realizing RT methodology is to derive an efficient and correct ASMD description
of an algorithm. Once this is accomplished, the VHDL derivation is more or less a mechanical procedure. The most subtle part of the ASMD derivation is using a register in Boolean
expressions of the decision boxes. We intentionally avoided this issue in the ASMD of
Figure 11.6 and used the somewhat vague a_is_0, b_is_0 and count_0 status signals inside the decision boxes. A more descriptive way is to express the Boolean conditions with
registers or input signals.

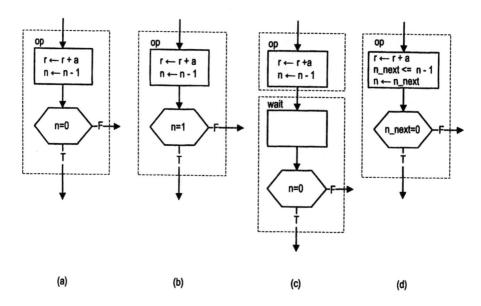

Figure 11.9 Register used in a decision box.

In the second decision box, the a_is_0=1 or b_is_0=1 expression can easily be trans-lated into a_in=0 or b_in=0. In the third decision box, the condition for the count_0=1 expression is more subtle. The n register is used as a counter to keep track of the number of iterations. The iteration stops when n reaches 0. In pseudocode, it is expressed as

```
n = n - 1;
if (n = 0) then{
    goto stop;}
else{
    goto op;}
```

Since the execution is sequential, the n variable is updated in the n = n - 1 statement, and then the new value is used in the n = 0 expression of the if statement.

In the corresponding ASMD chart, the $n \leftarrow n - 1$ operation and the decision box are in the same ASMD block. Since n is updated when the FSMD exits the block, the old value of n is used in decision box. If we write the condition as n = 0 inside the decision box, as in Figure 11.9(a), one extra iteration is introduced and thus the result is not correct.

One way to fix the problem is to use the condition of the previous iteration, n = 1, to terminate the loop, as in Figure 11.9(b). This approach may not work for other algorithms when the condition of the previous iteration cannot be determined in advance. The dis-crepancy between the pseudocode and the ASMD chart also makes the ASMD chart less intuitive.

One clumsy way to solve the problem is to insert an artificial wait state so that the content of n can be updated before it is used in a decision box. This approach is shown in Figure 11.9(c). While this makes the ASMD looks like the original algorithm, the wait state introduces one extra clock cycle in the iteration and thus severely degrades the performance.

A better way is to use the *next value* of the n register in the Boolean expression of the decision box. Since the next value is calculated during the op state, it is available at the end of the clock cycle and can be used in the decision box. Note that the previous VHDL code actually uses this value to generate the count_0 status signal:

```
count_0 <= '1' when n_next=0 else '0';
```

To express this idea in the ASMD chart, we have to split the RT operation $r \leftarrow f(\cdot)$ into two parts:

- r_next <= $f(\cdot)$
- $r \leftarrow$ r_next;

The first part means that the next value of the r register is calculated and updated within the current clock cycle. We use the signal assignment notation, <=, to emphasize that the assignment is independent of the clock. The second part indicates that the r_next signal is then assigned to r at the exit of the current state, as a regular RT operation. We can use this notation to replace the count_0=0 expression of the ASMD chart, as shown in Figure 11.9(d). This approach is the preferred method since it does not use the condition of the previous iteration, maintains consistency with the original sequential algorithm and introduces no performance penalty.

11.3.5 Four- and two-segment VHDL descriptions of FSMD

The previous multi-segment description follows the detailed FSMD block diagram. For a simple design, some blocks are very straightforward, and partitioning the VHDL code into so many code segments is overkill. We can merge some blocks to make the code more compact.

For the FSMD block diagram in Figure 11.5, we can merge the combinational circuits of the data path and control path respectively, and divide the code into four segments: the data path registers, data path combinational circuit, control path register and control path combinational circuit. The detailed VHDL code is shown in Listing 11.2. Some duplicated segments are omitted. Note that we eliminate the a_is_0, b_is_0 and count_0 status signals, and use the a_in, b_in and n_next signals directly in the Boolean conditions of the control path.

Listing 11.2   Four-segment description of a repetitive-addition multiplier

```
      architecture four_seg_arch of seq_mult is
              -- declarations same as mult_seg_arch , omitted
              . . .
      begin
  5           -- control path: state register
              -- same as mult_seg_arch , omitted
              . . .
              -- control path: combinational logic
              process (start , state_reg , a_in , b_in , n_next )
 10           begin
                  ready <='0';
                  case state_reg is
                      when idle =>
                          if start='1' then
 15                           if (a_in="00000000" or b_in="00000000") then
                                  state_next <= ab0;
                              else
                                  state_next <= load;
```

```
                              end if;
20             else
                         state_next <= idle;
                 end if;
                 ready <='1';
           when ab0 =>
25             state_next <= idle;
           when load =>
               state_next <= op;
           when op =>
               if (n_next="00000000") then
30                 state_next <= idle;
               else
                     state_next <= op;
               end if;
         end case;
35   end process;
     -- data path: data register
       -- same as mult_seg_arch , omitted
       . . .
     -- data path: combinational circuit
40   process(state_reg,a_reg,n_reg,r_reg,a_in,b_in)
     begin
         -- default value
         a_next <= a_reg;
         n_next <= n_reg;
45       r_next <= r_reg;
         case state_reg is
             when idle =>
             when ab0 =>
                 a_next <= unsigned(a_in);
50               n_next <= unsigned(b_in);
                 r_next <= (others=>'0');
             when load =>
                 a_next <= unsigned(a_in);
                 n_next <= unsigned(b_in);
55               r_next <= (others=>'0');
             when op =>
                 n_next <= n_reg - 1;
                 r_next <= ("00000000" & a_reg) + r_reg;
         end case;
60   end process;
       . . .
   end four_seg_arch;
```

Since the data registers and the state register are synchronized by the same clock signal, we can merge them into a single code segment. Similarly, since the descriptions of both combinational circuits are based on the state of the FSM, we can merge them into one segment. The resulting code consists of only two segments, one for the registers and one for the combinational circuits. The VHDL code is shown in Listing 11.3.

Listing 11.3   Two-segment description of a repetitive-addition multiplier

```vhdl
architecture two_seg_arch of seq_mult is
    -- declarations same as mult_seg_arch, omitted
    . . .
begin
    -- state and data registers
    process(clk,reset)
    begin
        if reset='1' then
            state_reg <= idle;
            a_reg <= (others=>'0');
            n_reg <= (others=>'0');
            r_reg <= (others=>'0');
        elsif (clk'event and clk='1') then
            state_reg <= state_next;
            a_reg <= a_next;
            n_reg <= n_next;
            r_reg <= r_next;
        end if;
    end process;
    -- combinational circuit
    process(start,state_reg,a_reg,n_reg,r_reg,a_in,b_in,
            n_next)
    begin
        -- default value
        a_next <= a_reg;
        n_next <= n_reg;
        r_next <= r_reg;
        ready <='0';
        case state_reg is
            when idle =>
                if start='1' then
                    if (a_in="00000000" or b_in="00000000") then
                        state_next <= ab0;
                    else
                        state_next <= load;
                    end if;
                else
                        state_next <= idle;
                end if;
                ready <='1';
            when ab0 =>
                a_next <= unsigned(a_in);
                n_next <= unsigned(b_in);
                r_next <= (others=>'0');
                state_next <= idle;
            when load =>
                a_next <= unsigned(a_in);
                n_next <= unsigned(b_in);
                r_next <= (others=>'0');
                state_next <= op;
            when op =>
                n_next <= n_reg - 1;
```

```
                    r_next <= ("00000000" & a_reg) + r_reg;
                    if (n_next="00000000") then
55                      state_next <= idle;
                    else
                        state_next <= op;
                    end if;
            end case;
60      end process;
        r <= std_logic_vector(r_reg);
    end two_seg_arch;
```

The combinational segment basically follows the ASMD chart. It uses a case statement to list the states of the ASMD chart and specifies the actions needed in each state, which include the RT operations to be performed in the data path, the next state of the control path and the external status signal of the control path.

In the beginning of the process, we use the default signal assignment statements:

```
a_next <= a_reg;
n_next <= n_reg;
r_next <= r_reg;
ready <='0';
```

These imply that registers will keep their previous values and the output signal will be unasserted if they are not assigned in a branch of the case statement. Use of the default signal assignment statements is consistent with our notation of the ASMD chart, in which only the non-default RT operations and asserted output signals are listed inside a state box. Following the two-segment coding style, we can derive the VHDL code directly from an ASMD chart and quickly realize it in hardware.

The four- and two-segment coding styles are just some possible ways to merge the blocks of an FSMD. Since an FSMD is a sequential circuit, it is a good practice to separate the registers from the combinational circuit. Other than that, we can combine or isolate combinational blocks as needed and exercise different degrees of control over the underlying hardware configuration. In an FSMD design, the functional units of the data path are normally the most complex components and are the dominant factor in circuit size and system performance. We should pay more attention to these parts and may need to separate them from the remaining code to achieve the desired area or performance constraints. The other portion of the combinational circuit can be treated as "random logic" and will be optimized during logic synthesis.

11.3.6 One-segment coding style and its deficiency

In VHDL, it is possible to combine the registers and combinational circuit into a single segment. This style may introduce some subtle mistakes, as discussed in Section 8.7. We can use this style to code FSMD as well and it allows us to translate an ASMD chart directly into one-segment VHDL code. Although this approach seems to be quick and compact at first glance, it may again introduce many subtle problems and is not recommended. The following example illustrates some of the problems. The one-segment VHDL code of the repetitive-addition multiplier is shown in Listing 11.4.

Listing 11.4   One-segment description of a repetitive-addition multiplier

```
architecture one_seg_arch of seq_mult is
   constant WIDTH: integer:=8;
   type state_type is (idle, ab0, load, op);
   signal state_reg: state_type;
5  signal a_reg, n_reg: unsigned(WIDTH-1 downto 0);
   signal r_reg: unsigned(2*WIDTH-1 downto 0);
begin
   process(clk,reset)
      variable n_next: unsigned(WIDTH-1 downto 0);
10 begin
      if reset='1' then
         state_reg <= idle;
         a_reg <= (others=>'0');
         n_reg <= (others=>'0');
15       r_reg <= (others=>'0');
      elsif (clk'event and clk='1') then
         case state_reg is
            when idle =>
               if start='1' then
20                if (a_in="00000000" or b_in="00000000") then
                     state_reg <= ab0;
                  else
                     state_reg <= load;
                  end if;
25             end if;
            when ab0 =>
               a_reg <= unsigned(a_in);
               n_reg <= unsigned(b_in);
               r_reg <= (others=>'0');
30             state_reg <= idle;
            when load =>
               a_reg <= unsigned(a_in);
               n_reg <= unsigned(b_in);
               r_reg <= (others=>'0');
35             state_reg <= op;
            when op =>
               n_next := n_reg - 1;
               n_reg <= n_next;
               r_reg <= ("00000000" & a_reg) + r_reg;
40             if (n_next="00000000") then
                  state_reg <= idle;
               end if;
         end case;
      end if;
45 end process;
   ready <='1' when (state_reg=idle) else '0';
   r <= std_logic_vector(r_reg);
end one_seg_arch;
```

There are several subtle problems in the code. First, since a register is inferred for any signal within the clk'event **and** clk='1' branch, the next value of a data register cannot be referred by a signal. To overcome this, we must define n_next as a variable for immediate assignment. Note that the variable here is used to achieve the effect of immediate assignment and has nothing to do with the variables used in the pseudocode. Second, to avoid the unnecessary output buffer, the ready output signal has to be moved outside the process and be coded as a separate segment. The problems encountered in the one-segment coding style usually require more attention and offset the original hope for quick and clear coding. We avoid this coding style in this book.

11.4   ALTERNATIVE DESIGN OF A REPETITIVE-ADDITION MULTIPLIER

After studying the basic FSMD construction and VHDL coding of the repetitive-addition algorithm, we examine two variations in this section. The variations introduce the concept of sharing and Mealy-controlled RT operation.

11.4.1   Resource sharing via FSMD

We discussed combinational resource sharing in Section 7.2. It can be applied only to few restricted scenarios. Since an FSMD provides a mechanism to schedule RT operations, sharing can be achieved in a *time-multiplexed* fashion; i.e., we can assign the same functional unit in different states (i.e., different clock cycles) and use it repeatedly. For example, if an algorithm needs to perform three additions, instead of using three adders to perform the three additions at the same time, we can use one adder and schedule the additions in three states. The FSMD allows us to have another dimension of flexibility to obtain a good trade-off between the circuit size and performance.

When we convert an algorithm into an FSMD, the functional units of the data path are usually the most complex components. Since many RT operations perform the same or similar functions, some functional units can be shared as long as these operations are scheduled in different states. In the previous repetitive-addition multiplier implementation, the function units include a 16-bit adder and an 8-bit decrementor. In the original ASMD of Figure 11.6, both addition and decrementing RT operations are scheduled in the op state, and thus no sharing is possible. If we wish to reduce the circuit size, one possibility is to split the operations and schedule them into two states. This idea is shown in the revised ASMD chart in Figure 11.10, in which the original op state is split into the op1 and op2 states. Note that an iteration now travels through two states and thus requires two clock cycles. Since the main calculation of the algorithm is done through the iterations, it takes almost twice the number of clock cycles to complete the same task.

The block diagram of the revised data path is shown in Figure 11.11. Note that there is only one adder. Two 2-to-1 multiplexers route the desired inputs to the adder. The inputs can be either a_reg and r_reg, or n_reg and "11···11", which is −1 in 2's-complement format. The former will be routed to the adder only if the current state of the control path is op1. Since there are already multiplexing circuits for the registers' inputs, no special routing circuit is needed for the output of the adder. Note that the output of the adder, add_out, is routed to the op1 port of the r_reg register's input multiplexer and to the op2 port of the n_reg register's input multiplexer.

As we discussed in Chapter 6, synthesis software is weak in performing RT-level optimization. If we use the two-segment coding style, the software may not be able to detect the intended sharing in the data path. To ensure the proper hardware construction, we can explicitly specify the desired functional unit sharing in VHDL code. The revised VHDL code is shown in Listing 11.5. It basically uses the two-segment coding style but isolates

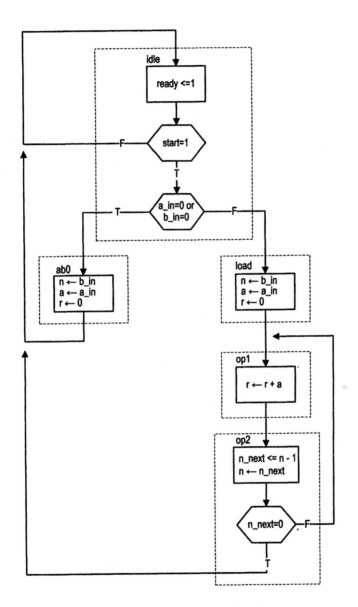

Figure 11.10   ASMD chart with sharing.

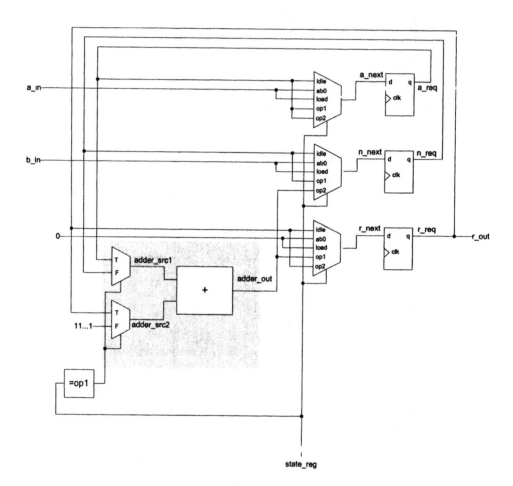

Figure 11.11   Conceptual block diagram of a sharing data path.

the functional unit from the remaining code. Note that the n register is only 8 bits wide and some adjustments are made in code to accommodate the 16-bit adder.

Listing 11.5   Sharing on a repetitive-addition multiplier

```
   architecture sharing_arch of seq_mult is
      constant WIDTH: integer:=8;
      type state_type is (idle, ab0, load, op1, op2);
      signal state_reg, state_next: state_type;
5     signal a_reg, a_next: unsigned(WIDTH-1 downto 0);
      signal n_reg, n_next: unsigned(WIDTH-1 downto 0);
      signal r_reg, r_next: unsigned(2*WIDTH-1 downto 0);
      signal adder_src1,adder_src2: unsigned(2*WIDTH-1 downto 0);
      signal adder_out: unsigned(2*WIDTH-1 downto 0);
10 begin
      -- state and data registers
      process(clk,reset)
      begin
         if reset='1' then
15          state_reg <= idle;
            a_reg <= (others=>'0');
            n_reg <= (others=>'0');
            r_reg <= (others=>'0');
         elsif (clk'event and clk='1') then
20          state_reg <= state_next;
            a_reg <= a_next;
            n_reg <= n_next;
            r_reg <= r_next;
         end if;
25    end process;
      -- next-state logic/ouput logic and data path routing
      process(start,state_reg,a_reg,n_reg,r_reg,a_in,b_in,
              adder_out,n_next)
      begin
30       -- defaut value
         a_next <= a_reg;
         n_next <= n_reg;
         r_next <= r_reg;
         ready <='0';
35       case state_reg is
            when idle =>
               if start='1' then
                  if (a_in="00000000" or b_in="00000000") then
                     state_next <= ab0;
40                else
                     state_next <= load;
                  end if;
               else
                  state_next <= idle;
45             end if;
               ready <='1';
            when ab0 =>
               a_next <= unsigned(a_in);
               n_next <= unsigned(b_in);
```

```
50            r_next <= (others=>'0');
              state_next <= idle;
           when load =>
              a_next <= unsigned(a_in);
              n_next <= unsigned(b_in);
55            r_next <= (others=>'0');
              state_next <= op1;
           when op1 =>
              r_next <= adder_out;
              state_next <= op2;
60         when op2 =>
              n_next <= adder_out(WIDTH-1 downto 0);
              if (n_next="00000000") then
                 state_next <= idle;
              else
65               state_next <= op1;
              end if;
        end case;
     end process;
     -- data path input routing and functional units
70   process(state_reg,r_reg, a_reg, n_reg)
     begin
        if (state_reg=op1) then
           adder_src1 <= r_reg;
           adder_src2 <= "00000000" & a_reg;
75      else   -- for op2 state
           adder_src1 <= "00000000" & n_reg;
           adder_src2 <= (others=>'1');
        end if;
     end process;
80   adder_out <= adder_src1 + adder_src2;
     -- output
     r <= std_logic_vector(r_reg);
  end sharing_arch;
```

Because the 8-bit decrementor is a relatively simple functional unit, the new design will not reduce the circuit size significantly, and the sharing is probably overkill for this particular example. Clearly, the sharing will become more predominant if complex functional units, such as combinational multipliers, are involved.

11.4.2 Mealy-controlled RT operations

In Section 10.4, we discussed the difference between a Mealy output and a Moore output. A Mealy output is preferred for an edge-sensitive control signal because it responds faster and requires fewer states in an FSM. Since the control path and data path are synchronized by the same clock signal, the control signals connected to the data path are edge-sensitive, and thus the Mealy output can be used. In terms of the FSMD, this means that we can specify RT operations in a conditional output box of an ASMD chart.

A representative ASMD block with a conditional output box is shown in Figure 11.12(a). The conditional output box indicates that the r2 ← r3 + r4 operation will be performed if the a > b condition is true. If the condition is false, r2 remains unchanged, which

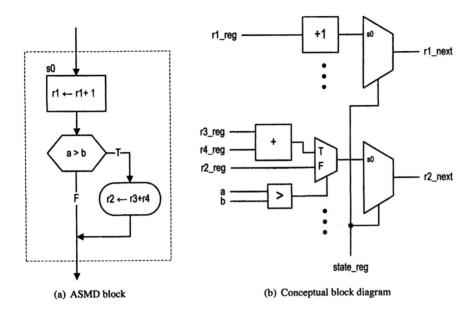

(a) ASMD block   (b) Conceptual block diagram

Figure 11.12 ASMD block with a conditional output box.

means that the r2 ← r2 operation will be performed. For comparison purposes, a Moore output-controlled operation, r1 ← r1 + 1, is included in the state box.

If this is a regular flowchart, the condition a > b is first evaluated and, if the condition is met, the r2 ← r3 + r4 operation will be performed accordingly. However, in an ASMD chart, all operations inside an ASMD block are evaluated in parallel. When the FSMD is in the s0 state, evaluations of a > b, r3 + r4 and r1 + 1 are performed at the same time. At the end of the clock cycle, the FSMD checks the result of a > b and stores the value of r2 or the result of r3 + r4 to r2 accordingly.

When an RT operation is specified inside a state box, as in r1 ← r1 + 1, there is only one possible next value (i.e., r1 + 1) in the s0 state. On the other hand, when a conditional output box exists, there are several possible next values (i.e., r2 or r3 + r4). This implies that an additional multiplexing circuit is needed. The corresponding conceptual block diagram is shown in Figure 11.12(b). An additional 2-to-1 multiplexer is added to handle the conditional output box. The result of the a > b operation is used as a selection signal and routes the desired next value to the s0 port of the abstract multiplexer.

We can apply this idea to the repetitive-addition multiplier. The original ASMD chart in Figure 11.6 is actually somewhat awkward. In the idle state, the start, a_in and b_in signals are used in the decision box, and thus they have to be available at the exit of the idle state. If the start signal is asserted, a_in and b_in will be loaded into the a and n registers in the load or ab0 state. Because of the delayed store, the actual sampling of the a_in and b_in signals occurs when the FSMD exits the load or ab0 state, and thus the a_in and b_in signals must be available at this clock edge again. For an external system that uses the multiplication circuit, this means that it has to place the two operands on a_in and b_in ports for two consecutive clocks. To release the external system from this artificial timing constraint, a better design should be able to sample the start, a_in and b_in signals at the same time and at only one clock edge.

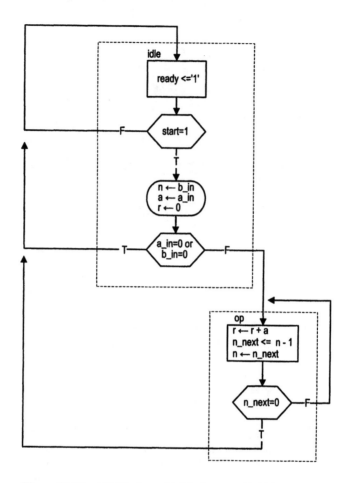

Figure 11.13   ASMD chart with Mealy-controlled RT operations.

This design can be achieved by using Mealy-controlled RT operations. The revised
ASMD chart is shown in Figure 11.13. It merges the ab0 and load states into the idle
state and moves the corresponding RT operations into a conditional output box. In addition
to relaxing the timing constraint on the external system, the revised design reduces the
number of states from four to two and improves the overall performance. The VHDL code
is shown in Listing 11.6. It uses the two-segment coding style. Note that some next-value
statements, such as a_next <= unsigned(a_in), are within the then branch of the if
statement, which corresponds to the conditional output box of the ASMD chart.

Listing 11.6   Mealy-controlled RT operations for a repetitive-addition multiplier

```
architecture mealy_arch of seq_mult is
    constant WIDTH: integer:=8;
    type state_type is (idle, op);
    signal state_reg, state_next: state_type;
    signal a_reg, a_next: unsigned(WIDTH-1 downto 0);
    signal n_reg, n_next: unsigned(WIDTH-1 downto 0);
    signal r_reg, r_next: unsigned(2*WIDTH-1 downto 0);
begin
```

```vhdl
    -- state and data registers
10  process(clk,reset)
    begin
       if reset='1' then
          state_reg <= idle;
          a_reg <= (others=>'0');
15        n_reg <= (others=>'0');
          r_reg <= (others=>'0');
       elsif (clk'event and clk='1') then
          state_reg <= state_next;
          a_reg <= a_next;
20        n_reg <= n_next;
          r_reg <= r_next;
       end if;
    end process;
    -- combinational circuit
25  process(start,state_reg,a_reg,n_reg,r_reg,a_in,b_in,
             n_next)
    begin
       a_next <= a_reg;
       n_next <= n_reg;
30     r_next <= r_reg;
       ready <='0';
       case state_reg is
          when idle =>
             if start='1' then
35              a_next <= unsigned(a_in);
                n_next <= unsigned(b_in);
                r_next <= (others=>'0');
                if a_in="00000000" or b_in="00000000" then
                   state_next <= idle;
40              else
                   state_next <= op;
                end if;
             else
                state_next <= idle;
45           end if;
             ready <='1';
          when op =>
             n_next <= n_reg - 1;
             r_next <= ("00000000" & a_reg) + r_reg;
50           if (n_next="00000000") then
                state_next <= idle;
             else
                state_next <= op;
             end if;
55     end case;
    end process;
    r <= std_logic_vector(r_reg);
  end mealy_arch;
```

11.5   TIMING AND PERFORMANCE ANALYSIS OF FSMD

An FSMD is a synchronous circuit and thus is subject to similar setup and hold time constraints. The setup time constraint, in turn, imposes the maximal clock rate. Unlike a regular sequential circuit, an algorithm described by an FSMD requires a sequence of RT operations to complete. Thus, in addition to the clock rate, the total computation time of an FSMD depends on the number of clock cycles needed to complete the computation as well. The following subsections discuss these issues.

11.5.1   Maximal clock rate

We analyzed timing of a regular sequential circuit and an FSM in Chapters 8 and 10. Both analyses are based on the basic block diagram shown in Figure 8.5. The basic diagram of an FSMD, shown in Figure 11.5, is somewhat different. It has two separate but interactive feedback loops, one for the control path and one for the data path. In theory, we can merge the two feedback loops, convert the FSMD block diagram into the standard diagram, and then analyze it as an ordinary sequential circuit. Because of the interaction between the two loops, it will be difficult to manually analyze the merged combinational circuit. We must rely on a software tool to do the timing analysis and determine the maximal clock rate.

Although the manual analysis cannot determine the exact maximal clock rate, it is possible to determine the boundaries of the rate. This analysis provides more insights into the FSMD operation and helps us to derive a more efficient design. The basic FSDM block diagram of Figure 11.5 has two feedback loops. The data path loop is based on the data register, and the control path loop is based on the state register. The two loops are not independent but interact via the control signals and status signals. For example, a function unit in the data path cannot operate until the control signals set the selection signal of the input multiplexer, and the next-state logic in the control path cannot proceed until the status signals are available. The exact maximal clock rate depends on where the control signals are needed and where the status signals are generated. This depends on the individual implementation and cannot be generalized. Our analysis considers the best- and worst-case scenarios and thus determines the boundaries of the maximal clock rate.

The control path is the bottom part of Figure 11.5. Its timing parameters are the same as those of an FSM. They are defined as follows:

- $T_{cq(state)}$: clock-to-q delay of the state register.
- $T_{setup(state)}$: setup time of the state register.
- T_{next}: maximal propagation delay of the next-state logic of the control path FSM.
- T_{output}: maximal propagation delay of the output-state logic of the control path FSM.

The conceptual diagram of the data path is shown at the top of Figure 11.5. The relevant timing parameters are:

- $T_{cq(data)}$: clock-to-q delay of the data register.
- $T_{setup(data)}$: setup time of the data register.
- T_{func}: maximal propagation delay of the functional units.
- T_{route}: maximal propagation delay of the routing multiplexing circuit.
- T_{dp}: maximal propagation delay of the combinational circuit of the data path, which is the sum of T_{func} and $2T_{route}$.

As before, we use T_c for the the clock period. In a normal design, T_{func} is likely to be the largest and the most dominant of all timing parameters. We use this assumption in the analysis.

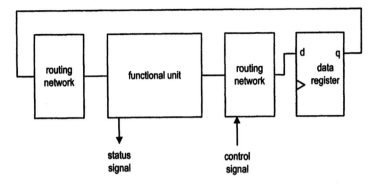

status
signal

control
signal

(a) Conceptual data path diagram

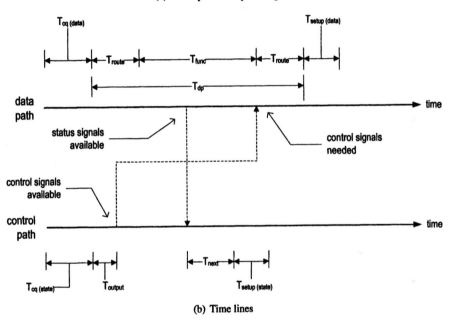

(b) Time lines

Figure 11.14 Time lines for the best-case scenario.

We first consider the best-case scenario. In this scenario, the control signals are required at a late stage of data path operation, and the status signals are generated in an early stage of data path operation, as shown in the conceptual data path diagram in Figure 11.14(a). The time lines of the data- and control-path operations are shown in Figure 11.14(b). They start with the rising edge of the clock signal. Since the data path uses control signals in the late stage, the operation of the output logic overlaps with the operation of data path and thus contributes no extra delay for the data path loop. Similarly, since the status signal is available at an early stage, the operation of the next-state logic of the control path and the computation of data path are done in parallel. When the data path computation is complete, the next-state value is also ready in the control path. The time lines show that the minimal clock period of the FSMD is the same as the clock period of the data path, which is

$$T_c = T_{cq(data)} + T_{dp} + T_{setup(data)}$$

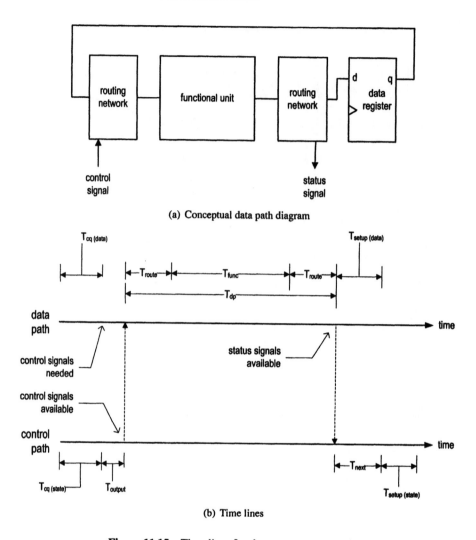

(a) Conceptual data path diagram

(b) Time lines

Figure 11.15   Time lines for the worst-case scenario.

The worst-case scenario reverses the conditions of the best-case scenario. In this scenario, control signals are required at the beginning of data path operation, and the status signals are generated at the end of data path operation. The conceptual diagram of the data path and the time lines are shown in Figure 11.15. The data path must wait for the FSM to generate the output signals, and the control path must wait for status signals to generate the next-state value. Except for the register, there is no overlapped operation between the control path and the data path. The minimal clock period can be found by following the time lines, and it includes the propagation delays of all combinational components:

$$T_c = T_{cq(state)} + T_{output} + T_{dp} + T_{next} + T_{setup(state)}$$

Assume that the state register and data register have similar timing characteristics, and clock-to-q delay and setup time are T_{cq} and T_{setup} respectively. From the two extreme scenarios, we can establish the boundaries of the minimal clock period:

$$T_{cq} + T_{dp} + T_{setup} \leq T_c \leq T_{cq} + T_{output} + T_{dp} + T_{next} + T_{setup}$$

Consequently, the maximal clock rate is bound by

$$\frac{1}{T_{cq} + T_{output} + T_{dp} + T_{next} + T_{setup}} \leq f \leq \frac{1}{T_{cq} + T_{dp} + T_{setup}}$$

For a design with a wide, complex data path, T_{dp} will be much larger than T_{next} and T_{output}, and thus variation in the minimal clock period is relatively small. For a circuit with a complex control path, we may need to minimize T_{next} and T_{output} to obtain better performance. For this kind of design, we can isolate the control path FSM in VHDL code, as in the multi- or four-segment coding styles, and apply special FSM optimization software to obtain a more efficient FSM implementation.

11.5.2 Performance analysis

In an FSMD, computation is performed in a sequence of steps, and it usually takes many clock cycles to complete a task. Thus, the total required time becomes

$$T_{total} = K * T_c$$

where K is the number of clock cycles and T_c is the clock period. K is determined by the algorithm, the width of the input and the value of the input. The determination of K is an ad hoc process and can sometimes be very difficult. For certain algorithms, K and T_c may work against each other. For example, we can merge more computation steps into a single state. This will reduce the number of states (and thus the clock cycles) but increase the clock period due to the larger data path propagation delay (T_{dp}). On the other hand, we can sometimes divide an operation into several smaller steps and schedule them in multiple clock cycles. This will decrease T_{dp} and the clock period, but requires more clock cycles to complete the same computation.

Consider the original ASMD design in Figure 11.6. The width of input operands is 8 bits. The K of this algorithm is not a constant but depends on the value of the b_in input. In the best case, b_in is 0, and the FSMD goes through the idle and ab0 states. The computation takes two clock cycles (i.e., $K=2$). In the worst case, b_in is 255 (i.e., 2^8-1) and a_in is not 0, the FSMD goes through the idle and load states once and loops the op state 255 times. K becomes 257. We can generalize this for n-bit input operands. In the worst case, the FSMD goes through the idle and load states once and loops the op state 2^n-1 times. Thus, it takes 2^n+1 clock cycles to complete the computation. We can apply the same analysis for the sharing ASMD design in Figure 11.10, in which the op state is split into two states. It requires two clock cycles for each loop iteration. For n-bit input operands, the worst-case K becomes $2 + 2(2^n - 1)$, which is 2^{n+1}. Whereas the data path size is smaller in this design, the required computation time is nearly doubled.

11.6 SEQUENTIAL ADD-AND-SHIFT MULTIPLIER

Although simple, the previous repetitive-addition algorithm is not practical since the required computation time is on the order of $O(2^n)$. This section introduces a more efficient sequential multiplication algorithm. The algorithm is based on the add-and-shift method

					a_3	a_2	a_1	a_0	multiplicand
$\times$					b_3	b_2	b_1	b_0	multiplier
				a_3b_0	a_2b_0	a_1b_0	a_0b_0	
			a_3b_1	a_2b_1	a_1b_1	a_0b_1		
		a_3b_2	a_2b_2	a_1b_2	a_0b_2			
$+$	a_3b_3	a_2b_3	a_1b_3	a_0b_3				
y_7	y_6	y_5	y_4	y_3	y_2	y_1	y_0	product

Figure 11.16 Multiplication as a summation of a_ib_j terms.

discussed in Section 7.5.4. The multiplication of two 4-bit numbers is illustrated in Figure 11.16. It includes three tasks:

1. Multiply the digits of the multiplier (b_3, b_2, b_1 and b_0) by the multiplicand (A) one at a time to obtain b_3*A, b_2*A, b_1*A and b_0*A. The b_i*A operation is bitwise and operation of b_i and the digits of A; that is,

$$b_i*A = (a_3 \cdot b_i, \ a_2 \cdot b_i, \ a_1 \cdot b_i, \ a_0 \cdot b_i)$$

2. Shift b_i*A to left i positions.
3. Add the shifted b_i*A terms to obtain the final product.

11.6.1 Initial design

The add-and-shift method can easily be converted into a sequential algorithm. We can process one digit of the multiplier (i.e., b_i) at a time and iterate through all digits of the multiplier (B). In each iteration, we calculate b_i*A, shift it to the left i positions, and then add it to the partial product. Since b_i is a binary digit, it can be either 0 or 1. Instead of computing b_i*A, we use an if statement to check the value of b_i and add the shifted A to the partial product when b_i is 1. Assume that the inputs are a_in and b_in. The pseudocode is

```
n = 0;
p = 0;
while (n!=8) {
    if (b_in(n)=1) then{
        p = p + (a_in << n);}
    n = n + 1;
}
r_out = p;
```

In hardware, it is expensive to do indexing (i.e., b_in(n)) and general shifting (i.e., a_in << n). To overcome the problem, we can "intelligently" shift a_in and b_in one position in each iteration. The pseudocode of this algorithm is

```
a = a_in;
b = b_in;
n = 8;
p = 0;
while (n!=0) {
```

```
          if (b(0)=1) then{
              p = p + a;}
          a = a << 1;
          b = b >> 1;
          n = n - 1;
      }
      r_out = p;
```

Four variables are used in the algorithm. The p variable is used to store the partial product, and the n variable is used to keep track of the number of iteration. Note that the counting direction of the n is reversed from the previous pseudocode to accommodate future hardware implementation. The a variable is used to store the shifted multiplicand (A), which is shifted left one position in each iteration. The b variable is used for the multiplier (B). It is shifted right one position in each iteration, and thus b_i of B becomes LSB (i.e., b(0)) of b in the ith iteration.

To facilitate development of an ASMD chart, we can convert the while loop into if and goto statements:

```
      a = a_in;
      b = b_in;
      n = 8;
      p = 0;
op:   if b(0)=1 then {
          p = p + a;}
      a = a << 1;
      b = b >> 1;
      n = n - 1
      if (n!=0) then{
          goto op;}
      r_out = p;
```

This pseudocode can easily be converted to an ASMD chart, as shown in Figure 11.17. The FSMD has three states. In the idle state, the FSMD checks the start signal. If it is asserted, the FSMD loads the initial values to registers and moves to either the add or shift state. If the corresponding bit of the multiplier is '1', the FSMD moves to the add state, in which the shifted multiplicand (A) is added to the partial product. Otherwise, the FSMD moves to the shift state, in which the multiplicand (A) is shifted left one position, the multiplier (B) is shifted right one position, and the counter is decremented by 1. The add-and-shift process continues to iterate until the counter reaches 0.

While the chart basically follows the pseudocode, there are two differences. First, since the two shift operations and the counter decrementing operation are independent, they are scheduled in the same state and performed in parallel. Second, due to the delayed store of RT operations, we use the next values of the registers in decision boxes. Note that b_next(0) and n_next are used in the decision boxes of the shift state, and b_in(0) is used in the decision box of the idle state.

After developing a correct, comprehensive ASMD chart, we can derive the VHDL description accordingly. The VHDL code is shown in Listing 11.7. Note that the two shifting operations are done by the concatenation operations (&).

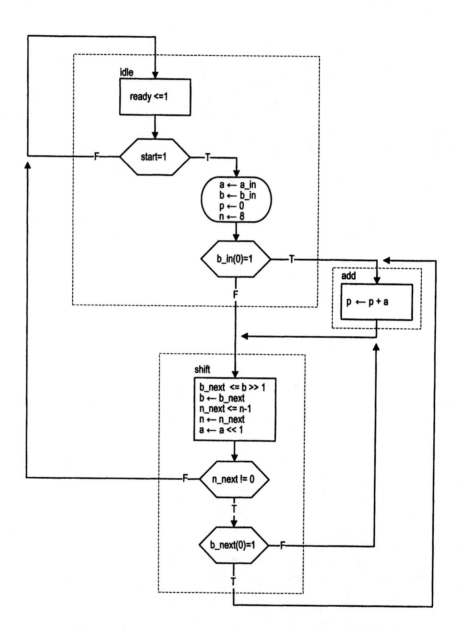

Figure 11.17 ASMD chart of the initial add-and-shift multiplier.

Listing 11.7 Initial description of an add-and-shift sequential multiplier

```vhdl
architecture shift_add_raw_arch of seq_mult is
   constant WIDTH: integer:=8;
   constant C_WIDTH: integer:=4; — width of the counter
   constant C_INIT: unsigned(C_WIDTH-1 downto 0):="1000";
5  type state_type is (idle, add, shift);
   signal state_reg, state_next: state_type;
   signal b_reg, b_next: unsigned(WIDTH-1 downto 0);
   signal a_reg, a_next: unsigned(2*WIDTH-1 downto 0);
   signal n_reg, n_next: unsigned(C_WIDTH-1 downto 0);
10   signal p_reg, p_next: unsigned(2*WIDTH-1 downto 0);
begin
   — state and data registers
   process(clk,reset)
   begin
15    if reset='1' then
         state_reg <= idle;
         b_reg <= (others=>'0');
         a_reg <= (others=>'0');
         n_reg <= (others=>'0');
20       p_reg <= (others=>'0');
      elsif (clk'event and clk='1') then
         state_reg <= state_next;
         b_reg <= b_next;
         a_reg <= a_next;
25       n_reg <= n_next;
         p_reg <= p_next;
      end if;
   end process;
   — combinational circuit
30   process(start,state_reg,b_reg,a_reg,n_reg,p_reg,
             b_in,a_in,n_next,a_next)
   begin
      b_next <= b_reg;
      a_next <= a_reg;
35     n_next <= n_reg;
      p_next <= p_reg;
      ready <='0';
      case state_reg is
         when idle =>
40          if start='1' then
               b_next <= unsigned(b_in);
               a_next <= "00000000" & unsigned(a_in);
               n_next <= C_INIT;
               p_next <= (others=>'0');
45            if b_in(0)='1' then
                  state_next <= add;
               else
                  state_next <= shift;
               end if;
50          else
               state_next <= idle;
            end if;
```

```
                    ready <='1';
                when add =>
55                  p_next <= p_reg + a_reg;
                    state_next <= shift;
                when shift =>
                    n_next <= n_reg - 1;
                    b_next <= '0' & b_reg (WIDTH-1 downto 1);
60                  a_next <= a_reg(2*WIDTH-2 downto 0) & '0';
                    if (n_next /= "0000") then
                        if a_next(0)='1' then
                            state_next <= add;
                        else
65                          state_next <= shift;
                        end if;
                    else
                        state_next <= idle;
                    end if;
70          end case;
        end process;
        r <= std_logic_vector(p_reg);
    end shift_add_raw_arch;
```

Recall that the functional units used in the data path are normally the most critical components in an FSMD, and understating their basic organization can help us to develop a more efficient design. The sketch of the data path is shown in Figure 11.18(a). To reduce the clutter, only functional units and major data flow are shown. Note that since the amount is fixed in two shift operations, the shifters require no real logic.

In the new algorithm, the number of iterations in the loop is equal to the width of the input operand. An iteration goes through the add and shift states if the corresponding multiplier bit is '1' and goes through only the shift state otherwise. For n-bit input operands, the computation requires $2n + 1$ clock cycles in the worst case (i.e., b_in is "1 $\cdots$ 1") and $n + 1$ clock cycles in the best case (i.e., b_in is "0 $\cdots$ 0"). It is far superior to the $2^n + 1$ clock cycles of the repetitive-addition algorithm.

11.6.2 Refined design

Our initial implementation of the add-and-shift multiplier closely follows the sequential pseudocode, which is presumed to be executed in a general-purpose processor. However, hardware implementation provides more flexibility and gives us an opportunity to further streamline the design. There are several possible improvements. We can first improve the efficiency of the ASMD chart. The main computation is done by iterating the add-and-shift loop, and each iteration may go through up to two states. If we examine the add and shift states closely, the RT operations in these states are independent. It is possible to merge the two states and utilize a conditional output box for the p ← p + a operation in the new state. The revised ASMD chart will require only one clock cycle for each iteration.

We can also improve the efficiency of the data path. Note that when a is added to the partial products, only the eight leftmost bits of the partial product are involved in the operation and the remaining trailing bits are kept unchanged. Instead of using a 16-bit adder, we can reduce the width of the adder to 9 bits (an 8-bit operand plus a 1-bit carry). This requires to selectively route a portion of the partial product to the adder. The "selective routing" involves complex multiplexing circuits and is not desirable. A better alternative

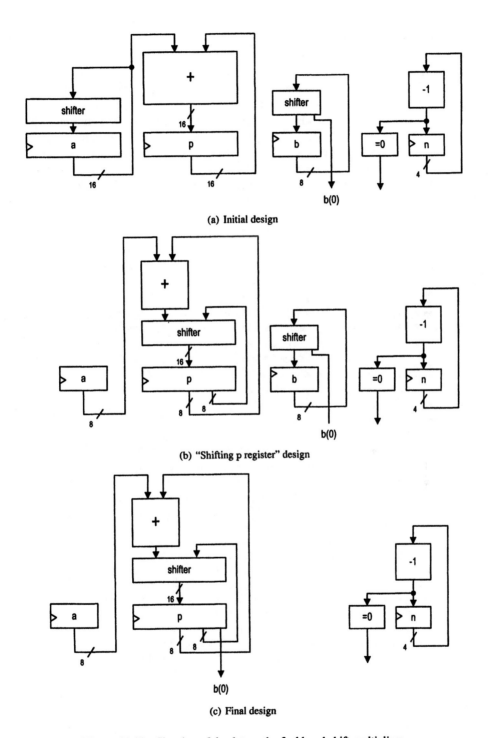

(a) Initial design

(b) "Shifting p register" design

(c) Final design

Figure 11.18   Sketches of the data path of add-and-shift multipliers.

is to shift the partial product to the right one position in each iteration, and thus the eight current leftmost bits are always connected to the input of the adder. This approach also eliminates the need of shifting multiplier (*A*) and reduces the width of the a register by half. The sketch of the revised data path is shown in Figure 11.18(b).

The circuit adds the upper half (the left half) of the p register and the a register and then combines the output of the adder with the original lower half (the right half) of the p register to form the new partial product. The p register is then shifted right one bit. Since we wish to merge the addition and shifting operations in the same state, there is no register between the adder and shifter and thus the two operations are performed in the same clock cycle. Because shifting right one bit involves only wiring, merging the two operations will not affect the critical path or increase the clock period.

Another minor improvement is to utilize the right unused portion of the p register. Note that initially only the left half of the p register contains the valid data. The valid portion expands to the right one position in each iteration when the shift-right operation is performed. On the other hand, the b register has eight valid bits initially. In each iteration, it shifts to the right one position and discards the LSB. Since the expansion of the left part of the p register matches the shrinkage of the b register, we can utilize the unused right part of the p register to function as the b register and eliminate the original b register. The sketch of the final data path is shown in Figure 11.18(c).

The refined and improved ASMD chart is shown in Figure 11.19. We use the notations pu and pl as the aliases for the upper half (left half) and lower half (right half) of the p register. Since the addition and shifting operations are performed in the same state, we must use the addition results for the following shift operation. To achieve the desired effect, we use the regular assignment notation (<=) instead of the RT notation (←) in the two conditional output boxes (i.e., pu_next <= pu and pu_next <= pu + a). The result (i.e., pu_next) is then used in a regular RT shift operation (i.e., p ← p_next >> 1). The VHDL code can be derived following the ASMD chart and is shown in Listing 11.8.

Listing 11.8   Refined description of an add-and-shift sequential multiplier

```
    architecture shift_add_better_arch of seq_mult is
       constant WIDTH: integer:=8;
       constant C_WIDTH: integer:=4;  -- width of the counter
       constant C_INIT: unsigned(C_WIDTH-1 downto 0):="1000";
 5     type state_type is (idle, add_shft);
       signal state_reg, state_next: state_type;
       signal a_reg, a_next: unsigned(WIDTH-1 downto 0);
       signal n_reg, n_next: unsigned(C_WIDTH-1 downto 0);
       signal p_reg, p_next: unsigned(2*WIDTH downto 0);
10     -- alias for the upper and lower parts of p_reg
       alias pu_next: unsigned(WIDTH downto 0) is
                      p_next(2*WIDTH downto WIDTH);
       alias pu_reg: unsigned(WIDTH downto 0) is
                     p_reg(2*WIDTH downto WIDTH);
15     alias pl_reg: unsigned(WIDTH-1 downto 0) is
                     p_reg(WIDTH-1 downto 0);
    begin
       -- state and data registers
       process(clk,reset)
20     begin
          if reset='1' then
             state_reg <= idle;
```

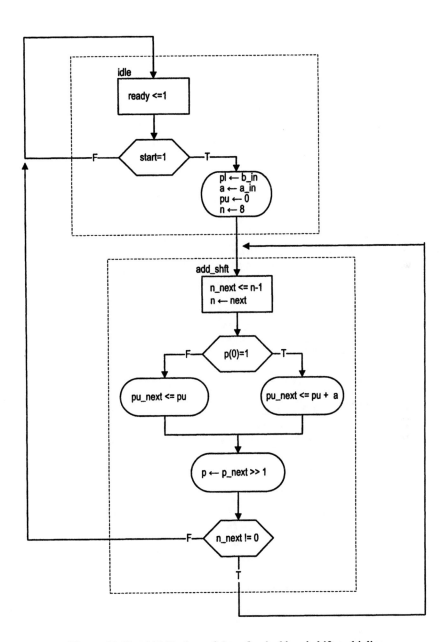

Figure 11.19 ASMD chart of the refined add-and-shift multiplier.

```vhdl
                a_reg <= (others=>'0');
                n_reg <= (others=>'0');
25              p_reg <= (others=>'0');
            elsif (clk'event and clk='1') then
                state_reg <= state_next;
                a_reg <= a_next;
                n_reg <= n_next;
30              p_reg <= p_next;
            end if;
        end process;
        -- combinational circuit
        process(start,state_reg,a_reg,n_reg,p_reg,
35              a_in,b_in,n_next,p_next)
        begin
            a_next <= a_reg;
            n_next <= n_reg;
            p_next <= p_reg;
40          ready <='0';
            case state_reg is
                when idle =>
                    if start='1' then
                        p_next <= "000000000" & unsigned(b_in);
45                      a_next <= unsigned(a_in);
                        n_next <= C_INIT;
                        state_next <= add_shft;
                      else
                        state_next <= idle;
50                  end if;
                    ready <='1';
                when add_shft =>
                    n_next <= n_reg - 1;
                    -- add if multiplier bit is '1'
55                  if (p_reg(0)='1') then
                        pu_next <= pu_reg + ('0' & a_reg);
                    else
                        pu_next <= pu_reg;
                    end if;
60                  --shift
                    p_next <= '0' & pu_next &
                                pl_reg(WIDTH-1 downto 1);
                    if (n_next /= "0000") then
                        state_next <= add_shft;
65                  else
                        state_next <= idle;
                    end if;
            end case;
        end process;
70      r <= std_logic_vector(p_reg(2*WIDTH-1 downto 0));
    end shift_add_better_arch;
```

Table 11.1   Comparison of performance and circuit complexity of three multipliers

Design method	# Clock cycles	Size of functional units	# Register bits
Repetitive-addition	2 to $2^n + 1$	$2n$-bit adder, n-bit decrementor	$4n$
Add-and-shift (initial)	$n + 1$ to $2n + 1$	$2n$-bit adder, $\lceil \log_2(n+1) \rceil$-bit dec	$5n + \lceil \log_2(n+1) \rceil$
Add-and-shift (refined)	$n + 1$	n-bit adder, $\lceil \log_2(n+1) \rceil$-bit dec	$3n + \lceil \log_2(n+1) \rceil + 1$

11.6.3 Comparison of three ASMD designs

We have examined several designs of a sequential multiplier. Table 11.1 summarizes the key characteristics of these designs, including the Mealy-based repetitive-addition design of Section 11.4.2 and two add-and-shift designs in this section. We assume that the width of the input operands is n bits. The table lists the range of the number of clock cycles to complete the multiplication, the size of the functional units in the data path, and the total number of bits in the data registers. Note that in add-and-shift designs, the counter counts from n to 0 for n-bit operands. There are $n + 1$ patterns in the counting sequence, and thus the width of the counter is $\lceil \log_2(n+1) \rceil$ bits.

The table shows that the hardware complexity of the repetitive-addition multiplier and the initial add-and-shift multiplier are comparable but that the latter significantly improves performance by reducing the worst-case clock cycles from about 2^n to $2n$. The refined add-and-shift design reduces the hardware complexity roughly by half and decreases the worst-case clock cycles from about $2n$ to n. The adder is the dominant part in the design and contributes most to the propagation delay of the data path. Because of the smaller adder, we can expect the refined add-and-shift design to have a smaller clock period as well.

After we become familiar with the process of converting an ASMD chart to VHDL code, it becomes more or less a mechanical procedure. The key is to find an efficient algorithm and researching an effective data path to support the RT operations in the algorithm. We then can derive the ASMD chart and VHDL code accordingly. As the sequential multiplier examples show, the effectiveness of a design ultimately relies on our understanding of the problem and hardware. No synthesis software can convert the repetitive-addition algorithm into an add-and-shift algorithm or convert the initial add-and-shift design to the refined design.

11.7 SYNTHESIS OF FSMD

The design methodology and VHDL coding style discussed in this chapter impose no new synthesis requirement. From the software's point of view, an FSMD code is just a code with both regular sequential circuits and an FSM and thus can be synthesized accordingly. We can separate the control path and data path in VHDL code if we want to use special FSM optimization software for the control path synthesis.

The synthesis of an algorithm can also be performed in a more abstract level, known as *high-level synthesis* or *behavioral synthesis*. The synthesis starts with abstract VHDL descriptions that are coded in pure sequential statements, similar to those used in the algorithm's pseudocode. The behavioral synthesis software converts the initial description into RT operations and automatically derives a control path and a data path. This kind of synthesis is limited to certain specialized applications. We discuss this in an example in Chapter 12.

11.8 SYNTHESIS GUIDELINES

An FSMD is a synchronous circuit with a regular sequential circuit (the data path) and an FSM (the control path). We should follow the guidelines from Chapters 8, 9 and 10. Few additional guidelines are related primarily to RT operations and construction of the data path:

- As any sequential circuit, the registers of the FSMD should be separated from the combinational circuits.

- Be aware that an RT operation exhibits a delayed-store behavior. Use of a register in a decision box should be carefully examined.

- The variables used in Boolean expressions of a pseudo algorithm normally correspond to the next values of the registers used in an ASMD.

- The function units are normally the most dominant components in a FSMD design. To exercise more control, we may need to isolate them from the rest of the code.

- Separate the control path from the code if the FSM optimization is needed later.

11.9 BIBLIOGRAPHIC NOTES

FSMD and ASMD chart provide a powerful methodology to realize sequential algorithms in hardware. The text, *Principles of Digital Design* by D. D. Gajski, has a comprehensive chapter on the representation and synthesis of FSMD. The text, *Verilog Digital Computer Design* by M. G. Arnold, applies RT methodology for computer design. As its name shows, Verilog is used for the text.

Problems

11.1   The ASMD chart in Figure 11.6 uses the n register to keep track of the number of iterations. It is initialized with b_in and counts down to 0. Alternatively, it can be initialized with 0 and counts up to b_in. From the implementation point of view, which method is better? Explain.

11.2   The ASMD chart in Figure 11.6 must return to the idle state after completion even when the main system is ready with a new set of inputs. An alternative is to allow the circuit to perform back-to-back operation in which the FSMD jumps to the ab0 or load state if the start signal is asserted while the current operation is completed.
  (a) Modify the ASMD chart to reflect the change.
  (b) Derive the VHDL code.

11.3   In the ASMD chart of Figure 11.13, what happens if we replace the a_in=0 or b_in=0 expression of the idle state with the n=0 or b=0 expression? Explain.

11.4   In Listing 11.4, what happens to the algorithm if n_next is declared as a signal?

11.5   Repetitive-subtraction division is an algorithm to implement division operation. Let y and d be the dividend and divisor respectively. This algorithm obtains the quotient (q) and the remainder (r) by subtracting d from y repeatedly until the remaining of y is smaller than d. Assume that all signals are 8 bits wide and interpreted as unsigned integers.

(a) Derive a pseudo algorithm.

(b) Convert the pseudo algorithm into an ASMD chart.

(c) Derive a detailed conceptual diagram.

(d) Derive the VHDL code according to the blocks of the conceptual diagram (i.e., in multi-segment style).

(e) Derive the VHDL code in two-segment style.

(f) Is this an efficient algorithm? Explain.

11.6   A leading-zero counting circuit counts the number of consecutive 0's from an input signal. We want to design a sequential version of this circuit. The design should check one bit of the input at a time and increment accordingly. The counting stops when the first '1' is encountered.

(a) Derive a pseudo algorithm.

(b) Convert the pseudo algorithm into an ASMD chart.

(c) Derive the VHDL code in two-segment style.

(d) Assume that the input is 16 bits wide. Synthesize both combinational and sequential versions in an ASIC technology. Compare the size and performance.

11.7   The Fibonacci function is defined as

$$fib(n) = \begin{cases} 0 & \text{if } n = 0 \\ 1 & \text{if } n = 1 \\ fib(n-1) + fib(n-2) & \text{if } n > 1 \end{cases}$$

We want to implement this function in hardware. Assume that n is a 6-bit input and interpreted as an unsigned integer. Note that $fib(63)$ is 6557470319842.

(a) Derive an ASMD chart.

(b) Derive the VHDL code.

11.8   In the ASMD chart of Figure 11.13, we can express the condition in the bottom decision box as n_next=0 or n=1. From the timing's point of view, which one can help to get a higher clock rate? Explain.

11.9   Synthesize the combinational multiplier in Section 7.5.4 and the sequential multiplier described by the ASMD chart of Figure 11.19 using an ASIC technology.

(a) Assume that the input operands are 8 bits wide. Compare the size and performance of the two circuits.

(b) Repeat part (a), but assume that the input operands are 16 bits wide.

11.10   For the sequential multiplier described by the ASMD chart of Figure 11.19, eight iterations of add-and-shift operation are needed. We can improve the design further by reducing the number of iterations to seven.

(a) Derive an ASMD chart.

(b) Derive the VHDL code.

(c) Assume that the width of the input operands is n. Calculate the relevant parameters of Table 11.1 for this improved design.

11.11   In the sequential add-and-shift multiplier, we can use a combinational circuit to process 2 bits at a time. Instead of adding 0 or shifted A to the partial product, we can add 0, shifted A, shifted $2A$, or shifted $3A$ to the partial product.

(a) Derive the revised ASMD chart for this circuit.

(b) Derive the VHDL code.

(c) Synthesize the design with an ASIC technology. Compare the size and performance of this design and the original design.

CHAPTER 12

REGISTER TRANSFER METHODOLOGY: PRACTICE

RT methodology is a powerful and versatile design technique. It can be applied to a wide variety of applications. In this chapter, we use several examples to illustrate how this methodology can be used in different types of problems and to highlight the design procedure and relevant issues.

12.1 INTRODUCTION

As discussed in Chapter 11, RT methodology can be thought of as a design technique that realizes an algorithm in hardware. The algorithm can be a complex process or just a simple sequential execution, and thus RT methodology is very flexible and versatile. We study five examples in this chapter, including a one-shot pulse generator, SRAM controller, universal asynchronous receiver and transmitter (UART), greatest common divisor (GCD) circuit, and square-root approximation circuit. The one-shot pulse generator is used to compare and contrast the differences among the regular sequential circuit, FSM and RT methodology. The SRAM controller illustrates the process of generating level-sensitive control signals to meet the timing requirement of a clockless device. The GCD circuit is another example of realizing a sequential algorithm in hardware. It shows how the hardware can be used to accelerate the performance. The UART receiver is a typical control-oriented application, which involves complex control structure and decision conditions. The square-root circuit, on the other hand, is a typical data-oriented application, which involves mainly arithmetic operations over data.

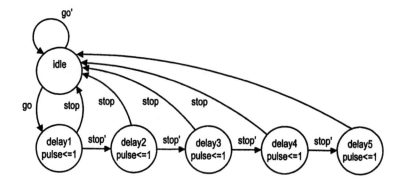

Figure 12.1 State diagram of a one-shot pulse generator.

12.2 ONE-SHOT PULSE GENERATOR

In Section 8.2.3, we divided sequential circuits into three categories based on the characteristics of the next-state logic:

- Regular sequential circuit. The next-state logic is regular.
- FSM. The next-state logic is random.
- RT methodology. The next-state logic consists of a regular part and a random part.

The RT methodology is the most flexible and capable scheme since it can accommodate both types of next-state logic.

The division is created to assist the circuit design and code development. There are no formal definitions of *regular* and *random*, and some applications can be designed as either type. In this section, we use a one-shot pulse generator as an example to illustrate the differences among the three types of circuits and to demonstrate the advantages and flexibility of the RT methodology.

A one-shot pulse generator is a circuit that generates a single fixed-width pulse upon activation of a trigger signal. We assume that the width of the pulse is five clock cycles. The detailed specifications are listed below.

- There are two input signals, go and stop, and one output signal, pulse.
- The go signal is the trigger signal that is usually asserted for only one clock cycle. During normal operation, assertion of the go signal activates the pulse signal for five clock cycles.
- If the go signal is asserted again during this interval, it will be ignored.
- If the stop signal is asserted during this interval, the pulse signal will be cut short and return to '0'.

Although the circuit is simple, it includes a regular part, which counts five clock cycles, and a random part, which keeps track of whether the circuit is idle or currently generating the pulse. Because of the simplicity, this circuit can be implemented as a pure regular sequential circuit, a pure FSM or a design using RT methodology.

12.2.1 FSM implementation

We first examine the FSM implementation. The state diagram is shown in Figure 12.1. The diagram consists of an idle state and five delay states, which activate the pulse signal for five clock cycles. The five delay states essentially function as a regular sequential circuit that counts for five clock cycles. The identical transition patterns of these five states hints at the "regularity" of this part of the operation. The corresponding VHDL code is shown in Listing 12.1.

Listing 12.1   FSM implementation of a one-shot pulse generator

```vhdl
   library ieee;
   use ieee.std_logic_1164.all;
   use ieee.numeric_std.all;
   entity pulse_5clk is
5     port(
          clk, reset: in std_logic;
          go, stop: in std_logic;
          pulse: out std_logic
       );
10 end pulse_5clk;

   architecture fsm_arch of pulse_5clk is
      type fsm_state_type is
          (idle, delay1, delay2, delay3, delay4, delay5);
15    signal state_reg, state_next: fsm_state_type;
   begin
      -- state register
      process(clk,reset)
      begin
20        if (reset='1') then
              state_reg <= idle;
          elsif (clk'event and clk='1') then
              state_reg <= state_next;
          end if;
25    end process;
      -- next-state logic & output logic
      process(state_reg,go,stop)
      begin
          pulse <= '0';
30        case state_reg is
              when idle =>
                  if go='1' then
                      state_next <= delay1;
                  else
35                    state_next <= idle;
                  end if;
              when delay1 =>
                  if stop='1' then
                      state_next <=idle;
40                else
                      state_next <=delay2;
                  end if;
                  pulse <= '1';
              when delay2 =>
45                if stop='1' then
                      state_next <=idle;
                  else
```

```
                               state_next <=delay3;
                           end if;
50                         pulse <= '1';
                       when delay3 =>
                           if stop='1' then
                               state_next <=idle;
                           else
55                             state_next <=delay4;
                           end if;
                           pulse <= '1';
                       when delay4 =>
                           if stop='1' then
60                             state_next <=idle;
                           else
                               state_next <=delay5;
                           end if;
                           pulse <= '1';
65                     when delay5 =>
                           state_next <=idle;
                           pulse <= '1';
               end case;
           end process;
70 end fsm_arch;
```

12.2.2 Regular sequential circuit implementation

We can also implement the pulse generator as a regular sequential circuit. It can be considered a mod-5 counter with a special control circuit to enable or disable the counting. To accommodate the generation of a single pulse, an additional FF is needed to flag whether the counter is active or idle. The VHDL code is shown in Listing 12.2.

Listing 12.2 Regular sequential circuit implementation of a one-shot pulse generator

```
architecture regular_seq_arch of pulse_5clk is
    constant P_WIDTH: natural:= 5;
    signal c_reg, c_next: unsigned(3 downto 0);
    signal flag_reg, flag_next: std_logic;
5 begin
    -- register
    process(clk,reset)
    begin
        if (reset='1') then
10          c_reg <= (others=>'0');
            flag_reg <= '0';
        elsif (clk'event and clk='1') then
            c_reg <= c_next;
            flag_reg <= flag_next;
15      end if;
    end process;
    -- next-state logic
    process(c_reg,flag_reg,go,stop)
    begin
20      c_next <= c_reg;
```

```
            flag_next <= flag_reg;
            if (flag_reg='0') and (go='1') then
                flag_next <= '1';
                c_next <= (others=>'0');
25          elsif (flag_reg='1') and
                    ((c_reg=P_WIDTH-1) or (stop='1')) then
                flag_next <= '0';
            elsif (flag_reg='1') then
                c_next <= c_reg + 1;
30          end if ;
        end process;
        -- output logic
        pulse <= '1' when flag_reg='1' else '0';
    end regular_seq_arch;
```

There are two registers. The c_reg register is used for the counter, and the flag_reg register indicates whether the counter is active. The critical part of the description is the if statement of the next-state logic. The first condition, (flag_reg='0') and (go='1'), indicates that the counter is currently idle and the go signal is asserted. Under this condition, the flag is asserted and the counter enters the active counting state at the next rising edge of the clock. The second condition indicates that the counter reaches 5 or the stop signal is asserted and the counting should stop. The last condition indicates that the counter is in the active state and should keep on counting.

In this code, the flag_reg register functions as some sort of state register to keep track of the current condition of the circuit. The state transitions are implicitly embedded in the if statement of the next-state logic.

12.2.3 Implementation using RT methodology

The RT methodology can separate the regular and random logic, and the ASMD chart is shown in Figure 12.2. Two states in the chart indicate whether the counter is active, and the arcs show the transitions under various conditions. The RT operation in the delay state specifies the desired increment of the counter. Following the ASMD chart, we can easily derive the VHDL code, as shown in Listing 12.3.

Listing 12.3 FSMD implementation of a one-shot pulse generator

```
    architecture fsmd_arch of pulse_5clk is
        constant P_WIDTH: natural:= 5;
        type fsmd_state_type is (idle, delay);
        signal state_reg, state_next: fsmd_state_type;
5       signal c_reg, c_next: unsigned(3 downto 0);
    begin
        -- state and data registers
        process(clk,reset)
        begin
10          if (reset='1') then
                state_reg <= idle;
                c_reg <= (others=>'0');
            elsif (clk'event and clk='1') then
                state_reg <= state_next;
15              c_reg <= c_next;
            end if;
```

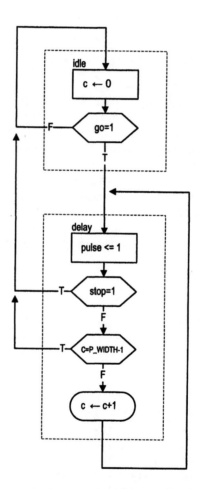

Figure 12.2   ASMD chart of a one-shot pulse generator.

```
      end process;
      — next—state  logic  &  data  path  functional  units/routing
      process(state_reg,go,stop,c_reg)
20    begin
         pulse <= '0';
         c_next <= c_reg;
         case state_reg is
            when idle =>
25                if go='1' then
                      state_next <= delay;
                   else
                      state_next <= idle;
                   end if;
30                c_next <= (others=>'0');
            when delay =>
                   if stop='1' then
                      state_next <=idle;
                   else
35                      if (c_reg=P_WIDTH-1) then
                         state_next <=idle;
                      else
                         state_next <=delay;
                         c_next <= c_reg + 1;
40                   end if;
                   end if;
                   pulse <= '1';
         end case;
      end process;
45 end fsmd_arch;
```

12.2.4 Comparison

The pulse generator example shows that we can use an FSM to emulate a regular sequential circuit, and vice versa. However, the emulation is cumbersome and convolved, and is only possible for a small design. On the other hand, the RT methodology can capture the essence of both regular and random logic, and the description is simple, flexible, clear and informative. That is why it is such a powerful methodology.

To further illustrate the capability of the RT methodology, let us consider an expanded programmable one-shot pulse generator. In this circuit, the width of the pulse can be programmed between 1 and 7. The "programming" is done as follows:

- The go and stop signals are asserted at the same time to indicate the beginning of the program mode.
- The desired value is shifted in via the go signal in the next three clock cycles.

With the RT methodology, we can easily incorporate the extension into the ASMD chart, as shown in Figure 12.3. The corresponding VHDL code is shown in Listing 12.4.

Listing 12.4 Programmable one-shot pulse generator

```
architecture prog_arch of pulse_5clk is
   type fsmd_state_type is (idle, delay, sh1, sh2, sh3);
   signal state_reg, state_next: fsmd_state_type;
   signal c_reg, c_next: unsigned(2 downto 0);
```

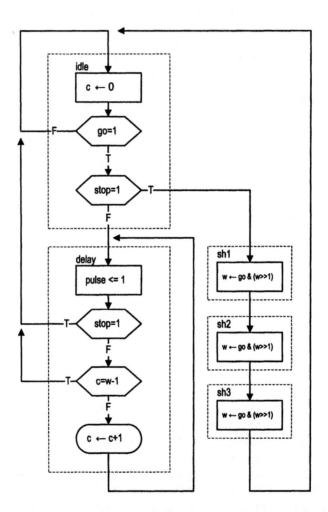

Figure 12.3 ASMD chart of a programmable one-shot pulse generator.

```vhdl
5    signal w_reg, w_next: unsigned(2 downto 0);
  begin
     -- state and data registers
     process(clk,reset)
     begin
10       if (reset='1') then
             state_reg <= idle;
             c_reg <= (others=>'0');
             w_reg <= "101";  -- default 5-cycle delay
         elsif (clk'event and clk='1') then
15           state_reg <= state_next;
             c_reg <= c_next;
             w_reg <= w_next;
         end if;
     end process;
20   -- next-state logic & data path functional units/routing
     process(state_reg,go,stop,c_reg,w_reg)
     begin
         pulse <= '0';
         c_next <= c_reg;
25       w_next <= w_reg;
         case state_reg is
             when idle =>
                 if go='1' then
                     if stop='1' then
30                       state_next <= sh1;
                     else
                         state_next <= delay;
                     end if;
                 else
35                   state_next <= idle;
                 end if;
                 c_next <= (others=>'0');
             when delay =>
                 if stop='1' then
40                   state_next <=idle;
                 else
                     if (c_reg=w_reg-1) then
                         state_next <=idle;
                     else
45                       c_next <= c_reg + 1;
                         state_next <=delay;
                     end if;
                 end if;
                 pulse <= '1';
50           when sh1 =>
                 w_next <= go & w_reg(2 downto 1);
                 state_next <= sh2;
             when sh2 =>
                 w_next <= go & w_reg(2 downto 1);
55               state_next <= sh3;
             when sh3 =>
                 w_next <= go & w_reg(2 downto 1);
```

```
                    state_next <= idle;
            end case;
60      end process;
    end prog_arch;
```

While we can implement the extended pulse generator as a pure FSM circuit or a pure regular sequential circuit in theory, the emulation becomes very involved and error-prone. It will require lots of effort to derive the code.

12.3 SRAM CONTROLLER

Random access memory (RAM) provides massive storage for digital systems. It is constructed as a two-dimensional array of memory cells. A cell is designed and optimized at the transistor level to achieve maximal efficiency. Since the silicon real estate is the primary concern, a memory cell is kept as simple as possible. Its control is level sensitive and uses no clock signal. To incorporate a RAM device into a synchronous digital system, we need a special circuit, known as a *memory controller*, to act as an interface to the synchronous system. Design of the memory controller illustrates control of a clockless subsystem.

12.3.1 Overview of SRAM

RAM is organized as a two-dimensional array of memory cells with special decoding and multiplexing circuits. The block diagram of a typical 2^{20}-by-1 static RAM (SRAM) is shown in Figure 12.4(a). It contains a 2^{10}-by-2^{10} cell array, two 10-to-2^{10} decoders and an I/O control circuit. The I/O of the SRAM includes a 20-bit address signal, ad, a 1-bit bidirectional data signal, d, and three control signals, ce, we and oe. The ad signal is split and connected to two decoders, which, in turn, enable the cell of the specified location. The three control signals are used to control SRAM operation. The chip select signal, cs, specifies whether to enable the SRAM. The output enable signal, oe, and the write enable signal, we, choose between write and read modes and control the direction of data flow. The function table is shown in Figure 12.4(b). Note that these signals are active low.

Because of the lack of a clock signal, SRAM timing is quite involved. A set of minimum and maximum timing constraints has to be satisfied to ensure proper operation. We first examine the timing of a read operation. There are two methods to read data. In the first method, both the oe and cs signals are already activated (i.e., '0') and the address signal is used to access the desired data. It is known as an *address-controlled* read cycle and the timing diagram is shown in Figure 12.5(a). In the second method, the address signal is already stable and the cs signal already activated, and the oe signal is used to initiate the read operation. It is known as an *oe-controlled* read cycle, and the timing diagram is shown in Figure 12.5(b). Note the activities of the tri-state data bus when the oe signal is activated and deactivated.

The relevant timing parameters associated with a read cycle are:

- T_{aa}: address access time, the required time to obtain stable output data after an address change. It is somewhat like the propagation delay of the read operation and is used to characterize the speed of an SRAM, as in "50-ns SRAM."
- T_{oh}: output hold from address change time, the time that the output data remains valid after the address changes. This should not be confused with the hold time of an edge-triggered FF, which is a constraint for the d input.

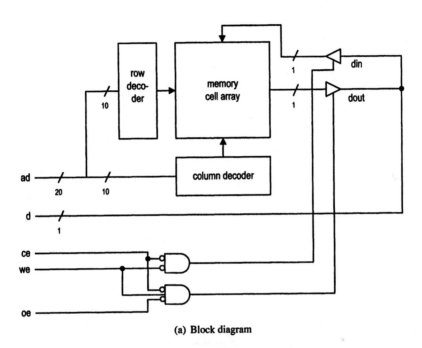

(a) Block diagram

ce	we	oe	**Operation**	**Data pin** d
1	-	-	no operation	Z
0	0	-	write	data in
0	1	0	read	data out
0	1	1	no operation	Z

(b) Function table

Figure 12.4   Block diagram and functional table of a 2^{20}-by-1 SRAM.

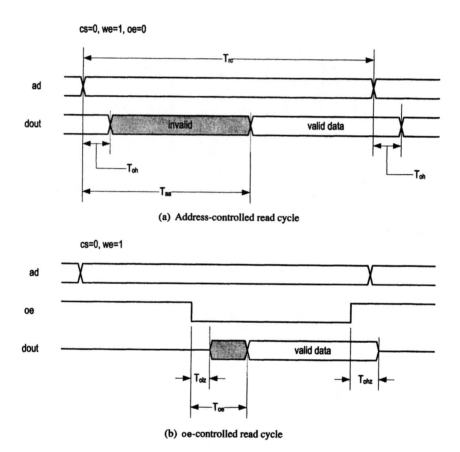

(a) Address-controlled read cycle

(b) oe-controlled read cycle

Figure 12.5   Timing diagrams of an SRAM read cycle.

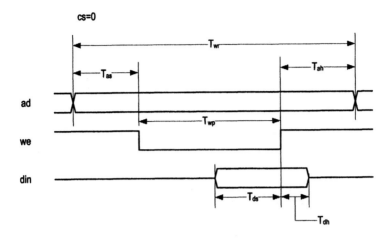

Figure 12.6 Timing diagram of an SRAM write cycle.

- T_{olz}: output enable to output in low-impedance time, the time for the tri-state buffer to leave from the high-impedance state after oe is activated. Note that even when the output is no longer in the high-impedance state, the data is still invalid.
- T_{oe}: output enable to output valid time, the time required to obtain valid data after oe is activated.
- T_{ohz}: output to Z time, the time for the tri-state buffer to enter the high-impedance state.
- T_{rc}: read cycle time, the minimal elapsed time between two read operations. It is about the same as T_{aa} for SRAM.

The write cycle is more complex. The timing diagram of a write cycle is shown in Figure 12.6. The key to understanding the write cycle timing is the assertion of the we signal, which latches the input data into the designated memory cell and plays a key role in the write operation. There are three major constraints:

- To latch data into the designated memory cell, the we signal must be activated (i.e., being '0') for a certain amount of time. This is specified by T_{wp}.
- The address needs to be stable for the entire write operation. Actually, it must be stable before we is activated and remain stable for a small amount of time after we is deactivated. The two time intervals are specified by T_{as} and T_{ah}.
- The input data must be stable in a small window when it is latched. The latch operation occurs at the edge when we transits from '0' to '1'. The input data has to be stable before and after the edge for a small amount of time. The two time intervals are specified by T_{ds} and T_{dh}. This constraint is somewhat like the constraint imposed on the d signal of a D FF at the rising edge of the clock.

These timing parameters are formally defined as follows:

- T_{wp}: write pulse width, the minimal time that the we signal must be activated.
- T_{as}: address setup time, the minimal time that the address must be stable before we is activated.
- T_{ah}: address hold time, the minimal time that the address must be stable after we is deactivated.

Table 12.1   Timing parameters of two SRAMs

Parameter	120-ns SRAM	20-ns SRAM
T_{aa} (max)	120 ns	20 ns
T_{oh} (min)	10 ns	3 ns
T_{olz} (min)	10 ns	0 ns
T_{oe} (max)	80 ns	9 ns
T_{ohz} (max)	40 ns	9 ns
T_{rc} (min)	120 ns	20 ns
T_{wp} (min)	70 ns	12 ns
T_{as} (min)	20 ns	0 ns
T_{ah} (min)	5 ns	0 ns
T_{ds} (min)	35 ns	1 ns
T_{dh} (min)	5 ns	0 ns
T_{wr} (min)	120 ns	20 ns

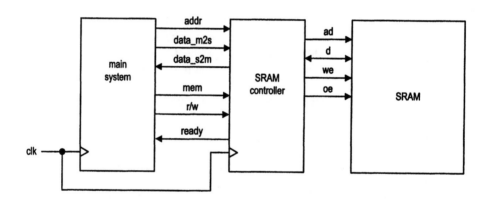

Figure 12.7   Role of an SRAM controller.

- T_{ds}: data setup time, the minimal time that data must be stable before the latching edge (the edge in which we moves from '0' to '1').
- T_{dh}: data hold time, the minimal time that data must be stable after the latching edge.
- T_{wr}: write cycle time, the minimal elapsed time between two write operations.

While there has been little change in the basic SRAM architecture over the years, its capacity and speed have improved significantly. The address access time (T_{aa}) can range from a few nanoseconds to several hundred nanoseconds. The typical timing parameters of an older, slow 120-ns SRAM and a more recent 20-ns SRAM are shown in Table 12.1.

12.3.2   Block diagram of an SRAM controller

The purpose of a memory controller is to interface the clockless memory and a synchronous system. The role of an SRAM controller is shown in Figure 12.7. It takes command from the main system and generates proper signals to store data into or retrieve data from the SRAM. The main system is a synchronous system. There are two command signals, mem and rw, and one status signal, ready. The main system activates the mem signal when a

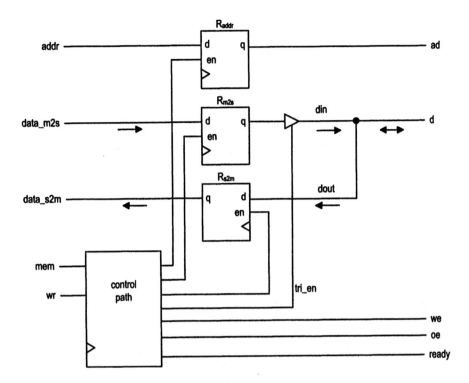

Figure 12.8   Block diagram of an SRAM controller.

memory operation is required and uses rw to specify the type of operation ('0' for write and '1' for read). The SRAM controller uses the ready signal to indicate whether it is ready for the operation. The addr signal is the address used to indicate the location of the memory. The data_m2s and data_s2m signals are the data transferred from the main system to the SRAM and from the SRAM to the main system respectively.

The main system treats the memory operation as a synchronous operation. For a write operation, it activates mem, makes rw '0', and places the address on addr and data on data_m2s for one clock cycle. At the rising edge of the clock, this information will be sampled by the SRAM controller, which, in turn, initiates an SRAM write cycle and generates proper control signals. It may take several clock cycles to complete an operation. For a read operation, the main system activates mem, makes rw '1', and places the address on addr for one clock cycle. Again, this information will be sampled by the SRAM controller at the rising edge of the clock, and an SRAM read cycle is initiated. After a predetermined number of clock cycles, the SRAM controller will put the data on data_s2m and make the data available to the main system.

Note that the main system and memory controller are controlled by the same clock. From the main system's point of view, the memory operation is completely synchronous. The combined memory controller and SRAM function somewhat like the register file of Section 9.3.1. However, whereas accessing a location in a register file can be done in one clock cycle, it takes many clock cycles to complete an SRAM read or write operation.

The block diagram of the SRAM controller is shown in Figure 12.8. The data path contains three registers, R_{addr}, R_{m2s} and R_{s2m}, which are used to store the address, the data from the main system to the SRAM, and the data from the SRAM to the main system respectively. Since the data input of the SRAM is bidirectional, a tri-state buffer is used to avoid conflict. The output from the register R_{m2s} will be placed in the data line, d, when the tri-state buffer is enabled.

The control path coordinates the overall SRAM access and generates the control signals, which include the we and oe signals of the SRAM and the enable signals of tri-state buffer and registers in the data path. There are several requirements for these control signals. First, the signals must be activated in the order specified in the read and write cycles. Second, the signals must meet various timing constraints of the SRAM. Finally, the signals need to ensure that there is no conflict (i.e., fighting) on the bidirectional data line.

12.3.3 Control path of an SRAM controller

We design the control path in two steps:

- Derive a sketch of an FSM according to the activities in read and write cycles.
- Refine the FSM with the actual SRAM timing parameters and clock period.

In the first step, we derive a sketch of an FSM that can activate and deactivate various signals in the desired order. This can be done by dividing the read or write cycles into multiple parts according to the activities of the signals and assigning a state for each part. For example, the write cycle can be divided into five parts, as shown in Figure 12.9(a).

A segment of an FSM can be constructed accordingly, as shown in Figure 12.10(a). We assume that the address and data are stored into the registers before the FSM moves to the s1 state. The data can be placed on the bidirectional line by activating the tri_en signal. The task of the FSM is essentially to generate two output signals, we and tri_en, in the following order: "10", "00", "01", "11" and "10".

Closer examination of the SRAM's timing specifications can help us to simplify the FSM. For example, T_{wp} is normally much larger than T_{ds} in most SRAMs, and there is no harm in placing data on the din line earlier. Thus, we can merge the s2 and s3 states into a single state. Also, since there in no constraint specified between the order of deactivation of address and data, we can merge the s4 and s5 states. The revised division and FSM segment are shown in Figures 12.9(b) and 12.10(b) respectively.

There are two issues with the initial sketch. First, the length of a state in the FSM corresponds to the period of the clock signal. The period must be large enough to accommodate the most strenuous timing parameter. Since T_{wp} is much larger than other parameters, the time allocated to T_{as} and T_{dh} in the ss1 and ss3 states are unnecessarily inflated. Second, in a practical design, the memory controller is usually a part of a larger system, and the clock rate is determined by the main system. The memory controller cannot have a separate clock and must work with the system clock.

In the second step, we refine the FSM according to the system clock period and SRAM timing parameters. The SRAM's access time and the main system's clock rate are the two key factors in the final design of the control path. The following examples illustrate the design and relevant issues for a slow SRAM and a fast SRAM.

Control path for a slow SRAM   The term *slow* here means that the SRAM's address access time (T_{aa}) is relatively large to the main system's clock period. For example, if we assume that the main system's clock period is 25 ns (i.e., the clock rate is 40 MHz), the 120-ns SRAM shown in Table 12.1 will be considered as a slow SRAM to this system since it takes about five clock cycles to complete a memory operation.

Because of the slow SRAM speed, it takes five (i.e., $\lceil \frac{120}{25} \rceil$) clock cycles to cover T_{aa} and three (i.e., $\lceil \frac{70}{25} \rceil$) cycles to cover T_{wp} We need to use multiple states in the FSM to

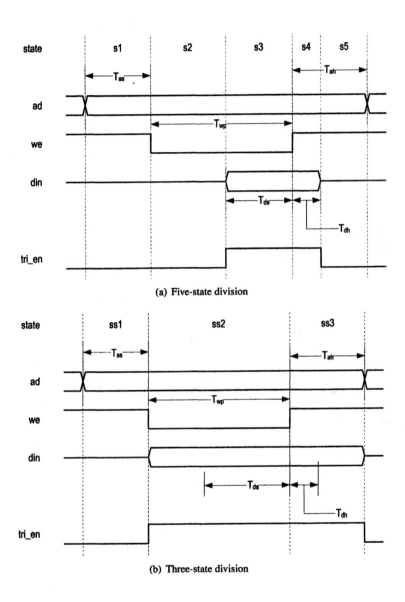

(a) Five-state division

(b) Three-state division

Figure 12.9   Divisions of a write cycle.

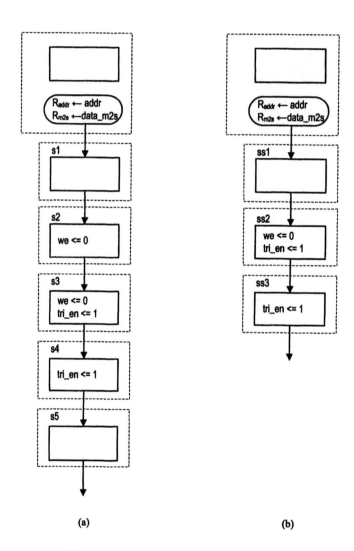

(a)   (b)

Figure 12.10 FSM segments for a write cycle.

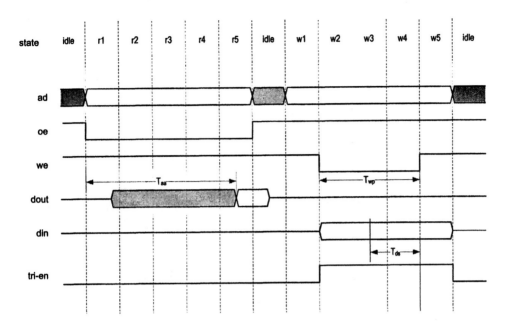

Figure 12.11 Division of read and write cycles of a slow SRAM.

accommodate the timing requirement. Figure 12.11 shows the division in the write and read cycles. An extra clock cycle, which represents the idle state, is inserted between the two operations. A read or write operation takes six clock cycles (i.e., 150 ns) to complete. The periods include one for the idle state and five for a read or write cycle. We can do a quick check on the SRAM timing parameters:

- T_{aa}: 120 ns < 125 ns (5*25 ns)
- T_{wp}: 70 ns < 75 ns (3*25 ns)
- T_{as}: 20 ns < 25 ns
- T_{ah}: 5 ns < 25 ns
- T_{ds}: 35 ns < 75 ns (3*25 ns)
- T_{dh}: 5 ns < 25 ns

It is clear that all timing parameters are satisfied and there is a margin of at least 5 ns.

The quick check is based on an ideal FSMD. To obtain more detailed timing information, we also need to consider the various propagation delays introduced by the data path and control path of the memory controller. For example, the oe signal is disabled in the end of the r5 state and the data on the d line is sampled and stored when the FSM moves from the r5 state to the idle state. We must perform a detailed timing analysis to ensure that there is no setup or hold time violation for the R_{s2m} register. The detailed timing diagram is shown in Figure 12.12. The read operation progresses as follows. At t_1, the FSM moves to the r1 state. After the T_{ctrl} delay (at t_2), the oe signal is activated and the SRAM starts the read operation. After T_{aa} (at t_3), the data is available. At t_4, the FSM moves from the r5 state to the idle state, and the memory controller samples and stores the data into the R_{s2m} register. After the T_{ctrl} delay (at t_5), the oe signal is deactivated. The data line (dout) of the SRAM returns to the high-impedance state after the T_{oz} delay (at t_6).

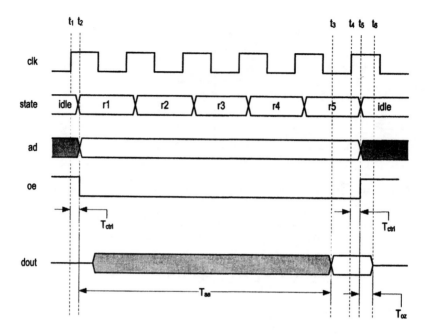

Figure 12.12 Detailed timing diagram of the read cycle.

To avoid the setup time violation, the data has to be stable before T_{setup} of the rising edge of the clock; that is,

$$T_{setup} < 5T_c - T_{ctrl} - T_{aa}$$

We can use the look-ahead output buffer to minimize T_{ctrl} and reduce it to the clock-to-q delay (T_{cq}) of the FF. With a 25-ns clock and 120-ns SRAM, the inequality becomes

$$T_{setup} + T_{cq} < 5 \text{ ns}$$

This condition can be met by most of today's device technology.

To avoid the hold time violation, the data has to be stable after T_{hold} of the rising edge of the clock:

$$T_{hold} < T_{ctrl} + T_{oz}$$

Since T_{ctrl} is T_{cq}, this condition can be easily satisfied.

Other timing requirements, such as the data bus conflict, the exact timing on the SRAM's T_{ds} and T_{dh} requirement, can be analyzed in a similar way. Because of the relatively large safety margin of this design, the initial checking should still be valid.

Following the division and signal activation, we can derive the ASMD chart accordingly, as shown in Figure 12.13. The VHDL code of the complete memory controller is shown in Listing 12.5. It includes both the data path and control path. Note that we use the look-ahead output buffer scheme for the we, oe and tri_en signals to ensure that the signals are glitch-free and to minimize the clock-to-output delay.

Listing 12.5 Memory controller of a slow SRAM

```
library ieee;
use ieee.std_logic_1164.all;
entity sram_ctrl is
```

Default: oe <= 1; we <= 1; tri_en <= 0; ready <= 0

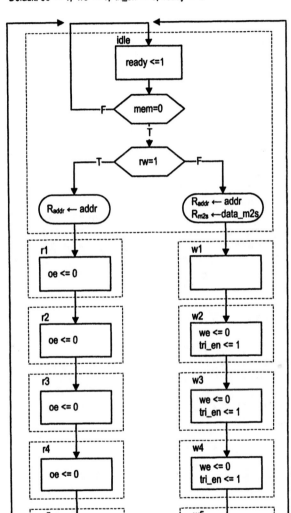

Figure 12.13   ASMD chart for a slow SRAM controller.

```vhdl
    port(
5       clk, reset: in std_logic;
        mem: in std_logic;
        rw: in std_logic;
        addr: in std_logic_vector(19 downto 0);
        data_m2s: in std_logic;
10      we, oe: out std_logic;
        ready: out std_logic;
        data_s2m: out std_logic;
        d: inout std_logic;
        ad: out std_logic_vector(19 downto 0)
15  );
  end sram_ctrl;

  architecture arch of sram_ctrl is
    type state_type is
20      (idle, r1, r2, r3, r4, r5, w1, w2, w3, w4, w5);
    signal state_reg, state_next: state_type;
    signal data_m2s_reg, data_m2s_next: std_logic;
    signal data_s2m_reg, data_s2m_next: std_logic;
    signal addr_reg, addr_next: std_logic_vector(19 downto 0);
25  signal tri_en_buf, we_buf, oe_buf: std_logic;
    signal tri_en_reg, we_reg, oe_reg: std_logic;
  begin
    -- state & data registers
    process(clk,reset)
30    begin
        if (reset='1') then
            state_reg <= idle;
            addr_reg <= (others=>'0');
            data_m2s_reg <= '0';
35          data_s2m_reg <= '0';
            tri_en_reg <= '0';
            we_reg <= '1';
            oe_reg <='1';
        elsif (clk'event and clk='1') then
40          state_reg <= state_next;
            addr_reg <= addr_next;
            data_m2s_reg <= data_m2s_next;
            data_s2m_reg <= data_s2m_next;
            tri_en_reg <= tri_en_buf;
45          we_reg <= we_buf;
            oe_reg <= oe_buf;
        end if;
    end process;
    -- next-state logic & data path functional units/routing
50    process(state_reg,mem,rw,d,addr,data_m2s,
            data_m2s_reg,data_s2m_reg,addr_reg)
    begin
        addr_next <= addr_reg;
        data_m2s_next <= data_m2s_reg;
55      data_s2m_next <= data_s2m_reg;
        ready <= '0';
```

```vhdl
                   case state_reg is
                      when idle =>
                         if mem='0' then
60                          state_next <= idle;
                         else
                            if rw='0' then  ――write
                               state_next <= w1;
                               addr_next <= addr;
65                             data_m2s_next <= data_m2s;
                            else ―― read
                               state_next <= r1;
                               addr_next <= addr;
                            end if;
70                       end if;
                         ready <= '1';
                      when w1 =>
                         state_next <= w2;
                      when w2 =>
75                       state_next <= w3;
                      when w3 =>
                         state_next <= w4;
                      when w4 =>
                         state_next <= w5;
80                    when w5 =>
                         state_next <= idle;
                      when r1 =>
                         state_next <= r2;
                      when r2 =>
85                       state_next <= r3;
                      when r3 =>
                         state_next <= r4;
                      when r4 =>
                         state_next <= r5;
90                    when r5 =>
                         state_next <= idle;
                         data_s2m_next <= d;
                   end case;
                end process;
95        ―― look―ahead output logic
          process(state_next)
          begin
             tri_en_buf <='0';
             oe_buf <= '1';
100          we_buf <= '1';
             case state_next is
                when idle =>
                when w1 =>
                when w2 =>
105                we_buf <= '0';
                   tri_en_buf <= '1';
                when w3 =>
                   we_buf <= '0';
                   tri_en_buf <= '1';
```

```
110          when w4 =>
                 we_buf <= '0';
                 tri_en_buf <= '1';
             when w5 =>
                 tri_en_buf <= '1';
115          when r1 =>
                 oe_buf <= '0';
             when r2 =>
                 oe_buf <= '0';
             when r3 =>
120              oe_buf <= '0';
             when r4 =>
                 oe_buf <= '0';
             when r5 =>
                 oe_buf <= '0';
125       end case;
       end process;
       --   output
       we <= we_reg;
       oe <= oe_reg;
130    ad <= addr_reg;
       d <= data_m2s_reg when tri_en_reg ='1' else 'Z';
       data_s2m <= data_s2m_reg;
    end arch;
```

Control path for a fast SRAM   The major problem with the previous memory system is its speed. Since it takes six clock cycles to read or write a data item from the SRAM, it can be used only if the main system accesses the memory sporadically. One way to improve the memory performance is to use a faster SRAM. For example, we can use the 20-ns SRAM of Table 12.1, whose address access time (T_{aa}) is smaller than the 25-ns clock period of the main system. Figure 12.14 shows the timing of one possible design, in which a read cycle and a write cycle are done in one clock cycle. We can again check the division against the SRAM timing parameters:

- T_{aa}: 20 ns < 25 ns
- T_{wp}: 12 ns < 25 ns
- T_{as}: 0 ns $\leq$ 0 ns
- T_{ah}: 0 ns $\leq$ 0 ns
- T_{ds}: 12 ns < 25 ns
- T_{dh}: 0 ns $\leq$ 0 ns

Although all constraints are satisfied, the timing is very tight. The timing of T_{as}, T_{ah} and T_{dh} just meet the specification and there is no safety margin. The propagation delays of the control path and data path may cause timing violations. We may need to manually fine-tune the propagation delays of various signals to ensure correct operation.

In this design, a read or write operation requires two clock cycles because the FSM must return to the idle state after each operation. Since performance is the goal of a fast SRAM controller, it is desirable to perform back-to-back memory operations without returning to the idle state. This requires an ad hoc circuit to generate a we pulse whose activation time is only a fraction of a clock period and more manual fine-tuning on propagation delays to avoid data bus fighting.

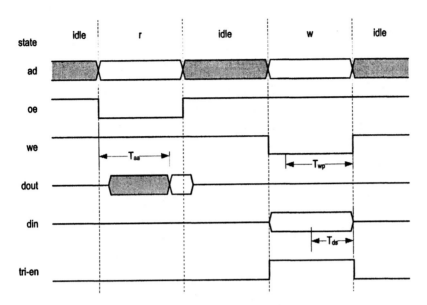

Figure 12.14   Division of read and write cycles of a fast SRAM.

In summary, while it is possible to perform single-clock back-to-back memory operations, the design imposes very strict and tricky timing requirements on the control signals. These requirements are delay-sensitive and cannot be expressed or implemented by a regular FSM. This kind of circuit is not suitable for RT-level synthesis. To implement the controller, we need to manually derive the schematic using cells from the device library and even to manually do the placement and routing. Many device manufacturers have recognized the design difficulty and incorporated the memory controller into a memory chip. This kind of device is known as *synchronous memory*. Since the main system only needs to issue commands, place address and data, or retrieve data at rising edges of the clock, this type of device greatly simplifies the memory interface to a synchronous system.

12.4 GCD CIRCUIT

The gcd(a, b) function returns the greatest common divisor (GCD) of two positive integers, a and b. For example, gcd($1, 10$) is 1 and gcd($12, 9$) is 3. The gcd function can be obtained by using subtraction, which is based on the equation

$$\gcd(a, b) = \begin{cases} a & \text{if } a = b \\ \gcd(a - b, b) & \text{if } a > b \\ \gcd(a, b - a) & \text{if } a < b \end{cases}$$

Assume that a_in and b_in are positive nonzero integers and their GCD is r. The equation can easily be converted into the following pseudocode:

```
a = a_in;
b = b_in;
while (a /= b) {
    if (a > b) then
        a = a - b;
```

```
        else
            b = b - a;
        end if
    }
    r = a;
```

To make the pseudocode more compatible with the ASMD chart, we convert the while loop into a goto statement and use a swap operation to reduce the number of required subtractions. The revised pseudocode becomes

```
            a = a_in;
            b = b_in;
    swap:   if (a = b) then
                goto stop;
            else
                if (a < b) then  — swap a and b
                    a = b;       — assume the two operations
                    b = a;       — can be done in parallel
                end if;
                a = a - b;
                goto swap;
            end if;
    stop:   r = a;
```

The code first moves the larger value into a and then performs a single subtraction of a - b. This code can easily be converted into an ASMD chart, as shown in Figure 12.15. As the sequential multiplier circuit of Chapter 11, the start and ready signals are added to interface external systems. The corresponding VHDL code is shown in Listing 12.6.

Listing 12.6   Initial implementation of a GCD circuit

```
    library ieee;
    use ieee.std_logic_1164.all;
    use ieee.numeric_std.all;
    entity gcd is
5       port(
            clk, reset: in std_logic;
            start: in std_logic;
            a_in, b_in: in std_logic_vector(7 downto 0);
            ready: out std_logic;
10          r: out std_logic_vector(7 downto 0)
        );
    end gcd ;

    architecture slow_arch of gcd is
15      type state_type is (idle, swap, sub);
        signal state_reg, state_next: state_type;
        signal a_reg, a_next, b_reg, b_next: unsigned(7 downto 0);
    begin
        — state & data registers
20      process(clk,reset)
        begin
            if reset='1' then
                state_reg <= idle;
                a_reg <= (others=>'0');
```

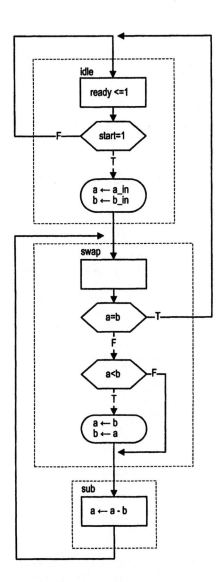

Figure 12.15   ASMD chart of the initial GCD circuit.

```
25          b_reg <= (others=>'0');
         elsif (clk'event and clk='1') then
            state_reg <= state_next;
            a_reg <= a_next;
            b_reg <= b_next;
30       end if;
      end process;
      -- next-state logic & data path functional units/routing
      process(state_reg,a_reg,b_reg,start,a_in,b_in)
      begin
35       a_next <= a_reg;
         b_next <= b_reg;
         case state_reg is
            when idle =>
               if start='1' then
40                a_next <= unsigned(a_in);
                  b_next <= unsigned(b_in);
                  state_next <= swap;
               else
                  state_next <= idle;
45             end if;
            when swap =>
               if (a_reg=b_reg) then
                  state_next <= idle;
               else
50                if (a_reg < b_reg) then
                     a_next <= b_reg;
                     b_next <= a_reg;
                  end if;
                  state_next <= sub;
55             end if;
            when sub =>
               a_next <= a_reg - b_reg;
               state_next <= swap;
         end case;
60    end process;
      -- output
      ready <= '1' when state_reg=idle else '0';
      r <= std_logic_vector(a_reg);
   end slow_arch;
```

As discussed in Section 11.5, one factor in the performance of an FSMD is the number of clock cycles required to complete the computation. In this design, the input values are subtracted successively until the a_reg=b_reg condition is reached. The number of clock cycles required to complete computation of this GCD circuit depends on the input values. It requires more time if only a small value is subtracted each time. The calculation of $\gcd(1, 2^8 - 1)$ represents the worst-case scenario. The loop has to be repeated $2^8 - 1$ times until the two values are equal. For a circuit with an N-bit input, the computation time is on the order of $O(2^N)$, and thus this is not an effective design.

One way to improve the design is to take advantage of the binary number system. For a binary number, we can tell whether it is odd or even by checking the LSB. Based on the

LSBs of two inputs, several simplification rules can be applied in the derivation of the GCD function:

- If both a and b are even, $\gcd(a, b) = 2\gcd(\frac{a}{2}, \frac{b}{2})$.
- If a is odd and b is even, $\gcd(a, b) = \gcd(a, \frac{b}{2})$.
- If a is even and b is odd, $\gcd(a, b) = \gcd(\frac{a}{2}, b)$.

Since the divided-by-2 operation corresponds to shifting right one position, it can be implemented easily in hardware. The previous equation can be extended:

$$
\gcd(a, b) = \begin{cases}
a & \text{if } a = b \\
2\gcd(\frac{a}{2}, \frac{b}{2}) & \text{if } a \neq b \text{ and } a, b \text{ even} \\
\gcd(a, \frac{b}{2}) & \text{if } a \neq b \text{ and } a \text{ odd}, b \text{ even} \\
\gcd(\frac{a}{2}, b) & \text{if } a \neq b \text{ and } a \text{ even}, b \text{ odd} \\
\gcd(a - b, b) & \text{if } a > b \text{ and } a, b \text{ odd} \\
\gcd(a, b - a) & \text{if } a < b \text{ and } a, b \text{ odd}
\end{cases}
$$

To incorporate the new rules into the algorithm, the main issue is how to handle computation of $2\gcd(\frac{a}{2}, \frac{b}{2})$. One way is ignoring the factor 2 in initial iterations and using an additional register, n, to keep track of the number of occurrences in which both operands are even. The final GCD value can be restored by multiplying the initial result by 2^n, which corresponds to shifting the initial result left n positions.

The expanded ASMD chart is shown in Figure 12.16. It has several modifications. In the swap state, the LSBs of the a and b registers are checked. The register is shifted right one position (i.e., divided by 2) if it is even. Furthermore, the n register is incremented if both are even. If the a and b registers are odd, they are compared and, if necessary, swapped, and the FSM moves to the sub state. An extra state, labeled res (for "restore"), is added to restore the final GCD value. The initial result in a is shifted left repeatedly until the n counter reaches 0. The corresponding VHDL code is shown in Listing 12.7.

Listing 12.7 More efficient implementation of a GCD circuit

```
    architecture fast_arch of gcd is
        type state_type is (idle, swap, sub, res);
        signal state_reg, state_next: state_type;
        signal a_reg, a_next, b_reg, b_next: unsigned(7 downto 0);
 5      signal n_reg, n_next: unsigned(2 downto 0);
    begin
        --- state & data registers
        process(clk, reset)
        begin
10          if reset='1' then
                state_reg <= idle;
                a_reg <= (others=>'0');
                b_reg <= (others=>'0');
                n_reg <= (others=>'0');
15          elsif (clk'event and clk='1') then
                state_reg <= state_next;
                a_reg <= a_next;
                b_reg <= b_next;
                n_reg   <= n_next;
20          end if;
        end process;
        --- next-state logic & data path functional units/routing
```

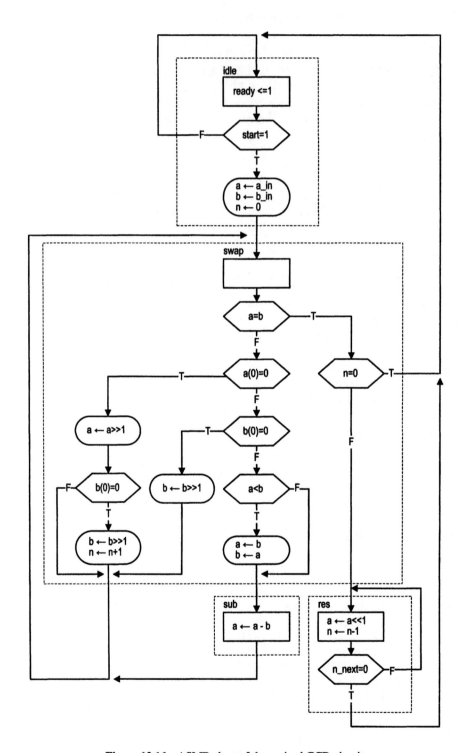

Figure 12.16   ASMD chart of the revised GCD circuit.

```
    process(state_reg,a_reg,b_reg,n_reg,start,a_in,b_in,n_next)
    begin
25      a_next <= a_reg;
        b_next <= b_reg;
        n_next <= n_reg;
        case state_reg is
            when idle =>
30              if start='1' then
                    a_next <= unsigned(a_in);
                    b_next <= unsigned(b_in);
                    n_next <= (others=>'0');
                    state_next <= swap;
35              else
                    state_next <= idle;
                end if;
            when swap =>
                if (a_reg=b_reg) then
40                  if (n_reg=0) then
                        state_next <= idle;
                    else
                        state_next <= res;
                    end if;
45              else
                    if (a_reg(0)='0') then  -- a_reg even
                        a_next <= '0' & a_reg(7 downto 1);
                        if (b_reg(0)='0') then   -- both even
                            b_next <= '0' & b_reg(7 downto 1);
50                          n_next <= n_reg + 1;
                        end if;
                        state_next <= swap;
                    else  -- a_reg odd
                        if (b_reg(0)='0') then  -- b_reg even
55                          b_next <= '0' & b_reg(7 downto 1);
                            state_next <= swap;
                        else  -- both a_reg and b_reg odd
                            if (a_reg < b_reg) then
                                a_next <= b_reg;
60                              b_next <= a_reg;
                            end if;
                            state_next <= sub;
                        end if;
                    end if;
65              end if;
            when sub =>
                a_next <= a_reg - b_reg;
                state_next <= swap;
            when res =>
70              a_next <= a_reg(6 downto 0) & '0';
                n_next <= n_reg - 1;
                if (n_next=0) then
                    state_next <= idle;
                else
75                  state_next <= res;
```

```
                    end if;
                 end case;
              end process;
              —output
80            ready <= '1' when state_reg=idle else '0';
              r <= std_logic_vector(a_reg);
           end fast_arch;
```

Now let us consider the number of clock cycles needed to complete one computation. Assume that the width of the input operand is N bits. The algorithm gradually reduces the values in the a_reg and b_reg until they are equal. In the worst case, there are $2N$ bits to be processed initially. If a value is even, the LSB is shifted out and thus the number of bits is reduced by 1. If both values are odd, a subtraction is performed and the difference is even, and the number of bits can be reduced by 1 in the next iteration. In the most pessimistic scenario, the $2N$ bits can be processed in $2 * 2N$ iterations, and the required computation time is on the order of $O(N)$, which is much better than the $O(2^N)$ of the original algorithm.

Because of the flexibility of hardware implementation, it is possible to invest extra hardware resources to improve the performance. For example, instead of handling the data bit by bit in the swap and res states, we can use more sophisticated combinational circuits to process the data in parallel. In the swap state, the circuit checks and shifts out the trailing 0's of a and b. In the res state, a shift-left barrel shifter restores the final result in a single step. The revised VHDL code is shown in Listing 12.8.

Listing 12.8 Performance-oriented implementation of a GCD circuit

```
    architecture fastest_arch of gcd is
       type state_type is (idle, swap, sub, res);
       signal state_reg, state_next: state_type;
       signal a_reg, a_next, b_reg, b_next: unsigned(7 downto 0);
5      signal n_reg, n_next, a_zero, b_zero: unsigned(2 downto 0);
    begin
       —  state & data registers
       process(clk,reset)
       begin
10         if reset='1' then
              state_reg <= idle;
              a_reg <= (others=>'0');
              b_reg <= (others=>'0');
              n_reg <= (others=>'0');
15         elsif (clk'event and clk='1') then
              state_reg <= state_next;
              a_reg <= a_next;
              b_reg <= b_next;
              n_reg   <= n_next;
20         end if;
       end process;
       —  next-state logic & data path functional units/routing
       process(state_reg,a_reg,b_reg,n_reg,start,
               a_in,b_in,a_zero,b_zero)
25     begin
          a_next <= a_reg;
          b_next <= b_reg;
          n_next <= n_reg;
```

```vhdl
        a_zero <= (others=>'0');
30      b_zero <= (others=>'0');
        case state_reg is
            when idle =>
                if start='1' then
                    a_next <= unsigned(a_in);
35                  b_next <= unsigned(b_in);
                    n_next <= (others=>'0');
                    state_next <= swap;
                else
                    state_next <= idle;
40              end if;
            when swap =>
                if (a_reg=b_reg) then
                    if (n_reg=0) then
                        state_next <= idle;
45                  else
                        state_next <= res;
                    end if;
                else
                    if (a_reg(0)='1' and b_reg(0)='1') then  -- swap
50                      if (a_reg < b_reg) then
                            a_next <= b_reg;
                            b_next <= a_reg;
                        end if;
                        state_next <= sub;
55                  else
                        -- shift out 0s of a_reg
                        if (a_reg(0)='1') then
                            a_zero <="000";
                        elsif (a_reg(1)='1') then
60                          a_next <= "0" & a_reg(7 downto 1);
                            a_zero <="001";
                        elsif (a_reg(2)='1') then
                            a_next <= "00" & a_reg(7 downto 2);
                            a_zero <="010";
65                      elsif (a_reg(3)='1') then
                            a_next <= "000" & a_reg(7 downto 3);
                            a_zero <="011";
                        elsif (a_reg(4)='1') then
                            a_next <= "0000" & a_reg(7 downto 4);
70                          a_zero <="100";
                        elsif (a_reg(5)='1') then
                            a_next <= "00000" & a_reg(7 downto 5);
                            a_zero <="101";
                        elsif (a_reg(6)='1') then
75                          a_next <= "000000" & a_reg(7 downto 6);
                            a_zero <="110";
                        else   -- a_reg(7)='1'
                            a_next <= "0000000" & a_reg(7);
                            a_zero <="111";
80                      end if;
                        -- shift out 0s of b_reg
```

```
               if (b_reg(0)='1') then
                   b_zero <="000";
               elsif (b_reg(1)='1') then
85                 b_next <= "0" & b_reg(7 downto 1);
                   a_zero <="001";
               elsif (b_reg(2)='1') then
                   b_next <= "00" & b_reg(7 downto 2);
                   b_zero <="010";
90             elsif (b_reg(3)='1') then
                   b_next <= "000" & b_reg(7 downto 3);
                   b_zero <="011";
               elsif (b_reg(4)='1') then
                   b_next <= "0000" & b_reg(7 downto 4);
95                 b_zero <="100";
               elsif (b_reg(5)='1') then
                   b_next <= "00000" & b_reg(7 downto 5);
                   b_zero <="101";
               elsif (b_reg(6)='1') then
100                b_next <= "000000" & b_reg(7 downto 6);
                   b_zero <="110";
               else   -- b_reg(7)='1'
                   b_next <= "0000000" & b_reg(7);
                   b_zero <="111";
105            end if;
               -- find common number of 0s
               if (a_zero > b_zero) then
                   n_next <= n_reg + b_zero;
               else
110                n_next <= n_reg + a_zero;
               end if;
               state_next <= swap;
            end if;
         end if;
115   when sub =>
         a_next <= a_reg - b_reg;
         state_next <= swap;
      when res =>
         case n_reg is
120         when "000" =>
               a_next <= a_reg;
            when "001" => a_next <=
               a_reg(6 downto 0) & '0';
            when "010" =>
125            a_next <= a_reg(5 downto 0) & "00";
            when "011" =>
               a_next <= a_reg(4 downto 0) & "000";
            when "100" => a_next <=
               a_reg(3 downto 0) & "0000";
130         when "101" =>
               a_next <= a_reg(2 downto 0) & "00000";
            when "110" =>
               a_next <= a_reg(1 downto 0) & "000000";
            when others =>
```

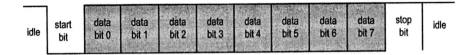

Figure 12.17   Transmission of a byte.

```
135                        a_next <= a_reg(0) & "0000000";
                   end case;
                   state_next <= idle;
               end case;
           end process;
140        -- output
           ready <= '1' when state_reg=idle else '0';
           r <= std_logic_vector(a_reg);
       end fastest_arch;
```

12.5   UART RECEIVER

Universal asynchronous receiver and transmitter (UART) is a scheme that sends bytes of data through a serial line. The transmission of a single byte is shown in Figure 12.17. The serial line is in the '1' state when it is idle. The transmission is started with a *start bit*, which is '0', followed by eight data bits and ended with a *stop bit*, which is '1'. It is also possible to insert an optional parity bit in the end of the data bits to perform error detection. Before the transmission starts, the transmitter and receiver must agree on a set of parameters in advance, which include the baud rate (i.e., number of bits per second), the number of data bits, and use of the parity bit.

The UART transmitter is essentially a shift register that shifts out data bits at a specific rate. Construction of a UART receiver is more involved since no clock information is conveyed through the serial line. The receiver can retrieve the data bits only by using the predetermined parameters. It uses an oversampling scheme to ensure that the data bits are retrieved at the correct point. This scheme utilizes a high-frequency sampling signal to estimate the middle point of a data bit and then retrieve data bits at these points. For example, assume that the sampling rate is 16 times the baud rate (i.e., there are 16 sampling pulses for each bit). The incoming stream can be recovered as follows:

1. When the incoming line becomes '0' (i.e., the beginning of the start bit), initiate the sampling pulse counter.
2. When the counter reaches 7, clear it to 0 and restart. At this point, the incoming signal reaches about a half of the start bit (i.e., the middle point of the start bit).
3. When the counter reaches 15, clear it to 0 and restart. At this point, the incoming signal progresses for one bit and reaches the middle of the first data bit. The data in the serial line should be retrieved and shifted into a register.
4. Repeat Step 3 seven times to retrieve the remaining seven data bits.
5. Repeat Step 3 one more time but without shifting. The incoming signal should reach the middle of the stop bit at this point, and its value should be '1'.

The idea behind this scheme is to use oversampling to overcome the uncertainty of the initiation of the start bit. Even when we don't know the exact onset point of the start bit, it can be off by at most $\frac{1}{16}$. The subsequent data bit retrievals are off by at most $\frac{1}{16}$ from the middle point as well.

With understanding of the oversampling procedure, we can derive the ASMD chart accordingly. One issue is the creation of sampling pulses. The easiest way is to treat the UART as a separate subsystem that utilizes a clock signal whose frequency is just 16 times that of the baud rate. This approach violates the synchronous design principle and should be avoided. A better alternative is to use a single-clock enable pulse that is synchronized with the system clock, as discussed in Section 9.1.3. Assume that the system clock is 1 MHz and the baud rate is 1200 baud. The frequency of the sampling enable pulse should be $16 * 1200$, which can be obtained by a mod-52 counter (note that $\frac{1,000,000}{16*2000} = 52$). It can easily be coded in VHDL:

```
process(clk,reset)
begin
   if reset='1' then
      clk16_reg <= (others=>'0');
   elsif (clk'event and clk='1') then
      clk16_reg <= clk16_next;
   end if;
end process;
— next-state/output logic
clk16_next <= (others=>'0') when clk16_reg=51 else
              clk16_reg + 1 ;
s_pulse <= '1' when clk16_reg=0 else '0';
```

The ASMD chart of a simplified UART receiver is shown in Figure 12.18. The chart follows the previous steps and includes three major states, start, data and stop, which represent the processing of the start bit, data bits and stop bit respectively. The s_pulse signal is the enable pulse whose frequency is 16 times that of the baud rate. Note that the FSMD stays in the same state unless the s_pulse signal is activated. There are two counters, represented by the s and n registers. The s register keeps track of the number of sampling pulses and counts to 7 in the start state and to 15 in the data and stop states. The n register keeps track of the number of data bits received in the data state. The retrieved bits are shifted into and reassembled in the b register. The corresponding VHDL code is shown in Listing 12.9. We assume that the system clock is 1 MHz and the baud rate is 1200 baud.

Listing 12.9   Simplified UART receiver

```
   library ieee;
   use ieee.std_logic_1164.all;
   use ieee.numeric_std.all;
   entity uart_receiver is
5     port(
         clk, reset: in std_logic;
         rx: in std_logic;
         ready: out std_logic;
         pout: out std_logic_vector(7 downto 0)
10    );
   end uart_receiver ;

   architecture arch of uart_receiver is
      type state_type is (idle, start, data, stop);
```

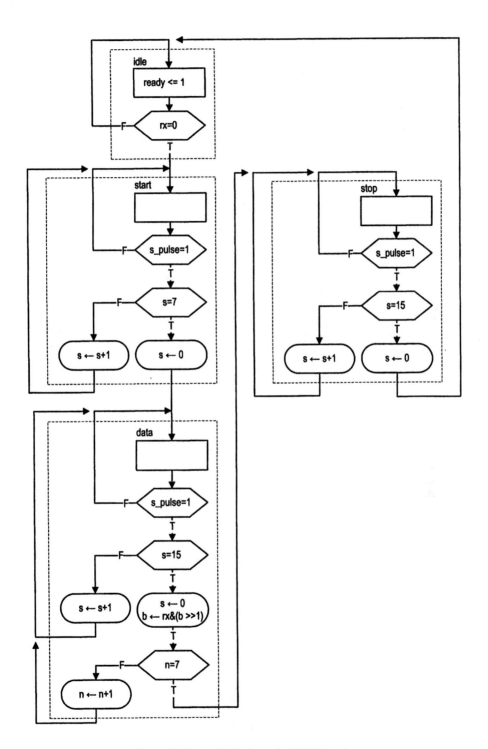

Figure 12.18   ASMD chart of a UART receiver.

```
15      signal state_reg, state_next: state_type;
        signal clk16_next, clk16_reg: unsigned(5 downto 0);
        signal s_reg, s_next: unsigned(3 downto 0);
        signal n_reg, n_next: unsigned(2 downto 0);
        signal b_reg, b_next: std_logic_vector(7 downto 0);
20      signal s_pulse: std_logic;
        constant DVSR: integer := 52;
     begin
        -- free-running mod-52 counter, independent of FSMD
        process(clk,reset)
25      begin
           if reset='1' then
              clk16_reg <= (others=>'0');
           elsif (clk'event and clk='1') then
              clk16_reg <= clk16_next;
30         end if;
        end process;
        -- next-state/output logic
        clk16_next <= (others=>'0') when clk16_reg=(DVSR-1) else
                         clk16_reg + 1 ;
35      s_pulse <= '1' when clk16_reg=0 else '0';

        -- FSMD state & data registers
        process(clk,reset)
        begin
40         if reset='1' then
              state_reg <= idle;
              s_reg <= (others=>'0');
              n_reg <= (others=>'0');
              b_reg <= (others=>'0');
45         elsif (clk'event and clk='1') then
              state_reg <= state_next;
              s_reg <= s_next;
              n_reg <= n_next;
              b_reg <= b_next;
50         end if;
        end process;
        -- next-state logic & data path functional units/routing
        process(state_reg,s_reg,n_reg,b_reg,s_pulse,rx)
        begin
55         s_next <= s_reg;
           n_next <= n_reg;
           b_next <= b_reg;
           ready <='0';
           case state_reg is
60            when idle =>
                 if rx='0' then
                    state_next <= start;
                 else
                    state_next <= idle;
65               end if;
                 ready <='1';
              when start =>
```

```
                 if (s_pulse = '0') then
                    state_next <= start;
70               else
                    if s_reg=7 then
                       state_next <= data;
                       s_next <= (others=>'0');
                    else
75                     state_next <= start;
                       s_next <= s_reg + 1;
                    end if;
                 end if;
              when data =>
80               if (s_pulse = '0') then
                    state_next <= data;
                 else
                    if s_reg=15 then
                       s_next <= (others=>'0');
85                     b_next <= rx & b_reg(7 downto 1);
                       if n_reg=7 then
                          state_next <= stop ;
                          n_next <= (others=>'0');
                       else
90                        state_next <= data;
                          n_next <= n_reg + 1;
                       end if;
                    else
                       state_next <= data;
95                     s_next <= s_reg + 1;
                    end if;
                 end if;
              when stop =>
                 if (s_pulse = '0') then
100                 state_next <= stop;
                 else
                    if s_reg=15 then
                       state_next <= idle;
                       s_next <= (others=>'0');
105                 else
                       state_next <= stop;
                       s_next <= s_reg + 1;
                    end if;
                 end if;
110        end case;
        end process;
        pout <= b_reg;
     end arch;
```

Several extensions are possible for this UART receiver, including adding a parity bit to detect the transmission error, checking the stop bit for the framing error, and making the baud rate adjustable. The main problem with the UART scheme is its performance. Because of the oversampling, the baud rate can be only a small fraction of the system clock rate, and thus this scheme can be used only for a low data rate.

12.6   SQUARE-ROOT APPROXIMATION CIRCUIT

The previous UART example is a typical *control-oriented* application, which is character-ized by the dominance of the sophisticated decision conditions and branching structures in the algorithm. The opposite type is a *data-oriented* application, which involves mainly data manipulation and arithmetic operations. It is also known as a *computation-intensive* application.

Although a data-oriented application can be implemented by a combinational circuit in theory, the approach uses a large number of functional units and thus requires a significant amount of hardware resources. RT methodology allows us to share functional units in a time-multiplexed fashion, and we can schedule the operations sequentially to achieve the desired trade-off between performance and circuit complexity. A square-root approximation circuit in this section illustrates the design procedure and relevant issues of data-oriented applications.

The square-root approximation circuit uses simple adder-type components to obtain the approximate value of $\sqrt{a^2 + b^2}$, where a and b are signed integers. The approximation is obtained by the following formula:

$$\sqrt{a^2 + b^2} \approx \max(((x - 0.125x) + 0.5y), x)$$
$$\text{where } x = \max(|a|, |b|) \text{ and } y = \min(|a|, |b|)$$

Note that the $0.125x$ and $0.5y$ operations correspond to shift x right three positions and shift y right one position, and that no actual multiplication circuit is needed. The equation can be coded in a traditional programming language. Let the two input operands be a_in and b_in and the output be r. One possible pseudocode is

```
a = a_in;
b = b_in;
t1 = abs(a);
t2 = abs(b);
x = max(t1, t2);
y = min(t1, t2);
t3 = x*0.125;
t4 = y*0.5;
t5 = x - t3;
t6 = t4 + t5;
t7 = max(t6, x)
r = t7;
```

To help VHDL conversion, we intentionally avoid reuse of the same variable name on the left-hand side of the statements. Because of the lack of control structure, the pseudocode can be translated to synthesizable VHDL code directly. The corresponding code is shown in Listing 12.10.

Listing 12.10   Square-root approximation circuit using direct dataflow description

```
library ieee;
use ieee.std_logic_1164.all;
use ieee.numeric_std.all;
entity sqrt is
    port(
        a_in, b_in: in std_logic_vector(7 downto 0);
        r: out std_logic_vector(8 downto 0)
```

```
          );
      end sqrt;
10
      architecture comb_arch of sqrt is
          constant WIDTH: natural:=8;
          signal a, b, x, y: signed(WIDTH downto 0);
          signal t1, t2, t3, t4, t5, t6, t7: signed(WIDTH downto 0);
15    begin
          a <= signed(a_in(WIDTH-1) & a_in);  — signed extension
          b <= signed(b_in(WIDTH-1) & b_in);
          t1 <= a when a > 0 else
                  0 - a;
20        t2 <= b when b > 0 else
                  0 - b;
          x <= t1 when t1 - t2 > 0 else
                  t2;
          y <= t2 when t1 - t2 > 0 else
25                t1;
          t3 <= "000" & x(WIDTH downto 3);
          t4 <= "0" & y(WIDTH downto 1);
          t5 <= x - t3;
          t6 <= t4 + t5;
30        t7 <= t6 when t6 - x > 0 else
                  x;
          r <= std_logic_vector(t7);
      end comb_arch;
```

Note that the code consists only of concurrent statements, and thus their order does not matter. The original sequential execution is embedded in the interconnection of components and the flow of data. The VHDL code consists of seven arithmetic components, including one adder and six subtractors. Since the addition and subtractions are not mutually exclusive, sharing is not possible.

For a data-oriented application, it will be helpful to examine the dependency and movement of the data. This information can be visualized by a *dataflow graph*, in which an operation is represented by a node (a circle), and its input and output variables are represented by the incoming and outgoing arcs. The dataflow graph of the square-root approximation algorithm is shown in Figure 12.19.

The graph shows that the algorithm has only a limited degree of parallelism since at most only two operations can be executed concurrently. The seven arithmetic components of the previous VHDL code cannot significantly increase the performance, and most hardware resources are wasted. Thus, while the code is simple, it is not very efficient. RT methodology is a better alternative.

To transform a dataflow chart to an ASMD chart, we need to specify when and how operations in the dataflow graph are executed. The transformation include two major tasks: scheduling and binding. *Scheduling* specifies *when* a function (i.e., a circle) can start execution, and *binding* specifies *which* functional unit is assigned to perform the execution. One important design constraint is the number of functional units allowed to be used in a design. We can allocate a minimal number of functional units to reduce the circuit size, allocate a maximal number of units to exploit full potential parallelism, or find a specific number to achieve the desired trade-off between performance and circuit size. Obtaining

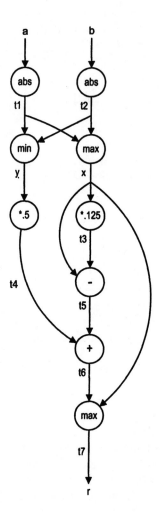

Figure 12.19   Dataflow graph.

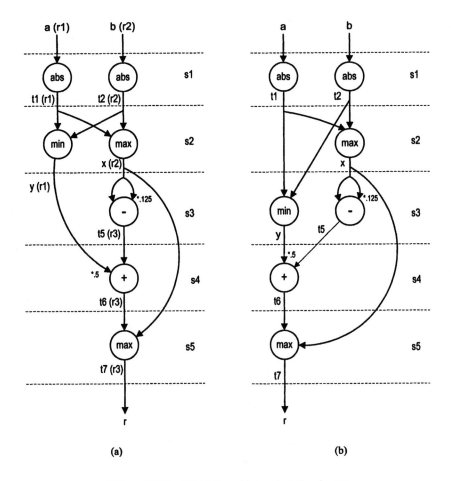

Figure 12.20   Schedules with two functional units.

an optimal schedule involves sophisticated algorithms and is a difficult task. Specialized EDA software tools are needed for a complex dataflow graph.

The dataflow graph of the square-root approximation algorithm involves a variety of operations. The ∗.125 and ∗.5 operations can be implemented by fixed-amount shifting circuits, which require no physical logic and thus should not be considered in the scheduling process. The other operations can be constructed by adders with some "glue" and routing logic. Thus, we can assume that the adder/subtractor is the only functional unit type required for the algorithm. Because at most two operations can be executed in parallel, the ASMD design can only utilize up to two functional units.

One possible schedule is shown in Figure 12.20(a). Note that the ∗.125 and ∗.5 operations are removed from the graph. The parentheses associated with the variables will be explained later. The dataflow graph is divided into five time intervals, which are later mapped into five states of an ASMD chart. It utilizes two units. One possible binding is to assign the two operations in the left column to one unit and the five operations in the right column to another unit. An alternative schedule and binding is shown in Figure 12.20(b), which requires the same amount of time to complete the computation. A schedule that uses only

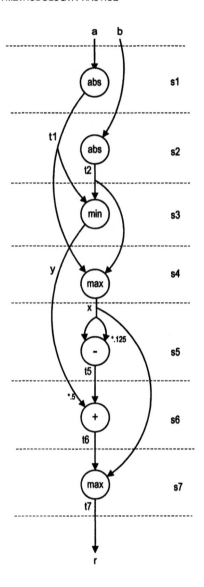

Figure 12.21   Schedule with one functional unit.

one functional unit is shown in Figure 12.21. It needs two extra time intervals to complete the operation.

Once the scheduling and binding are done, the dataflow graph can be transformed into an ASMD chart. Since each time interval represents a state in the chart, a register is needed when a signal is passed through the state boundary. The corresponding ASMD chart of Figure 12.20(a) is shown in Figure 12.22(a). The variables in the graph are mapped into the registers of the ASMD chart. There are two operations in the s1 and s2 states and one operation in the s3, s4 and s5 states. The start and ready signals and an additional idle state are included to interface the circuit with an external system.

Additional optimization schemes can be applied to reduce the number of registers and to simplify the routing structure. For example, instead of creating a new register for each

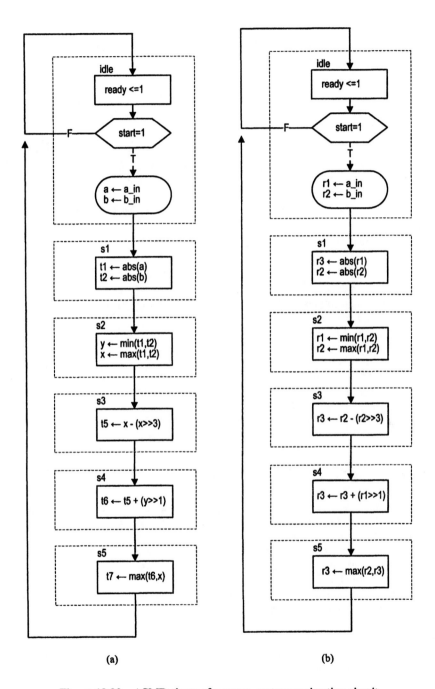

Figure 12.22   ASMD charts of a square-root approximation circuit.

variable, we can reuse an existing register if its value is no longer needed. This corresponds to properly renaming the variables in the dataflow graph. Close examination of Figure 12.20(a) shows that we can use three variables to cover the entire operation. The relationship between the new registers and the original registers is:

- Use r1 to replace a, t1 and y.
- Use r2 to replace b, t2 and x.
- Use r3 to replace t5, t6 and t7.

The replacement variables are shown in parentheses in Figure 12.20(a). The revised ASMD chart is shown in Figure 12.22(b). The number of the registers is reduced from seven to three.

The VHDL code can be derived according to the ASMD chart and is shown in Listing 12.11. To ensure proper sharing, the two functional units are isolated from the other description and coded as two separated segments. The first unit uses a single subtractor to perform the max and abs functions. The second unit uses a single adder to perform the abs and max functions as well as addition and subtraction. For clarity, we use the + operator for the carry-in signal. The synthesis software should be able to map it to the carry-in port of the adder rather than inferring another adder.

Listing 12.11   Square-root approximation circuit using RT methodology

```vhdl
library ieee;
use ieee.std_logic_1164.all;
use ieee.numeric_std.all;
entity sqrt is
    port(
        clk, reset: in std_logic;
        start: in std_logic;
        a_in, b_in: in std_logic_vector(7 downto 0);
        ready: out std_logic;
        r: out std_logic_vector(8 downto 0)
    );
end sqrt;

architecture seq_arch of sqrt is
    constant WIDTH: integer:=8;
    type state_type is (idle, s1, s2, s3, s4, s5);
    signal state_reg, state_next: state_type;
    signal r1_reg, r2_reg, r3_reg: signed(WIDTH downto 0);
    signal r1_next, r2_next, r3_next: signed(WIDTH downto 0);
    signal sub_op0, sub_op1, diff, au1_out:
        signed(WIDTH downto 0);
    signal add_op0, add_op1, sum, au2_out:
        signed(WIDTH downto 0);
    signal add_carry: integer ;
begin
    -- state & data registers
    process(clk,reset)
    begin
        if reset='1' then
            state_reg <= idle;
            r1_reg <= (others=>'0');
            r2_reg <= (others=>'0');
```

```
                    r3_reg <= (others=>'0');
                elsif (clk'event and clk='1') then
35                  state_reg <= state_next;
                    r1_reg <= r1_next;
                    r2_reg <= r2_next;
                    r3_reg <= r3_next;
                end if;
40      end process;
        -- next-state logic and data path routing
        process(start,state_reg,r1_reg,r2_reg,r3_reg,
                a_in,b_in,au1_out,au2_out)
        begin
45          r1_next <= r1_reg;
            r2_next <= r2_reg;
            r3_next <= r3_reg;
            ready <='0';
            case state_reg is
50              when idle =>
                    if start='1' then
                        r1_next <= signed(a_in(WIDTH-1) & a_in);
                        r2_next <= signed(b_in(WIDTH-1) & b_in);
                        state_next <= s1;
55                   else
                        state_next <= idle;
                    end if;
                    ready <='1';
                when s1 =>
60                  r1_next <= au1_out; -- t1 = |a|
                    r2_next <= au2_out; -- t2 = |b|
                    state_next <= s2;
                when s2 =>
                    r1_next <= au1_out; -- y=min(t1,t2)
65                  r2_next <= au2_out; -- x=max(t1,t2)
                    state_next <= s3;
                when s3 =>
                    r3_next <= au2_out; -- t5=x-0.125x
                    state_next <= s4;
70              when s4 =>
                    r3_next <= au2_out; -- t6=0.5y+t5
                    state_next <= s5;
                when s5 =>
                    r3_next <= au2_out; -- t7=max(t6,x)
75                  state_next <= idle;
            end case;
        end process;
        -- arithmetic unit 1
        -- subtractor
80      diff <= sub_op0 - sub_op1;
        -- input routing
        process(state_reg,r1_reg,r2_reg)
        begin
            case state_reg is
85              when s1 => -- 0-a
```

```
                          sub_op0 <= (others=>'0');
                          sub_op1 <= r1_reg; — a
                      when others =>   — s2: t2−t1
                          sub_op0 <= r2_reg; — t2
90                        sub_op1 <= r1_reg; — t1
                  end case;
              end process;
              — output routing
              process(state_reg,r1_reg,r2_reg,diff)
95            begin
                  case state_reg is
                      when s1 => —|a|
                          if diff(WIDTH)='0' then   — (0−a)>0
                              au1_out <= diff; — − a
100                       else
                              au1_out <= r1_reg; — a
                          end if;
                      when others =>   — s2: min(a,b)
                          if diff(WIDTH)='0' then —(t2−t1)>0
105                           au1_out <= r1_reg; — t1
                          else
                              au1_out <= r2_reg; — t2
                          end if;
                  end case;
110           end process;
              — arithmetic unit 2
              — adder
              sum <= add_op0 + add_op1 + add_carry;
              — input routing
115           process(state_reg,r1_reg,r2_reg,r3_reg)
              begin
                  case state_reg is
                      when s1 => — 0−b
                          add_op0 <= (others=>'0'); —0
120                       add_op1 <= not r2_reg;   — not b
                          add_carry <= 1;
                      when s2 => — t1−t2
                          add_op0 <= r1_reg; —t1
                          add_op1 <= not r2_reg; —not t2
125                       add_carry <= 1;
                      when s3 => — −− x−0.125x
                          add_op0 <= r2_reg; —x
                          add_op1 <= not ("000" & r2_reg(WIDTH downto 3));
                          add_carry <= 1;
130                   when s4 => — 0.5*y + t5
                          add_op0 <= "0" & r1_reg(WIDTH downto 1);
                          add_op1 <= r3_reg;
                          add_carry <= 0;
                      when others => — t6 − x
135                       add_op0 <= r3_reg; —t1
                          add_op1 <= not r2_reg; —not x
                          add_carry <= 1;
                  end case;
```

```
      end process;
140   -- output routing
      process(state_reg,r1_reg,r2_reg,r3_reg,sum)
      begin
         case state_reg is
            when s1 => -- |b|
145            if sum(WIDTH)='0' then    -- (0-b)>0
                  au2_out <= sum;  -- -b
               else
                  au2_out <= r2_reg; -- b
               end if;
150         when s2 =>
               if sum(WIDTH)='0' then
                  au2_out <= r1_reg;
               else
                  au2_out <= r2_reg;
155            end if;
            when s3|s4 => -- +,-
               au2_out <= sum;
            when others => -- s5
               if sum(WIDTH)='0' then
160               au2_out <= r3_reg;
               else
                  au2_out <= r2_reg;
               end if;
         end case;
165   end process;
      -- output
      r <= std_logic_vector(r3_reg);
   end seq_arch;
```

12.7 HIGH-LEVEL SYNTHESIS

The square-root approximation circuit of Section 12.6 shows that deriving the optimal RT design for data-oriented applications is by no means a simple task. The procedure is complex and involves many sophisticated algorithms. Derivation of this type of circuit belongs to a specific class of design, known as *high-level synthesis* or as somewhat misleading *behavioral synthesis*.

The synthesis starts with a set of constraints and an abstract VHDL description similar to the algorithm's pseudocode. The high-level synthesis software converts the initial description into an FSMD and automatically derives code for the control path and data path. In other words, the high-level synthesis software basically transforms from code in the form of Listing 12.10 to code in the form of Listing 12.11. The main task of the synthesis is to find an optimal schedule and binding to minimize the required hardware resources, to maximize performance or to obtain the best trade-off within a given constraint.

High-level synthesis is best for data-oriented, computation-intensive applications, such as those encountered in signal processing. It requires a separate software package, and its output is fed to regular synthesis software.

12.8 BIBLIOGRAPHIC NOTES

High-level synthesis covers primarily algorithms to perform the *binding* and *scheduling* of hardware resources, with emphasis on functional units. The treatment is normally very theoretical. The texts, *Synthesis and Optimization of Digital Circuits* by G. De Micheli, and *High-Level Synthesis: Introduction to Chip and System Design* by D. D. Gajski et al., provide good coverage on this topic. The square-root approximation circuit is adopted from the text, *Principles of Digital Design* by D. D. Gajski, which uses the circuit to demonstrate the procedures and various optimization algorithms of high-level synthesis.

Problems

12.1   In the ASMD chart of the programmable one-shot pulse generator of Section 12.2, shifting the desired values requires three states. This operation can be done by using a single state and a counter.

(a) Revise the ASMD chart to accommodate the change.

(b) Derive the VHDL code of the revised chart.

12.2   Redesign the programmable one-shot pulse generator of Section 12.2 as a pure regular sequential circuit. Derive the VHDL code.

12.3   Redesign the programmable one-shot pulse generator of Section 12.2 as a pure FSM.

(a) Derive the state diagram.

(b) Derive the VHDL code.

12.4   For the memory controller in Section 12.3, assume that the period of the system clock is 50 ns. Redesign the circuit for the 120-ns SRAM. The design should use a minimal number of states in the FSMD.

(a) Derive the revised ASMD chart.

(b) Derive the VHDL code.

(c) Determine the required time to perform a read operation.

12.5   Repeat the Problem 12.4 with a system clock of 15 ns.

12.6   The memory controller of Listing 12.5 must return to the idle state after each operation. We can improve performance by skipping this state when back-to-back memory operations are issued.

(a) Derive the revised ASMD chart.

(b) Derive the VHDL code.

(c) When a read operation follows immediately after a write operation, the direction of data flow in the bidirectional d line changes. Do a detailed timing analysis to examine whether a conflict can occur. We can assume that the timing parameters of the tri-state buffer in the data path are similar to those of the SRAM.

(d) Repeat part (c) for a write operation immediately following a read operation.

12.7   The FIFO buffer of Section 9.3.2 uses a register file as temporary storage. Revise the design to use an SRAM device for storage. Assume that the 120-ns SRAM is used and the system clock is 25 ns. We wish to design a FIFO controller for this system. Since it takes several clock cycles to complete a memory operation, the controller should have an additional status signal, ready, to indicate whether the SRAM is currently in operation.

(a) Derive the ASMD chart for the FIFO controller.

(b) Derive the VHDL code.

12.8   Repeat Problem 12.7 for a stack controller.

12.9   Consider the GCD circuit in Section 12.4. Assume that inputs are N-bit unsigned integers. For each architecture:

(a) Determine the input pattern that leads to the maximal number of clock cycles.

(b) Calculate the exact number of clock cycles for the pattern.

12.10   For the fast_arch architecture of the GCD circuit, we can combine the subtraction with the comparison and merge the sub state into the swap state.

(a) Derive the revised ASMD chart.

(b) Derive the VHDL code.

(c) Assume that the clock period is doubled because of the merge. Discuss whether the merge actually increases the performance (i.e., completes the computation in less time).

(d) Repeat part (c), but assume that the clock period is increased by only 50%.

12.11   In the fastest_arch architecture of the GCD circuit, up to two shifting operations are performed in the swap state and one is performed in the res state. Since a barrel shifter is a complex circuit, we want the three operations to share one unit.

(a) Derive the VHDL code of a barrel shifter that can perform both shift-right and shift-left operations.

(b) Derive the revised ASMD chart.

(c) Derive the VHDL code.

12.12   Design a transmitter for the UART discussed in Section 12.5.

(a) Derive the ASMD chart.

(b) Derive the VHDL code.

12.13   Expand the UART receiver of Section 12.5 to make the baud rate adjustable. Assume that there is an additional 2-bit control signal, baud_sel, which specifies the desired baud rate, which can be 1200, 2400, 4800 or 9600 baud.

12.14   Revise the UART of Section 12.5 to include an even-parity bit. The length of the data is now 7 bits, and the eighth bit is the parity bit. The parity bit is asserted when there is an odd number of 1's in data bits (and thus makes the 8 received bits always have an even number of 1's). Design a transmitter and a receiver for the modified UART.

(a) Derive the ASMD chart for the transmitter.

(b) Derive the VHDL code for the transmitter.

(c) Derive the ASMD chart for the receiver. The receiver should include an extra output signal to indicate the occurrence of a parity error.

(d) Derive the VHDL code for the receiver.

12.15   Expand the UART receiver of Section 12.5 to accommodate different parity schemes. Assume that there is an extra 2-bit control signal, pairty_sel, which selects the desired parity scheme, which can be odd parity, even parity or no parity.

12.16   Consider a UART that can communicate at four baud rates: 1200, 2400, 4800 and 9600 baud. Assume that the actual baud rate is unknown but the transmitter always sends a "11111111" data byte at the beginning of the session. Design a circuit that can automatically determine the baud rate and derive the VHDL code.

12.17 Consider the schedule in Figure 12.20(b).

  (a) Map the variables into a minimum number of registers.

  (b) Derive the ASMD chart for the schedule. Recall that two arithmetic units are used in this schedule.

  (c) Derive the VHDL code.

12.18 Repeat Problem 12.17 for the schedule in Figure 12.21. Note that only one arithmetic unit is used in this schedule.

12.19 Multiplication can be implemented by performing additions of shifted bit-products, as discussed in Section 7.5.4. Let $p_7, p_6, \ldots, p_1, p_0$ be the shifted bit-products of an 8-bit multiplier. The final product can be expressed as

$$y = p_7 + p_6 + \cdots + p_1 + p_0$$

  (a) Derive the dataflow graph for the expression. Arrange the additions as a tree to exploit parallelism.

  (b) Assume that only one adder is provided. Derive a schedule.

  (c) Derive the ASMD chart for the schedule. Use a minimal number of registers in the chart.

  (d) Derive the VHDL code.

  (e) Discuss the difference between this design and the sequential multiplier discussed in Section 11.6.

12.20 Repeat parts (b), (c) and (d) of Problem 12.19, but use two adders to accelerate the operation.

12.21 Repeat parts (b), (c) and (d) of Problem 12.19, but use three adders to accelerate the operation.

CHAPTER 13

HIERARCHICAL DESIGN IN VHDL

As the size of a digital system increases, its complexity grows accordingly. One method of managing the complexity is to describe the system in a hierarchical structure, in which the system is gradually divided into smaller parts. With a hierarchy, we only need to focus on a small, manageable part at a time. One of the goals of VHDL is to facilitate the development and modeling of large digital systems. It consists of versatile mechanisms and language constructs to specify and configure a design hierarchy and to organize design information and files. This chapter provides an overview of constructs relevant to the RT-level design and synthesis.

13.1 INTRODUCTION

Hierarchical design is a methodology that divides a system recursively into small modules and then constructs each module independently. The term *recursively* means that the division process can be applied repeatedly and the modules can be further decomposed. For example, consider the sequential multiplier of Section 11.3.3. One possible design hierarchy is shown in Figure 13.1. The system is first divided into a control path and a data path. The control path is then divided into the next-state logic and the state register, and the data path is divided into a routing circuit, functional units and a data register. The functional units are then further decomposed into an adder and a decrementor. If needed, we can continue the process and further refine the leaf modules. The sequential multiplier can also be part of a larger system. For example, it can be a module of an arithmetic unit, which in turn, can be a module of a processor.

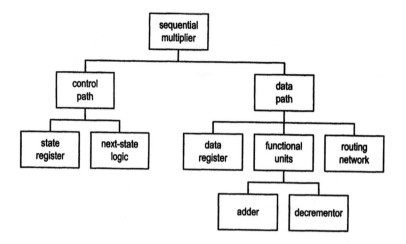

Figure 13.1   Hierarchical description of a sequential multiplier.

13.1.1 Benefits of hierarchical design

There are two major benefits of using the hierarchy: *complexity management* and *design reuse*. As the size of transistor continues to decrease, more functionality can be included in a device and the digital system grows larger and more complex. Managing the complexity becomes a key challenge in today's design. Hierarchical design methodology allows us to apply the divide-and-conquer strategy and break a system into smaller modules. This approach helps us to manage a large design in several ways:

- Instead of looking at the entire system, we can focus on a manageable portion of the system, and analyze, design and verify each module in isolation.
- Once the hierarchy and modules are specified, a large system can be constructed in stages by a designer or concurrently by a team of designers.
- The synthesis software may require a significant amount of memory space and computation time to synthesize a large system. Breaking the system into smaller modules and synthesizing them independently can make the process more effective.

Hierarchical design methodology also helps to facilitate design reuse:

- Some predesigned modules or third-party cores (i.e., IPs) may exist and can be used in the system. Therefore, we don't need to construct every system from scratch.
- Many systems contain some common or similar functionalities. After we design and verify a module in a system, the same module can be used in future design.
- Some design may contain certain device-dependent components, such as an SRAM module. To achieve portability, we can isolate these components in the top level of the hierarchy and substitute them according to the target technology.

13.1.2 VHDL constructs for hierarchical design

One objective of VHDL is to facilitate the modeling and developments of complex digital systems. Many language constructs are designed for this purpose. These include the following:

- Component
- Generic

- Configuration
- Library
- Package
- Subprogram

The *component, generic* and *configuration* constructs provide flexible and versatile mechanism to describe a hierarchical design. These constructs are discussed in Sections 13.2, 13.3 and 13.4. The *library, package,* and *subprogram* help the management of complicated code and are briefly reviewed in Section 13.5. To take full advantage of the hierarchical design methodology, we have to develop general and flexible modules. This issue is discussed in Chapters 14 and 15.

13.2 COMPONENTS

Hierarchical design methodology basically divides a system into smaller modules and then constructs the modules accordingly. Although not stated explicitly, our previous derivations generally followed this approach. We usually started with a top-level diagram with several major parts and then derived the VHDL code according to the diagram, with a VHDL segment for each part. The VHDL component construct provides a formal and explicit way to describe a hierarchical design.

We examined the VHDL component construct briefly in Section 2.2.2. It is the mechanism used to describe a digital system in a structural view. Recall that a structural view is essentially a block diagram, in which we specify the types of parts used and the interconnections among these parts. While the component construct is supported in both VHDL 87 and VHDL 93, the syntax of VHDL 93 is much simpler. However, since the IEEE RTL synthesis standard is based on VHDL 87, it follows the old syntax. To obtain maximal portability, our discussion mainly follows the IEEE RTL synthesis standard (i.e., VHDL 87). In Section 13.4.4, we briefly examine the newer version.

In VHDL 87, using a component involves two steps. The first step is *component declaration*, in which a component is "make known" to an architecture. The second step is *component instantiation*, in which an instance of the component is created and its external I/O interface is specified.

13.2.1 Component declaration

Component declaration provides information about the external interface of a component, which includes the input and output ports and relevant parameters. The information is similar to that provided in an entity declaration. The simplified syntax of component declaration is as follows:

```
component component_name
   generic(
      generic_declaration;
      generic_declaration;
      . . .
   );
   port(
      port_declaration;
      port_declaration;
      . . .
   );
```

```
end component;
```

The generic portion is optional and consists of relevant parameters to be passed into the component. It is discussed in the next section. The port portion consists of port declarations, which are similar to those in an entity declaration. Note that the **is** keyword is in the entity declaration but not in the component declaration (the **is** keyword is allowed in VHDL 93). As in entity declaration, no information about internal implementation is specified in component declaration.

Assume that we have already designed a decade (i.e., mod-10) counter and that its entity declaration is

```
entity dec_counter is
   port(
      clk, reset: in std_logic;
      en: in std_logic;
      q: out std_logic_vector(3 downto 0);
      pulse: out std_logic
   );
end dec_counter;
```

If we want to use it as a component in other designs, the easiest way is to declare a component that has the same name and ports:

```
component dec_counter
   port(
      clk, reset: in std_logic;
      en: in std_logic;
      q: out std_logic_vector(3 downto 0);
      pulse: out std_logic
   );
end component;
```

Note that the same information is used in the entity declaration and the corresponding component declaration. Graphically, a component can be thought of as a circuit part with properly named input and output ports. The conceptual diagram of the dec_counter component is shown in Figure 13.2(a).

During the elaboration process, the component eventually has to be bound with an architecture body. The complete VHDL code of the decade counter is shown in Listing 13.1. The en input functions as an enable signal. The pulse output is a status signal and is asserted when the counter reaches 9 and is ready to warp around.

Listing 13.1   Decade counter

```
   library ieee;
   use ieee.std_logic_1164.all;
   use ieee.numeric_std.all;
   entity dec_counter is
5    port(
        clk, reset: in std_logic;
        en: in std_logic;
        q: out std_logic_vector(3 downto 0);
        pulse: out std_logic
10   );
   end dec_counter;
```

```
   architecture up_arch of dec_counter is
      signal r_reg: unsigned(3 downto 0);
15    signal r_next: unsigned(3 downto 0);
      constant TEN: integer:= 10;
   begin
      -- register
      process(clk,reset)
20    begin
         if (reset='1') then
            r_reg <= (others=>'0');
         elsif (clk'event and clk='1') then
            r_reg <= r_next;
25       end if;
      end process;
      -- next-state logic
      process(en,r_reg)
      begin
30       r_next <= r_reg;
         if (en='1') then
            if r_reg=(TEN-1) then
               r_next <= (others=>'0');
            else
35             r_next <= r_reg + 1;
            end if;
         end if;
      end process;
      -- output logic
40    q <= std_logic_vector(r_reg);
      pulse <= '1' when r_reg=(TEN-1) else
               '0';
   end up_arch;
```

In a VHDL code, a component declaration is included in the declaration part of an architecture body, as shown in Listing 13.2. If it is used in multiple architecture bodies, the declaration may be placed in a package. Use of packages is discussed in Section 13.5.3.

13.2.2 Component Instantiation

Once a component is declared, its instance can be created inside the architecture body. The simplified syntax of component instantiation is

```
instance_label: component_name
   generic map(
      generic_association;
      generic_association;
      . . .
   )
   port map(
      port_association;
      port_association;
      . . .
   );
```

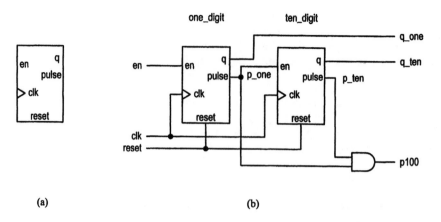

(a)   (b)

Figure 13.2 Block diagram of a two-digit decimal counter.

In the first line, component_name specifies the component to be used, and instance_label assigns the instance with a unique label for identification. The **generic map** portion assigns the actual values to the generics. This portion, which is optional, is discussed in Section 13.3. Note that there is no semicolon after the generic map portion. The **port map** portion specifies the connections (i.e., "association") between the component's ports (known as *formal signals*) and the external signals (known as *actual signals*). The port_association term has the general format

```
port_name => signal_name
```

This is known as the *named association*.

The use of component instantiation can best be explained by an example. Assume that we want to implement a two-digit decimal counter, which counts up in BCD format (i.e., from 00 to 99) and wraps around. One possible implementation is to cascade two decade counters. The diagram is shown in Figure 13.2(b). The left decade counter represents the digit in the one's place. Its pulse port is connected to an external wire, which is labeled as the p_one signal, which in turn is connected to the en port of the right decade counter, which represents the digit in the ten's place. If the two-digit decimal counter is enabled, the left decade counter asserts p_one signal every 10 clock cycles and wraps around. The right decade counter is controlled by the p_one signal and thus counts only once for every 10 clock cycles. The p100 output is a pulse to indicate that the two-digit decimal counter reaches 99 and is ready to wrap around.

To describe this diagram in VHDL, we need to create two instances of the dec_counter component and specify the relevant I/O connections. The main task in component instantiation is to specify the mapping between the formal signals and the actual signals. This is a tedious and error-prone task. The best way to do it is to draw a properly labeled block diagram and then derive the VHDL code following the connections of the diagram. The diagram should contain necessary information, which includes the component names, instance labels, and properly labeled ports and connection signals. The block diagram in Figure 13.2(b) is created for this purpose. In our convention, the information from the component declaration, which includes the component name and port names (i.e., the formal signals), is placed inside the block. The external signal names (i.e., actual signals) and instance names, on the other hand, are placed outside the block. Note that a formal signal name and an actual signal name can be the same.

Following the diagram, we can derive the code segments for the two instances:

```
one_digit: dec_counter
    port map (clk=>clk, reset=>reset, en=>en,
              pulse=>p_one, q=>q_one);
ten_digit: dec_counter
    port map (clk=>clk, reset=>reset, en=>p_one,
              pulse=>p_ten, q=>q_ten);
```

The port mapping used here is known as the *named association* because both the name of the formal signal and the name of the actual signal are listed in each port association. The order of the port associations does not matter. For example, the first instance can also be written as

```
one_digit: dec_counter
    port map (pulse=>p_one, reset=>reset, en=>en,
              q=>q_one, clk=>clk);
```

The complete VHDL code is shown in Listing 13.2.

Listing 13.2 Two-digit decimal counter using decade counters

```
library ieee;
use ieee.std_logic_1164.all;
entity hundred_counter is
    port(
5       clk, reset: in std_logic;
        en: in std_logic;
        q_ten, q_one: out std_logic_vector(3 downto 0);
        p100: out std_logic
    );
10 end hundred_counter;

architecture vhdl_87_arch of hundred_counter is
    component dec_counter
        port(
15          clk, reset: in std_logic;
            en: in std_logic;
            q: out std_logic_vector(3 downto 0);
            pulse: out std_logic
        );
20      end component;
    signal p_one, p_ten: std_logic;
    begin
    one_digit: dec_counter
        port map (clk=>clk, reset=>reset, en=>en,
25                pulse=>p_one, q=>q_one);
    ten_digit: dec_counter
        port map (clk=>clk, reset=>reset, en=>p_one,
                  pulse=>p_ten, q=>q_ten);
    p100 <= p_one and p_ten;
30 end vhdl_87_arch;
```

In VHDL, component instantiation is just another concurrent statement and thus can be mixed with other statements. For example, a simple signal assignment statement is used to derive the p100 signal.

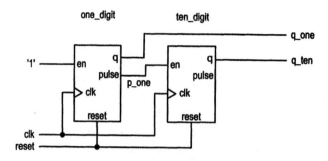

Figure 13.3 Block diagram of a free-running two-digit decimal counter.

13.2.3 Caveats in component instantiation

As long as we derive a proper block diagram, the use of components is straightforward. There are two caveats about component instantiation. One is the use of position association in port mapping, and the other is the handling of unused ports.

So far, we have used the *named association* method for port mapping. Alternatively, we can omit the formal signal names and place the actual names according to the positions of the formal signals. This is known as *positional association*. For example, the component declaration of dec_counter shows that the order of the ports is

```
clk, reset, en, q, pulse
```

In the previous vhdl_87_arch architecture, we can put the actual signals in this order in component instantiation. The VHDL code becomes

```
one_digit: dec_counter
    port map (clk, reset, en, q_one, p_one);
ten_digit: dec_counter
    port map (clk, reset, p_one, q_ten, p_ten);
```

At first glance this method may seem to be more compact, but it can cause problems in the long run, especially for a component with many I/O ports. For example, we may revise the port declaration of dec_counter later and switch the order of the clk and reset signals:

```
. . .
port(
    reset, clk: in std_logic;
    . . .
```

The modification has no effect for the dec_counter code in Listing 13.1 but introduces a serious problem for code that instantiates dec_counter with positional association. To make the code more reliable, it is good practice to use named association in port and generic mapping.

When we instantiate a component, some ports may not be needed to connect to actual signals. For example, assume that we wish to design a free-running two-digit decimal counter, in which the en and p100 signals are removed. The modified block diagram is shown in Figure 13.3, in which the en signal of the one_digit instance is tied to logic '1', and the pulse signal of the ten_digit instance is left unconnected. To describe the mapping of en, we can simply use a constant expression to replace the actual signal and the association becomes en=>'1'. Some synthesis software may not accept the constant expression. To overcome this, we can create a signal, assign it with the desired constant and then use it as the actual signal in port mapping.

To specify the unused port, we can associate the port with the **open** keyword and the association becomes `pulse=>`**open**. Good synthesis software should know that the port is not used, backtrack the corresponding circuit and remove the unneeded circuit from implementation. The **open** keyword can also be associated with an input port, and the association means that the initial value in port declaration will be used for the port. Since it is not good practice to assign an initial value to a signal or port in synthesis, this should be avoided.

The VHDL code for the free-running two-digit decimal counter is shown in Listing 13.3.

Listing 13.3   Free-running two-digit decimal counter

```
   library ieee;
   use ieee.std_logic_1164.all;
   entity free_run_hundred_counter is
      port(
5        clk, reset: in std_logic;
         q_ten, q_one: out std_logic_vector(3 downto 0)
      );
   end free_run_hundred_counter;

10 architecture vhdl_87_arch of free_run_hundred_counter is
      component dec_counter
         port(
            clk, reset: in std_logic;
            en: in std_logic;
15          q: out std_logic_vector(3 downto 0);
            pulse: out std_logic
         );
      end component;
      signal p_one: std_logic;
20 begin
      one_digit: dec_counter
         port map (clk=>clk, reset=>reset, en=>'1',
                   pulse=>p_one, q=>q_one);
      ten_digit: dec_counter
25       port map (clk=>clk, reset=>reset, en=>p_one,
                   pulse=>open, q=>q_ten);
   end vhdl_87_arch;
```

In named association, a formal port can be omitted in the list and VHDL assumes that it is mapped to **open** by default. To make the code reliable, it is good practice to list all ports in port map and explicitly associate the unused output ports with **open**.

13.3   GENERICS

The *generic* construct of VHDL is a mechanism to pass information into an entity and a component. Generics are like parameters. They are first declared in entity and component declaration and later assigned a value during component instantiation.

The use of generics starts with the entity declaration by adding a generic declaration section. The simplified syntax is

```
entity entity_name is
   generic(
      generic_names: data_type;
      generic_names: data_type;
      . . .
   );
   port(
      port_names: mode data_type;
      ...
   );
end entity_name;
```

Once a generic is declared, it can be used in subsequent port declarations and associated architecture bodies. For example, consider the free-running binary counter of Section 8.5.4, which has a fixed width of 4 bits. We can modify it to a more versatile parameterized free-running binary counter by defining a WIDTH generic to specify the desired width (i.e., number of bits). The modified entity declaration becomes

```
entity para_binary_counter  is
   generic(WIDTH: natural);
   port(
      clk, reset: in std_logic;
      q: out std_logic_vector(WIDTH-1 downto 0)
   );
end para_binary_counter;
```

Note that the range of the q output is not fixed, but is expressed in terms of the WIDTH generic, as in std_lgic_vector(WIDTH-1 downto 0).

After the declaration, the generic can be used in the associated architecture bodies. A generic cannot be modified inside the architecture body and thus functions like a constant. It is sometimes referred to as a *generic constant*. As a constant, we use uppercase letters for the generics in the book.

The corresponding architecture body of the binary counter is

```
architecture arch of para_binary_counter is
   signal r_reg, r_next: unsigned(WIDTH-1 downto 0);
begin
   process(clk,reset)
   begin
      if (reset='1') then
         r_reg <= (others=>'0');
      elsif (clk'event and clk='1') then
         r_reg <= r_next;
      end if;
   end process;
   r_next <= r_reg + 1;
   q <= std_logic_vector(r_reg);
end arch;
```

Again, note that the WIDTH generic is used to specify the range of internal signals.

To use the parameterized free-running binary counter in a hierarchical design, a similar component declaration should be included in the architecture declaration. The generic can then be assigned a value in the generic mapping section when a component instance is instantiated. An example code is shown in Listing 13.4. The code creates a 4-bit counter and a 12-bit counter.

Listing 13.4   Example of the use of generics

```
library ieee;
use ieee.std_logic_1164.all;
entity generic_demo is
    port(
        clk, reset: in std_logic;
        q_4: out std_logic_vector(3 downto 0);
        q_12: out std_logic_vector(11 downto 0)
    );
end generic_demo;

architecture vhdl_87_arch of generic_demo is
    component para_binary_counter
        generic(WIDTH: natural);
        port(
            clk, reset: in std_logic;
            q: out std_logic_vector(WIDTH-1 downto 0)
        );
    end component;
begin
    four_bit: para_binary_counter
        generic map (WIDTH=>4)
        port map (clk=>clk, reset=>reset, q=>q_4);
    twe_bit: para_binary_counter
        generic map (WIDTH=>12)
        port map (clk=>clk, reset=>reset, q=>q_12);
end vhdl_87_arch;
```

In the second example, we consider the design of a parameterized mod-n counter, in which n can be specified as a parameter. The counter counts from 0 to $n - 1$ and then wraps around. To count to n patterns, the counter needs at least $\lceil \log_2 n \rceil$ bits. Our first design uses two generics, the N generic for n and the WIDTH generic for the number of bits in the counter. The VHDL code is shown in Listing 13.5. It is patterned after the decade counter of Listing 13.1 and includes an en control signal and a pulse output signal, which is asserted when the counter reaches $n - 1$.

Listing 13.5   Parameterized mod-n counter

```
library ieee;
use ieee.std_logic_1164.all;
use ieee.numeric_std.all;
entity mod_n_counter is
    generic(
        N: natural;
        WIDTH: natural
    );
    port(
        clk, reset: in std_logic;
        en: in std_logic;
        q: out std_logic_vector(WIDTH-1 downto 0);
        pulse: out std_logic
    );
end mod_n_counter;
```

```
   architecture arch of mod_n_counter is
      signal r_reg: unsigned(WIDTH-1 downto 0);
      signal r_next: unsigned(WIDTH-1 downto 0);
20 begin
      -- register
      process(clk,reset)
      begin
         if (reset='1') then
25          r_reg <= (others=>'0');
         elsif (clk'event and clk='1') then
            r_reg <= r_next;
         end if;
      end process;
30    -- next-state logic
      process(en,r_reg)
      begin
         r_next <= r_reg;
         if (en='1') then
35          if r_reg=(N-1) then
               r_next <= (others=>'0');
            else
               r_next <= r_reg + 1;
            end if;
40       end if;
      end process;
      -- output logic
      q <= std_logic_vector(r_reg);
      pulse <= '1' when r_reg=(N-1) else
45             '0';
   end arch;
```

Note that WIDTH is not an independent parameter. It can be derived from N and thus is not actually required. Section 13.5 shows how to achieve this.

We can redesign the two-digit decimal counter of Section 13.2.2 by replacing the two decade counters with two parameterized mod-n counters. To do this, we simply assign 10 to N and 4 to WIDTH in component instantiation and thus customize the mod-n counter as a mod-10 counter. The corresponding VHDL code is shown in Listing 13.6.

Listing 13.6   Two-digit decimal counter using parameterized mod-n counters

```
   architecture generic_arch of hundred_counter is
      component mod_n_counter
         generic(
            N: natural;
5           WIDTH: natural
         );
         port(
            clk, reset: in std_logic;
            en: in std_logic;
10          q: out std_logic_vector(WIDTH-1 downto 0);
            pulse: out std_logic
         );
      end component;
```

```
      signal p_one, p_ten: std_logic;
 15 begin
      one_digit: mod_n_counter
         generic map (N=>10, WIDTH=>4)
         port map (clk=>clk, reset=>reset, en=>en,
                   pulse=>p_one, q=>q_one);
 20   ten_digit: mod_n_counter
         generic map (N=>10, WIDTH=>4)
         port map (clk=>clk, reset=>reset, en=>p_one,
                   pulse=>p_ten, q=>q_ten);
      p100 <= p_one and p_ten;
 25 end generic_arch;
```

The two examples show the potential of combining the generics and component. Instead of creating an array of counters with different widths, we can use a single parameterized module and customize it to the desired width. This makes the module more flexible and more versatile, and greatly enhances its chance to be reused. The next two chapters provide comprehensive coverage of the design of parameterized modules.

Another major application of generics is to pass delay information in modeling and simulation. For example, we can define the Tpd generic for the propagation delay. It can then be used in a statement like

```
y <= a + b after Tpd ns;
```

This allows us to pass delay information into the model when it becomes available.

13.4 CONFIGURATION

13.4.1 Introduction

When a component is declared and instantiated, as the example in Listing 13.2, only basic generic and port information is provided. *Configuration* of VHDL is the process of *binding* a component with a design entity and architecture. The process includes two parts:

1. Bind a component with a design entity.
2. Bind the design entity with an architecture body.

The configuration of VHDL is very flexible, and thus its detailed syntax and use are quite complex. However, most features are not needed in RT-level design and synthesis. The IEEE RTL synthesis standard supports only the second part of the process, the binding of entity and architecture. We discuss only default binding, and top-level entity and architectural body binding in the book.

Explicit configuration is not always required for a component. For example, the codes in previous sections use no configuration constructs. In this case, the component is bound by the *default binding*, which is processed as follows:

- The component is bound to an entity with the identical name.
- Component ports are bound to entity ports of the same names.
- The most recently analyzed architecture body is bound to the entity declaration.

In RT-level design, the hierarchy is normally simple, and only one architecture body exists. The default binding should be satisfactory most of the time, and no explicit configuration statement is needed.

Figure 13.4   Block diagram of a testbench.

In synthesis, multiple architectures may be needed for several reasons. First, there is frequently a trade-off between area and performance in a digital circuit. A complex circuit, such as a multiplier, may have several implementations, each with a unique area–delay characteristics. Each implementation represents an architecture body. Second, we sometimes need to adjust certain circuit characteristics to fit a specific application. For example, we may need to force a counter to circulate different patterns. One way to accomplish this is to use a separate architecture body for each pattern.

In modeling and simulation, having multiple architectures is more common. For example, we discussed the concept of testbench in Section 2.2.4. The diagram of a basic testbench is shown in Figure 13.4. A complex design is normally first specified in an abstract behavioral description, converted to an RT-level description, and then synthesized to a cell-level structural description. Each description represents an architecture body. As the design progresses, a more detailed architecture body becomes available. We can use configuration to bind the new description for verification.

There are two ways to specify the configuration. It can be described in an independent design unit, which is known as *configuration declaration*, or included in the declaration section of the architecture body, which is known as *configuration specification*. The two methods are discussed in the following subsections. The IEEE RTL standard supports only the configuration declaration method.

13.4.2 Configuration declaration

In the configuration declaration method, we create a new kind of design unit, known as *configuration*, to specify the binding of a component. A configuration unit is an independent design unit in VHDL, just like an entity declaration and an architecture body. It is analyzed and stored independently when the VHDL code is processed. The simplified syntax of a configuration unit is

```
configuration conf_name of entity_name is
  for archiecture_name
    for instance_label: component_name
      use entity lib_name.bound_entity_name(bound_arch_name);
    end for;
    for instance_label: component_name
      use entity lib_name.bound_entity_name(bound_arch_name);
    end for;
    . . .
  end for;
end;
```

The conf_name term is the unique identifier for this configuration unit. The entity_name and architecture_name terms identify the entity and the architecture for which the configuration is intended. The instance_label term specifies a specific component instance, and the following "**use** ..." clause indicates the entity and architecture to be bound to the instance. The lib_name term is the name of the library in which the entity and architecture reside. The library is discussed in Section 13.5. In the place of instance_label, we can use the **all** keyword to represent all instances of this particular component, or use **others** in the end to represent all the unbound instances of the component.

To demonstrate the use of configuration, we create a second architecture, down_arch, for the dec_counter entity of Section 13.2.1. The VHDL code is shown in Listing 13.7. It counts down from 9 to 0 and then wraps around. The pulse output is asserted when the counter reaches 0 and is ready to wrap around. If we use this architecture in the vhdl_87_arch architecture of the two-digit decimal counter of Section 13.2.2, the two-digit counter counts down from 99 to 00 and then wraps around.

Listing 13.7   Decade counter with a count-down sequence

```
architecture down_arch of dec_counter is
    signal r_reg: unsigned(3 downto 0);
    signal r_next: unsigned(3 downto 0);
    constant TEN: integer := 10;
5 begin
    -- register
    process(clk,reset)
    begin
        if (reset='1') then
10          r_reg <= (others=>'0');
        elsif (clk'event and clk='1') then
            r_reg <= r_next;
        end if;
    end process;
15  -- next-state logic
    process(en,r_reg)
    begin
        r_next <= r_reg;
        if (en='1') then
20          if r_reg=0 then
                r_next <= to_unsigned(TEN-1,4);
            else
                r_next <= r_reg - 1;
            end if;
25      end if;
    end process;
    -- output logic
    q <= std_logic_vector(r_reg);
    pulse <= '1' when r_reg=0 else
30              '0';
end down_arch;
```

Depending on the requirement of the direction of a two-digit decimal counter, we can create a configuration unit to specify the desired binding. The VHDL in Listing 13.8 binds the two instances with the down_arch architecture.

Listing 13.8 Configuration for a two-digit decimal counter

```
configuration count_down_config of hundred_counter is
   for vhdl_87_arch
      for one_digit: dec_counter
         use entity work.dec_counter(down_arch);
 5    end for;
      for ten_digit: dec_counter
         use entity work.dec_counter(down_arch);
      end for;
   end for;
10 end;
```

Note that the work library is the default library used in VHDL. It represents the current working library.

13.4.3 Configuration specification

The configuration declaration is general and flexible. However, for a simple design, creating a new design unit for this purpose is somewhat cumbersome. An alternative is to specify the relevant configuration in the declaration section of the architecture body. This is known as a *configuration specification*. The simplified syntax is

```
for instance_label: component_name
   use entity lib_name.bound_entity_name(bound_arch_name);
for instance_label: component_name
   use entity lib_name.bound_entity_name(bound_arch_name);
. . .
```

For example, the configuration declaration in Listing 13.8 can also be specified by a configuration specification. We simply revise the vhdl_87_arch by adding the relevant configuration information to the declaration section:

```
architecture vhdl_87_config_arch of hundred_counter is
   component dec_counter
      port(
         clk, reset: in std_logic;
         en: in std_logic;
         q: out std_logic_vector(3 downto 0);
         pulse: out std_logic
      );
   end component;
   for one_digit: dec_counter
      use entity work.dec_counter(down_arch);
   for ten_digit: dec_counter
      use entity work.dec_counter(down_arch);
   signal p_one, p_ten: std_logic;
begin
   . . .
```

13.4.4 Component instantiation and configuration in VHDL 93

Components and configuration are flexible in VHDL, but its syntax is involved and tedious. Since RT-level design uses relatively simple component instantiation and binding, the syntactic constructs becomes cumbersome. For example, consider the previous vhdl_87_config_arch architecture. We need a relatively lengthy declaration to use and bind the two component instances in design. VHDL 93 provides a much simpler mechanism. It allows a component to be bound directly to an entity and an architecture in component instantiation. The simplified syntax is

```
instance_label:
    entity lib_name.bound_entity_name(bound_arch_name)
        generic map (. . .)
        port map (. . .);
```

The **entity** lib_name.bound_entity_name(bound_arch_name) clause specifies the associated entity and architecture, and no component declaration or any additional configuration construct is needed. The (bound_arch_name) term is optional. If it is omitted, the most recently analyzed architecture will be bound to the entity.

Consider the two-digit decimal counter of Section 13.2.2. With this mechanism, a more compact code can be derived, as shown in Listing 13.9.

Listing 13.9 Two-digit decimal counter with direct entity binding

```
architecture vhdl_93_arch of hundred_counter is
    signal p_one, p_ten: std_logic;
begin
    one_digit: entity work.dec_counter(up_arch)
5       port map (clk=>clk, reset=>reset, en=>en,
                  pulse=>p_one, q=>q_one);
    ten_digit: entity work.dec_counter(up_arch)
        port map (clk=>clk, reset=>reset, en=>p_one,
                  pulse=>p_ten, q=>q_ten);
10  p100 <= p_one and p_ten;
end vhdl_93_arch;
```

Since this kind of instantiation is valid only in VHDL 93, it is not supported by the IEEE RTL synthesis standard. However, some software does accept this type of component instantiation.

13.5 OTHER SUPPORTING CONSTRUCTS FOR A LARGE SYSTEM

13.5.1 Library

As we discussed in Section 3.2.5, a VHDL program is analyzed and stored as individual design units, which include entity declaration, architecture body, configuration declaration, and so on. A VHDL *library* is the virtual repository that stores the analyzed design units. VHDL does not define the physical location of a library. Most software maps a library to a physical directory in a hard disk. By default, the current design units are stored in a working library named work. For example, the work library is used in the previous component instantiation:

```
    . . .
    one_digit: entity work.dec_counter(up_arch)
    . . .
```

For a complex design, there may exist a large number of design units. It is desirable to organize these units and store them in separate places. Also, we may have a collection of commonly used design units that are shared by many different designs. It is more effective to save these units in a common library rather than duplicating them in every design directory.

To access the content of a library, we must first make it known by using a library statement. The syntax is

```
library lib_name, lib_name, ... , lib_name;
```

For example, assume that we create a library named c_lib and save the previous dec_counter entity and relevant architectures in the library. The count_down_config configuration discussed in Section 13.4.2 must be revised accordingly:

```
library c_lib;  — make c_lib visible
configuration clib_config of hundred_counter is
   for vhdl_87_arch
      for one_digit: dec_counter
         use entity c_lib.dec_counter(down_arch);  — c_lib
      end for;
      for ten_digit: dec_counter
         use entity c_lib.dec_counter(down_arch);  — c_lib
      end for;
   end for;
end;
```

Note that the work library of the original code is replaced with c_lib.

If a design unit is accessed frequently, we can make it visible by adding a **use** clause. The syntax is

```
use lib_name.unit_name;
```

The unit can then be accessed directly without referring to the library. The **all** keyword can be used in place of unit_name to make all units of the library visible. For example, the previous code can be revised as:

```
library c_lab;
use c_lib.dec_counter;  — make dec_counter visible
configuration clib_config of hundred_counter is
   for vhdl_87_arch
      for one_digit: dec_counter
         use entity dec_counter(down_arch);  — lib dropped
      end for;
      for ten_digit: dec_counter
         use entity dec_counter(down_arch);
      end for;
   end for;
end;
```

Note that the library name is dropped from the "**use entity** dec_counter(down_arch);" statement.

The work library is declared implicitly by VHDL definition, and thus there is no need for the "**library work;**" statement.

13.5.2 Subprogram

Subprograms in VHDL include *functions* and *procedures*. Their bodies are made of sequential statements, and their behaviors are similar to those in traditional programming languages. Unlike entity and architecture, procedures and functions are not design units and thus cannot be processed independently. For example, we cannot isolate a function from the code and synthesize it separately. Therefore, while the functions and procedures are basic building blocks of software hierarchy, they are not adequate to describe the hardware hierarchy.

VHDL functions are more versatile and useful than procedures, and thus our discussion focuses mainly on functions. In synthesis, functions should not be used to specify the design hierarchy, but should be treated as a shorthand for simple, repeatedly used operations. Functions are also needed to perform certain house-keeping tasks, such as data type conversion or operator overloading in IEEE packages.

A VHDL function takes several parameters and returns a single value. It must first be declared in the declaration section and then can be called later. A function can be thought of as an extension of the expression and can be "called" wherever an expression is used. The simplified syntax of a function is

```
function func_name(parameter_list) return data_type is
   declarations;
begin
   sequential statement;
   sequential statement;
   . . .
   return(expression);
end;
```

The following examples illustrate the construction of a function. The first example is a function that performs a majority function. It returns '1' if two or more input parameters, a, b and c, are '1'. The function can be treated as a shorthand for the $a \cdot b + a \cdot c + b \cdot c$ expression.

```
function maj(a, b, c: std_logic) return std_logic is
   variable result: std_logic;
begin
   result := (a and b) or (a and c) or (b and c);
   return result;
end maj;
```

The maj function must be declared and then can be invoked. The following code segment illustrates its use:

```
architecture arch of . . .
   -- declaration
   function maj(a, b, c: std_logic) return std_logic is
      variable result: std_logic;
   begin
      result := (a and b) or (a and c) or (b and c);
      return result;
   end maj;
   signal i1, i2, i3, i4, x, y: std_logic;
begin
   . . .
```

```
x <= maj(i1, i2, i3) or i4;
y <= i1 when maj(i2, i3, i4)='1' else
    . . .
```

Note that the entire function definition is included in the declaration section of the architecture body. This may become cumbersome. An alternative is to declare the function in a package, which is discussed in the next subsection.

The second example is a function that performs data type conversion. It converts the std_logic data type to the boolean data type.

```
function to_boolean(a: std_logic) return boolean is
    variable result: boolean;
begin
    if a='1' then
        result := true;
    else
        result := false;
    end if;
    return result;
end to_boolean;
```

If this function is declared, we can use to_boolean(a) to replace the a='1' expression.

The last example is a function that performs $\lceil \log_2 n \rceil$, which is frequently needed in calculating the width of data signals.

```
function log2c(n: integer) return integer is
    variable m, p: integer;
begin
    m := 0;
    p := 1;
    while p < n loop
        m := m + 1;
        p := p * 2;
    end loop;
    return m;
end log2c;
```

13.5.3 Package

As a system becomes complex, more information is included in the declaration section. The declaration section of an architecture body may consist of the declarations of constants, data types, components, functions and so on. When a system is divided into several smaller subsystems, some declarations must be duplicated in many different design units. The VHDL *package* construct is a method of organizing declarations. We can gather the commonly used declarations in a design, group them together and store them in a package. A design unit just needs to include a use clause to access these declarations.

A VHDL package is divided into *package declaration* and *package body*. The declaration items are placed in a package declaration. If an item is a subprogram, only the declaration of the subprogram is included. The body (i.e., the implementation) of the subprogram is placed in the associated package body. The package body is optional and is needed only when subprograms exist.

Package declaration and package body are design units of VHDL. They are analyzed independently and stored in a library. To make a declaration item visible, a use clause is needed. Its syntax is

```
use lib_name.package_name.item_name;
```

Most of the time, we use the **all** keyword in place of item_name to make all items of the named package visible.

Many extensions to VHDL are done by defining additional packages, such as the IEEE std_logic_1164 and numeric_std packages. Almost all of our VHDL codes include the statement

```
use ieee.std_logic_1164.all;
```

It makes all of the declaration items of the predefined std_logic_1164 package visible, and thus we can use the std_logic and std_logic_vector data types in VHDL code.

We can also define our own package. The syntax of a package declaration is very simple:

```
package package_name is
    declaration item;
    declaration item;
    . . .
end package_name;
```

If the declaration items include subprograms, an associated package body is needed. Its syntax is

```
package body package_name is
    subprogram;
    subprogram;
    . . .
end package_name;
```

An example of package declaration is shown in the first part of Listing 13.10. It consists of the definition of the std_logic_2d data type, which is a two-dimensional array with element of std_logic data type, and the declaration of the log2c function. Note that the package also invokes the IEEE std_logic_1164 package so that the std_logic data type can be used. The corresponding package body is shown in the second part of Listing 13.10, which is the implementation of the log2c function.

Listing 13.10   Example of a package

```
— package declaration
library ieee;
use ieee.std_logic_1164.all;
package util_pkg is
5    type std_logic_2d is
        array(integer range <>, integer range <>) of std_logic;
    function log2c (n: integer) return integer;
end util_pkg ;

10 —package body
  package body util_pkg is
      function log2c(n: integer) return integer is
          variable m, p: integer;
      begin
```

```
15        m := 0;
          p := 1;
          while p < n loop
             m := m + 1;
             p := p * 2;
20        end loop;
          return m;
      end log2c;
   end util_pkg;
```

For the parameterized mod-n counter of Section 13.3, one drawback is that we must use a redundant WIDTH generic to specify the width of the output signal. To overcome the problem, we need a previously defined function to calculate WIDTH from N. The log2c function of the util_pkg package can be used for this purpose. We can invoke this package before the entity declaration. Assume that the package is saved in the same working directory. The improved code is shown in Listing 13.11.

Listing 13.11   Improved parameterized mod-n counter

```
  library ieee;
  use ieee.std_logic_1164.all;
  use ieee.numeric_std.all;
  use work.util_pkg.all;
5 entity better_mod_n_counter is
     generic(N: natural);
     port(
        clk, reset: in std_logic;
        en: in std_logic;
10      q: out std_logic_vector(log2c(N)-1 downto 0);
        pulse: out std_logic
     );
  end better_mod_n_counter;

15 architecture arch of better_mod_n_counter is
     constant WIDTH: natural := log2c(N);
     signal r_reg: unsigned(WIDTH-1 downto 0);
     signal r_next: unsigned(WIDTH-1 downto 0);
  begin
20   -- register
     process(clk,reset)
     begin
        if (reset='1') then
           r_reg <= (others=>'0');
25      elsif (clk'event and clk='1') then
           r_reg <= r_next;
        end if;
     end process;
     -- next-state logic
30   process(en,r_reg)
     begin
        r_next <= r_reg;
        if (en='1') then
           if r_reg=(N-1) then
35            r_next <= (others=>'0');
```

```
                else
                    r_next <= r_reg + 1;
                end if;
            end if;
40      end process;
        -- output logic
        q <= std_logic_vector(r_reg);
        pulse <= '1' when r_reg=(N-1) else
                    '0';
45 end arch;
```

We add the statement

```
use work.util_pkg.all;
```

to make the log2c function visible. The function can then be used in the range specification for the q output, as in std_logic_vector(log2c(N)-1 **downto** 0), as well as the calculation of the internal WIDTH constant.

13.6 PARTITION

VHDL provides powerful mechanisms and versatile language constructs to support hierarchical design methodology and to manage the design of large systems. To apply these features, we must first determine the design hierarchy and divide the system into smaller parts. The process is sometimes known as *design partition*.

For synthesis, the design partition can be viewed from two perspectives: physical partition and logical partition. *Physical partition* is the division of the physical implementation. It specifies how the circuit is divided during synthesis. *Logical partition* is imposed by human designers. The goal is to make the design, development and verification of a system manageable. The two kinds of partitions are correlated but not necessarily identical.

13.6.1 Physical partition

A digital system can be described by a hierarchy of an arbitrary number of levels. The circuit parts becomes simpler as we traverse down the hierarchy. In VHDL code, the circuit parts are described as component instances. When the code is processed, the components are replaced by the actual architecture bodies level by level, and the hierarchy is gradually converted to a flattened description. One way to perform synthesis is to collapse the entire hierarchy into a one big, flattened circuit and then to synthesize the circuit accordingly. More sophisticated software provides mechanisms to selectively flatten the hierarchy and preserve some high-level components. The software will synthesize and optimize these components separately and then merge the resulting netlists (i.e., the cell level descriptions) to form the final circuit.

An important issue is the size of the preserved components. Since synthesis involves many sophisticated algorithms, the required computation time and memory space are normally much worse than the linear order, $O(n)$, where n is the size of the circuit. This implies that synthesizing a large circuit will take much more computation time and memory space than that required by several smaller circuits. For example, assume that an algorithm is on the order of $O(n^3)$. If it requires 1 second to synthesize a 1000-gate circuit, it will take 125 seconds (i.e., 5^3 seconds) to synthesize a 5000-gate circuit and 35 hours (i.e., 50^3 seconds) to synthesize a 50,000-gate circuit. However, if we break the 50,000-gate circuit into ten 5000-gate circuits, it requires only about 21 minutes (i.e., $10*5^3$ seconds) to synthesize the ten small parts.

On the other hand, when we preserve a component, it implicitly forms a "synthesis boundary," and optimization can only be performed within the boundary. This prevents synthesis software from exploring optimization opportunities that exist between components. Thus, small component size hinders the optimization process and leads to less efficient overall implementation. Many software tools suggest that the maximal gate count for synthesis is between 5000 and 50,000 gates.

13.6.2 Logical partition

The logical partition is determined by human designers. A good partition simplifies and streamlines the development and verification process, and makes the code reliable and portable. To facilitate the synthesis, a "logical circuit part" in the hierarchy should be within the range of the maximal gate count recommended by the synthesis software. On the other hand, since synthesis software can flatten and collapse a part of the hierarchy, smaller "logical parts" can be used in the design hierarchy.

In addition to synthesis concerns, partition should also be used to help develop reliable design, and portable and reusable code. We should pay particular attention to the circuits that may hinder portability or introduce problems in development flow. It is a good idea to separate these circuits from ordinary logic and instantiate them as components in a design hierarchy. Two types of circuits of concern are:

- Device-dependent circuits
- Non-Boolean circuits

Device-dependent circuits   Device-dependent circuits are those not synthesized by generic logic gates. They are predesigned or even prefabricated for a specific device technology. For example, most device technology has various types of prefabricated memory modules. These circuits are inferred by component instantiation and require no synthesis. Once a device-dependent circuit is used, the VHDL code becomes device dependent. To maintain portability, one way is to isolate these circuits in the top-level hierarchy and instantiate them as individual components. If the VHDL code is used later for a different device technology, we need only substitute these components with equivalent circuits of the new technology and keep the remaining code intact.

Non-Boolean circuits   Digital system design is primarily based on a mathematical model of Boolean algebra and its derivations. The algorithms in analysis, synthesis, verification and testing are developed within this framework. If a circuit does not follow the basic mathematical model, we call it a *non-Boolean* circuit. Some examples are listed below.

- *Tri-state buffer.* It has a third possible value, high impedance, in its output. The high impedance cannot be optimized or propagated as regular logic values.
- *Delay-sensitive circuit.* It uses logic gates to introduce a specific amount of propagation delay. The function of the circuit relies on the delay characteristics, not on Boolean algebra manipulation.
- *Clock distribution circuit.* It distributes the clock signal to the connected FFs. The circuit functions as a current amplifier and performs no logic operation.
- *Synchronization circuit.* It uses FFs to resolve the metastable condition, not for regular storage. This is discussed in Chapter 16.

These circuits should be isolated in the hierarchy so that later they can be processed independently.

13.7 SYNTHESIS GUIDELINES

- Use components, not subprograms, to specify the design hierarchy.

- Use the `std_logic` and `std_logic_vector` data types in the ports of components to maintain portability.

- Use named association, not positional association, in port mapping and generic mapping.

- List all ports of a component in port mapping and use **open** for unused output ports.

- For synthesis, partition the system into 5000- to 50,000-gate modules. Collapse and flatten low-level hierarchy if the components are too small.

- Separate device-dependent parts from ordinary logic and instantiate them as components in a hierarchy.

- Separate non-Boolean circuits from ordinary logic and instantiate them as components in a hierarchy.

13.8 BIBLIOGRAPHIC NOTES

This chapter provides a detailed discussion of VHDL components and gives a brief review of many other language constructs. The review is aimed primarily at synthesis of the RT-level system. Comprehensive coverage and the syntax of these constructs can be found in VHDL texts, such as *The Designer's Guide to VHDL, 2nd edition*, by P. J. Ashenden.

Problems

13.1   Consider a three-digit decimal counter that counts from 000 to 999 and wraps around. Use the `dec_counter` of Section 13.2.1 as a component to design this circuit.
  (a) Derive the block diagram and properly label the formal and actual signals.
  (b) Follow the block diagram to derive the VHDL code.
  (c) Use a configuration specification for configuration.
  (d) Same as part (c), but use a configuration declaration for configuration.

13.2   Redesign the three-digit decimal counter of Problem 13.1 using the mod-n counter of Section 13.3.
  (a) Derive the block diagram and properly label the formal and actual signals.
  (b) Follow the block diagram to derive the VHDL code.

13.3   We want to design a timer that counts from 00 to 59 seconds and then wraps around. Assume that the system clock is 1 MHz. Use the mod-n counter of Section 13.3 as a component to design this circuit.
  (a) Derive the block diagram and properly label the formal and actual signals.
  (b) Follow the block diagram to derive the VHDL code.

13.4   Consider a counter that counts from m to n and then wraps around. Derive VHDL code for the counter. Use generics, M and N, for m and n of the counter.

13.5   Divide the FIFO control circuit in Figure 9.14 into a hierarchy of two counters and two comparison circuits.

    (a) Derive VHDL entities and architectures for the counter and comparison circuits.

    (b) Follow the diagram in Figure 9.14 to derive the VHDL code.

13.6   Some synthesis software does not accept the ** operator. Derive a VHDL function, power2, that implements the function $f(n) = 2^n$.

13.7   Derive a function that converts the boolean data type into the std_logic data type. The true and false values of the boolean type are converted to '1' and '0' of the std_logic data type respectively.

CHAPTER 14

PARAMETERIZED DESIGN: PRINCIPLE

Design reuse is one of the major goals in developing VHDL code. Ideally, we want to design some common modules that can be shared by many applications. Since every application is different, it is desirable that a module can be customized to some degree to meet the specific need of an application. Customization is normally specified by explicit or implicit parameters, and we call this *parameterized design*. The most important parameter is the "width" of the module, which describes the number of bits of the data signal, as in a 24-bit adder. VHDL provides several mechanisms to pass and infer parameters and includes several language constructs to describe the replicated structure. In this chapter, we examine these basic mechanisms and constructs and use simple examples to illustrate their use. More detailed and comprehensive parameterized designs and case studies are discussed in Chapter 15.

14.1 INTRODUCTION

As the size of digital systems continues to grow, designing every system from scratch requires a tremendous amount of time and effort. One way to increase productivity and efficiency is design reuse. Many applications use parts of common functionalities. We can design and verify these parts once, store them in a library and then reuse them in other applications. As we discussed in Chapter 13, VHDL provides a versatile and powerful framework to facilitate the hierarchical design methodology and to accommodate predesigned components. Thus, once the commonly used parts are developed, design reuse can readily be incorporated into the VHDL environment.

While circuits share some parts of common functionalities, the exact specification of the part differs. For example, many applications need a binary counter. The basic construction of counters is similar, but the numbers of bits and the direction of the counting sequence depend on the need of a specific application. The chance that a fixed-size counter, say, an 11-bit up counter, will be reused is very small. On the other hand, if we develop a counter module that can be customized with different numbers of bits and counting directions, it can be utilized by many applications. The customization is normally done by describing certain circuit aspects with external parameters, and thus we call this *parameterized design*.

VHDL supports parameterized design in several ways. First, it provides mechanisms to pass parameters into an entity and to extract information from objects inside the entity. Second, most operators of VHDL and overloaded operators of the std_1164 and numeric_std packages are defined over *unconstrained arrays*, which are "implicitly parameterized." Finally, VHDL has two language constructs, *for generate* and *for loop*, that can be used to describe replicated structures. The desired circuit width can be obtained by properly specifying the index range of these constructs.

14.2   TYPES OF PARAMETERS

In a parameterized design, we can broadly divide the parameters into *width parameters* and *feature parameters*. They are discussed in the following subsections.

14.2.1   Width parameters

For design-reuse purposes, we can classify a system's input and output signals into data signals and non-data signals. The clearest example is an FSMD system. The external signals that flow into and out of the data path are the data signals, and the clock and reset signals as well as the command and status signals are the non-data signals. For example, consider the sequential multiplier of Section 11.3.3. The a_in, b_in and r are the data signals and the clk, reset, start and ready are the non-data signals. Some combinational circuits, such as a multiplier or a barrel shifter, contain only data signals.

The widths of data signals normally can be modified to meet different requirements whereas the non-data signals need little or no revision. Again, consider the sequential binary multiplier. We can modify the design to process 16-, 24- or 32-bit operands. The width of the data signals (i.e., a_in, b_in and r) as well as the internal signals and registers will change accordingly. On the other hand, the non-data signals (i.e., clk, reset, start and ready) remain the same.

The *width parameters* of a parameterized design specify the sizes (i.e., number of bits) of the relevant data signals. A system may need one or more parameters to describe the sizes of input and output signals as well as the sizes of internal signals and registers. For example, the sequential binary multiplier requires one independent width parameter to specify the size of the operands (and the size of the product can be derived accordingly). The FIFO buffer requires two independent width parameters, one for the number of bits in a word and one for the number of words in a buffer.

The main goal of parameterized design is to describe the desired design in terms of the width parameters so that the same VHDL description can be used for applications with different size requirements. Since the sizes of the data signals can be increased or decreased, we also call this *scalable design*.

14.2.2   Feature parameters

In addition to width, we can use parameters to specify the structure or organization of a design. We call these *feature parameters*. The feature parameters are defined on an ad hoc basis. We normally use feature parameters to include or exclude certain functionalities (i.e., features) from the implementation or to select one particular version of the implementation.

A feature parameter is generally used to specify small variations within a design. For example, we can specify whether to include an output buffer for the output signal of an FSM, or whether to use a synchronous or asynchronous reset signal for a counter.

In theory, we can also use the feature parameters to select totally different implementations. For example, a counter may have several possible implementations, and we can use a parameter to choose binary counter–based implementation, Gray counter–based implementation, or LFSR-based implementation. To accommodate this, the corresponding VHDL code almost has the description of three independent designs. It may be better to code the three implementations in three separate architecture bodies and use a configuration to instantiate the desired implementation.

There is no definite rule about the use of the feature parameters and the configuration. When a feature parameter leads to significant modification or addition of the non-feature code, it is probably time to use separate architecture bodies and configurations. An example is given in Section 14.6.3.

14.3   SPECIFYING PARAMETERS

A parameterized design needs a mechanism to specify the parameters. There are several ways to do this in VHDL, including *generics, array attribute* and *unconstrained array*. Generics behave somewhat like parameters passing between the main program and a routine in a traditional programming language. Array attribute and unconstrained array derive the needed parameter values indirectly from a signal or port declaration.

14.3.1   Generics

We discussed generics in Section 13.3. They can be thought of as symbolic constants that are passed into the entity declaration. When the entity is used later as a component, the generics are assigned values during component instantiation.

Although a generic can assume any data type, only the `integer` data type is allowed in the IEEE 1076.6 RTL synthesis standard. While the `integer` data type is used mainly with a width parameter, we can also utilize it as a flag to specify the desired feature. For example, we can use the values of 0 and 1 to specify whether a buffer is needed for an output signal. Since the width parameter cannot be negative, we sometimes use the `natural` data type, which is a subtype of `integer`, for a generic.

In this chapter, we use the reduced-xor circuit to illustrate various concepts. The reduced-xor circuit applies xor operation over the elements of an array. For example, assume that the input signal is $a_3 a_2 a_1 a_0$. The reduced-xor circuit performs the $a_3 \oplus a_2 \oplus a_1 \oplus a_0$ operation. In Section 5.6.2, this circuit was implemented by using a for loop statement. Since the original code was written with reuse in mind, it can easily be converted to parameterized design.

To utilize a generic, we need to replace the constant declaration in the original code with a generic declaration in the entity declaration. The parameterized code is shown in Listing 14.1.

Listing 14.1 Parameterized reduced-xor circuit using a generic

```vhdl
library ieee;
use ieee.std_logic_1164.all;
entity reduced_xor is
   generic(WIDTH: natural);   -- generic declaration
5  port(
      a: in std_logic_vector(WIDTH-1 downto 0);
      y: out std_logic
   );
end reduced_xor;

10
architecture loop_linear_arch of reduced_xor is
   signal tmp: std_logic_vector(WIDTH-1 downto 0);
begin
   process(a,tmp)
15 begin
      tmp(0) <= a(0);   -- boundary bit
      for i in 1 to (WIDTH-1) loop
         tmp(i) <= a(i) xor tmp(i-1);
      end loop;
20 end process;
   y <= tmp(WIDTH-1);
end loop_linear_arch;
```

14.3.2 Array attribute

A VHDL *attribute* provides information about a named item, such as a data type or a signal. We have used the 'event attribute, as in clk'event, to express the changing edge of the clk signal. There is a set of attributes associated with an object of an array data type. Let s be a signal with an array data type. The following attributes provide some information about the array:

- s'left, s'right: the left and right bounds of the index range of s.
- s'low, s'high: the lower and upper bounds of the index range of s.
- s'length: the length of the index range of s.
- s'range: the index range of s.
- s'reverse_range: the reversed index range of s.

Recall that the std_logic_vector, unsigned and signed data types are defined as array types. The attributes can be applied to the signals defined with these data types. For example, consider the following signals:

```vhdl
signal s1: std_logic_vector(31 downto 0);
signal s2: std_logic_vector(8 to 15);
```

The attributes of s1 are

- s1'left = 31; s1'right = 0;
- s1'low = 0; s1'high = 31;
- s1'length = 32;
- s1'range = 31 **downto** 0
- s1'reverse_range = 0 **to** 31

The attributes of s2 are

- s2'left = 8; s2'right = 15;
- s2'low = 8; s2'high = 15;
- s2'length = 8;
- s2'range = 8 **to** 15
- s2'reverse_range = 15 **downto** 8

These attributes provide information about the width and boundary of a signal. This information can be used as parameters in VHDL code. For example, we can rewrite the reduced-xor code in Listing 14.1 using the 'length attribute, as shown in Listing 14.2. The a'length returns the size of the a signal and plays the role of the previous WIDTH generic.

Listing 14.2   Parameterized reduced-xor circuit using an attribute

```
architecture attr_arch of reduced_xor is
    signal tmp: std_logic_vector(a'length-1 downto 0);
begin
    process(a,tmp)
5   begin
        tmp(0) <= a(0);
        for i in 1 to (a'length-1) loop
            tmp(i) <= a(i) xor tmp(i-1);
        end loop;
10  end process;
    y <= tmp(a'length-1);
end attr_arch;
```

The range of the for loop can also be expressed in other attributes:

- **for** i **in** a'low+1 **to** a'high **loop**
- **for** i **in** a'right+1 **to** a'left **loop**

The last signal assignment statement of the code accesses the leftmost bit of the tmp signal. We can use the 'left attribute to obtain the left bound of the signal and rewrite the statement as

```
y <= tmp(tmp'left);
```

Since the WIDTH generic is included in the entity declaration, the relevant boundaries can be expressed clearly and concisely by the WIDTH generic, as in Listing 14.1. Use of the attributes is somewhat redundant and even cumbersome in this example. The real application of the array attributes is with the unconstrained array, which is discussed in the next subsection.

14.3.3 Unconstrained array

The std_logic_vector, unsigned and signed data types are the three main array types used in this book. They are defined as an *unconstrained array* internally. For example, in the std_logic_1164 package, the std_logic_vector data type is defined as follows:

```
type std_logic_vector is array(natural range <>)
    of std_logic;
```

It indicates that the data type of the index value must be natural, but it does not specify the exact bounds. If an object is declared with an unconstrained array data type, we must specify its index range (i.e., a constraint) when the data type is used, as 15 **downto** 0 in

```
signal x: std_logic_vector(15 downto 0);
```

The port declaration is considered a special case. The unconstrained array can be declared without specifying the range. For example, we can describe a register with no explicit range, as shown in Listing 14.3.

Listing 14.3   Unconstrained D FF

```
   library ieee;
   use ieee.std_logic_1164.all;
   entity unconstrain_dff is
      port(
5        clk: std_logic;
         d: in std_logic_vector;
         q: out std_logic_vector
      );
   end unconstrain_dff;
10
   architecture arch of unconstrain_dff is
   begin
      process(clk)
      begin
15       if (clk'event and clk='1') then
            q <= d;
         end if;
      end process;
   end arch;
```

Note that the data type for the d and q ports is std_logic_vector and no range is specified. The actual range of the std_logic_vector data type is inferred when an instance of unconstrain_dff is instantiated. The ranges of the actual signals become the ranges of the d and q signals. For example, the dff16 instance is instantiated as a 16-bit register in the code segment shown below.

```
   . . .
   signal din, qout: std_logic_vector(15 downto 0);
   signal clk: std_logic;
   . . .
   dff16: unconstrain_dff
      port map(clk=>clk, d=>din, q=>qout);
   . . .
```

In this mechanism, we can think that the width parameter is embedded in the actual signal and passed to the entity declaration when the corresponding component is instantiated.

Since no range is specified for d and q, the boundaries of the two signals will not be checked in the analysis stage. The following code segment is syntactically correct:

```
   . . .
   signal din: std_logic_vector(15 downto 0);
   signal qout: std_logic_vector(7 downto 0);
   . . .
   dff_error: unconstrain_dff
      port map(clk=>clk, d=>din, q=>qout);
   . . .
```

The error can only be detected during the elaboration or execution stage of the code. To make the design more robust, we may need to add error-checking code in the unconstrain_dff description to ensure that d and q have the same range when the component is instantiated.

The previous reduced-xor circuit can also be described without using an explicit range in the a signal. The VHDL code is shown in Listing 14.4. The description is basically patterned after the code in Listing 14.1. In the new code, the generic declaration is removed and the range of the a signal is omitted. The width parameter is inferred from the 'length attribute of the a signal and then declared as a constant in the declaration of the architecture body.

Listing 14.4   Parameterized reduced-xor circuit using an unconstrained array

```
library ieee;
use ieee.std_logic_1164.all;
entity unconstrain_reduced_xor is
    port(
5       a: in std_logic_vector;
        y: out std_logic
    );
end unconstrain_reduced_xor;

10 architecture arch of unconstrain_reduced_xor is
    constant WIDTH : natural:= a'length;
    signal tmp: std_logic_vector(WIDTH-1 downto 0);
begin
    process(a,tmp)
15  begin
        tmp(0) <= a(0);
        for i in 1 to (WIDTH-1) loop
            tmp(i) <= a(i) xor tmp(i-1);
        end loop;
20      end process;
    y <= tmp(WIDTH-1);
end arch;
```

The code appears to be correct at first glance. For example, if we map the a signal to an actual signal with the data type of std_logic_vector(7 **downto** 0) during component instantiation, the code functions as expected. However, since the range of the a signal is inferred from the actual signal, it is same as the actual signal. For an 8-bit actual signal, the following range specification formats are possible:

- std_logic_vector(7 **downto** 0)
- std_logic_vector(0 **to** 7)
- std_logic_vector(15 **downto** 8)
- std_logic_vector(8 **to** 15)

The code does not work properly for the last two formats.

One way to fix the problem is to assign the a signal to an internal signal of known format and use that signal in the code. This scheme is shown in Listing 14.5. We first assign a into an internal aa signal, whose range is specified as WIDTH-1 **downto** 0, and use it in the remaining architecture body.

Listing 14.5 Improved parameterized reduced-xor circuit using an unconstrained array

```
architecture better_arch of unconstrain_reduced_xor is
    constant WIDTH : natural:= a'length;
    signal tmp: std_logic_vector(WIDTH-1 downto 0);
    signal aa: std_logic_vector(WIDTH-1 downto 0);
5 begin
    aa <= a;
    process(aa,tmp)
    begin
        tmp(0) <= aa(0);
10      for i in 1 to (WIDTH-1) loop
            tmp(i) <= aa(i) xor tmp(i-1);
        end loop;
    end process;
    y <= tmp(WIDTH-1);
15 end better_arch;
```

14.3.4 Comparison between a generic and an unconstrained array

A generic and an unconstrained array are two mechanisms to convey width parameter information. The unconstrained array mechanism uses attributes to infer the relevant information from the actual signals. Since the width parameter is derived automatically, this mechanism is more general and flexible than the generic mechanism. However, the flexibility also introduces more opportunities for errors, as shown in the examples in Section 14.3.3.

To develop robust and reliable code for an unconstrained array, we must consider the different formats of range specifications and the potential width mismatch between various signals. These require comprehensive error-checking code to cover possible erroneous conditions. This code may become very involved and unnecessarily complicate the developing and coding process and even overshadow the real design issues. Unless a module is extremely general and widely used, the generic mechanism is satisfactory. We prefer to use the generic mechanism in this book. When the mechanism is more rigid, it clearly specifies the range, direction and width of each signal and avoids many subtle erroneous conditions. This allows us to focus on development of real hardware rather than on error checking.

14.4 CLEVER USE OF AN ARRAY

The logical, relational and arithmetic operators of VHDL and the overloaded operators of the std_logic_1164 and numeric_std packages are defined over unconstrained arrays and thus can be applied to arrays of any size. We can think that they are "implicitly parameterized." For example, consider the following code segment:

```
r <= a - b when a > b else
        a + b;
```

Since the +, - and > operators can accommodate any array sizes, the code is implicitly parameterized.

In a more sophisticated code, an element or a slice of array may be referred and a signal may be assigned or compared with a constant vector value. One key to developing parameterized design is to refrain from fixed-size references. Instead, the references should be expressed in terms of attributes or width parameters. We actually have followed this practice from the beginning of the book. One early coding guideline is to use symbolic constants instead of hard literals. In a properly coded program, we can convert a regular design into a parameterized design by replacing symbolic constants with expressions derived from attributes and generics. The following subsections discuss techniques to achieve this goal and present several examples to illustrate use of these techniques. Lots of regular code can be modified and converted to parameterized descriptions by cleverly using the array data type.

14.4.1 Description without fixed-size references

As in input and output ports, we can classify the internal signals as data and non-data signals. Data signals normally have an array data type, such as std_logic_vector, unsigned or signed. To achieve parameterized design, we should try to use the width parameters or attributes to describe operations that involve data signals. Following are some techniques to avoid a fixed-size description.

Using named association for aggregates   A signal or variable is frequently assigned with a fixed value, as in the initiation of a sequential system. For example, the following is the initialization statement of an 8-bit counter:

```
q_reg <= "00000000";
```

This statement must be revised every time when the width of the counter is modified. A better alternative is to use named association:

```
q_reg <= (others => '0');
```

This statement will remain the same regardless of the width of the counter. Other frequently used constant aggregates include all 1's (i.e., "11...11"):

```
q_reg <= (others => '1');
```

and a single 1 in the LSB (i.e., "00...01"):

```
q_reg <= (0=>'1', others => '0');
```

The aggregate has to be assigned to an object of known size and cannot be used in an expression. For example, the following code segment attempts to check whether a is all-zero and is invalid:

```
signal a: std_logic_vector(WIDTH-1 downto 0)
. . .
x <= '1' when (a=(others=>'0')) else ...
```

One way to correct the problem is to use the 'range attribute to provide the size information:

```
x <= '1' when (a=(a'range=>'0')) else ...
```

Another somewhat cumbersome, but more descriptive way is to define a constant for the all-zero conditions:

```
constant ZERO std_logic_vector(WIDTH-1 downto 0)
            :=(others=>'0');
signal a: std_logic_vector(WIDTH-1 downto 0)
. . .
x <= '1' when (a=ZERO)) else ...
```

Using an integer and conversion function in an expression   If an object is with
the unsigned or signed data types, we can express a constant in integer format since the
relational and arithmetic operators are overloaded with the integer or natural data type.
For example, assume that the a signal is with the unsigned data type. Instead of using a
constant in the unsigned data type, as in

```
x <= '1' when (a="00000110") else ...
```

we can express the constant in the natural data type, as in

```
x <= '1' when (a=6) else ...
```

The constant 6 will be converted to the proper number of bits, and thus no revision is needed
when the width of the a signal changes.

   If a constant is assigned to an object, we can convert the integer to the designated data
type by the width parameter. For example, assume that the x signal is with the unsigned
data type of WIDTH bits. A constant, say 6, can be assigned to x as

```
x <= to_unsigned(6,WIDTH);
```

When the integer 6 is converted to the unsigned type, the number of bits is automatically
adjusted with the WIDTH parameter.

   We can do this for an object with the std_logic_vector data type with additional type
casting. For example, if we assume that the x signal is with the std_logic_vector data
type, the statement becomes

```
x <= std_logic_vector(to_unsigned(6,WIDTH));
```

   Similarly, the previous a=ZERO expression can also be written as follows without using
the constant declaration

```
a=std_logic_vector(to_unsigned(0,WIDTH))
```

Of course, the numeric_std package has to be invoked to use the unsigned data type.

Using the width parameter to refer to a slice or element in an array   Some
VHDL code must make reference to a single element or a slice of an array. The reference is
frequently the MSB or the LSB and is sometimes dependent on the width of the array. For
example, assume that src is with the signed(7 **downto** 0) data type and we use alias to
refer to the sign of the src signal:

```
alias sign: std_logic := src(7);
```

Instead of a hard literal, we can use an attribute to refer to the MSB bit:

```
alias sign: std_logic := src(src'left);
```

If the data type of src is already parameterized as signed(WIDTH-1 **downto** 0), we can
code the statement as

```
alias sign: std_logic := src(WIDTH-1);
```

   Similarly, instead of expressing rotating right 1 bit as

```
dest <= src(0) & src(7 downto 1);
```

we can write

```
dest <= src(src'right) & src(src'left downto src'right+1);
```

This statement can work only if the size of src is larger than 2 and its range is in descending (i.e., **downto**) order. If the data type of src is already parameterized as signed(WIDTH-1 **downto** 0), the rotation operation can be coded in a more descriptive fashion with the WIDTH parameter:

```
dest <= src(0) & src(WIDTH-1 downto 1);
```

14.4.2 Examples

Reduced-xor circuit Several codes were developed for the fixed-size reduced-xor circuit in Section 7.4.1. In Listing 7.17, we used an auxiliary internal signal to represent the intermediate results and described the circuit in a compact, array format. The code can easily be converted to a parameterized design by replacing the constant with a generic. The entity declaration is the same as the one shown in Listing 14.1, and the architecture body is shown in Listing 14.6.

Listing 14.6 Parameterized reduced-xor circuit using a clever array representation

```
architecture array_arch of reduced_xor is
    signal tmp: std_logic_vector(WIDTH-1 downto 0);
begin
    tmp <= (tmp(WIDTH-2 downto 0) & '0') xor a;
5   y <= tmp(WIDTH-1);
end array_arch;
```

Reduced-and circuit The reduced-and circuit applies and operation to the elements of an array. For example, if the input signal is $a_3a_2a_1a_0$, the reduced-and circuit generates the result of $a_3 \cdot a_2 \cdot a_1 \cdot a_0$.

While the reduced-and circuit can be implemented using methods similar to those of the reduced-xor circuit, we use a different approach in this example. The design is based on the observation that the reduced-and circuit returns '1' only when all the inputs are '1'. The code is shown in Listing 14.7. The key is the Boolean condition a=(a'range=>'1'). We use the 'range attribute to obtain the range of the a signal and then construct an aggregate (a'range=>'1'), in which all elements are assigned to '1' (i.e., "1...1"). The a=(a'range=>'1') expression returns true only when the a signal consists of only 1's.

Listing 14.7 Parameterized reduced-and circuit using a clever array representation

```
library ieee;
use ieee.std_logic_1164.all;
entity reduced_and is
    generic(WIDTH: natural);
5   port(
        a: in std_logic_vector(WIDTH-1 downto 0);
        y: out std_logic
    );
end reduced_and;
10
    architecture array_arch of reduced_and is
    begin
        y <= '1' when a=(a'range=>'1') else
             '0';
15 end array_arch;
```

Serial-to-parallel converter   A serial-to-parallel converter accepts input data serially and stores the data in a shift register. Since the output of the register can be accessed simultaneously, the serial data is converted into parallel format. A parameterized design is shown in Listing 14.8.

Listing 14.8   Parameterized serial-to-parallel converter using a clever array representation

```
   library ieee;
   use ieee.std_logic_1164.all;
   entity s2p_converter is
      generic(WIDTH: natural);
 5    port(
         clk: in std_logic;
         si: in std_logic;
         q: out std_logic_vector(WIDTH-1 downto 0)
      );
10 end s2p_converter;

   architecture array_arch of s2p_converter is
      signal q_reg, q_next: std_logic_vector(WIDTH-1 downto 0);
   begin
15    process(clk)
      begin
         if (clk'event and clk='1') then
            q_reg <= q_next;
         end if;
20    end process;
      q_next  <= si & q_reg(WIDTH-1 downto 1);
      q <= q_reg;
   end array_arch;
```

Adder with status circuit   In Section 7.5.3, we discussed a general adder circuit that contains a carry-in signal and various status output signals. To process these extra signals, the adder is expanded by 2 bits internally. We can convert this circuit into a parameterized design, as shown in Listing 14.9. Note that the WIDTH generic is used to access the various elements of the array.

Listing 14.9   Parameterized adder with status circuit

```
   library ieee;
   use ieee.std_logic_1164.all;
   use ieee.numeric_std.all;
   entity para_adder_status is
 5    generic(WIDTH: natural);
      port(
         a, b: in std_logic_vector(WIDTH-1 downto 0);
         cin: in std_logic;
         sum: out std_logic_vector(WIDTH-1 downto 0);
10       cout, zero, overflow, sign: out std_logic
      );
   end para_adder_status;

   architecture arch of para_adder_status is
15    signal a_ext, b_ext, sum_ext: signed(WIDTH+1 downto 0);
```

```
    signal ovf: std_logic;
    alias sign_a: std_logic is a_ext(WIDTH);
    alias sign_b: std_logic is b_ext(WIDTH);
    alias sign_s: std_logic is sum_ext(WIDTH);
20 begin
    a_ext <= signed('0' & a & '1');
    b_ext <= signed('0' & b & cin);
    sum_ext <= a_ext + b_ext;
    ovf <= (sign_a and sign_b and (not sign_s)) or
25           ((not sign_a) and (not sign_b) and sign_s);
    cout <= sum_ext(WIDTH+1);
    sign <= sign_s when ovf='0' else
              not sign_s;
    zero <= '1' when (sum_ext(WIDTH downto 1)=0
30                        and ovf='0') else
             '0';
    overflow <= ovf;
    sum <= std_logic_vector(sum_ext(WIDTH downto 1));
   end arch;
```

Ring counter   We studied two possible implementations of an 8-bit ring counter in Section 9.2.2. The first implementation uses the reset signal to initialize the counter to "000000001". The parameterized version is shown in the reset_arch architecture of Listing 14.10. The second implementation is a self-correcting design. In the original code, a '1' is inserted into the serial input when all 7 MSBs are 0's. For the parameterized version, the WIDTH-1 MSBs need to be checked. The VHDL code is shown in the self_correct_arch architecture of Listing 14.10. We create the r_high alias to represent the WIDTH-1 MSBs and use the r_high=(r_high'range=>'0') expression to check the all-zero condition.

Listing 14.10   Parameterized ring counter

```
    library ieee;
    use ieee.std_logic_1164.all;
    entity para_ring_counter is
       generic(WIDTH: natural);
5      port(
          clk, reset: in std_logic;
          q: out std_logic_vector(WIDTH-1 downto 0)
       );
    end para_ring_counter;
10
    — architecture using asynchronous initialization
    architecture reset_arch of para_ring_counter is
       signal r_reg: std_logic_vector(WIDTH-1 downto 0);
       signal r_next: std_logic_vector(WIDTH-1 downto 0);
15 begin
       — register
       process(clk,reset)
       begin
          if (reset='1') then
20           r_reg <= (0=>'1',others=>'0');
          elsif (clk'event and clk='1') then
```

```
            r_reg <= r_next;
        end if;
    end process;
25  — next—state logic
    r_next <=r_reg(0) & r_reg(WIDTH-1 downto 1);
    — output logic
    q <= r_reg;
  end reset_arch;
30
  — architecture using self—correcting circuit
  architecture self_correct_arch of para_ring_counter is
    signal r_reg: std_logic_vector(WIDTH-1  downto 0);
    signal r_next: std_logic_vector(WIDTH-1  downto 0);
35  signal s_in: std_logic;
    alias r_high: std_logic_vector(WIDTH-2  downto 0) is
                  r_reg(WIDTH-1 downto 1)   ;
  begin
    — register
40  process(clk,reset)
    begin
        if (reset='1') then
            r_reg <= (others=>'0');
        elsif (clk'event and clk='1') then
45          r_reg <= r_next;
        end if;
    end process;
    — next—state logic
    s_in <= '1' when r_high=(r_high'range=>'0') else
50          '0';
    r_next <= s_in & r_reg(WIDTH-1 downto 1);
    — output logic
    q <= r_reg;
  end self_correct_arch;
```

14.5 FOR GENERATE STATEMENT

The generate statements are concurrent statements with embedded internal concurrent statements, which can be interpreted as a circuit part. There are two types of generate statements. The first type is the *for generate statement*, which is used to create a circuit by replicating the hardware part. The second type is the *conditional* or *if generate statement*, which is used to specify whether or not to create an optional hardware part. The generate statements are especially useful for the parameterized design. This section discusses the for generate statement and the next section covers the conditional generate statement.

Many digital circuits can be implemented as a repetitive composition of basic building blocks. They frequently exhibit a regular structure, such as a one-dimensional cascading chain, a tree-shaped connection or a two-dimensional mesh. Since we can easily expand the structure by increasing the number of iterations, these circuits are natural for the parameterized design. The for generate statement is used to describe this kind of circuit.

14.5.1 Syntax

The simplified syntax of the for generate statement is

```
gen_label:
for loop_index in loop_range generate
    concurrent statements;
end generate;
```

The for generate statement is somewhat similar to the basic for loop statement discussed in Chapter 4. The for generate statement repeats the loop body of concurrent statements for a fixed number of iterations. The loop_range term specifies a range of values between the left and right bounds. The range has to be static, which means that it has to be determined by the time of execution (synthesis). It is normally specified by the width parameters. The loop_index term is used to keep track of the iteration and takes a successive value from loop_range in each iteration, starting from the leftmost value. The index automatically takes the data type of loop_range's element and does not need to be declared. The gen_label term is mandatory. It is the label used to identify to this particular generate statement.

The loop body contains a collection of concurrent statements, which may include other generate statements. The concurrent statements describe a *stage* of the iterative circuit. A stage description is composed of two main ingredients. One is the description of the basic building block and the other is the input–output connection pattern between the blocks. The connection pattern is normally specified by a collection of internal signals, which are represented as a one- or two-dimensional array with loop_index in their index expression.

The key to designing an iterative circuit is to identify the basic block and connection pattern of a stage. To determine the connection pattern and to describe the relationship between the input and output signals of successive stages, we can first draw a small-scale circuit diagram, label a few specific connection signals and then derive the general relationship.

14.5.2 Examples

Binary decoder A binary n-to-2^n decoder is a circuit that asserts one of the 2^n possible output signals. The codes of a 2-to-2^2 decoder are shown in Chapters 4 and 5. While these codes are simple, none can be modified for parameterized design.

One way to view the binary decoder is to treat each bit of the decoded output as the result of a constant comparator. The decoded bit is asserted when the value of the input signal matches the hardwired constant value. The block diagram of a 2-to-2^2 decoder is shown in Figure 14.1.

Note that since one input of the comparator is a constant (i.e., hardwired), it can be simplified during synthesis. This diagram can easily be replicated with a different input width. In the ith stage, code(i) is asserted when the binary value of the a input is equal to i. This can be translated into the VHDL statement:

```
code(i) <= '1' when a=std_logic_vector(unsigned(i)) else
           '0';
```

The parameterized VHDL code using the for generate statement is shown in Listing 14.11.

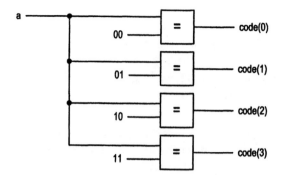

Figure 14.1 Block diagram of a 2-to-4 decoder.

Listing 14.11 Parameterized binary decoder using a for generate statement

```
library ieee;
use ieee.std_logic_1164.all;
use ieee.numeric_std.all;
entity bin_decoder is
   generic(WIDTH: natural);
   port(
      a: in std_logic_vector(WIDTH-1 downto 0);
      code: out std_logic_vector(2**WIDTH-1 downto 0)
   );
end bin_decoder;

architecture gen_arch of bin_decoder is
begin
   comp_gen:
   for i in 0 to (2**WIDTH-1) generate
      code(i) <= '1' when i=to_integer(unsigned(a)) else
                 '0';
   end generate;
end gen_arch;
```

Note that we have to include the numeric_std package to use the unsigned data type.

Reduced-xor circuit As discussed in Section 7.4.1, the reduced-xor circuit can be implemented as a cascading chain or a tree. The block diagram of an 8-bit cascading-chain implementation is shown again in Figure 14.2. The diagram exhibits a regular, iterative pattern and thus is a good match for the for generate statement description. The building block is the xor gate. We divide the chain into stages and number the stages from left to right, starting at the 0th stage. The internal signals connect the output of the current stage to the input of the next stage. These signals are arranged as an array and the tmp(i) signal represents the output of the $(i-1)$th stage and the input to the ith stage, as shown in Figure 14.2. From the diagram, we can see that there is a clear relationship among the two input signals and the output signal of an xor gate. For the ith stage, the three signals can be expressed as

```
tmp(i+1) <= tmp(i) xor a(i+1);
```

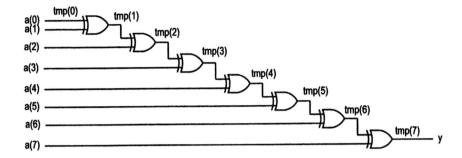

Figure 14.2   Block diagram of a reduced-xor circuit.

Note that the statement shows the basic building block (i.e., xor gate) and the interconnection between blocks.

Base on the equation, we can derive the VHDL code using a for generate statement. The code is shown in Listing 14.12. The loop body iterates WIDTH−1 times and thus infers WIDTH−1 xor gates.

Listing 14.12   Parameterized reduced-xor circuit using a for generate statement

```
architecture gen_linear_arch of reduced_xor is
    signal tmp: std_logic_vector(WIDTH-1 downto 0);
begin
    tmp(0) <= a(0);
5   xor_gen:
    for i in 1 to (WIDTH-1) generate
        tmp(i) <= a(i) xor tmp(i-1);
    end generate;
    y <= tmp(WIDTH-1);
10 end gen_linear_arch;
```

In an iterative structure, the boundary stages interface to the external input and output signals, and sometimes their connections are different from the regular blocks. Note that we use two special statements to handle the boundary signals of the leftmost and rightmost stages.

The code here and the array-based code in Listing 14.6 actually specify the same circuit structure. While the former describes the design stage by stage, the latter lumps the signals together in a single array.

Serial-to-parallel converter   We discussed a simple serial-to-parallel converter in Section 14.4.2. It is composed of a series of cascading D FFs, and the conceptual diagram of a 4-bit implementation is shown in Figure 14.3. Each stage consists of a D FF and next-state logic, which is a wire that connects the output of the previous D FF to the input of the current D FF.

The first VHDL description is shown in Listing 14.13. To accommodate the naming convention, we extend the q_reg signal by one extra bit and assign the external si signal to q_reg(WIDTH). Since q_reg(WIDTH) is not assigned inside the for generate statement, only WIDTH−1 D FFs will be inferred. The loop body consists of two concurrent statements. One is the process for the D FF and the other is the next-state logic. Even the next-state

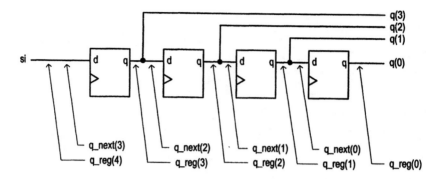

Figure 14.3   Block diagram of a serial-to-parallel converter.

logic is very simple; we still follow synchronous design practice and separate it from the memory element.

Listing 14.13   Parameterized serial-to-parallel converter using a for generate statement

```vhdl
architecture gen_proc_arch of s2p_converter is
    signal q_next: std_logic_vector(WIDTH-1 downto 0);
    signal q_reg: std_logic_vector(WIDTH downto 0);
begin
    q_reg(WIDTH) <= si;
    dff_gen:
    for i in (WIDTH-1) downto 0 generate
        -- D FF
        process(clk)
        begin
            if (clk'event and clk='1') then
                q_reg(i) <= q_next(i);
            end if;
        end process;
        -- next-state logic
        q_next(i) <= q_reg(i+1);
    end generate;
    --output
    q <= q_reg(WIDTH-1 downto 0);
end gen_proc_arch;
```

Alternatively, we can also define the D FF as an entity and use it through component instantiation. The VHDL codes for the D FF and the alternative description are shown in Listing 14.14. The code is essentially the structural description of the block diagram in Figure 14.3.

Listing 14.14   Alternative serial-to-parallel converter using a for generate statement

```vhdl
-- D FF
library ieee;
use ieee.std_logic_1164.all;
entity dff is
    port(
        clk: in std_logic;
```

```
          d: in std_logic;
          q: out std_logic
       );
10 end dff;

   architecture arch of dff is
   begin
       process(clk)
15     begin
          if (clk'event and clk='1') then
             q  <= d;
          end if;
       end process;
20 end arch;

   -- architecture using component instantiation
   architecture gen_comp_arch of s2p_converter is
       signal q_reg: std_logic_vector(WIDTH downto 0);
25     component dff
          port(
             clk: in std_logic;
             d: in std_logic;
             q: out std_logic
30        );
       end component;
   begin
       q_reg(WIDTH) <= si;
       dff_gen:
35     for i in (WIDTH-1) downto 0 generate
          dff_array: dff
             port map (clk=>clk, d=>q_reg(i+1), q=>q_reg(i));
       end generate;
       q <= q_reg(WIDTH-1 downto 0);
40 end gen_comp_arch;
```

14.6 CONDITIONAL GENERATE STATEMENT

14.6.1 Syntax

The conditional generate statement is used to specify an optional circuit that can be included or excluded in the final implementation. It can be used to realize the feature parameters of a parameterized design. The simplified syntax of the conditional generate statement is

```
gen_label:
if boolean_exp generate
   concurrent statements;
end generate;
```

The boolean_exp is an expression that returns a value with the boolean data type. If it is true, the internal concurrent statements are invoked, which means that the circuit described by the concurrent statements will be included in the implementation. If the expression is false, no concurrent statement is invoked, and thus the corresponding circuit is excluded from the implementation. Note that there is no else branch. If we want to include one of the two possible circuits in an implementation, we must use two separate if generate statements. The gen_label term is the label and is mandatory.

For synthesis purposes, the boolean_exp expression must be static so that synthesis software knows whether the corresponding concurrent statements should be included in the physical implementation. The expression is normally described in terms of generics.

14.6.2 Examples

Reduced-xor circuit revisited One common use of the conditional generate statement is to describe the "irregular" stages in a for generate statement. Consider the VHDL code for the reduced-xor circuit in Listing 14.12. The first and last stages are different from others because they interface with the external input and output signals, which use different name conventions. Two statements are used to rename the signals:

```
tmp(0)  <= a(0);
y  <= tmp(WIDTH-1);
```

To eliminate these statements, we can use conditional generate statements inside the for generate statement. Each Boolean expression of a conditional generate statement represents a specific condition and specifies what kind of circuit should be generated for the corresponding stages. In this design, there are three kinds of stages: the leftmost stage, regular middle stages and the rightmost stage. The if generate statement can check the stage number and then generate a circuit that matches the naming convention accordingly. The VHDL code is shown in Listing 14.15.

Listing 14.15 Parameterized reduced-xor circuit with a conditional generate statement

```
architecture gen_if_arch of reduced_xor is
    signal tmp: std_logic_vector(WIDTH-2 downto 1);
begin
    xor_gen:
 5  for i in 1 to (WIDTH-1) generate
        — leftmost stage
        left_gen: if i=1 generate
            tmp(i) <= a(i) xor a(0);
        end generate;
10      — middle stages
        middle_gen: if (1 < i) and (i < (WIDTH-1)) generate
            tmp(i) <= a(i) xor tmp(i-1);
        end generate;
        — rightmost stage
15      right_gen: if i=(WIDTH-1) generate
            y <= a(i) xor tmp(i-1);
        end generate;
    end generate;
end gen_if_arch;
```

Up-or-down free-running binary counter An up-or-down binary counter is a counter that can be instantiated in a specific mode. Note that the "or" here means that only one mode of operation, either counting up or counting down but not both, can be implemented in the final circuit. We use the UP generic as the feature parameter to specify the desired mode.

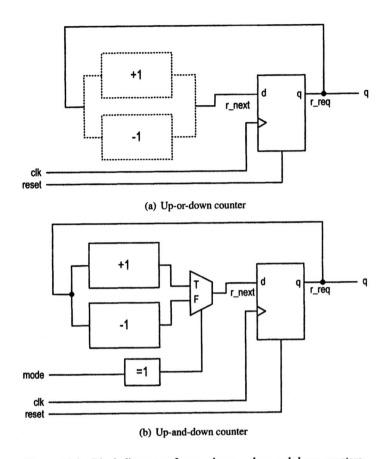

(a) Up-or-down counter

(b) Up-and-down counter

Figure 14.4   Block diagrams of up-or-down and up-and-down counters.

The counter counts up if UP is 1 and counts down otherwise. Since there are two possible features, the boolean data type will be more appropriate for the UP generic. However, since the IEEE RTL synthesis standard and some software accept only the integer data type and its subtypes, the natural type is used.

The conceptual block diagram of this counter is shown in Figure 14.4(a). We use dashed blocks to indicate the optional features of a circuit, such as the incrementor and decrementor in the diagram. In this particular example, only one of the dashed blocks will be used in synthesis, and thus there is no output confliction for the r_next signal. The VHDL code is shown in Listing 14.16.

Listing 14.16   Up-or-down free-running binary counter

```
library ieee;
use ieee.std_logic_1164.all;
use ieee.numeric_std.all;
entity up_or_down_counter is
    generic(
        WIDTH: natural;
        UP: natural
    );
    port(
```

```
10        clk, reset: in std_logic;
          q: out std_logic_vector(WIDTH-1 downto 0)
      );
   end up_or_down_counter;

15 architecture arch of up_or_down_counter is
      signal r_reg: unsigned(WIDTH-1 downto 0);
      signal r_next: unsigned(WIDTH-1 downto 0);
   begin
      -- register
20    process(clk,reset)
      begin
         if (reset='1') then
            r_reg <= (others=>'0');
         elsif (clk'event and clk='1') then
25          r_reg <= r_next;
         end if;
      end process;
      -- next-state logic
      inc_gen:  -- incrementor
30    if UP=1 generate
         r_next <= r_reg + 1;
      end generate;
      dec_gen:  --decrementor
      if UP/=1 generate
35       r_next <= r_reg - 1;
      end generate;
      -- output logic
      q <= std_logic_vector(r_reg);
   end arch;
```

The two next-state logics are described by the two separated if generate statements. Note that the Boolean expressions of the two statements are complementary, and thus only one circuit will be generated.

For comparison purposes, let us examine a dual-mode binary counter that counts in both up and down directions. The mode is specified by an additional mode input signal. This implementation includes an incrementor and a decrementor, and uses a multiplexer to select the desired result, as shown in Figure 14.4(b). The corresponding VHDL code is listed in Listing 14.17.

Listing 14.17 Up-and-down free-running binary counter

```
library ieee;
use ieee.std_logic_1164.all;
use ieee.numeric_std.all;
entity up_and_down_counter is
5    generic(WIDTH: natural);
     port(
        clk, reset: in std_logic;
        mode: in std_logic;
        q: out std_logic_vector(WIDTH-1 downto 0)
10   );
   end up_and_down_counter;
```

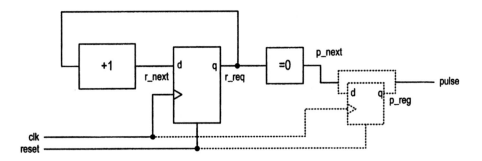

Figure 14.5 Block diagram of a counter with an optional output buffer.

```vhdl
   architecture arch of up_and_down_counter is
      signal r_reg: unsigned(WIDTH-1 downto 0);
15    signal r_next: unsigned(WIDTH-1 downto 0);
   begin
      -- register
      process(clk,reset)
      begin
20       if (reset='1') then
            r_reg <= (others=>'0');
         elsif (clk'event and clk='1') then
            r_reg <= r_next;
         end if;
25    end process;
      -- next-state logic
      r_next <= r_reg + 1 when mode='1' else
                r_reg - 1;
      -- output logic
30    q <= std_logic_vector(r_reg);
   end arch;
```

Counter with an optional output buffer   An output buffer can remove glitches from the signal. Since the buffer is only needed for certain applications, it will be convenient to include the buffer as an optional part of the circuit. This can be achieved by using a feature parameter and conditional generate statements. Consider a binary counter that has a pulse output signal that is activated when the counter reaches 0. We can use the BUFF generic to indicate whether a buffer should be inserted. The conceptual diagram is shown in Figure 14.5 and the VHDL code is shown in Listing 14.18.

Listing 14.18   Counter with an optional output buffer

```vhdl
   library ieee;
   use ieee.std_logic_1164.all;
   use ieee.numeric_std.all;
   entity op_buf_counter  is
5     generic(
         WIDTH: natural;
         BUFF: natural
      );
      port(
```

```
10        clk, reset: in std_logic;
          pulse: out std_logic
       );
    end op_buf_counter;

15 architecture arch of op_buf_counter is
       signal r_reg: unsigned(WIDTH-1 downto 0);
       signal r_next: unsigned(WIDTH-1 downto 0);
       signal p_next, p_reg: std_logic;
    begin
20     -- register
       process(clk,reset)
       begin
          if (reset='1') then
             r_reg <= (others=>'0');
25        elsif (clk'event and clk='1') then
             r_reg <= r_next;
          end if;
       end process;
       -- next-state logic
30     r_next <= r_reg + 1;
       -- output logic
       p_next <= '1' when r_reg=0 else '0';
       buf_gen:    -- with buffer
       if BUFF=1 generate
35        process(clk,reset)
          begin
             if (reset='1') then
                p_reg <= '0';
             elsif (clk'event and clk='1') then
40              p_reg <= p_next;
             end if;
          end process;
          pulse <= p_reg;
       end generate;
45     no_buf_gen:   -- without buffer
       if BUFF/=1 generate
          pulse <= p_next;
       end generate;
    end arch;
```

FSM with a selectable clear signal In a sequential circuit, we usually include a clear signal to perform system initialization. The clear signal can be either synchronous or asynchronous, and the choice sometimes depends on the target device technology. To make the code portable, it is beneficial to include a generic to specify the type of clear signal to be synthesized. Implementing the asynchronous clear is straightforward. We just replace the state register by a register with an asynchronous reset signal. Implementing the synchronous clear needs to revise the next-state logic of the sequential circuit. To minimize the modification, we can wrap the original next-state logic with a 2-to-1 multiplexer. The conceptual diagram is shown in Figure 14.6. The initial state value (assume that it is idle) will be routed to the register if the synchronous clear signal is asserted.

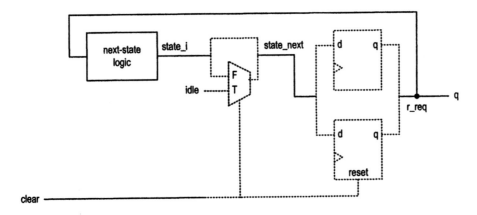

Figure 14.6   Block diagram of FSM with a selectable clear signal.

We use the memory controller FSM of Chapter 10 to demonstrate the design. To accommodate the "clear" feature, we must selectively generate circuits in two places, as shown in Figure 14.6. One is the register, which can be a register with or without the asynchronous reset signal. The other is the optional multiplexer to route the `idle` value to the `state_next` signal. We introduce the SYNC generic to specify the desired type of clear and use two if generate statements to create the corresponding circuits. The VHDL code is shown in Listing 14.19.

Listing 14.19   Memory controller FSM with a selectable clear signal

```
   library ieee;
   use ieee.std_logic_1164.all;
   entity clr_mem_fsm is
      generic(SYNC: integer);
 5    port(
         clk, clear: in std_logic;
         mem, rw, burst: in std_logic;
         oe, we: out std_logic
      );
10 end clr_mem_fsm ;

   architecture mult_seg_arch of clr_mem_fsm is
      type mc_state_type is
           (idle, read1, read2, read3, read4, write);
15    signal state_reg, state_i, state_next: mc_state_type;
   begin
      —— state register
      reset_ff_gen:   —— register with asynchronous clear
      if SYNC/=1 generate
20       process(clk,clear)
         begin
            if (clear='1') then
               state_reg <= idle;
            elsif (clk'event and clk='1') then
25             state_reg <= state_next;
            end if;
```

```
            end process;
        end generate;
        no_reset_ff_gen: -- register without asynchronous clear
30      if SYNC=1 generate
            process(clk)
            begin
                if (clk'event and clk='1') then
                    state_reg <= state_next;
35              end if;
            end process;
        end generate;
        -- next-state logic
        process(state_reg,mem,rw,burst)
40      begin
            case state_reg is
              when idle =>
                  if mem='1' then
                      if rw='1' then
45                        state_i <= read1;
                      else
                          state_i <= write;
                      end if;
                  else
50                    state_i <= idle;
                  end if;
              when write =>
                  state_i <= idle;
              when read1 =>
55                if (burst='1') then
                      state_i <= read2;
                  else
                      state_i <= idle;
                  end if;
60            when read2 =>
                  state_i <= read3;
              when read3 =>
                  state_i <= read4;
              when read4 =>
65                state_i <= idle;
            end case;
        end process;
        no_sync_clr_gen:  -- without mux
        if SYNC/=1 generate
70          state_next <= state_i;
        end generate;
        sync_clr_gen:      -- with mux
        if SYNC=1 generate
            state_next <= idle when clear='1' else
75                          state_i;
        end generate;
        -- Moore output logic
        process(state_reg)
        begin
```

```
80        we <= '0';
          oe <= '0';
          case state_reg is
              when idle =>
              when write =>
85                we <= '1';
              when read1 =>
                  oe <= '1';
              when read2 =>
                  oe <= '1';
90            when read3 =>
                  oe <= '1';
              when read4 =>
                  oe <= '1';
          end case;
95    end process;
   end mult_seg_arch;
```

14.6.3   Comparisons with other feature-selection methods

In addition to the conditional generate statement, there are two other methods to create a circuit with a selectable feature. One is to create a full-featured circuit and then connect some input control signals to constant values to permanently enable the desired feature. The other is to use the configuration construct. Their uses and differences are discussed in the following subsections.

Comparison to a full-featured circuit   The full-featured scheme can be explained best by an example. Consider the up-and-down counter from Listing 14.17. The counter has an external control signal, mode, which specifies the direction of the counting. The code implies that the next-state logic consists of an incrementor, a decrementor and a multiplexer, as shown in Figure 14.4(b). Although there is no feature parameter in this design, we can imitate the UP generic of the up-or-down counter by connecting the mode signal to a constant value.

For example, assume that we need a 16-bit up counter in a design. To use the parameterized up-or-down counter, we can use the following component instantiation to create the instance:

```
count16up: up_or_down_counter
    generic map(WIDTH=>16, UP=>1);
    port(clk=>clk, reset=>reset, q=>q);
```

To create the same counter instance using an the up-and-down counter, we can map the mode signal to '1'. The component instantiation becomes

```
count16up: up_and_down_counter
    generic map(WIDTH =>16);
    port(clk=>clk, reset=>reset, mode=>'1', q=>q);
```

Since the mode signal is tied to '1', the counter always counts up, just as in the previous up-or-down counter instance.

Although the two instances have the same functionality, they are two different circuits. The up-or-down counter instance creates a circuit with only the needed features. The up-

and-down counter instance creates a circuit that consists of all features and uses an external control signal to selectively enable a portion of the circuit.

This difference will also be reflected in the processing of the VHDL program. Recall that the processing is divided into analysis, elaboration and execution (synthesis) stages. The conditional generate statement is processed in the elaboration stage and the unneeded circuit is removed. The synthesis software only needs to synthesize the selected portion. On the other hand, while the code from the full-featured scheme is processed, the entire VHDL code will be passed to the synthesis stage. It is the synthesis software's responsibility to propagate the constant signal through the circuits and eliminate the unused portion through logic optimization. This will increase the processing time. For a complex description, the software may not be able to eliminate all the unneeded logic in the final implementation.

In general, use of the feature parameters and conditional generate statements is better than the full-featured approach because it clearly identifies the optional part, and the unused portion of the circuit is removed before synthesis.

Comparison to configuration   The selected hardware creation can also be achieved by configuration. We can construct multiple architecture bodies, each containing a specific feature. Instead of using the feature generic, we can select the desired feature by configuring the entity with a proper architecture body.

For example, for the previous up-or-down free-running counter, we can eliminate the UP generic and construct one architecture body with a counting-up sequence and another with a counting-down sequence. The VHDL code is shown in Listing 14.20.

Listing 14.20   Up-or-down counter with two architecture bodies

```vhdl
library ieee;
use ieee.std_logic_1164.all;
use ieee.numeric_std.all;
entity updown_counter is
    generic(WIDTH: natural);
    port(
        clk, reset: in std_logic;
        q: out std_logic_vector(WIDTH-1 downto 0)
    );
end updown_counter;

    -- architecture for the count-up sequence
    architecture up_arch of updown_counter is
        signal r_reg: unsigned(WIDTH-1 downto 0);
        signal r_next: unsigned(WIDTH-1 downto 0);
    begin
        -- register
        process(clk,reset)
        begin
            if (reset='1') then
                r_reg <= (others=>'0');
            elsif (clk'event and clk='1') then
                r_reg <= r_next;
            end if;
        end process;
        -- next-state logic
        r_next <= r_reg + 1;
        -- output logic
```

```
        q <= std_logic_vector(r_reg);
30  end up_arch;

    -- architecture for the count-down sequence
    architecture down_arch of updown_counter is
        signal r_reg: unsigned(WIDTH-1 downto 0);
35      signal r_next: unsigned(WIDTH-1 downto 0);
    begin
        -- register
        process(clk,reset)
        begin
40          if (reset='1') then
                r_reg <= (others=>'0');
            elsif (clk'event and clk='1') then
                r_reg <= r_next;
            end if;
45      end process;
        -- next-state logic
        r_next <= r_reg - 1;
        -- output logic
        q <= std_logic_vector(r_reg);
50  end down_arch;
```

We can create a configuration declaration unit or add the configuration specification to bind the architecture body with the desired feature. This can also be done via the component instantiation in VHDL 93. For example, we can create a 16-bit up counter as follows:

```
count16up: work.updown_counter(up_arch)
    generic map(WIDTH =>16);
    port(clk=>clk, reset=>reset, q=>q);
```

Conversely, we can merge the logic from several architecture bodies into a single body and use a feature generic and conditional generate statements to select the desired portion. This essentially replaces the configuration with a feature parameter.

There is no rule about when to use a feature parameter and when to use a configuration construct. In general, code with a feature parameter is more difficult to develop and comprehend because we are essentially describing several different versions of the circuit in the same code. The code of the two architecture bodies of the previous example is clearer and more descriptive than the code with the UP generic. On the other hand, if we use a separate architecture body for each distinctive feature, the number of architecture bodies will grow exponentially and becomes difficult to manage. For example, if we want a binary counter to count up or down, to be equipped with either synchronous or asynchronous clear, and to include a buffered and unbuffered output pulse, we must create eight architecture bodies to cover all possible combinations.

In general, when a feature parameter leads to significant modification or addition of the no-feature code and starts to make the code incomprehensible, it is probably a good idea to use separate architecture bodies and the configuration construct.

14.7 FOR LOOP STATEMENT

14.7.1 Introduction

The for loop statement is a sequential statement and is the only sequential loop construct that can be synthesized. The simplified syntax of the for loop statement is

```
for index in loop_range loop
   sequential statements;
end loop;
```

The syntax and operation of the for loop statement are similar to those of the generate loop statement except that the loop body is composed of sequential statements. As in the generate loop statement, the loop_range must be static.

The for loop statement is more general and flexible because of the sequential statements. In addition to the statements discussed in Chapter 5, the sequential statements also include the *exit statement*, which skips the remaining iterations of the loop, and the *next statement*, which skips the remaining part of the current iteration. The exit and next statements are discussed in the next section.

The basic way to synthesize a for loop statement is to unroll or flatten the loop. *Unrolling a loop* means to replace the loop structure by explicitly listing all iterations. Since the range is static, the number of iterations is fixed. Once a for loop statement is unrolled, the code is converted to a sequence of regular sequential statements, which can be synthesized accordingly. To derive an effective design, we need to know the implication of various language constructs on the underlying hardware. The examples in the next subsection show the implementation issues of the for loop statement.

14.7.2 Examples of a simple for loop statement

Binary decoder The structure of the binary decoder is discussed in Section 14.5.2. We can use the diagram in Figure 14.1 as a reference and derive a for loop statement to describe the basic building block and interconnection pattern. The code is very similar to the for generate version in Listing 14.11 and is shown in Listing 14.21. Note that the conditional signal assignment statement in Listing 14.11 is replaced by the if statement.

Listing 14.21 Parameterized binary decoder using a for loop statement

```
architecture loop_arch of bin_decoder is
begin
   process(a)
   begin
      for i in 0 to (2**WIDTH-1) loop
         if i=to_integer(unsigned(a)) then
            code(i) <= '1';
         else
            code(i) <= '0';
         end if;
      end loop;
   end process;
end loop_arch;
```

Reduced-xor circuit In Section 14.3.1, the reduced-xor circuit is described using a for loop statement in Listing 14.1. The code is patterned after the version that uses the for generate statement in Listing 14.12.

Serial-to-parallel converter The serial-to-parallel converter discussed in Section 14.5.2 can also be described using a for loop statement. The first version is shown in Listing 14.22. It is patterned after the code using the for generate statement in Listing 14.13.

Listing 14.22   Parameterized serial-to-parallel converter using a for loop statement

```
architecture loop1_arch of s2p_converter is
    signal q_next: std_logic_vector(WIDTH-1 downto 0);
    signal q_reg: std_logic_vector(WIDTH downto 0);
begin
5   q_reg(WIDTH) <= si;
    process(clk,q_reg)
    begin
        for i in WIDTH-1 downto 0 loop
            -- D FF
10          if (clk'event and clk='1') then
                q_reg(i) <= q_next(i);
            end if;
            -- next-state logic
            q_next(i) <= q_reg(i+1);
15      end loop;
    end process;
    q <= q_reg(WIDTH-1 downto 0);
end loop1_arch;
```

Since the for loop statement is a sequential statement, it must be enclosed within a process. The loop body, which contains both the D FF and next-state logic, is enclosed within the same process accordingly. Note that the both clk and q_reg signals are listed in the sensitivity list. An alternative is to split the register and the next-state logic and describe the structure in two separate for loop statements. The revised code is shown in Listing 14.23.

Listing 14.23   Parameterized serial-to-parallel converter using separate for loop statements

```
architecture loop2_arch of s2p_converter is
    signal q_reg, q_next: std_logic_vector(WIDTH-1 downto 0);
begin
    -- registers
5   process(clk)
    begin
        for i in WIDTH-1 downto 0 loop
            if (clk'event and clk='1') then
                q_reg(i) <= q_next(i);
10          end if;
        end loop;
    end process;
    -- next-state logic
    process(si,q_reg)
15  begin
        q_next(WIDTH-1) <= si;
```

```
        for i in WIDTH-2 downto 0 loop
            q_next(i) <= q_reg(i+1);
        end loop;
20    end process;
        q <= q_reg;
    end loop2_arch;
```

In Section 14.5.2, the code in Listing 14.14 uses component and instantiation to describe the structure of the serial-to-parallel converter. Since the component instantiation statement is a concurrent statement, this approach cannot be duplicated for the for loop statement.

14.7.3 Examples of a loop body with multiple signal assignment statements

Only sequential signal assignment statements can be used inside the loop body of a for loop statement. Recall that a signal can be assigned multiple times inside a process and only the last assignment takes effect. We can use this property to develop more abstract description. Two examples are shown below.

Priority encoder Recall that a priority encoder is a circuit that returns the binary code of the highest-priority request. In the parameterized version, we assume that the request signals are arranged as an array of $r(WIDTH-1$ **downto** $0)$, and priority is given in descending order (i.e., the $r(WIDTH-1)$ signal has the highest priority). In addition to the binary code, the bcode signal, the output includes the valid signal, which is asserted when at least one request signal is activated. One possible VHDL code of a parameterized priority encoder is shown in Listing 14.24.

Listing 14.24 Parameterized priority encoder using a for loop statement

```
    library ieee;
    use ieee.std_logic_1164.all;
    use ieee.numeric_std.all;
    use work.util_pkg.all;
5 entity prio_encoder is
        generic(WIDTH: natural);
        port(
            r: in std_logic_vector(WIDTH-1 downto 0);
            bcode: out std_logic_vector(log2c(WIDTH)-1 downto 0);
10        valid: out std_logic
        );
    end prio_encoder;

    architecture loop_linear_arch of prio_encoder is
15    constant B: natural := log2c(WIDTH);
        signal tmp: std_logic_vector(WIDTH-1 downto 0);
    begin
        -- binary code
        process(r)
20    begin
            bcode <= (others=>'0');
            for i in 0 to (WIDTH-1) loop
                if r(i)= '1' then
                    bcode <= std_logic_vector(to_unsigned(i,B));
```

```
25          end if ;
         end loop ;
      end process ;
      -- reduced-or circuit
      process (r , tmp)
30    begin
         tmp (0) <= r (0) ;
         for i in 1 to (WIDTH-1) loop
            tmp (i) <= r (i) or tmp (i-1) ;
         end loop ;
35    end process ;
      valid <= tmp (WIDTH-1) ;
   end loop_linear_arch ;
```

For an input with n request signals, the number of bits of the binary code is $\lceil \log_2 n \rceil$. To express the range of the bcode signal, we can use the log2c function derived in Chapter 12. We assume that it is stored in the util_pkg package and include the use statement to make it visible.

The major part of the program is the first for loop statement, which iterates from the lowest index to the highest index. When the corresponding request is asserted, its binary code will be assigned to the bcode signal. Recall that the last signal assignment takes effect in a process. The bcode signal is assigned with the binary code of the highest index, which represents the highest-priority request. If none of the request is asserted, the bcode assumes the default assignment of all 0's.

Unlike the previous binary decoder and reduced-xor examples, in which the program codes are derived from the actual circuit structures, this program is based on an abstract, behavioral description of a priority encoding circuit. We drive the circuit structure from the VHDL code, but not the other way around.

To derive the conceptual implementation, we first need to unroll the loop. Assume that WIDTH is 4. The flattened code becomes

```
bcode <= "00"
if r (0)= '1' then
    bcode <= "00" ;
end if ;
if r (1)= '1' then
    bcode <= "01" ;
end if ;
if r (2)= '1' then
    bcode <= "10" ;
end if ;
if r (3)= '1' then
    bcode <= "11" ;
end if ;
```

The code performs a sequence of assignments to the same signal. As discussed in Section 5.4.1, this kind of code is equivalent to an if statement with multiple elsif branches, which implies a priority routing network. The conceptual diagram of the flattened code can be derived using the procedure in Section 5.4.1 and is shown in Figure 14.7. It is basically a cascading chain of 2-to-1 multiplexers.

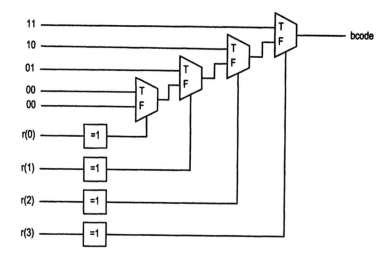

Figure 14.7   Block diagram of a priority encoder.

The valid signal is obtained by performing an or operation on all request signals. It is essentially a reduced-or circuit, and its implementation is similar to that of the reduced-xor circuit.

Multiplexer   A multiplexer routes the designated input signal to the output port. In the parameterized version, we assume that the input signals are arranged as an array, from a(WIDTH-1) to a(0), and the selection signal of the multiplexer uses the index of the array to specify the designated signal. One possible VHDL code is shown in Listing 14.25.

Listing 14.25   Parameterized multiplexer using a for loop statement

```
library ieee;
use ieee.std_logic_1164.all;
use ieee.numeric_std.all;
use work.util_pkg.all;
5 entity mux1 is
    generic(WIDTH: natural);
    port(
        a: in std_logic_vector(WIDTH-1 downto 0);
        sel: in std_logic_vector(log2c(WIDTH)-1 downto 0);
10      y: out std_logic
    );
  end mux1;

    architecture loop_linear_arch of mux1 is
15 begin
    process(a,sel)
    begin
        y <='0';
        for i in 0 to (WIDTH-1) loop
20          if i= to_integer(unsigned(sel)) then
                y <= a(i);
            end if;
```

```
            end loop;
         end process;
25 end loop_linear_arch;
```

The code is based on an abstract behavioral description. It infers a cascading priority routing networks, similar to that of the previous priority encoder. Although the code is simple and clear, it leads to bulky and inefficient hardware implementation. A better alternative is shown in Chapter 15.

14.7.4 Examples of a loop body with variables

Variables can be used in sequential statements and thus can be used in the body of a for loop statement. Since a variable assignment takes effect immediately, it is useful when an object inside the loop needs to be updated in each iteration. Two examples are shown below.

Reduced-xor circuit with variables The reduced-xor circuit can be described using a variable. The VHDL code is in Listing 14.26.

Listing 14.26 Parameterized reduced-xor circuit using a variable

```
architecture loop_linear_var_arch of reduced_xor is
begin
    process(a)
        variable tmp: std_logic;
5   begin
        tmp := a(0);
        for i in 1 to (WIDTH-1) loop
            tmp := a(i) xor tmp;
        end loop;
10        y <= tmp;
    end process;
end loop_linear_var_arch;
```

The code is more like a program in a traditional programming language. Although it is more abstract and descriptive, deriving the conceptual implementation for this code requires more effort. The key is to convert the variables into constructs that can be mapped into a hardware entity. We first unroll the loop and then rename the variable to avoid self-reference.

Assume that WIDTH is 4. The flattened code is

```
tmp := a(0);
tmp := a(1) xor tmp;
tmp := a(2) xor tmp;
tmp := a(3) xor tmp;
y <= tmp;
```

To avoid self-reference, the variable is given a new name in each statement and the new names are propagated through subsequent statements:

```
tmp0 := a(0);
tmp1 := a(1) xor tmp0;
tmp2 := a(2) xor tmp1;
tmp3 := a(3) xor tmp2;
y <= tmp3;
```

We can now interpret that each variable is a connection wire and derive the conceptual diagram accordingly.

For comparison purposes, we also unroll the code of Listing 14.1, which uses signals in the body of the for loop statement. The code becomes

```
tmp(0) <= a(0);
tmp(1) <= a(1) xor tmp(0);
tmp(2) <= a(2) xor tmp(1);
tmp(3) <= a(3) xor tmp(2);
y <= tmp3;
```

Note that the appearances of the two flattened codes are very similar. The tmp variable can be thought of as shorthand to replace an array of signals, which are needed to express the intermediate values.

Population counter   The population counter counts the number of 1's from the elements of an array input signal. A fixed-size circuit was discussed in Section 7.5.5. One way to derive the parameterized version is to use a for loop statement and a variable to keep track of the occurrences of 1's. The abstract VHDL code is shown in Listing 14.27.

Listing 14.27   Parameterized population counter

```
library ieee;
use ieee.std_logic_1164.all;
use ieee.numeric_std.all;
use work.util_pkg.all;
5 entity popu_count is
    generic(WIDTH: natural);
    port(
        a: in std_logic_vector(WIDTH-1 downto 0);
        count: out std_logic_vector(log2c(WIDTH)-1 downto 0)
10   );
  end popu_count;

  architecture loop_linear_arch of popu_count is
  begin
15   process(a)
        variable sum: unsigned(log2c(WIDTH)-1 downto 0);
      begin
        sum := (others=>'0');
        for i in 0 to (WIDTH-1) loop
20           if a(i)= '1' then
                sum := sum + 1;
            end if;
        end loop;
        count <= std_logic_vector(sum);
25   end process;
  end loop_linear_arch;
```

Deriving the conceptual implementation of this code involves more work. Assume that WIDTH is 3. The loop can be unrolled into three iterations:

```
sum := 0;
if a(0)= '1' then
    sum := sum + 1;
```

```
end if ;
if a(1)= '1' then
    sum := sum + 1;
end if ;
if a(2)= '1' then
    sum := sum + 1;
end if ;
count <= sum ;
```

Unlike the previous reduced-xor example, the following simple renaming will not work properly:

```
sum0 := 0;
if a(0)= '1' then
    sum1 := sum0 + 1;
end if ;
if a(1)= '1' then
    sum2 := sum1 + 1;
end if ;
if a(2)= '1' then
    sum3 := sum2 + 1;
end if ;
count <= sum3 ;
```

To correctly rename the signals, we must include the else branch, which implies that the sum variable remains unchanged. The revised, unrolled code is

```
sum := 0;
--- 1st stage
if a(0)= '1' then
    sum := sum + 1;
else
    sum := sum;
end if ;
--- 2nd stage
if a(1)= '1' then
    sum := sum + 1;
else
    sum := sum;
end if ;
--- 3rd stage
if a(2)= '1' then
    sum := sum + 1;
else
    sum := sum;
end if ;
count <= sum ;
```

We can easily rename the variables now and the code segment becomes

```
sum0 := 0;
--- 1st stage
if a(0)= '1' then
    sum1 := sum0 + 1;
else
    sum1 := sum0;
```

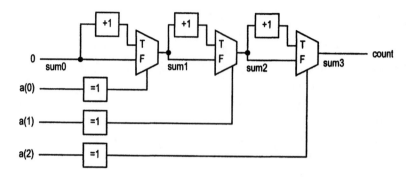

Figure 14.8   Block diagram of population counter.

```
end if;
-- 2nd stage
if a(1)= '1' then
    sum2 := sum1 + 1;
else
    sum2 := sum1;
end if;
-- 3rd stage
if a(2)= '1' then
    sum3 := sum2 + 1;
else
    sum3 := sum2;
end if;
count <= sum3;
```

The basic building block of each stage is now clear. The corresponding diagram of the flattened and renamed code is shown in Figure 14.8. Note that the diagram also exhibits a cascading-chain structure.

14.7.5   Comparison of the for generate and for loop statements

Both the for generate and for loop statements are used to describe replicated structures. The major difference is the type of statements in their loop bodies. The for generate statement can only use concurrent statements, and the for loop statement can only use sequential statements.

Because there is a clear mapping between concurrent statements and hardware parts, the circuit involved in a stage can be easily visualized. When a for generate statement is used, we frequently start a design with a conceptual diagram of a few stages. The diagram is used to identify the basic building block and connection pattern, and then the code of the loop body is derived accordingly. We sometimes create an entity for the basic building block and then describe the replicated structure by instantiating the component instances.

On the other hand, because of the sequential semantics and the existence of variables, the body of the for loop statement can be more general and versatile. While this allows us to develop more abstract code, it may also lead to an unnecessarily complex implementation or even an unsynthesizable description. The synthesis issue is discussed in Section 14.9.

14.8 EXIT AND NEXT STATEMENTS

The exit and next statements are sequential statements used inside a for loop statement to alter the regular iterations of the loop. The exit statement stops execution of a for loop statement and exits the loop immediately. The remaining iterations will be skipped. The next statement stops execution of the current iteration and jumps to the beginning of the next iteration. The remaining statements of the iteration will be skipped. While the two statements can sometimes be useful for modeling, they are difficult to synthesize. Many synthesis software packages do not support these two statements. The following subsections discuss the use and conceptual implementation of the two statements and the alternative coding style.

14.8.1 Syntax of the exit statement

The syntax of the exit statement is

```
exit when boolean_expr;
```

The `boolean_expr` term is a Boolean expression indicating the exiting condition. When it is evaluated as `true`, the exit takes effect and the execution skips the remaining iterations of the loop.

Note that the **when** `boolean_expr` portion is optional. It is not needed if the exit statement is associated with a condition of an if statement, as in the following code segment:

```
if (boolean_expr) then
    . . .
    exit;
else
    . . .
```

14.8.2 Examples of the exit statement

reduced-and circuit The reduced-and circuit was discussed in Section 14.4.2. We use a different approach in this example. One property of the and operation is that $x \cdot 0 = 0$. Thus, the reduced-and operation can return '0' as soon as the first '0' element of the input array is found. The VHDL code in Listing 14.28 is based on this observation. The code uses a for loop statement to check the values of the array's elements and uses the exit statement to terminate the loop after the first '0' element is found.

Listing 14.28 Parameterized reduced-and circuit using an exit statement

```
   architecture exit_arch of reduced_and is
   begin
      process(a)
         variable tmp: std_logic;
5     begin
         tmp := '1';    -- default output
         for i in 0 to (WIDTH-1) loop
            if a(i)='0' then
               tmp := '0';
10             exit;
            end if;
```

```
              end loop;
              y <= tmp;
          end process;
      15 end exit_arch;
```

If this code is developed as a routine in a traditional programming language, it is an effective approach since the execution does not need to go through all iterations of the loop and can cut one half of the execution time in average. However, in synthesis, we cannot dynamically create the circuit according to the input pattern. Instead, the synthesized circuit must accommodate all possible input combinations. Using the exit statement actually introduces additional hardware overhead and complicates the synthesis process. This is discussed in more detail in the next subsection.

Leading-zero counting circuit   The leading-zero counting circuit is a combinational circuit that counts the number of leading 0's in front of an input signal. One possible implementation is to use a for loop statement to keep track of the number of consecutive 0's and use an exit statement to terminate the loop when the first '1' is encountered. The VHDL code is shown in Listing 14.29.

Listing 14.29   Parameterized leading-zero counting circuit using an exit statement

```
    library ieee;
    use ieee.std_logic_1164.all;
    use ieee.numeric_std.all;
    use work.util_pkg.all;
  5 entity leading0_count is
        generic(WIDTH: natural);
        port(
            a: in std_logic_vector(WIDTH-1 downto 0);
            zeros: out std_logic_vector(log2c(WIDTH)-1 downto 0)
 10     );
    end leading0_count;

    architecture exit_arch of leading0_count is
    begin
 15     process(a)
            variable sum: unsigned(log2c(WIDTH)-1 downto 0);
        begin
            sum := (others=>'0');     -- initial value
            for i in WIDTH-1 downto 0 loop
 20             if a(i)='1' then
                    exit;
                else
                    sum := sum + 1;
                end if;
 25         end loop;
            zeros <= std_logic_vector(sum);
        end process;
    end exit_arch;
```

This code infers multiple adders and is not an efficient design. It is only for demonstration of the use of the exit statement.

14.8.3 Conceptual implementation of the exit statement

Since the hardware cannot expand or shrink dynamically to accommodate the input pattern, the exit statement cannot be implemented directly in hardware. However, we can emulate the effect of the exit statement using a special circuit to bypass some iterations.

The "bypassing" can best be explained by an example. Consider the VHDL code of the leading-zero counting circuit in Listing 14.29. The exit statement specifies that the remaining iterations of the loop should be skipped if the condition a(i)='1' is met. We can achieve the same effect using an array of flags, which indicate whether the execution of the corresponding stages should be bypassed. The revised VHDL code is shown in Listing 14.30.

Listing 14.30 Parameterized leading-zero counting circuit using a bypass flag

```
   architecture bypass_arch of leading0_count is
      signal bypass: std_logic_vector(WIDTH downto 0);
   begin
      process(a,bypass)
5         variable sum: unsigned(log2c(WIDTH)-1 downto 0);
      begin
         -- initial value
         sum := (others=>'0');
         bypass(WIDTH) <= '0';
10        -- bypass flags
         for i in WIDTH-1 downto 0 loop
            if a(i)='1' then
               bypass(i) <= '1';
            else
15             bypass(i) <= bypass(i+1);
            end if;
         end loop;
         -- counting 1's
         for i in WIDTH-1 downto 0 loop
20          if bypass(i)='0' then
               if a(i)='0' then
                  sum := sum + 1;
               end if;
            end if;
25       end loop;
         zeros <= std_logic_vector(sum);
      end process;
   end bypass_arch;
```

The flags are implemented by the bypass signal. The leftmost element, bypass(WIDTH), is set to '0'. An element of the bypass signal is set to '1' when the condition a(i)='1' is met, and the condition will be propagated to the subsequent elements. For example, if bypass(3) is set to '1', bypass(2), bypass(1) and bypass(0) will be set to '1' as well.

The bypass signal is then used to control the increment operation of each stage. In the ith stage, the increment operation is performed only if the corresponding bypass(i) is not set (i.e., is '0'). Once an element of the bypass signal is set to '1', all the subsequent increment operations will be skipped and the values of the sum variable will remain unchanged in the remaining iterations. This essentially achieves the effect of the exit statement without actually using it inside the for loop statement.

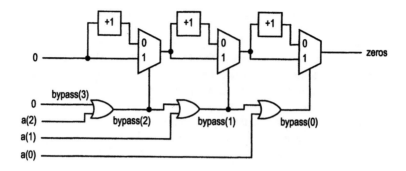

Figure 14.9   Block diagram of a leading-zero counting circuit.

Note that when the bypass(i) signal is '0', the a(i) must be '0'. Thus, checking the a(i)='0' condition is not needed in the second for loop statement, and it can be simplified to

```
for i in WIDTH-1 downto 0 loop
    if bypass(i)='0' then
        sum := sum + 1;
    end if;
end loop;
```

Since there is no exit statement in the revised code, we can derive the conceptual diagram by unrolling the two for loops. Assume that WIDTH is 3. The diagram is shown in Figure 14.9. The upper loop can be simplified to a chain of or gates, as shown at the bottom of the diagram. The bottom loop is similar to the population counter of Figure 14.8, as shown at the top of the diagram. The bypass(i) signal specifies whether to skip the incrementing operation by routing either the original or the incremented input to the next stage. Note that once a bypass value is set to '1' in a stage, the '1' will be propagated through the descending stages, and thus all subsequent incrementing operations are skipped.

The example shows that the use of the exit statement actually introduces additional "bypass overhead" to the original circuit. This overhead makes synthesis more difficult and may increase the size of the final implementation. Some synthesis software packages may not be able to handle a for loop with the exit statement.

14.8.4   Next statement

The syntax of the next statement is

```
next when boolean_expr;
```

When the boolean_exp term is evaluated as true, the next statement takes effect and the execution skips the remaining statements of the iteration. Like the exit statement, the **when** boolean_expr portion is not needed if the next statement is used with an if statement.

The VHDL code of the population counter of Section 14.7.4 can be revised by using the next statement, as shown in Listing 14.31. When a(i) is '0', the next statement skips the remaining statements of the loop (i.e., the sum := sum + 1 statement).

Listing 14.31   Parameterized population counter using a next statement

```
architecture next_arch of popu_count is
begin
   process(a)
      variable sum: unsigned(log2c(WIDTH)-1 downto 0);
   begin
      sum := (others=>'0');
      for i in 0 to (WIDTH-1) loop
         next when a(i)='0';
         sum := sum + 1;
      end loop;
      count <= std_logic_vector(sum);
   end process;
end next_arch;
```

The next statement is somewhat similar to the "opposite" of an if statement with only a then branch. The former skips some statements when the corresponding condition is met, while the latter executes some statements when the corresponding condition is met. Based on this observation, we can convert the next statement to a modified if statement. For example, consider the following VHDL segment:

```
for ... loop
   sequential statement 1;
   next when boolean_exp;
   sequential statement 2;
end loop;
```

It can be rewritten as

```
for ... loop
   sequential statement 1;
   if (not boolean_exp) then
      sequential statement 2;
   end if
end loop;
```

The if statement is preferred because it is more descriptive and modular. The revised code also shows that the implementation of the next statement should be similar to that of an if statement without an else branch, as in Listing 14.27.

14.9   SYNTHESIS OF ITERATIVE STRUCTURE

VHDL provides a variety of mechanisms to describe the iterative structure. From the synthesis's point of view, the key is to identify the circuit involved in a stage. Once it is done, the circuit can be replicated to a specific number set by the width parameters. The synthesis software can process the flattened description as a regular circuit.

When deriving code with for generate statements, we normally first draw a sketch diagram of the hardware and then derive the VHDL description accordingly. This is partially due to the semantics of the concurrent statements, which prevents us from thinking in terms of sequential programming constructs. Ideally, we should apply the same approach when using for loop statements. Unfortunately, since sequential statements are more abstract and closer to the statements of traditional programming languages, it is easy to use for loop statements to write abstract, sequential codes. This frequently leads to bulky, unnecessarily complex implementation. Thus, when a for loop statement is used, we should be conscious of the implications on the underlying hardware.

The for loop and for generate statements provide an easy mechanism to describe the iterative structure. Unfortunately, the simple description does not always lead to efficient implementation. The examples of this chapter utilize a single-level for generate or for loop statement, which translates into a one-dimensional structure. Except for special cases, such as the binary decoder, the one-dimensional structure leads to a cascading-chain type of circuit. This kind of topology is difficult to handle during placement and routing and may introduce a large propagation delay, especially when the chain is very long. A more effective two-dimensional tree- or mesh-shaped structure is more desirable. This issue is discussed in the Chapter 15.

14.10   SYNTHESIS GUIDELINES

- Use a generic to specify the width parameter.

- If an unconstrained array is used for parameterized design, take the range and direction of the array into consideration.

- Use the if generate statement for small feature variation.

- The for loop statement should be considered as a construct to describe a circuit with a replicated structure.

- A single-level for generate or for loop statement normally leads to a one-dimensional cascading-chain structure.

14.11   BIBLIOGRAPHIC NOTES

While many texts cover the VHDL generate and loop constructs, few literatures provide in-depth discussions of parameterized design. One place to find good examples is in the package body of the IEEE numerid_std package. The source file can normally be found in the directory where the IEEE library resides. The functions and overloaded operators of the package are defined over the unsigned and signed data types with no explicit range specification, and the needed parameters are derived from attributes. Because the codes include comprehensive error-checking, they are quite complex. Another source is the VHDL code of "Library of Parameterized Modules" (LPM). It is an early, not-so-successful attempt to develop parameterized device-independent VHDL modules. The VHDL package should still be available via internet search.

Problems

14.1   Consider a 1-bit incrementor cell that adds 1 to the input operand. It has two 1-bit input signals, a and cin, which represent the input operand and carry-in respectively, and two 1-bit output signals, s and cout, which represent the sum and carry-out respectively.

   (a) Derive the function table for this circuit.
   (b) Derive the VHDL code for this circuit using only simple signal assignment statements and logical operators.

(c) Derive the block diagram of a 4-bit incrementor using four incrementor cells.

14.2 Follow the block diagram of Problem 14.1(c) to design a parameterized incrementor in which the width of the input operand is specified by a generic. Derive the VHDL code using the for generate statement. Use a simple signal assignment statement in the loop body, and no VHDL arithmetic operator is allowed.

14.3 Repeat Problem 14.2, but create the 1-bit incrementor cell as a component and use component instantiation in the loop body.

14.4 Repeat Problem 14.2, but use conditional generate statements for the boundary cells.

14.5 Repeat Problem 14.2, but use the for loop statement.

14.6 Repeat Problem 14.2, but apply the clever-use-of-array techniques discussed in Section 14.4. No for generate or for loop statement is allowed.

14.7 Repeat Problem 14.2, but use no generic. Declare the data type of the input port as std_logic_vector with no explicit range specification. Make sure that the code can work with different formats of specification when the component is instantiated.

14.8 Follow the technique of the reduced-and circuit of Listing 14.4.2, and derive a parameterized VHDL code for the reduced-or circuit.

14.9 For the memory controller FSM circuit in Section 10.7.2, the output signal can be unbuffered or buffered. The buffered output uses the look-ahead output buffer scheme. Derive a VHDL code that includes both schemes and use the BUF generic as a feature parameter to specify which buffer scheme to use.

14.10 Consider the priority encoder code of Listing 14.24. Rewrite the code using a for generate statement.

14.11 Consider the population counter code of Listing 14.27. Rewrite the code using a for generate statement.

14.12 Consider the reduced-and code of Listing 14.28. Follow the conceptual implementation procedure discussed in Section 14.8.3 to replace the exit statement with flag signals.
   (a) Derive the VHDL code.
   (b) Draw the conceptual diagram.
   (c) Prove that the conceptual diagram actually performs the reduced-and operation.

CHAPTER 15

PARAMETERIZED DESIGN: PRACTICE

After learning the basic language constructs for the parameterized design, we study more sophisticated circuit examples in this chapter. In addition to parameterization, the emphasis is on the efficiency and performance of the circuits. The main focus is on the derivation of efficient parameterized RT-level modules that can be used as building blocks of larger systems.

15.1 INTRODUCTION

Parameterization is not directly related to the efficiency of a digital circuit. However, it frequently leads to inefficient design for several reasons. First, a parameterized description relies on a small set of language constructs, mainly the for generate and for loop statements. We have less freedom to describe the intended circuit structure. Second, because the for loop statement is similar to the loop constructs found in the traditional programming languages, we tend to develop behavioral descriptions and become less aware of the underlying hardware organization.

We constructed several parameterized modules in Chapter 14. Because of the regular, repetitive nature of these circuits, they are described by a for loop or for generate statement. While the code is simple and easy to understand, a single for loop or for generate statement generally describes a one-dimensional cascading structure. This kind of structure introduces a long propagation delay and thus penalizes the performance. Since synthesis software can only perform certain local transformation, it is not able to restructure and optimize the entire chain. This is particularly problematic for parameterized description since we may instantiate a module with a large parameter.

To derive efficient parameterized modules, we must pay particular attention to the topology of the underlying structure, as discussed in Section 7.4. The description should help the synthesis process to derive a more effective implementation. In this chapter, we discuss the data types used to describe scalable two-dimensional structure, illustrate the design and description of various RT-level components, and study several more difficult application examples.

15.2   DATA TYPES FOR TWO-DIMENSIONAL SIGNALS

One-dimensional arrays, which include std_logic_vector of the std_logic_1164 package, and unsigned and signed of the numeric_std package, are the primary data types used in our codes. They are natural matches to represent a multiple-bit signal. To enhance portability and improve readability, we prefer to use these predefined data types and avoid the multidimensional array in general. In a parameterized design, a circuit may exhibit a scalable two-dimensional structure and require two-dimensional data types to represent the internal signals or I/O ports. While the data types of VHDL are flexible and versatile, no two-dimensional array data type is defined in the std_logic_1164 or numeric_std package. Thus, we must create a user-defined data type for two-dimensional signals.

Because of the lack of a common synthesis standard, software support varies and multidimensional array data types are not accepted by all software tools. This section discusses use of genuine VHDL two-dimensional data types, and two work-arounds, which are the *array-of-arrays* and *emulated two-dimensional array* data types. We can choose the scheme that is supported by the software in hand.

15.2.1   Genuine two-dimensional data type

The array data type in VHDL is defined as a collection of elements of the same data types. Its definition is very general. The simplified syntax is

```
type data_type_name is array (range_1, range_2, ...)
    of element_data_type;
```

The range terms inside the parentheses specify the boundaries of the array, and the number of range terms corresponds to the dimensions of the array. The data type is known as a *constrained array* if the ranges are fixed and is known as an *unconstrained array* otherwise.

Constrained array   The definition and use of a user-defined two-dimensional data type can best be explained by an example. Consider a 4-by-6 SRAM (i.e., an SRAM that contains four words and each word is 6 bits wide). The structure of the SRAM is a natural match for a two-dimensional data type:

```
constant ROW natural:=4;
constant COL natural:=6;
type sram_4_by_6_type is
    array (ROW-1 downto 0, COL-1 downto 0) of std_logic;
```

This is a constrained array since its ranges are fixed.

We can assign a two-dimensional constant to a signal with this data type. As in a one-dimensional array, both positional and named association can be used for the aggregate. Some examples are

```
signal t1, t2, t3: sram_4_by_6_type;
. . .
— positional association
t1 <= ("000000",
       "010101",
       "000111",
       "111111");
— "101010" to all rows
t2 <= (others=>"101010");
— all 0's
t3 <= (others=>(others=>'0'));
```

We can use the two indexes, in the form of (i,j), to access a particular element of the array, as in the following examples:

```
signal t4: sram_4_by_6_type;
signal e1, e2, e3: std_logic;
. . .
t4(0,0) <= '1';
t4(1,2) <= e1 and e2;
e3 <= t4(3,5);
```

In an SRAM, we frequently want to access a word, which corresponds to a row in the two-dimensional data type. Unfortunately, there is no build-in VHDL mechanism to specify a particular dimension. The work-around is to use a for loop or for generate statement to iterate through the individual elements. An example of using the for loop statement is

```
signal t5: sram_4_by_6_type;
signal v1: std_logic_vector(COL-1 downto 0);
. . .
process(...)
begin
   for i in COL-1 downto 0 loop
      v1(i) <= t5(1,i);
   end loop;
. . .
```

Unconstrained array   An unconstrained array does not specify the boundary of the range when the data type is defined. This information is provided later when the data type is used. An unconstrained two-dimensional array of element type of std_logic can be defined as

```
type std_logic_2d is
   array (natural range <>, natural range <>) of std_logic;
```

This is more effective since it can accommodate two-dimensional arrays of various sizes. For example, assume that three different sizes are used in a design. If the constrained array is used, we need three data types:

```
type array_4_by_6_type is
   array (3 downto 0, 5 downto 0) of std_logic;
type array_16_by_32_type is
   array (15 downto 0, 31 downto 0) of std_logic;
type array_8_by_2_type is
   array (7 downto 0, 1 downto 0) of std_logic;
```

```
signal s1: array_4_by_6_type;
signal s2, s3: array_16_by_32_type;
signal s4, s5: array_8_by_2_type;
```

On the other hand, the code will be much simpler if an unconstrained array is used:

```
type std_logic_2d is
   array (natural range <>, natural range <>) of std_logic;
signal s1: std_logic_2d(3 downto 0, 5 downto 0);
signal s2, s3: std_logic_2(15 downto 0, 31 downto 0);
signal s4, s5: std_logic_2(7 downto 0, 1 downto 0);
```

If we include the std_logic_2d data type in a package, this data type can be used in VHDL code after the package is invoked. In fact, the std_logic_vector, unsigned and signed data types are defined as an unconstrained one-dimensional array. The utility package discussed in Section 13.5.3 actually follows this practice and includes the two-dimensional data type in the package declaration. The definition of the std_logic_2d data type is very general, and its dimension can be specified as generic parameters in the port declaration of an entity and in the signal declaration of an architecture body. A simple example is

```
. . .
use work.util_pkg.all;
entity . . .
   generic(
      ROW: natural;
      COL: natural
   );
   port(
      p1, p2: in
         std_logic_2d(ROW-1 downto 0, COL-1 downto 0);
      . . .
   );
architecture . . .
   signal sig1, sig2:
      std_logic_2d(ROW-1 downto 0, COL-1 downto 0)
   . . .
```

While this is an elegant scheme, the std_logic_2d data type may not be accepted by some synthesis software because of the lack of support for multidimensional arrays.

15.2.2 Array-of-arrays data type

The array in VHDL is very flexible, and the data type of the element of an array can also be an array. Thus, a two-dimensional structure can be defined as a one-dimensional array whose element's data type is also a one-dimensional array. We call this an *array-of-arrays* data type. For example, we can replace the previous sram_4_by_6_type data type with an array-of-arrays data type:

```
constant ROW natural:=4;
constant COL natural:=6;
type sram_aoa_46_type is array (ROW-1 downto 0) of
      std_logic_vector(COL-1 downto 0);
```

This data type is a one-dimensional array with four elements whose data type is a six-element one-dimensional array (i.e., std_logic_vector(5 downto 0)).

The structures of sram_4_by_6_type and sram_aoa_46_type are very similar. In fact, the constant assignment for the two data types are identical:

```
signal t1, t2, t3: sram_aoa_46_type;
. . .
—— positional association
t1 <= ("000000",
       "010101",
       "000111",
       "111111");
—— "101010" to all rows
t2 <= (others=>"101010");
—— all 0's
t3 <= (others=>(others=>'0'));
```

We can also access a single bit in an array-of-arrays data type. Instead of (i,j), the index is in the form of (i)(j). The example in Section 15.2.1 becomes

```
signal t4: sram_aoa_46_type;
signal e1, e2, e3: std_logic;
. . .
t4(0)(0) <= '1';
t4(1)(2) <= e1 and e2;
e3 <= t4(3)(5);
```

Since a row of an array-of-arrays data type is an element of a one-dimensional array, to access a row is much easier. For example, to retrieve a word from the previous SRAM, we can just use the first index:

```
signal t5: sram_aoa_46_type;
signal v1: std_logic_vector(COL-1 downto 0);
. . .
v1 <= t5(1);
. . .
```

Thus, for this particular application, an array-of-arrays data type is more natural than a genuine two-dimensional data type.

The major limitation of the array-of-arrays data type is that the data type of its element must be a constrained array. At best, we can only define a data type as

```
type std_logic_aoa_N is
    array (natural range <>) of std_logic_vector(N-1 downto 0);
```

In other words, only the first dimension (i.e., the number of rows) can be left unconstrained.

If an array-of-arrays data type is defined and used inside the architecture body, it is still possible to pass the two-dimensional parameters via generics. A simple example is

```
. . .
entity . . .
   generic(
      ROW: natural;
      COL: natural
   );
. . .
architecture . . .
   type aoa_RC_type is
```

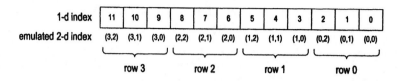

Figure 15.1   Emulation of a two-dimensional 4-by-3 array with a one-dimensional array.

```
array(ROW-1 downto 0) of std_logic_vector(COL-1 downto 0);
signal sig1, sig2: aoa_RC_type;
    . . .
```

However, if an array-of-arrays data type is used for a port declaration, only the first dimension can be parameterized. Since the size of the second dimension must be fixed in port declaration, the array-of-arrays data type is not as flexible as the std_logic_2d data type. This is the most severe constraint of using an array-of-arrays data type in a parameterized design.

Because an array-of-arrays data type is a one-dimensional array, more synthesis software accepts this form.

15.2.3   Emulated two-dimensional array

Emulated two-dimensional array is a scheme that imitates a two-dimensional structure using a one-dimensional array. In this scheme, we introduce no new data type but cleverly interpret a one-dimensional array as a two-dimensional structure. For example, consider a 4-by-3 two-dimensional array. We can enumerate the four rows in a single list, as shown in Figure 15.1. The relationship between the one-dimensional index, n, and the two-dimensional indexes, i and j, is characterized by a simple equation:

$$n = i * 3 + j$$

Since the regular one-dimensional array data type is used to describe a two-dimensional structure, we call it an *emulated two-dimensional array*.

Let us consider the 4-by-6 SRAM example again. Although no new data type is needed, it will be handy to define a function to calculate the indexes. The code is

```
constant ROW natural:=4;
constant COL natural:=6;
--- data type is std_logic_vector(ROW*COL-1 downto 0);
function ix(r,c: natural) return natural is
begin
    return (r*COL + c);
end ix;
```

To access a single bit in the emulated array, we can invoke the ix function to calculate the corresponding position in the one-dimensional array. We can replace the index (i,j) used in a genuine two-dimensional array with the indexing function, ix(i,j). The indexing example in Section 15.2.1 becomes

```
signal t4: std_logic_vector(ROW*COL-1 downto 0);
signal e1, e2, e3: std_logic;
    . . .
```

```
t4(ix(0,0)) <= '1';
t4(ix(1,2)) <= e1 and e2;
e3 <= t4(ix(3,5));
```

A row in the SRAM corresponds to a slice in the one-dimensional array. We can access the entire row after determining the upper and lower boundaries of the row. For the ith row, the index of the upper boundary is `ix(i,COL-1)` and the lower boundary is `ix(i,0)`, and thus we can use the range `ix(i,COL-1)` **downto** `ix(i,0)` to access the row. The example in Section 15.2.1 retrieves the first word of the SRAM. It can be modified as follows

```
signal t5: std_logic_vector(ROW*COL-1 downto 0);
signal v1: std_logic_vector(COL-1 downto 0);
. . .
v1 <= t5(ix(1,COL-1) downto ix(1,0));
. . .
```

Assigning a constant to the emulated array is just assigning a constant to a regular one-dimensional array. We can use the concatenation operator to make the code clearer and consistent with other schemes. The constant expression in Section 15.2.1 can be modified as follows:

```
signal t1, t2, t3: std_logic_vector(ROW*COL-1 downto 0);
. . .
t1 <= "000000" &
      "010101" &
      "000111" &
      "111111";
— "101010" to all rows
t2 <= "101010" & "101010" & "101010" & "101010";
— all 0's
t3 <= (others=>'0');
```

Because the emulated array uses a predefined one-dimensional array data type, the parameters for two dimensions can be passed via generics for the port declaration and signal declaration. An example is shown below.

```
. . .
entity . . .
   generic(
      ROW: natural;
      COL: natural
   );
   port(
      p1, p2: in std_logic_vector(ROW*COL-1 downto 0);
      . . .);
architecture . . .
   function ix(r,c: natural) return natural is
   begin
      return (r*COL + c);
   end ix;
   signal sig1, sig2: in std_logic_vector(ROW*COL-1 downto 0);
   . . .
```

The emulated array involves numerous calculations to map a two-dimensional index into a one-dimensional index. However, these calculations are static and thus can be determined when the VHDL code is elaborated. No physical circuit will be inferred for this purpose.

15.2.4 Example

In Chapter 14, we presented a parameterized multiplexer in Listing 14.25. In this design, the number of input ports is specified by a generic but the number of bits per port is fixed (i.e., 1 bit). A more general description should add an additional parameter to specify the number of bits of a port as well. Let the number of input ports be P and the number of bits per port be B. The a input signal now represents a two-dimensional P*B-bit signal. The following codes illustrate how to use the three two-dimensional data types to implement the new multiplexer.

Implementation with a genuine two-dimensional array   The first description uses the genuine two-dimensional `std_logic_2d` data type. The VHDL code is shown in Listing 15.1. The `util_pkg` package is needed for the `std_logic_2d` data type and the `log2c` function.

Listing 15.1   Parameterized two-dimensional multiplexer using a genuine array

```
   library ieee;
   use ieee.std_logic_1164.all;
   use ieee.numeric_std.all;
   use work.util_pkg.all;
 5 entity mux2d is
      generic(
         P: natural; — number of input ports
         B: natural  — number of bits per port
      );
10    port(
         a: in std_logic_2d(P-1 downto 0, B-1 downto 0);
         sel: in std_logic_vector(log2c(P)-1 downto 0);
         y: out std_logic_vector(B-1 downto 0)
      );
15 end mux2d;

   architecture two_d_arch of mux2d is
   begin
      process(a,sel)
20    begin
         y <=(others=>'0');
         for i in 0 to (P-1) loop
            if i= to_integer(unsigned(sel)) then
               for j in 0 to (B-1) loop — B—bits of the port
25                y(j) <= a(i,j);
               end loop;
            end if;
         end loop;
      end process;
30 end two_d_arch;
```

The code is basically patterned after the one-dimensional multiplexer code in Listing 14.25. An extra inner for loop statement is added to route B bits from an input port to the output.

Implementation with an emulated array   The second description uses an emulated array, and the VHDL code is shown in Listing 15.2. The code also includes the util_pkg package since the log2c function is needed. It follows the description in Listing 15.1 but with several simple modifications:

- Use the regular std_logic_vector data type.
- Define the ix function.
- Use a(ix(i,j)) to replace a(i,j).

Listing 15.2   Parameterized two-dimensional multiplexer using an emulated array

```
library ieee;
use ieee.std_logic_1164.all;
use ieee.numeric_std.all;
use work.util_pkg.all;
5 entity mux_emu_2d is
     generic(
          P: natural; — number of input ports
          B: natural  — number of bits per port
     );
10    port(
          a: in std_logic_vector(P*B-1 downto 0);
          sel: in std_logic_vector(log2c(P)-1 downto 0);
          y: out std_logic_vector(B-1 downto 0)
     );
15 end mux_emu_2d;

   architecture emu_2d_arch of mux_emu_2d is
     function ix(r,c: natural) return natural is
          begin
20            return (r*B + c);
          end ix;
   begin
     process(a,sel)
     begin
25      y <=(others=>'0');
        for r in 0 to (P-1) loop
          if r= to_integer(unsigned(sel)) then
            for c in 0 to (B-1) loop — B−bits of the port
               y(c) <= a(ix(r,c));
30          end loop;
          end if;
        end loop;
     end process;
   end emu_2d_arch;
```

Since we can specify a slice of array in the emulated array scheme, the inner loop

```
for c in 0 to (B-1) loop
   y(c) <= a(ix(r,c));
end loop;
```

can be replaced by

```
y <= a(ix(r,B-1) downto ix(r,0));
```

Implementation with an array of arrays   Because the element data type of an array of arrays must be a constrained array, the array-of-arrays data type is not general enough to be used in the port declaration of the two-dimensional multiplexer. However, this data type can still be used inside the architecture body. We can use the previous emulated array in the entity declaration and then convert the input into the array-of-arrays data type in the architecture body. The VHDL code of the architecture body is shown in Listing 15.3.

Listing 15.3   Parameterized two-dimensional multiplexer using an array of arrays

```
architecture a_of_a_arch of mux_emu_2d is
   type std_aoa_type is
      array(P-1 downto 0) of std_logic_vector(B-1 downto 0);
   signal aa: std_aoa_type;
5 begin
   -- convert to array-of-arrays data type
   process(a)
   begin
      for r in 0 to (P-1) loop
10          for c in 0 to (B-1) loop
               aa(r)(c) <= a(r*B+c);
            end loop;
         end loop;
   end process;
15  -- mux
   process(aa,sel)
   begin
      y <=(others=>'0');
      for i in 0 to (P-1) loop
20          if i= to_integer(unsigned(sel)) then
               y <= aa(i);
            end if;
         end loop;
   end process;
25 end a_of_a_arch;
```

The first process is for type conversion. It is static, and no physical circuit should be inferred. The second process describes the actual multiplexer. The code is identical to one-dimensional multiplexer code in Listing 14.25. The only difference is that the element data type of the aa signal is `std_logic_vector(B-1 downto 0)`, and the element data type of the a signal of the one-dimensional multiplexer is `std_logic`. From this point of view, the array-of-arrays data type is the most concise representation of the underlying circuit structure.

15.2.5 Summary

Ideally, we wish to select a two-dimensional representation that can effectively describe the underlying circuit structure and be universally accepted by synthesis software. It cannot be easily achieved due to the intrinsic limitation of VHDL and the variation on synthesis software support. However, since these representations describe the same two-dimensional structure, conversion between the representations is fairly straightforward. We should select a scheme that works with the synthesis software in hand and properly document the use of these data types in the VHDL code so that they can be easily modified when needed.

In the remainder of this chapter, we use the `std_logic_2d` data type in general and use the array-of-arrays data type if it closely matches the underlying structure.

15.3 COMMONLY USED INTERMEDIATE-SIZED RT-LEVEL COMPONENTS

We discussed the level of abstraction in Section 1.4. The focus of this book is on the RT level, in which the main parts are intermediate-sized components. Most synthesis software contains predesigned modules for relational operators and addition and subtraction operators, and these modules are inferred and instantiated during synthesis. There are many other intermediate-sized RT-level components that are frequently encountered in a large design, including reduction circuit, decoder, encoder, multiplexer, barrel shifter and multiplier. Since these components are common building parts that are needed in many applications, they are good candidates to be parameterized.

As discussed in Section 7.4, the efficiency of a circuit relies heavily on its basic structure and underlying topology. A good description helps the synthesis process to derive a more effective implementation. To describe a parameterized multidimensional circuit is more involved. The key to designing this type of circuit is to identify a general pattern and then use for loop or for generate statements to describe the desired connection pattern. The following procedure helps us to achieve this goal:

- Draw a small-scale diagram with basic building blocks.
- Derive a proper index for the connection signals in each stage.
- Identify the general relationship between the signals in successive stages.
- Identify the connection patterns between boundary stages and I/O ports.
- Derive the VHDL code accordingly.

The remaining section illustrates the design and derivation of several RT-level components.

15.3.1 Reduced-xor circuit

In Chapter 14, we constructed a parameterized reduced-xor circuit using various VHDL language constructs, as in Listings 14.1, 14.6 and 14.12. These codes essentially describe the same cascading circuit of Figure 14.2. For an n-bit input, the critical path includes n xor gates. We can rearrange the cascading chain into a tree-shaped structure, as discussed in Section 7.4.1, and reduce the critical path to $\log_2 n$ xor gates.

For a non-parameterized design, we can use parentheses to force the desired order of evaluation and thus implicitly construct a tree-shaped circuit, as shown in Listing 7.18. Translating this approach into a parameterized description is not feasible. We need to explicitly specify the connection pattern in VHDL code. The circuit diagram of a tree-shaped eight-input reduced-xor circuit is shown in Figure 15.2. This is a two-dimensional structure. We first divide the tree into stages and number the stages from right to left. Each stage now contains multiple xor gates. We treat each xor gate as a row and number the rows from top to bottom. An xor gate can be identified with a two dimensional index (s, r), which represents the rth row of the sth stage. The corresponding output signals of the xor gate is named $p_{s,r}$. We can label all the interconnection signals according to this naming convention, as shown in Figure 15.2. Note that the input signals to the leftmost stage are also named following the same convention to make a homogeneous diagram.

The key to describing a repetitive structure is to identify the relationship of the signals between successive stages. Let us examine the xor gate in the rth row of the sth stage. Its two inputs are from the the $2r$th row and $(2r+1)$th row of the left stage (i.e., the $(s+1)$th stage).

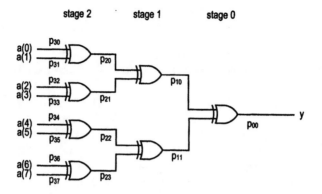

Figure 15.2   Tree-shaped reduced-xor circuit.

The factor 2 in a row's index reflects the fact that the number of rows is reduced by half in each stage. The input–output relationship of this xor gate can be described as

$$p_{s,r} = p_{s+1,2r} \oplus p_{s+1,2r+1}$$

After identifying the key relationship, we can convert the circuit into VHDL code. The two-dimensional structure implies that we need a two-dimensional data type for the p signal and a nested generate statement for the structure, with the outer statement for iteration in terms of the stages and the inner statement for iteration in terms of the rows. Since an xor gate has two inputs, the number of rows is reduced by half at each stage. For an input of n bits, the implementation needs $\log_2 n$ stages and there are 2^s rows in the sth stage.

The VHDL code is shown in Listing 15.4. The entity declaration is the same as the one in Chapter 14 and is included for clarity. We assume that the width of the input is in a power of 2. The code uses a nested two-level for generate statement for the general structure and an additional for generate statement to convert the input signal to the internal naming convention.

Listing 15.4   Parameterized tree-shaped reduced-xor circuit with input of 2^n bits

```
  library ieee;
  use ieee.std_logic_1164.all;
  use work.util_pkg.all;
  entity reduced_xor is
5     generic(WIDTH: natural);
      port(   .
          a: in std_logic_vector(WIDTH-1 downto 0);
          y: out std_logic
      );
10 end reduced_xor;

  architecture gen_tree_arch of reduced_xor is
      constant STAGE: natural:= log2c(WIDTH);
      signal p:
15        std_logic_2d(STAGE downto 0, WIDTH-1 downto 0);
  begin
      -- rename input signal
      in_gen: for i in 0 to (WIDTH-1) generate
```

```
              p(STAGE,i) <= a(i);
20      end generate;
        -- replicated structure
        stage_gen:
        for s in (STAGE-1) downto 0 generate
           row_gen:
25         for r in 0 to (2**s-1) generate
              p(s,r) <= p(s+1,2*r) xor p(s+1,2*r+1);
           end generate;
        end generate;
        -- rename output signal
30      y <= p(0,0);
     end gen_tree_arch;
```

If the number of input bits is not a power of 2, the input stage may appear irregular.
One way to handle the input of arbitrary width is to create a full-sized reduced-xor tree and
tie the unused inputs to 0's. Since $x \oplus 0 = x$, there is no effect on functionality. These
0 inputs are static, and the redundant xor gates will be removed during synthesis. Thus, the
padding 0's should have no adverse impact on the physical implementation. The revised
VHDL code is shown in Listing 15.5. An if generate statement is added. The input to the
leftmost stage will be padded with 0's if its number is not a power of 2.

Listing 15.5 Parameterized tree-shaped reduced-xor circuit with input of arbitrary bits

```
architecture gen_tree2_arch of reduced_xor is
    constant STAGE: natural:= log2c(WIDTH);
    signal p:
        std_logic_2d(STAGE downto 0, 2**STAGE-1 downto 0);
5 begin
    -- rename input signal
    in_gen:
    for i in 0 to (WIDTH-1) generate
       p(STAGE,i) <= a(i);
10  end generate;
    -- padding 0's
    pad0_gen:
    if WIDTH < (2**STAGE) generate
       zero_gen:
15     for i in WIDTH to (2**STAGE-1) generate
          p(STAGE,i) <= '0';
       end generate;
    end generate;
    -- replicated structure
20  stage_gen:
    for s in (STAGE-1) downto 0 generate
       row_gen:
       for r in 0 to (2**s-1) generate
          p(s,r) <= p(s+1,2*r) xor p(s+1,2*r+1);
25     end generate;
    end generate;
    -- rename output signal
    y <= p(0,0);
  end gen_tree2_arch;
```

The design can also be coded with a for loop statement, as shown in Listing 15.6.

Listing 15.6 Parameterized tree-shaped reduced-xor circuit using for loop statement

```
architecture loop_tree_arch of reduced_xor is
   constant STAGE: natural:= log2c(WIDTH);
   signal p:
      std_logic_2d(STAGE downto 0, 2**STAGE-1 downto 0);
5 begin
   process(a,p)
   begin
      for i in 0 to (2**STAGE-1) loop
         if i < WIDTH then
10          p(STAGE,i) <= a(i); -- rename input signal
         else
            p(STAGE,i) <= '0'; -- padding 0's
         end if;
      end loop;
15       -- replicated structure
      for s in (STAGE-1) downto 0 loop
         for r in 0 to (2**s-1) loop
            p(s,r) <= p(s+1,2*r) xor p(s+1, 2*r+1);
         end loop;
20       end loop;
   end process;
   -- rename output signal
   y <= p(0,0);
end loop_tree_arch;
```

15.3.2 Binary decoder

We discussed the design of a parameterized binary decoder in Section 14.7.2. The code in Listing 14.21 represents a one-dimensional vertical structure, as shown in Figure 14.1. Since the decoding of each output bit is done in parallel, the code is better than the codes of a cascading chain. However, the parallel vertical structure introduces a large number of input signals and may hinder the placement and routing process.

An alternative is to construct a larger decoder with a collection of smaller decoders that are arranged as a two-dimensional tree. This example illustrates the construction with 1-to-2^1 decoders. The block diagram and the function table of the 1-to-2^1 decoder are shown in Figure 15.3(a). An enable signal, en, is added to the decoder to accommodate the construction. When it is not asserted, the decoder is disabled with an all-zero output. The logic equations for this circuit are very simple:

$$y_0 = en \cdot a'$$
$$y_1 = en \cdot a$$

The block diagram of a 3-to-2^3 decoder with 1-to-2^1 decoders is shown in Figure 15.3(b). In this scheme, the input signal is decoded in stages, from the MSB to the LSB. The leftmost stage (i.e., stage 2) decodes the a_2 bit, and its output enables either the top or bottom part of the downstream decoding stages. The next stage decodes the a_1 bit and enables one-half of its downstream decoding stages. Thus, after two stages, only one-fourth of the downstream

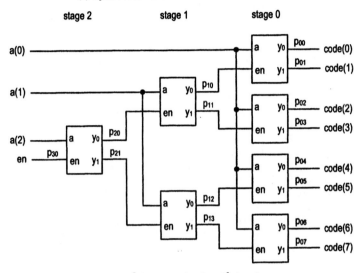

input		output	
en	a	y_1	y_0
0	-	0	0
1	0	0	1
1	1	1	0

(a) Symbol and function table of a 1-to-2^1 decoder

(b) 3-to-2^3 decoder using 1-to-2^1 decoders

Figure 15.3   Tree-shaped binary decoder.

decoding stages is enabled. For an n-to-2^n decoder, this operation repeats for each bit until all the bits are decoded and one out of 2^n output bits is asserted.

Note that there is an additional enable signal, en, in the input of the parameterized module. If the en signal is not asserted, it disables the leftmost 1-to-2^1 decoder, which, in turn, disables all downstream 1-to-2^1 decoders. None of the output bits will be asserted.

The VHDL description is shown in Listing 15.7, and the entity declaration of Chapter 14 is included for clarity. It is coded with a nested two-level for loop statement. The two inner sequential signal assignments are based on the logic equations of the 1-to-2^1 decoder.

Listing 15.7   Parameterized tree-shaped binary decoder

```
library ieee;
use ieee.std_logic_1164.all;
use work.util_pkg.all;
entity tree_decoder is
   generic(WIDTH: natural);
   port(
      a: in std_logic_vector(WIDTH-1 downto 0);
      en:std_logic;
      code: out std_logic_vector(2**WIDTH-1 downto 0)
   );
```

```
      end tree_decoder;

      architecture loop_tree_arch of tree_decoder is
         constant STAGE: natural:= WIDTH;
15       signal p:
            std_logic_2d(STAGE downto 0, 2**STAGE-1 downto 0);
      begin
         process(a,p)
         begin
20          -- leftmost stage
            p(STAGE,0) <= en;
            -- middle stages
            for s in STAGE downto 1 loop
               for r in 0 to (2**(STAGE-s)-1) loop
25                p(s-1,2*r) <= (not a(s-1)) and p(s,r);
                  p(s-1,2*r+1) <= a(s-1) and p(s,r);
               end loop;
            end loop;
            -- last stage and output
30          for i in 0 to (2**STAGE-1) loop
               code(i) <= p(0,i);
            end loop;
         end process;
      end loop_tree_arch;
```

15.3.3 Multiplexer

A parameterized multiplexer was designed in Chapter 14 and the code is shown in Listing 14.25. The code represents a one-dimensional cascading priority routing network and thus is not an ideal structure.

Tree-shaped multiplexer   One scheme to derive a two-dimensional structure is to divide the multiplexing into stages that are controlled by the individual bits of the selection signal. The block diagram of an 8-to-1 multiplexer is shown in Figure 15.4. It consists of three stages of 2-to-1 multiplexers. At each stage, the selection signals of the 2-to-1 multiplexers are tied together and connected to a bit of the selection signal, sel, of the 8-to-1 multiplexer. The LSB of the sel signal is connected to the leftmost stage (i.e., stage 2). It selects one-half of the eight possible inputs and routes them to the next stage. The selection process repeats two more times until a single input is routed to the output.

The operation of this circuit can be understood by examining an example. Routing with the sel signal of "110" is shown in Figure 15.5. We use a "binary subscript" to make the routing process clearer. For example, the a_6 input is expressed as a_{110}. The routing is done as follows:

- Stage 2 (the leftmost stage): The LSB of the sel signal is '0' and thus input signals with index "xx0", which include a_{000}, a_{010}, a_{100} and a_{110}, are selected and routed to the next stage.
- Stage 1 (the middle stage): The second LSB of the sel signal is '1' and thus signals with index "x1x", which include a_{010}, and a_{110}, are selected and routed to the next stage.

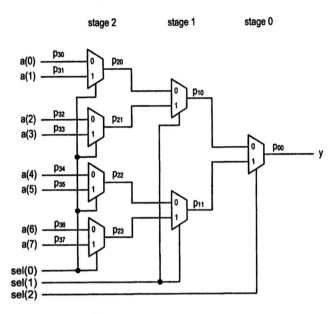

Figure 15.4   Tree-shaped 8-to-1 multiplexer.

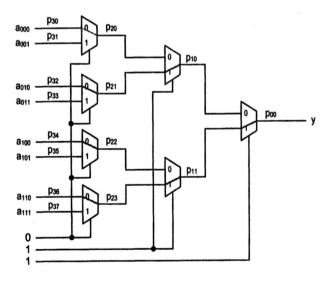

Figure 15.5   Routing with sel="110".

- Stage 0 (the rightmost stage): The MSB of the sel signal is '1' and thus the signal with index "1xx", which is a_{110}, is selected and routed to the output.

We can develop the VHDL code following the basic connection pattern of Figure 15.5. Note that the basic structure of the multiplexer is similar to the tree-shaped reduced-xor circuit of Section 15.3.1. Thus, the code of the reduced-xor circuit can be modified for the multiplexer. The VHDL code using the for loop statement is listed in Listing 15.8.

Listing 15.8 Parameterized tree-shaped multiplexer

```
library ieee;
use ieee.std_logic_1164.all;
use ieee.numeric_std.all;
use work.util_pkg.all;
5 entity mux1 is
    generic(WIDTH: natural);
    port(
        a: in std_logic_vector(WIDTH-1 downto 0);
        sel: in std_logic_vector(log2c(WIDTH)-1 downto 0);
10      y: out std_logic
    );
  end mux1;

  architecture loop_tree_arch of mux1 is
15    constant STAGE: natural:= log2c(WIDTH);
    signal p:
        std_logic_2d(STAGE downto 0, 2**STAGE-1 downto 0);
  begin
    process(a,sel,p)
20    begin
        for i in 0 to (2**STAGE-1) loop
            if i < WIDTH then
                p(STAGE,i) <= a(i); -- rename input signal
            else
25              p(STAGE,i) <= '0'; -- padding 0's
            end if;
        end loop;
        -- replicated structure
        for s in (STAGE-1) downto 0 loop
30          for r in 0 to (2**s-1) loop
                if sel((STAGE-1)-s)='0' then
                    p(s,r) <= p(s+1,2*r);
                else
                    p(s,r) <= p(s+1,2*r+1);
35              end if;
            end loop;
        end loop;
    end process;
    -- rename output signal
40    y <= p(0,0);
  end loop_tree_arch;
```

The code is identical to that in Listing 15.6 except that we replace the xor gate

```
p(s,r) <= p(s+1,2*r) xor p(s+1,2*r+1);
```

with a 2-to-1 multiplexer:

```
if sel((STAGE-1)-s)='0' then
    p(s,r) <= p(s+1,2*r);
else
    p(s,r) <= p(s+1,2*r+1);
end if;
```

Behavioral description If the input of a multiplexer is represented as an array, as in the code of Listing 15.8, the multiplexing can be considered as an indexing function that uses the `sel` signal as an index to select an element from the array. Based on this observation, we can derive the behaviorial VHDL code, as shown in Listing 15.9.

Listing 15.9 Behavioral description of a multiplexer

```
architecture beh_arch of mux1 is
begin
    y <= a(to_integer(unsigned(sel)));
end beh_arch;
```

We have used the complex index expressions before. However, these expressions are *static*, which means that their values are determined during the elaboration process, and no physical circuit will be inferred. On the other hand, the index expression in the `beh_arch` architecture depends on the `sel` input. This implies that the expression is *dynamic* and will infer a multiplexing circuit.

In the ideal case, the synthesis software recognizes this expression, and a predesigned, optimized multiplexer is inferred from the device library accordingly. We can use a simple one-line code to obtain an efficient implementation. However, not all synthesis software accepts the dynamic expression in array index, and thus the code is less portable.

Two-dimensional description In Section 15.2.4, we extended the multiplexer to accommodate two-dimensional input data. The code follows the cascading priority routing network of the one-dimensional design and suffers the same performance problem.

We can follow the process in Section 15.2.4 and extend the tree-shaped multiplexer to accept two-dimensional input data as well. The extension requires the use of a three-dimensional data type to represent the internal signal. This can be done by defining a new genuine data type like `std_logic_2d` or creating a new index function to emulate the three-dimensional data type with a one-dimensional array.

Alternatively, we can construct a two-dimensional multiplexer by duplicating the existing one-dimensional multiplexers. The VHDL code is shown in Listing 15.10. The a signal is converted into an array-of-arrays data type internally, and a for generate statement creates multiple instances of one-dimensional multiplexers.

Listing 15.10 Two-dimensional multiplexer using one-dimensional multiplexers

```
architecture from_mux1d_arch of mux2d is
    type aoa_transpose_type is
        array(B-1 downto 0) of std_logic_vector(P-1 downto 0);
    signal aa: aoa_transpose_type;
5   component mux1 is
        generic(WIDTH: natural);
        port(
            a: in std_logic_vector(WIDTH-1 downto 0);
```

Table 15.1   Function table of an 8-to-3 binary encoder

Input $a_7 a_6 \cdots a_1 a_0$	Encoded output $b_2 b_1 b_0$
0000 0001	000
0000 0010	001
0000 0100	010
0000 1000	011
0001 0000	100
0010 0000	101
0100 0000	110
1000 0000	111
others	don't-care

```
          sel: in std_logic_vector(log2c(WIDTH)-1 downto 0);
10        y: out std_logic
       );
    end component;
 begin
    -- convert to array-of-arrays data type
15  process(a)
    begin
       for i in 0 to (B-1) loop
          for j in 0 to (P-1) loop
             aa(i)(j) <= a(j,i);
20        end loop;
       end loop;
    end process;
    -- replicate 1-bit multiplexer B times
    gen_nbit: for i in 0 to (B-1) generate
25     mux: mux1
          generic map(WIDTH=>P)
          port map(a=>aa(i), sel=>sel, y=>y(i));
    end generate;
 end from_mux1d_arch;
```

15.3.4 Binary encoder

A binary encoder is a circuit that converts a one-hot input into a binary representation. The width of the input is normally a power of 2, and only 1 bit of the input is asserted. The function table of an 8-to-3 binary encoder is shown in Table 15.1. One unique characteristic of a binary encoder is the number of don't-care input combinations. For an n-bit input, $2^n - n$ combinations are not used. This can lead to significant circuit reduction.

The circuit can easily be constructed by observing the function table. The logic expressions of the previous 8-to-3 binary encoder are

$$b_2 = a_7 + a_6 + a_5 + a_4$$
$$b_1 = a_7 + a_6 + a_3 + a_2$$
$$b_0 = a_7 + a_5 + a_3 + a_1$$

Deriving an abstract parameterized code for the binary encoder is not very hard. However, this kind of description tends to "overspecify" the circuit. For example, the priority encoder code of Listing 14.24 can also be used to describe a binary encoder. Although the circuit functions correctly, the overspecification leads to unnecessary circuit complexity.

One way to describe a more efficient implementation is to follow the pattern of the previous or expressions. Close observation shows that the a_k bit will be included in the or expression of b_i if the following condition is met:

$$\frac{k}{2^i} \bmod 2 = 1$$

For example, let $i = 1$. For an 8-to-3 binary encoder, the range of k is between 0 and 7, and the condition is satisfied when k is 7, 6, 3 and 2. Thus, the or expression of b_1 can be written as $a_7 + a_6 + a_3 + a_2$.

To accommodate the condition, we create a mask table mirroring the desired patterns and apply the pattern to enable the desired bits. For example, the mask table of the previous 8-to-3 binary encoder is

```
"11110000"
"11001100",
"10101010",
```

To obtain b_2, we can perform the and operation between the a input and the first row of the mask table and then perform reduced-or operation over the result. This scheme is coded in Listing 15.11. We define a function, gen_or_mask, to generate the mask table with an array-of-arrays data type and then use it to disable the unneeded bits. The circuit is described by a nested two-level for loop statement. The outer loop iterates through the $\log_2 n$ output bits, and the inner loop performs the reduced-or operation over the masked input. The code for the reduced-or circuit represents a cascading structure. If needed, we can revise it to make a tree-shaped implementation, as the reduced-xor circuit in Section 15.3.1. This is probably not necessary since the synthesis software should be able to handle such a simple circuit.

Listing 15.11   Parameterized binary encoder

```
library ieee;
use ieee.std_logic_1164.all;
use work.util_pkg.all;
entity bin_encoder is
5     generic(N: natural);
      port(
          a: in std_logic_vector(N-1 downto 0);
          bcode: out std_logic_vector(log2c(N)-1 downto 0)
      );
10 end bin_encoder;

   architecture para_arch0 of bin_encoder is
      type mask_2d_type is array(log2c(N)-1 downto 0) of
          std_logic_vector(N-1 downto 0);
15    signal mask: mask_2d_type;
      function gen_or_mask return mask_2d_type is
          variable or_mask: mask_2d_type;
      begin
          for i in (log2c(N)-1) downto 0 loop
```

```
20              for k in (N-1) downto 0 loop
                  if (k/(2**i) mod 2)= 1 then
                      or_mask(i)(k) := '1';
                  else
                      or_mask(i)(k) := '0';
25                end if;
              end loop;
           end loop;
           return or_mask;
        end function;
30
   begin
      mask <= gen_or_mask;
      process(mask,a)
         variable tmp_row: std_logic_vector(N-1 downto 0);
35       variable tmp_bit: std_logic;
      begin
         for i in (log2c(N)-1) downto 0 loop
            tmp_row := a and mask(i);
            -- reduced or operation
40          tmp_bit := '0';
            for k in (N-1) downto 0 loop
               tmp_bit := tmp_bit or tmp_row(k);
            end loop;
            bcode(i) <= tmp_bit;
45       end loop;
      end process;
   end para_arch0;
```

Note that the gen_or_mask function and the mask operation are static. The masked bits will become 0's during elaboration process and be removed from the physical circuit during synthesis.

15.3.5 Barrel shifter

In Section 7.4.4, we studied the design of a fixed-size 8-bit rotating-right circuit. It consists of three stages of shifting–multiplexing circuits. According to the value of the control signal, the input can be either passed directly to the output or shifted by a fixed amount. The amount of shifting doubles in each stage, from 2^0 to 2^1 and 2^2. The 3-bit selection signal controls the three shifting–multiplexing circuits. After an input signal passes through three stages, the total shifted amount is the summation of the three individual stages set by the selection signal.

This is an efficient implementation for several reasons. First, as the number of inputs increases, the number of stages grows on the order of $O(\log_2 n)$. The length of the critical path grows in the same order, and thus its performance is much better than the cascading chain. Second, the circuit exhibits a regular two-dimensional structure and thus is easier for the synthesis and placement and routing software to obtain better results. Finally, recall that shifting a fixed amount requires only reconnection of the input and output signals. The shifting–multiplexing circuit is essentially a simple 2-to-1 multiplexer. Because of the regular structure, this scheme can be extended easily to accommodate parameterized design.

To make the parameterized shifting circuit more flexible, we include a feature parameter to indicate the type of shift operation, which can be shifting left, rotating left, shifting right and rotating right. The design starts with the shifting–multiplexing module. The basic block diagram is shown in Figure 15.6(a). The VHDL code of the parameterized shifting–multiplexing module is shown in Listing 15.12. The code includes three parameters. The WIDTH generic specifies the size of the circuit, the S_AMT generic specifies the amount to be shifted and the S_MODE generic specifies the type of shifting operation. Four if generate statements generate the desired amount of shifting or rotation, and the result is passed to a 2-to-1 multiplexer. Note that the shifted amount is determined by the S_AMT generic and thus is static. The shifting/rotation circuit involves only reconnection of the signals.

Listing 15.12   Parameterized fixed-size shifting–multiplexing module

```vhdl
library ieee;
use ieee.std_logic_1164.all;
use work.util_pkg.all;
entity fixed_shifter is
   generic(
      WIDTH: natural;
      S_AMT: natural;
      S_MODE: natural
   );
   port(
      s_in: in std_logic_vector(WIDTH-1 downto 0);
      shft: in std_logic;
      s_out: out std_logic_vector(WIDTH-1 downto 0)
   );
end fixed_shifter;

architecture para_arch of fixed_shifter is
   constant L_SHIFT: natural :=0;
   constant R_SHIFT: natural :=1;
   constant L_ROTAT: natural :=2;
   constant R_ROTAT: natural :=3;
   signal sh_tmp, zero: std_logic_vector(WIDTH-1 downto 0);
begin
   zero <= (others=>'0');
   -- shift left
   l_sh_gen:
   if S_MODE=L_SHIFT generate
      sh_tmp <= s_in(WIDTH-S_AMT-1 downto 0) &
                zero(WIDTH-1 downto WIDTH-S_AMT);
   end generate;
   -- rotate left
   l_rt_gen:
   if S_MODE=L_ROTAT generate
      sh_tmp <= s_in(WIDTH-S_AMT-1 downto 0) &
                s_in(WIDTH-1 downto WIDTH-S_AMT);
   end generate;
   -- shift right
   r_sh_gen:
   if S_MODE=R_SHIFT generate
      sh_tmp <= zero(S_AMT-1 downto 0) &
```

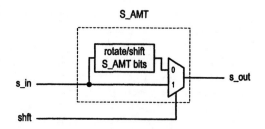

(a) Block diagram of a shifting–multiplexing module

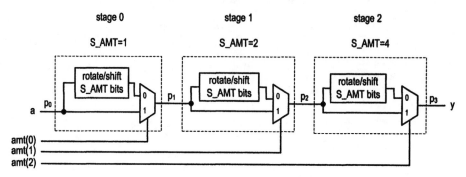

(b) Block diagram of an 8-bit three-stage barrel shifter

Figure 15.6   Parameterized barrel shifter.

```
                    s_in(WIDTH-1 downto S_AMT);
    end generate;
    -- rotate right
    r_rt_gen:
45  if S_MODE=R_ROTAT generate
        sh_tmp <= s_in(S_AMT-1 downto 0) &
                    s_in(WIDTH-1 downto S_AMT);
    end generate;
    -- 2-to-1 multiplexer
50  s_out <= sh_tmp when shft='1' else
                s_in;
    end para_arch;
```

The block diagram of a general 8-bit three-stage barrel shifter is shown in Figure 15.6(b). Each stage is a shifting–multiplexing module, and the ith bit of the amt signal is connected to the shft signal of the ith stage. The amount of shifting is determined by the stage and is 2^i for the ith stage. The VHDL code is shown in Listing 15.13. We assume that the value of input (i.e., the WIDTH parameter) is a power of 2.

Listing 15.13   Parameterized barrel shifter

```
library ieee;
use ieee.std_logic_1164.all;
use work.util_pkg.all;
entity barrel_shifter is
```

```
 5      generic(
            WIDTH: natural;
            S_MODE: natural
        );
        port(
10          a: in std_logic_vector(WIDTH-1 downto 0);
            amt: in std_logic_vector(log2c(WIDTH)-1 downto 0);
            y: out std_logic_vector(WIDTH-1 downto 0)
        );
    end barrel_shifter;
15
    architecture para_arch of barrel_shifter is
        constant STAGE: natural:= log2c(WIDTH);
        type std_aoa_type is array(STAGE downto 0) of
            std_logic_vector(WIDTH-1 downto 0);
20      signal p: std_aoa_type;
        component fixed_shifter is
            generic(
                WIDTH: natural;
                S_AMT: natural;
25              S_MODE: natural
            );
            port(
                s_in: in std_logic_vector(WIDTH-1 downto 0);
                shft: in std_logic;
30              s_out: out std_logic_vector(WIDTH-1 downto 0)
            );
        end component;
    begin
        p(0) <= a;
35      stage_gen:
        for s in 0 to (STAGE-1) generate
            shift_slice: fixed_shifter
                generic map(WIDTH=>WIDTH, S_MODE=>S_MODE,
                            S_AMT=>2**s)
40              port map(s_in=>p(s), s_out=>p(s+1), shft=>amt(s));
        end generate;
        y <= p(STAGE);
    end para_arch;
```

15.4 MORE SOPHISTICATED EXAMPLES

We study more sophisticated design examples in this section, including a reduced-xor-vector circuit and cell-based combinational multiplier, which exhibit more complex two-dimensional structures, and a priority encoder and FIFO, which are constructed using pre-designed parameterized RT-level components.

15.4.1 Reduced-xor-vector circuit

The reduced-xor-vector circuit was explained in Section 7.4.2. It performs the xor operation over successive ranges of the input. For example, for a 4-bit input $a_3 a_2 a_1 a_0$, the circuit returns four values: a_0, $a_1 \oplus a_0$, $a_2 \oplus a_1 \oplus a_0$ and $a_3 \oplus a_2 \oplus a_1 \oplus a_0$.

Cascading-chain structure   We discussed two implementations in Section 7.4.2. The linear cascading implementation requires a minimal number of gates, and its VHDL code is very simple. The code of Listing 7.21 takes advantage of the VHDL array construct and can easily be modified to accommodate a parameterized design. The revised code is shown in Listing 15.14.

Listing 15.14   Parameterized cascading-chain reduced-xor-vector circuit

```
library ieee;
use ieee.std_logic_1164.all;
use work.util_pkg.all;
entity reduced_xor_vector is
    generic(N: natural);
    port(
        a: in std_logic_vector(N-1 downto 0);
        y: out std_logic_vector(N-1 downto 0)
    );
end reduced_xor_vector;

architecture linear_arch of reduced_xor_vector is
    signal p: std_logic_vector(N-1 downto 0);
begin
    p <= (p(N-2 downto 0) & '0') xor a;
    y <= p;
end linear_arch;
```

The cascading structure experiences a large propagation delay. For an N-bit input, the critical path includes N xor gates.

Parallel-prefix structure   A more efficient structure was shown in Figure 7.8(b), which reduces the critical path to $\log_2 N$ xor gates and achieves the maximal amount of sharing. The interconnection is arranged according to a special class of structures based on the *parallel-prefix algorithm*.

The connection structure of this circuit is more involved. To better understand the connection pattern, we rename the signals in the circuit diagram of Figure 7.8(b) and add some pass-through boxes. The revised diagram is shown in Figure 15.7.

Assume that a reduced-xor-vector circuit has N-bit input and $N = 2^n$. The circuit can be divided into n stages, each containing 2^n blocks (rows). A block can be an xor gate or an empty pass-through box. We number the stages from left to right and the rows from top to bottom. For the ith row in the sth stage, its output is labeled as p_{si}. An 8-bit circuit is shown in Figure 15.7.

Closer observation of the diagram shows that it follows a simple pattern. Consider the sth stage:

- The stage is divided into 2^{n-s} modules. Each module contains 2^s blocks and is shown as a shaded rectangle in Figure 15.7.
- The top-half blocks of the module are pass-through boxes. The input of a box is connected to the output from the same row of the preceding stage.

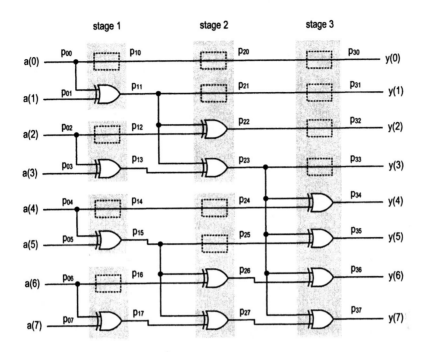

Figure 15.7   Parallel-prefix reduced-xor-vector circuit.

- The bottom-half blocks of the module are xor gates. One input of an xor gate is connected to the output from the same row of the preceding stage. The other input is the same for all xor gates in the module. It is from the output whose row index is one smaller than the index of the top xor gate in the module.

For example, consider the second stage in the diagram. We can divide it into two 2^2 modules. In the first module, the top half of the first module, whose outputs are labeled p_{20} and p_{21}, is connected to p_{10} and p_{11}. The outputs of the bottom half of the module are labeled p_{22} and p_{23}. In addition to the p_{12} and p_{13} signals, the xor gates share a common input, the p_{11} signal. The second module has a similar pattern. Note that the p_{15} signal is connected to the xor gates whose outputs are labeled p_{26} and p_{27}.

The VHDL code is shown in Listing 15.15. We assume that the number of elements of the a input is a power of 2.

Listing 15.15   Parameterized parallel-prefix reduced-xor-vector circuit

```
architecture para_prefix_arch of reduced_xor_vector is
   constant ST: natural:= log2c(N);
   signal p: std_logic_2d(ST downto 0, N-1 downto 0);
begin
   process(a,p)
   begin
      -- rename input
      for i in 0 to (N-1) loop
         p(0,i) <= a(i);
      end loop;
      -- main structure
      for s in 1 to ST loop
```

```
          for k in 0 to (2**(ST-s)-1) loop
              -- 1st half: pass-through boxes
15            for i in 0 to (2**(s-1)-1) loop
                  p(s, k*(2**s)+i) <= p(s-1, k*(2**s)+i);
              end loop;
              -- 2nd half: xor gates
              for i in (2**(s-1)) to (2**s-1) loop
20                p(s, k*(2**s)+i) <=
                      p(s-1, k*(2**s)+i) xor
                      p(s-1, k*(2**s)+2**(s-1)-1);
              end loop;
          end loop;
25    end loop;
      -- rename output
      for i in 0 to N-1 loop
          y(i) <= p(ST,i);
      end loop;
30  end process;
  end para_prefix_arch;
```

The main structure is described by a nested three-level for loop statement. The outer loop specifies the iterations over ST stages:

```
for s in 1 to ST loop
```

The middle loop iterates over the modules:

```
for k in 0 to (2**(ST-s)-1) loop
```

The two inner loops iterate over the blocks inside a module:

```
for i in 0 to (2**(s-1)-1) loop
    . . .
for i in 2**(s-1) to (2**s-1) loop
    . . .
```

The first inner loop iterates through the pass-through boxes and the second inner loop iterates through the xor gates. Note that the loop index represents half of the number of the blocks in a module.

15.4.2 Multiplier

Multiplication is a frequently needed arithmetic operation and its synthesis is not supported by all software. Two fixed-size implementations were discussed earlier, including an adder-based combinational multiplier in Section 11.6 and a sequential multiplier in Section 7.5.4. In this section, we convert the previous implementations to parameterized modules and also introduce a more efficient cell-based design.

Sequential multiplier The sequential multiplier utilizes a simple shift-and-add algorithm to iterate additions sequentially through a single adder. Since the algorithm can be applied for any input width, the design can be easily parameterized.

The original fixed-size 8-bit multiplier code is shown in Listing 11.8. Various array boundaries, initial values, and test conditions are based on the input width. To convert the code into a parameterized design, we just need to represent these values in terms of the WIDTH generic. The revised code is shown in Listing 15.16.

Listing 15.16 Parameterized sequential multiplier

```
library ieee;
use ieee.std_logic_1164.all;
use ieee.numeric_std.all;
use work.util_pkg.all;
5 entity seq_mult_para is
    generic(WIDTH: natural);
    port(
        clk, reset: in std_logic;
        start: in std_logic;
10      a_in, b_in: in std_logic_vector(WIDTH-1 downto 0);
        ready: out std_logic;
        r: out std_logic_vector(2*WIDTH-1 downto 0)
    );
  end seq_mult_para;
15
  architecture shift_add_better_arch of seq_mult_para is
    constant C_WIDTH: integer:=log2c(WIDTH)+1;
    constant C_INIT: unsigned(C_WIDTH-1 downto 0)
        :=to_unsigned(WIDTH,C_WIDTH);
20  type state_type is (idle, add_shft);
    signal state_reg, state_next: state_type;
    signal a_reg, a_next: unsigned(WIDTH-1 downto 0);
    signal n_reg, n_next: unsigned(C_WIDTH-1 downto 0);
    signal p_reg, p_next: unsigned(2*WIDTH downto 0);
25  -- alias for the upper part and lower parts of p_reg
    alias pu_next: unsigned(WIDTH downto 0) is
        p_next(2*WIDTH downto WIDTH);
    alias pu_reg: unsigned(WIDTH downto 0) is
        p_reg(2*WIDTH downto WIDTH);
30  alias pl_reg: unsigned(WIDTH-1 downto 0) is
        p_reg(WIDTH-1 downto 0);
  begin
    -- state and data registers
    process(clk,reset)
35  begin
        if reset='1' then
            state_reg <= idle;
            a_reg <= (others=>'0');
            n_reg <= (others=>'0');
40          p_reg <= (others=>'0');
        elsif (clk'event and clk='1') then
            state_reg <= state_next;
            a_reg <= a_next;
            n_reg <= n_next;
45          p_reg <= p_next;
        end if;
    end process;
    -- combinational circuit
    process(start,state_reg,a_reg,n_reg,p_reg,a_in,b_in,
50          n_next,p_next)
    begin
        a_next <= a_reg;
```

```
            n_next <= n_reg;
            p_next <= p_reg;
55          ready <='0';
            case state_reg is
                when idle =>
                    if start='1' then
                        p_next(WIDTH-1 downto 0) <= unsigned(b_in);
60                      p_next(2*WIDTH downto WIDTH) <= (others=>'0');
                        a_next <= unsigned(a_in);
                        n_next <= C_INIT;
                        state_next <= add_shft;
                     else
65                      state_next <= idle;
                    end if;
                    ready <='1';
                when add_shft =>
                    n_next <= n_reg - 1;
70                  -- add
                    if (p_reg(0)='1') then
                        pu_next <= pu_reg + ('0' & a_reg);
                    else
                        pu_next <= pu_reg;
75                  end if;
                    --shift
                    p_next <= '0' & pu_next & pl_reg(WIDTH-1 downto 1);
                    if (n_next /= 0) then
                        state_next <= add_shft;
80                  else
                        state_next <= idle;
                    end if;
            end case;
        end process;
85      r <= std_logic_vector(p_reg(2*WIDTH-1 downto 0));
    end shift_add_better_arch;
```

Adder-based combinational multiplier   The adder-based combinational multiplier uses an array of adders to perform additions in parallel, as discussed in Section 7.5.4. The revised block diagram of Section 9.4.3 illustrates the repetitive nature of this design. Our parameterized design is based on this structure. The block diagram is repeated in Figure 15.8. We modify the internal signal names to help us to identify the input and output relationships of each stage.

To increase the flexibility of this module, we include two parameters, N and WITH_PIPE, in this design. The N generic specifies the width of the operand, and the WITH_PIPE generic indicates whether to add a pipeline to the multiplier. If the pipeline is desired, registers will be inserted between the stages.

The VHDL code is shown in Listing 15.17. Two array-of-arrays data types are defined for the internal signals. The std_aoa_n_type data type is used for the propagated operands, and the std_aoa_2n_type data type is used to represent the partial product and the bit product. The code includes three major parts. The first part is composed of two if generate statements, which either generate buffer registers between stages or serve as a direct connection. The second part is the process that generates the bit product vector. The bit product in the *i*th

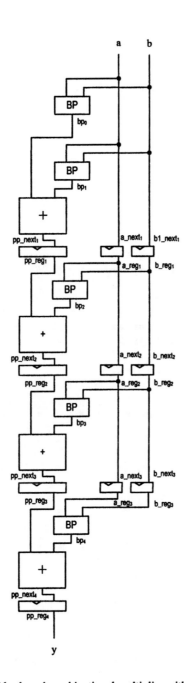

Figure 15.8   Adder-based combinational multiplier with new signal labels.

stage is represented by the bp(i) signal, which is in the form of $0 \cdots 0\, a_{n-1}b_i \cdots a_0b_i$ $0 \cdots 0$. There are $N - i$ and i padding 0's in the front and end respectively. The process includes two for loop statements, one for the two boundary bit products (i.e., bp(0) and bp(1)) and the other for regular stages. The third part specifies the addition operation in each stage. It includes a for generate statement for the middle stages and special signal connections for the first and the last stages.

Listing 15.17   Parameterized adder-based combinational multiplier

```
   library ieee;
   use ieee.std_logic_1164.all;
   use ieee.numeric_std.all;
   entity multn is
5     generic(
          N: natural;
          WITH_PIPE: natural
      );
      port(
10        clk, reset: std_logic;
          a, b: in std_logic_vector(N-1 downto 0);
          y: out std_logic_vector(2*N-1 downto 0)
      );
   end multn;
15
   architecture n_stage_pipe_arch of multn is
      type std_aoa_n_type is
          array(N-2 downto 1) of std_logic_vector(N-1 downto 0);
      type std_aoa_2n_type is
20        array(N-1 downto 0) of unsigned(2*N-1 downto 0);
      signal a_reg, a_next, b_reg, b_next: std_aoa_n_type;
      signal bp, pp_reg, pp_next: std_aoa_2n_type;

   begin
25    -- part 1
      -- without pipeline buffers
      g_wire:
      if (WITH_PIPE/=1) generate
          a_reg <= a_next;
30        b_reg <= b_next;
          pp_reg(N-1 downto 1) <= pp_next(N-1 downto 1);
      end generate;
      -- with pipeline buffers
      g_reg:
35    if (WITH_PIPE=1) generate
          process(clk,reset)
          begin
              if (reset ='1') then
                  a_reg <= (others=>(others=>'0'));
40                b_reg <= (others=>(others=>'0'));
                  pp_reg(N-1 downto 1) <= (others=>(others=>'0'));
              elsif (clk'event and clk='1') then
                  a_reg <= a_next;
                  b_reg <= b_next;
45                pp_reg(N-1 downto 1) <= pp_next(N-1 downto 1);
```

```
              end if;
          end process;
      end generate;
      — part 2
50    — bit–product generation
      process(a,b,a_reg,b_reg)
      begin
          — bp(0) and bp(1)
          for i in 0 to 1 loop
55            bp(i) <= (others=>'0');
              for j in 0 to N-1 loop
                  bp(i)(i+j) <= a(j) and b(i);
              end loop;
          end loop;
60        — regular bp(i)
          for i in 2 to (N-1) loop
              bp(i) <= (others=>'0');
              for j in 0 to (N-1) loop
                  bp(i)(i+j) <= a_reg(i-1)(j) and b_reg(i-1)(i);
65            end loop;
          end loop;
      end process;
      — part 3
      — addition of the first stage
70    pp_next(1) <= bp(0) + bp(1);
      a_next(1) <= a;
      b_next(1) <= b;
      — addition of the middle stages
      g1:
75    for  i in 2 to (N-2) generate
          pp_next(i) <= pp_reg(i-1) + bp(i);
          a_next(i) <= a_reg(i-1);
          b_next(i) <= b_reg(i-1);
      end generate;
80    — addition of the last stage
      pp_next(N-1) <= pp_reg(N-2) + bp(N-1);
      — rename output
      y <= std_logic_vector(pp_reg(N-1));
  end n_stage_pipe_arch;
```

Cell-based carry-ripple combinational multiplier

Cell-based carry-ripple combinational multiplier   The previous adder-based multiplier utilizes "coarse" RT-level parts, namely the $2N$-bit adders. The alternative is to use a 1-bit full-adder cell as the basic building block. This allows us to explore the "fine" structure of the multiplier and derive a more efficient circuit.

The multiplication of two 4-bit binary numbers is shown in Figure 15.9. The operation can be considered as the summation over the $a_i b_j$ terms, which are aligned in a specific two-dimensional pattern.

The $a_i b_j$ term returns a 1-bit value, and the addition of any two terms can be done by a 1-bit adder, which is commonly known as a *full adder*. The input of a full adder includes two 1-bit operands, ai and bi, and a 1-bit carry-in, ci, and the output includes a sum bit, so, and a carry-out, co. The gate-level VHDL description is shown in Listing 15.18. For

				a_3	a_2	a_1	a_0	multiplicand
×				b_3	b_2	b_1	b_0	multiplier
				a_3b_0	a_2b_0	a_1b_0	a_0b_0	
			a_3b_1	a_2b_1	a_1b_1	a_0b_1		
		a_3b_2	a_2b_2	a_1b_2	a_0b_2			
+		a_3b_3	a_2b_3	a_1b_3	a_0b_3			
y_7	y_6	y_5	y_4	y_3	y_2	y_1	y_0	product

Figure 15.9   Multiplication as a summation of a_ib_j terms.

most ASIC technologies, there is a predesigned full-adder cell in the device library, and it
will be inferred during synthesis.

Listing 15.18   1-bit full adder

```
library ieee;
use ieee.std_logic_1164.all;
entity fa is
   port(
      ai, bi, ci: in std_logic;
      so, co: out std_logic
   );
end fa;

architecture arch of fa is
begin
   so <= ai xor bi xor ci;
   co <= (ai and bi) or (ai and ci) or (bi and ci);
end arch;
```

To summate the a_ib_j terms, we can arrange the full-adder cells according to the two-
dimensional structure of multiplication operation in Figure 15.9. Two common arrange-
ments are *carry-ripple architecture* and *carry-save architecture*. We study the carry-ripple
multiplier in this subsection and the carry-save multiplier in the next subsection.

The block diagram of a 4-bit carry-ripple multiplier is shown in Figure 15.10. Because the
carry is propagated (i.e., rippled) from the LSB to the MSB stage by stage, this arrangement
is known as the *carry-ripple architecture*. In the diagram, each full adder cell is given an
index and expressed as FA_{ij}, indicating that the cell is located in the ith row and the jth
column. For a non-boundary cell, such as FA_{21} and FA_{22} in the diagram, the input and
output signals of the FA_{ij} cell follow a specific pattern:

- The ci port is connected to the $c_{i,j}$ signal.
- The co port is connected to the $c_{i+1,j}$ signal, which becomes the carry-in of the
  $FA_{i+1,j}$ cell.
- The so port is connected to the $s_{i,j}$ signal, which is connected to the bi port of the
  $FA_{i+1,j-1}$ cell.
- The ai port is connected to the a_ib_j term.
- The bi port is connected to the $s_{i-1,j+1}$ signal, which is the so signal of the $FA_{i-1,j+1}$
  cell.

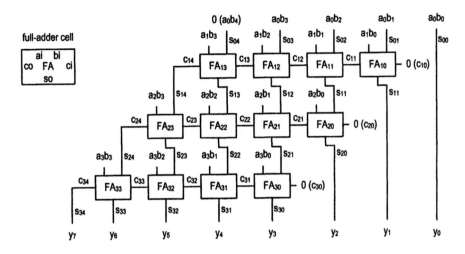

Figure 15.10   Cell-based carry-ripple combinational multiplier.

The boundary cells are located in the top and bottom rows, and the leftmost and rightmost columns. Their connections are modified as follows:

- *Top row*: The bi port of the FA_{1j} cell is connected to the $a_0 b_{j+1}$ term. Note that the b_4 bit does not exist and the leftmost term (i.e., $a_0 b_4$ in the diagram) is used for the naming convention. The $a_0 b_4$ term is actually connected to '0'.
- *Bottom row*: The so ports of the cells and the co port of the leftmost cell form the top portion of the final result.
- *Rightmost column*: The ci port of the FA_{i0} cell is connected to '0'. The so ports of the cells form the lower portion of the final result.
- *Leftmost column*: The bi port of the FA_{i4} cell is connected to the co port from the leftmost cell in the previous row. In other words, the $c_{i,3}$ signal is used in the place of the $s_{i,3}$ signal.

Once identifying the normal and boundary connection patterns and the signal naming convention, we can derive the VHDL description accordingly. The code is shown in Listing 15.19. We define an array-of-arrays type for the internal bit-product, carry and sum signals. The code is divided into several segments. The first segment is a nested two-level for generate statement that generates the ab signal, which consists of all $a_i \cdot b_j$ terms. The second segment specifies the connection patterns for the leftmost and rightmost columns. The third segment specifies the input signal of the top row. The fourth segment is a nested two-level for generate statement that instantiates the two-dimensional N-by-$(N-1)$ full-adder cells of the middle rows. The last segment uses the sum signals of the bottom row and rightmost column to form the final result.

Listing 15.19   Parameterized cell-based carry-ripple combinational multiplier

```
library ieee;
use ieee.std_logic_1164.all;
entity mult_array is
    generic(N: natural);
5   port(
        a_in, b_in: in std_logic_vector(N-1 downto 0);
        y: out std_logic_vector(2*N-1 downto 0)
```

```
      );
   end mult_array;
10
   architecture ripple_carry_arch of mult_array is
      type two_d_type is
         array(N-1 downto 0) of std_logic_vector(N downto 0);
      signal ab, c, s: two_d_type;
15    component fa
         port(
            ai, bi, ci: in std_logic;
            so, co: out std_logic
         );
20    end component;
   begin
      -- bit product
      g_ab_row:
      for i in 0 to N-1 generate
25       g_ab_col: for j in 0 to (N-1) generate
            ab(i)(j) <= a_in(i) and b_in(j);
         end generate;
      end generate;
      -- leftmost and rightmost columns
30    g_0_N_col:
      for i in 1 to (N-1) generate
         c(i)(0) <= '0';
         s(i)(N) <= c(i)(N); -- leftmost column
      end generate;
35    -- top row
      s(0) <= ab(0);
      ab(0)(N) <= '0';
      -- middle rows
      g_fa_row:
40    for i in 1 to (N-1) generate
         g_fa_col:
         for j in 0 to (N-1) generate
            u_middle: fa
               port map
45                (ai=>ab(i)(j), bi=>s(i-1)(j+1), ci=> c(i)(j),
                  so=>s(i)(j), co=>c(i)(j+1));
         end generate;
      end generate;
      -- bottom row and output
50    g_out:
      for i in 0 to (N-2) generate
         y(i) <= s(i)(0);
      end generate;
      y(2*N-1 downto N-1) <= s(N-1);
55 end ripple_carry_arch;
```

Although the appearance of this code is different from that of the previous adder-based code in Listing 15.17, the circuit it describes is very similar. Each row of the full-adder cells in Figure 15.10 forms a 4-bit ripple adder. Thus, this code actually describes a ripple adder-based combinational multiplier.

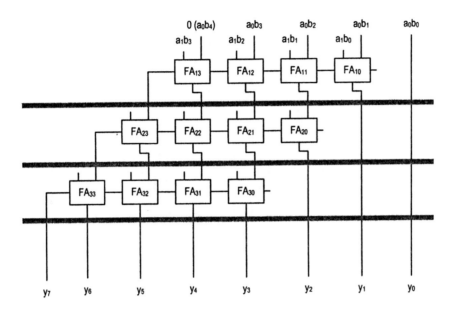

Figure 15.11   Non-optimal pipelined carry-ripple multiplier.

The fine granularity does provide more information about the underlying implementation and helps us better understand the operation of this circuit. For example, our previous pipelined design inserts pipeline registers for the sum output of the adders, as shown in Figure 15.11. These are not the optimal locations since no signal can be passed to the next row until the slowest carry bit (i.e., the MSB) becomes available.

A better division can be obtained by examining the signal propagation in the cell-level diagram. If we assume that the propagation delay of a full-adder cell is T_{fa} and the delay of obtaining $a_i \cdot b_j$ is negligible, the signal propagation from the LSB of the top row to the MSB of the bottom row is shown in Figure 15.12. The propagation is shown as a set of contour lines, each representing an increment of a delay of T_{fa}. Recall that a good pipelined design should divide the combinational circuit into stages of similar propagation delays. The pipeline registers should be inserted along these contour lines.

The contour lines also help us to identify the critical paths. One path is marked as a thick dashed line in Figure 15.12. For an N-bit multiplier, there are $N - 1$ rows, each consisting of N full-adder cells. The critical path includes N cells in the top row and two cells of each remaining $N - 2$ rows. Thus, the propagation delay is

$$NT_{fa} + 2(N - 2)T_{fa} = (3N - 4)T_{fa}$$

Cell-based carry-save combinational multiplier   The carries of the carry-ripple architecture form a cascading chain and introduce a large propagation delay. Instead of propagating the carry to the next cell in the same row, an alternative is to "save" the carry outputs and pass them to the cells in the next row, where they are summed in parallel. This is known as the *carry-save architecture*. The block diagram of a 4-bit carry-save combinational multiplier is shown in Figure 15.13. In the first three rows, a full-adder cell adds the $a_i b_j$ term and the sum bit (i.e., so) and the carry-out bit (i.e., co) from the previous row, and passes the results to the next row. The arrangement in each row represents a *carry-save adder*. The cells in the last row are arranged as a regular carry-ripple adder,

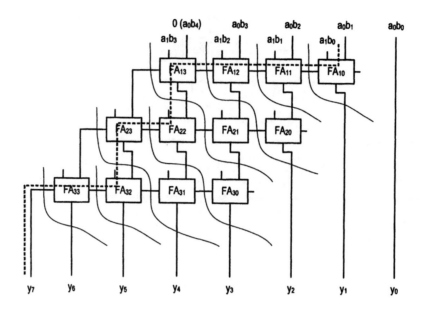

Figure 15.12 Propagation delay contour lines of a carry-ripple multiplier.

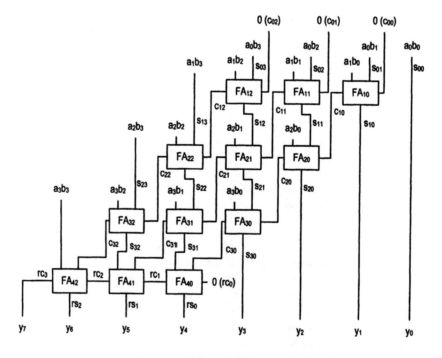

Figure 15.13 Cell-based carry-save multiplier.

which summates the carry-out signals from the last carry-save adder and forms the final result.

The derivation of the VHDL code is similar to that of the cell-based carry-ripple multiplier. We first identify the connection pattern of a non-boundary cell and then specify the special requirements for the cells in the first and last rows and the leftmost and rightmost columns. The complete VHDL code is shown in Listing 15.20.

Listing 15.20   Parameterized cell-based carry-save combinational multiplier

```vhdl
   architecture carry_save_arch of mult_array is
      type two_d_type is
         array(N-1 downto 0) of std_logic_vector(N-1 downto 0);
      signal ab, c, s: two_d_type;
 5    signal rs, rc: std_logic_vector(N-1 downto 0);
      component fa
         port(
            ai, bi, ci: in std_logic;
            so, co: out std_logic
10       );
      end component;
   begin
      -- bit product
      g_ab_row:
15    for i in 0 to N-1 generate
         g_ab_col: for j in 0 to (N-1) generate
            ab(i)(j) <= a_in(i) and b_in(j);
         end generate;
      end generate;
20    -- leftmost column
      g_N_col:
      for i in 1 to (N-1) generate
         s(i)(N-1) <= ab(i)(N-1);
      end generate;
25    -- top row
      s(0) <= ab(0);
      c(0) <= (others=>'0');
      -- middle rows
      g_fa_row:
30    for i in 1 to (N-1) generate
         g_fa_col: for j in 0 to (N-2) generate
            u_middle: fa
               port map
                  (ai=>ab(i)(j), bi=>s(i-1)(j+1), ci=> c(i-1)(j),
35                 so=>s(i)(j), co=>c(i)(j));
         end generate;
      end generate;
      -- bottom row ripple adder
      rc(0) <= '0';
40    g_acell_N_row:
      for j in 0 to (N-2) generate
         unit_N_row: fa
            port map (ai=>s(N-1)(j+1), bi=>c(N-1)(j), ci=> rc(j),
                      so=>rs(j), co=>rc(j+1));
45    end generate;
```

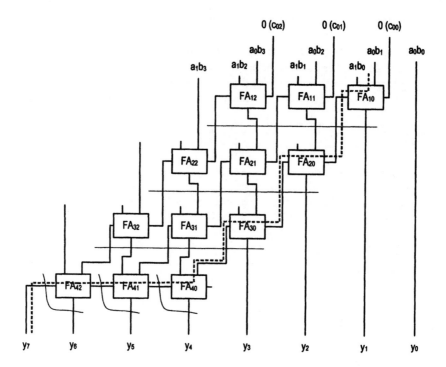

Figure 15.14   Propagation delay contour lines of a carry-save multiplier.

```
      — output signal
      g_out:
      for i in 0 to (N-1) generate
         y(i) <= s(i)(0);
50    end generate;
      y(2*N-2 downto N) <= rs(N-2 downto 0);
      y(2*N-1) <= rc(N-1);
   end carry_save_arch;
```

The propagation of the carries is much easier to trace for the carry-save multiplier. The propagation delay contour lines and the critical path are shown in Figure 15.14. For an N-bit multiplier, the critical path includes $N - 1$ cells in the bottom row and one cell of each remaining $N - 1$ rows. Thus, the propagation delay becomes

$$(N - 1)T_{fa} + (N - 1)T_{fa} = (2N - 2)T_{fa}$$

This value is about two-thirds of the delay of the previous ripple-carry multiplier. Furthermore, since the single ripple adder in the last row accounts for one-half of the delay, we can replace it with a faster adder architecture to further improve the performance.

Because of the clear propagation delay contour lines, we can easily divide the carry-save multiplier into stages of identical delays and convert it to a pipelined design. The sketch of the location of the pipeline registers is shown in Figure 15.15. The cells in the last row are rearranged for clarity. To reduce cluttering, the pipeline registers for the operands are not included.

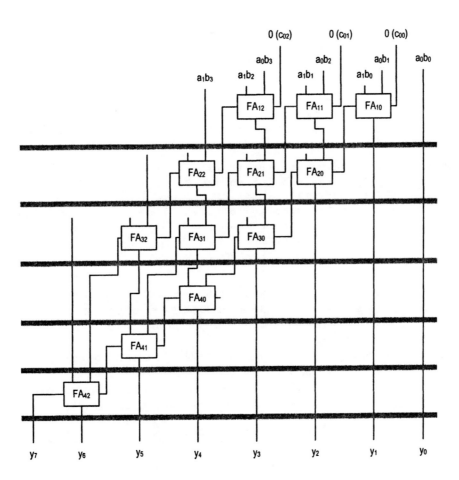

Figure 15.15 Pipelined carry-save multiplier.

15.4.3 Parameterized LFSR

The LFSR was discussed in Section 9.2.3. Its feedback circuit is simple and involves only one or three xor gates, as shown in Table 9.1. Despite its simplicity, the xor expression depends on the size of the shift register and is determined on an ad hoc basis. One way to parameterize the xor expression is to list all of the expressions in a table. Each row of the table corresponds to a specific size and indicates which register bits are needed in the expression. For example, the feedback expression of a 5-bit LFSR is $q_2 \oplus q_0$, and the corresponding row is "00101". The table can be considered as a mask table, and the pattern in each row can be used to enable or disable the corresponding bits. Consider the pervious example. The "00101" pattern can function as a mask. After performing a bitwise and operation between the mask pattern and $q_4q_3q_2q_1q_0$, we obtain $00q_20q_0$. The feedback circuit can be obtained by applying reduced-xor operation (i.e., $0 \oplus 0 \oplus q_2 \oplus 0 \oplus q_0$) over the result. Since $x \oplus 0 = x$, the 0's will be removed during synthesis, and the expression will be simplified to $q_2 \oplus q_0$.

There is no algorithm to generate the mask table. It must be exhaustively listed. Following Table 9.1, we can define the mask table as a constant of a two-dimensional array-of-arrays data type:

```
type tap_array_type is array(2 to MAX_N) of
   std_logic_vector(MAX_N-1 downto 0);
constant TAP_CONST_ARRAY: tap_array_type:=
   (2 => (1|0=>'1', others=>'0'),
    3 => (1|0=>'1', others=>'0'),
    4 => (1|0=>'1', others=>'0'),
    5 => (2|0=>'1', others=>'0'),
    . . .);
```

The MAX_N term is a constant. It specifies the maximal range of the parameter.

Section 9.2.3 shows that we can use additional logic in the feedback path to include the all-zero pattern and make an n-bit LFSR circulate through all 2^n states. This can be made as an option in a parameterized LFSR.

The complete VHDL code is shown in Listing 15.21. There are two generics: N, which specifies the size of the LFSR, and WITH_ZERO, which specifies whether the all-zero pattern should be included. The MAX_N is chosen to be 8, and thus the range of N is between 2 and 8. The MAX_N can be enlarged by adding additional rows to TAP_CONST_ARRAY.

Listing 15.21 Parameterized LFSR

```
library ieee;
use ieee.std_logic_1164.all;
use ieee.numeric_std.all;
entity lfsr is
s    generic(
        N: natural;
        WITH_ZERO: natural
     );
     port(
10       clk, reset: in std_logic;
        q: out std_logic_vector(N-1 downto 0)
     );
  end lfsr;
```

```
15 architecture para_arch of lfsr is
      constant MAX_N: natural:= 8;
      constant SEED: std_logic_vector(N-1 downto 0)
                   :=(0=>'1', others=>'0');
      type tap_array_type is array(2 to MAX_N) of
20       std_logic_vector(MAX_N-1 downto 0);
      constant TAP_CONST_ARRAY: tap_array_type:=
         (2 => (1|0=>'1', others=>'0'),
          3 => (1|0=>'1', others=>'0'),
          4 => (1|0=>'1', others=>'0'),
25        5 => (2|0=>'1', others=>'0'),
          6 => (1|0=>'1', others=>'0'),
          7 => (3|0=>'1', others=>'0'),
          8 => (4|3|2|0=>'1', others=>'0'));
      signal r_reg, r_next: std_logic_vector(N-1 downto 0);
30    signal fb, zero, fzero: std_logic;
   begin
      -- register
      process(clk,reset)
      begin
35       if (reset='1') then
            r_reg <= SEED;
         elsif (clk'event and clk='1') then
            r_reg <= r_next;
         end if;
40    end process;
      -- next-state logic
      process(r_reg)
         constant TAP_CONST: std_logic_vector(MAX_N-1 downto 0)
            := TAP_CONST_ARRAY(N);
45       variable tmp: std_logic;
      begin
         tmp := '0';
         for i in 0 to (N-1) loop
            tmp := tmp xor (r_reg(i) and TAP_CONST(i));
50       end loop;
         fb <= tmp;
      end process;
      -- with all-zero state
      gen_zero:
55    if (WITH_ZERO=1) generate
         zero <= '1' when r_reg(N-1 downto 1)=
                         (r_reg(N-1 downto 1)'range=>'0')
                    else
               '0';
60       fzero <= zero xor fb;
      end generate;
      -- without all-zero state
      gen_no_zero:
      if (WITH_ZERO/=1) generate
65       fzero <= fb;
      end generate;
      r_next <= fzero & r_reg(N-1 downto 1) ;
```

```
      — output logic
      q <= r_reg;
70 end para_arch;
```

The xor feedback circuit is implemented by a for loop statement, in which the reduced-xor operation is performed over the masked register output. The optional logic to process the all-zero pattern is implemented by two if generate statements. One statement generates the logic, and the other just reconnects the original feedback signal.

15.4.4 Priority encoder

A parameterized priority encoder was described in Listing 14.24. The code maps to a one-dimensional cascading priority routing network, and thus the performance suffers. One way to improve the performance is to construct the circuit using a collection of smaller priority encoders and multiplexers, as discussed in Section 7.4.3. The structure is quite complex.

An alternative way is to first convert the input into one-hot code and then pass the code into a regular binary encoder. For example, if an 8-bit input is "00110101", it will be converted to "0010000" and then encoded as a one-hot input. The conversion process can be explained by an example. Consider an 8-bit priority encoder whose input is $a_7, a_6, \ldots, a_0$ and a_7 has the highest priority. Let the corresponding one-hot code be $t_7, t_6, \ldots, t_0$. For the t_i bit to be asserted, the a_i bit must be '1' and all the upper bits, which include $a_7, a_6, \ldots, a_{i+1}$, must be '0'. This can be translated into a logic expression:

$$t_i = a_i \cdot a_7' \cdot a_6' \cdots a_{i+1}'$$

The logic expression represents a variant of *reduced-and* operations. As for the reduced-xor circuit, we can describe the reduced-and circuit as a tree to improve its performance. The specific pattern of the and operations also provides an opportunity for further optimization. Let us first list all logic expressions:

$$t_7 = a_7$$
$$t_6 = a_6 \cdot a_7'$$
$$t_5 = a_5 \cdot a_7' \cdot a_6'$$
$$t_4 = a_4 \cdot a_7' \cdot a_6' \cdot a_5'$$
$$t_3 = a_3 \cdot a_7' \cdot a_6' \cdot a_5' \cdot a_4'$$
$$t_2 = a_2 \cdot a_7' \cdot a_6' \cdot a_5' \cdot a_4' \cdot a_3'$$
$$t_1 = a_1 \cdot a_7' \cdot a_6' \cdot a_5' \cdot a_4' \cdot a_3' \cdot a_2'$$
$$t_0 = a_0 \cdot a_7' \cdot a_6' \cdot a_5' \cdot a_4' \cdot a_3' \cdot a_2' \cdot a_1'$$

If we ignore the first non-inverted element, the expressions become

$$a_7'$$
$$a_7' \cdot a_6'$$
$$a_7' \cdot a_6' \cdot a_5'$$
$$\cdots$$
$$a_7' \cdot a_6' \cdot a_5' \cdot a_4' \cdot a_3' \cdot a_2'$$
$$a_7' \cdot a_6' \cdot a_5' \cdot a_4' \cdot a_3' \cdot a_2' \cdot a_1'$$

The pattern is similar to the output of the reduced-xor-vector circuit discussed in Section 15.4.1. We can duplicate the code in Listing 15.15 to describe a reduced-and-vector circuit to take advantage of the sharing opportunity. The VHDL code is shown in Listing 15.22.

Listing 15.22   Parameterized parallel-prefix reduced-and-vector circuit

```
   library ieee;
   use ieee.std_logic_1164.all;
   use work.util_pkg.all;
   entity reduced_and_vector is
5      generic(N: natural);
       port(
           a: in std_logic_vector(N-1 downto 0);
           y: out std_logic_vector(N-1 downto 0)
       );
10 end reduced_and_vector;

   architecture para_prefix_arch of reduced_and_vector is
       constant ST: natural:= log2c(N);
       signal p: std_logic_2d(ST downto 0, N-1 downto 0);
15 begin
       process(a,p)
       begin
           -- rename input
           for i in 0 to (N-1) loop
20             p(0,i) <= a(i);
           end loop;
           -- main structure
           for s in 1 to ST loop
               for k in 0 to (2**(ST-s)-1) loop
                   -- 1st half: pass-through boxes
25                 for i in 0 to (2**(s-1)-1) loop
                       p(s, k*(2**s)+i) <= p(s-1, k*(2**s)+i);
                   end loop;
                   -- 2nd half: and gates
30                 for i in (2**(s-1)) to (2**s-1) loop
                       p(s, k*(2**s)+i) <=
                           p(s-1, k*(2**s)+i) and
                           p(s-1, k*(2**s)+2**(s-1)-1);
                   end loop;
35             end loop;
           end loop;
           -- rename output
           for i in 0 to (N-1) loop
               y(i) <= p(ST,i);
40         end loop;
       end process;
   end para_prefix_arch;
```

After developing the reduced-and-vector circuit, we can derive the VHDL code, as shown in Listing 15.23. The code uses the reduced-and-vector circuit and simple glue logic to generate the one-hot code and then pass it to a binary encoder. Two for loop statements are used to reverse the order of the input to match the convention used in the reduced-and-vector circuit. Since the critical paths of the parallel-prefix reduced-and-vector circuit and the optimized binary encoder circuits are on the order of $O(\log_2 n)$, the performance of this circuit is much better than that of the cascading design.

Listing 15.23   Parameterized priority encoder

```vhdl
library ieee;
use ieee.std_logic_1164.all;
use ieee.numeric_std.all;
use work.util_pkg.all;
entity prio_encoder is
   generic(N: natural);
   port(
      a: in std_logic_vector(N-1 downto 0);
      bcode: out std_logic_vector(log2c(N)-1 downto 0)
   );
end prio_encoder;

architecture para_arch of prio_encoder is
   component reduced_and_vector is
      generic(N: natural);
      port(
         a: in std_logic_vector(N-1 downto 0);
         y: out std_logic_vector(N-1 downto 0)
      );
   end component;
   component bin_encoder is
      generic(N: natural);
      port(
         a: in std_logic_vector(N-1 downto 0);
         bcode: out std_logic_vector(log2c(N)-1 downto 0)
      );
   end component;
   signal a_not_rev: std_logic_vector(N-1 downto 0);
   signal a_vec, a_vec_rev, t: std_logic_vector(N-1 downto 0);
begin
   -- reverse a
   gen_reverse_a:
   for i in 0 to (N-1) generate
      a_not_rev(i) <= not a(N-1-i);
   end generate;
   -- reduced and operation
   unit_token: reduced_and_vector
      generic map(N=>N)
      port map(a=>a_not_rev, y =>a_vec_rev);
   -- reverse the result
   gen_reverse_t:
   for i in 0 to (N-1) generate
      a_vec(i) <= a_vec_rev(N-1-i);
   end generate;
   -- form one-hot code
   t <= a and ('1' & a_vec(N-1 downto 1));
   -- regular binary encoder
   unit_bin_code: bin_encoder
```

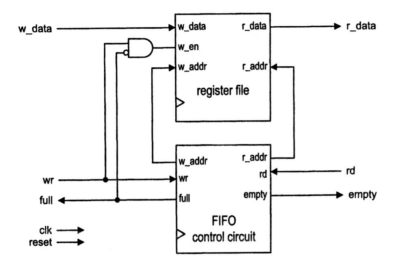

Figure 15.16   Block diagram of a FIFO buffer.

```
       generic map(N=>N)
   50      port map(a=>t, bcode=>bcode);
   end para_arch;
```

15.4.5 FIFO buffer

Implementation of a four-word FIFO buffer was discussed in Section 9.3.2. The code can be modified for a parameterized design. To achieve better performance, we use the previously developed modules to implement the circuit. The basic organization of the parameterized buffer is similar to that in Section 9.3.2, and its block diagram is shown in Figure 15.16. In the top level, the FIFO buffer is divided into a FIFO control circuit and a register file, which contains one write port and one read port. The control circuit contains two counters for the read and write pointers and the logic to generate full and empty status. The register file consists of a register array and a decoder to generate the proper enable signal and a multiplexer to route the desired value to output. The main components of the design hierarchy is shown in Figure 15.17.

For parameterized FIFO, we normally want to specify the width of a word (i.e., the number of bits in a word) and the size of the buffer (i.e., the number of words in the buffer). In our code, the B generic is used for the number of bits in a word. For simplicity, the buffer size is specified indirectly by the number of address bits of the buffer, represented by the W generic. To provide more flexibility and achieve better efficiency, we include a feature parameter, the CNT_MODE generic, to indicate whether binary or LFSR counters are used for the read and write pointers. Note that the sizes of the buffer for the binary and LFSR counter options are 2^W and $2^W - 1$ respectively.

The top-level VHDL code is shown in Listing 15.24. It is the instantiation of two components and a simple glue logic for the write enable signal of the register file. The codes of the register file and FIFO control circuit are discussed in the following two subsections.

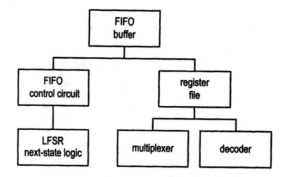

Figure 15.17   Design hierarchy of a FIFO buffer.

Listing 15.24   Parameterized FIFO buffer top-level instantiation

```
library ieee;
use ieee.std_logic_1164.all;
entity fifo_top_para is
   generic(
5       B: natural; — number of bits
        W: natural; — number of address bits
        CNT_MODE: natural — binary or LFSR
   );
   port(
10      clk, reset: in std_logic;
        rd, wr: in std_logic;
        w_data: in std_logic_vector (B-1 downto 0);
        empty, full: out std_logic;
        r_data: out std_logic_vector (B-1 downto 0)
15   );
 end fifo_top_para;

 architecture arch of fifo_top_para is
   component fifo_sync_ctrl_para
20      generic(
           N: natural;
           CNT_MODE: natural
        );
        port(
25         clk, reset: in std_logic;
           wr, rd: in std_logic;
           full, empty: out std_logic;
           w_addr, r_addr: out std_logic_vector(N-1 downto 0)
        );
30   end component;
    component reg_file_para
        generic(
           W: natural;
           B: natural
35      );
        port(
```

```
            clk, reset: in std_logic;
            wr_en: in std_logic;
            w_data: in std_logic_vector(B-1 downto 0);
40          w_addr, r_addr: in std_logic_vector(W-1 downto 0);
            r_data: out std_logic_vector(B-1 downto 0)
         );
      end component;
      signal r_addr : std_logic_vector(W-1 downto 0);
45    signal w_addr : std_logic_vector(W-1 downto 0);
      signal f_status, wr_fifo:std_logic;

   begin
      u_ctrl: fifo_sync_ctrl_para
50       generic map(N=>W, CNT_MODE=>CNT_MODE)
         port map(clk=>clk, reset=>reset, wr=>wr, rd=>rd,
                  full=>f_status, empty=>empty,
                  w_addr=>w_addr, r_addr=>r_addr);
      wr_fifo <= wr and (not f_status);
55    full <= f_status;
      u_reg_file: reg_file_para
         generic map(W=>W, B=>B)
         port map(clk=>clk, reset=>reset, wr_en=>wr_fifo,
                  w_data=>w_data, w_addr=>w_addr,
60               r_addr=> r_addr, r_data => r_data);
   end arch;
```

Register file   The operation and implementation of a fixed-size register file was discussed in Section 9.3.1. It consists of a register array, write-enable decoding logic and an output multiplexing circuit. The parameterized code can simply follow the skeleton of the fixed-size VHDL code in Listing 9.15 and replace the original segments with a parameterized register array and the predeveloped parameterized decoder and multiplexer. The array-of-arrays data type is a natural match for the register array. However, since the input data type of the parameterized multiplexer is a genuine two-dimensional array, the output of the register array must first be converted to the proper data type and then mapped to the input of the multiplexer. The complete VHDL code is shown in Listing 15.25.

Listing 15.25   Structural description of a parameterized register file

```
   library ieee;
   use ieee.std_logic_1164.all;
   use ieee.numeric_std.all;
   use work.util_pkg.all;
 5 entity reg_file_para is
      generic(
         W: natural;
         B: natural
      );
10    port(
         clk, reset: in std_logic;
         wr_en: in std_logic;
         w_data: in std_logic_vector(B-1 downto 0);
         w_addr, r_addr: in std_logic_vector(W-1 downto 0);
15       r_data: out std_logic_vector(B-1 downto 0)
```

```vhdl
          );
    end reg_file_para;

    architecture str_arch of reg_file_para is
20      component mux2d is
           generic(
              P: natural; — number of input ports
              B: natural  — number of bits per port
           );
25         port(
              a: in std_logic_2d(P-1 downto 0, B-1 downto 0);
              sel: in std_logic_vector(log2c(P)-1 downto 0);
              y: out std_logic_vector(B-1 downto 0)
           );
30      end component;
        component tree_decoder is
           generic(WIDTH: natural);
           port(
              a: in std_logic_vector(WIDTH-1 downto 0);
35            en: std_logic;
              code: out std_logic_vector(2**WIDTH-1 downto 0)
           );
        end component;
        constant W_SIZE: natural := 2**W;
40      type reg_file_type is array (2**W-1 downto 0) of
              std_logic_vector(B-1 downto 0);
        signal array_reg: reg_file_type;
        signal array_next: reg_file_type;
        signal array_2d: std_logic_2d(2**W-1 downto 0,B-1 downto 0);
45      signal en: std_logic_vector(2**W-1 downto 0);
    begin
        — register array
        process(clk,reset)
        begin
50          if (reset='1') then
               array_reg <= (others=>(others=>'0'));
            elsif (clk'event and clk='1') then
               array_reg <= array_next;
            end if;
55      end process;
        — enable decoding logic for register array
        u_bin_decoder: tree_decoder
           generic map(WIDTH=>W)
           port map(en=>wr_en, a=>w_addr, code=>en);
60      — next—state logic of register file
        process(array_reg,en,w_data)
        begin
           for i in (2**W-1) downto 0 loop
              if en(i)='1' then
65               array_next(i) <= w_data;
              else
                 array_next(i) <= array_reg(i);
              end if;
```

```
              end loop;
70     end process;
       -- convert to std_logic_2d
       process(array_reg)
       begin
           for r in (2**W-1) downto 0 loop
75             for c in 0 to (B-1) loop
                   array_2d(r,c)<=array_reg(r)(c);
               end loop;
           end loop;
       end process;
80     -- read port multiplexing circuit
       read_mux: mux2d
           generic map(P=>2**W, B=>B)
           port map(a=>array_2d, sel=>r_addr, y=>r_data);
   end str_arch;
```

Register file operation can be consider as accessing an array with a dynamic index (i.e., using a signal as an index), and some synthesis software may recognize this type of description. If this is the case, the behavioral VHDL code can be used for the register file, as shown in Listing 15.26.

Listing 15.26   Behavioral description of a parameterized register file

```
   architecture beh_arch of reg_file_para is
       type reg_file_type is array (2**W-1 downto 0) of
            std_logic_vector(B-1 downto 0);
       signal array_reg: reg_file_type;
5      signal array_next: reg_file_type;
   begin
       -- register array
       process(clk,reset)
       begin
10         if (reset='1') then
               array_reg <= (others=>(others=>'0'));
           elsif (clk'event and clk='1') then
               array_reg <= array_next;
           end if;
15     end process;
       -- next-state logic for register array
       process(array_reg,wr_en,w_addr,w_data)
       begin
           array_next <= array_reg;
20         if wr_en='1' then
               array_next(to_integer(unsigned(w_addr))) <= w_data;
           end if;
       end process;
       -- read port
25     r_data <= array_reg(to_integer(unsigned(r_addr)));
   end beh_arch;
```

FIFO Controller   We choose the look-ahead configuration of Section 9.3.2 for the parameterized FIFO controller because LFSR counters can be used to achieve better performance. The main task is to derive parameterized code to determine the counter's successive value.

Since the look-ahead configuration requires the next value of the counter, the predeveloped parameterized LFSR counter of Section 15.21 cannot be used directly. Instead, we must create a customized module for this purpose. This module is essentially the next-state logic of the parameterized LFSR of Listing 15.21. The VHDL code is shown in Listing 15.27.

Listing 15.27 Parameterized LFSR next-state logic

```
   library ieee;
   use ieee.std_logic_1164.all;
   entity lfsr_next is
      generic(N: natural);
5    port(
         q_in: in std_logic_vector(N-1 downto 0);
         q_out: out std_logic_vector(N-1 downto 0)
      );
   end lfsr_next;
10
   architecture para_arch of lfsr_next is
      constant MAX_N: natural:= 8;
      type tap_array_type is
         array(2 to MAX_N) of std_logic_vector(MAX_N-1 downto 0);
15    constant TAP_CONST_ARRAY: tap_array_type:=
         (2 => (1|0=>'1', others=>'0'),
          3 => (1|0=>'1', others=>'0'),
          4 => (1|0=>'1', others=>'0'),
          5 => (2|0=>'1', others=>'0'),
20        6 => (1|0=>'1', others=>'0'),
          7 => (3|0=>'1', others=>'0'),
          8 => (4|3|2|0=>'1', others=>'0'));
      signal fb: std_logic;
   begin
25    -- next-state logic
      process(q_in)
         constant TAP_CONST: std_logic_vector(MAX_N-1 downto 0)
            := TAP_CONST_ARRAY(N);
         variable tmp: std_logic;
30    begin
         tmp := '0';
         for i in 0 to (N-1) loop
            tmp := tmp xor (q_in(i) and TAP_CONST(i));
         end loop;
35       fb <= not(tmp); -- exclude all 1's
      end process;
      q_out <= fb & q_in(N-1 downto 1) ;
   end para_arch;
```

There is a minor modification over the original code. The feedback xor expression is inverted before it is appended to the MSB of the output. The purpose is to replace the all-zero state with the all-one state (i.e., the "11$\cdots$11" pattern, instead of the "00$\cdots$00" pattern, will be excluded from the circulation). This simplifies the system initialization.

The complete code of the parameterized FIFO controller is shown in Listing 15.28. It is similar to fixed-size code in Listing 9.16 except that two if generate statements are used to generate the desired successive value.

Listing 15.28   Parameterized FIFO control circuit

```
library ieee;
use ieee.std_logic_1164.all;
use ieee.numeric_std.all;
entity fifo_sync_ctrl_para is
    generic(
        N: natural;
        CNT_MODE: natural
    );
    port(
        clk, reset: in std_logic;
        wr, rd: in std_logic;
        full, empty: out std_logic;
        w_addr, r_addr: out std_logic_vector(N-1 downto 0)
    );
end fifo_sync_ctrl_para;

architecture lookahead_arch of fifo_sync_ctrl_para is
    component lfsr_next is
        generic(N: natural);
        port(
            q_in: in std_logic_vector(N-1 downto 0);
            q_out: out std_logic_vector(N-1 downto 0)
        );
    end component;
    constant LFSR_CTR: natural:=0;
    signal w_ptr_reg, w_ptr_next, w_ptr_succ:
        std_logic_vector(N-1 downto 0);
    signal r_ptr_reg, r_ptr_next, r_ptr_succ:
        std_logic_vector(N-1 downto 0);
    signal full_reg, empty_reg, full_next, empty_next:
            std_logic;
    signal wr_op: std_logic_vector(1 downto 0);
begin
    -- register for read and write pointers
    process(clk,reset)
    begin
        if (reset='1') then
            w_ptr_reg <= (others=>'0');
            r_ptr_reg <= (others=>'0');
        elsif (clk'event and clk='1') then
            w_ptr_reg <= w_ptr_next;
            r_ptr_reg <= r_ptr_next;
        end if;
    end process;
    -- statue FF
    process(clk,reset)
    begin
        if (reset='1') then
```

```vhdl
                    full_reg <= '0';
50                  empty_reg <= '1';
               elsif (clk'event and clk='1') then
                    full_reg <= full_next;
                    empty_reg <= empty_next;
               end if;
55        end process;
          -- successive value for LFSR counter
          g_lfsr:
          if (CNT_MODE=LFSR_CTR) generate
               u_lfsr_wr: lfsr_next
60                  generic map(N=>N)
                    port map(q_in=>w_ptr_reg, q_out=>w_ptr_succ);
               u_lfsr_rd: lfsr_next
                    generic map(N=>N)
                    port map(q_in=>r_ptr_reg, q_out=>r_ptr_succ);
65        end generate;
          -- successive value for binary counter
          g_bin:
          if (CNT_MODE/=LFSR_CTR) generate
               w_ptr_succ <= std_logic_vector(unsigned(w_ptr_reg) + 1);
70             r_ptr_succ <= std_logic_vector(unsigned(r_ptr_reg) + 1);
          end generate;
          -- next-state logic for read and write pointers
          wr_op <= wr & rd;
          process(w_ptr_reg,w_ptr_succ,r_ptr_reg,r_ptr_succ,wr_op,
75               empty_reg,full_reg)
          begin
               w_ptr_next <= w_ptr_reg;
               r_ptr_next <= r_ptr_reg;
               full_next <= full_reg;
80             empty_next <= empty_reg;
               case wr_op is
                    when "00" => -- no op
                    when "01" => -- read
                         if (empty_reg /= '1') then -- not empty
85                            r_ptr_next <= r_ptr_succ;
                              full_next <= '0';
                              if (r_ptr_succ=w_ptr_reg) then
                                   empty_next <='1';
                              end if;
90                       end if;
                    when "10" => -- write
                         if (full_reg /= '1') then -- not full
                              w_ptr_next <= w_ptr_succ;
                              empty_next <= '0';
95                            if (w_ptr_succ=r_ptr_reg) then
                                   full_next <='1';
                              end if;
                         end if;
                    when others => -- write/read;
100                      w_ptr_next <= w_ptr_succ;
                         r_ptr_next <= r_ptr_succ;
```

```
        end case;
      end process;
      — output
105   w_addr <= w_ptr_reg;
      full <= full_reg;
      r_addr <= r_ptr_reg;
      empty <= empty_reg;
    end lookahead_arch;
```

15.5 SYNTHESIS OF PARAMETERIZED MODULES

In a parameterized module, the parameter is assigned to a fixed value when the module is instantiated. At the time of synthesis, the software is always performing the synthesis of a fixed-size circuit. From this point of view, parameterized code imposes no additional requirement on actual synthesis.

On the other hand, to facilitate the use of parameters, the expressions tend to be more general and the parameterized code normally needs more "preparation" work, including the flattening of the multidimensional array and the processing and optimization of static expressions. Recall that a static expression is an expression whose value can be calculated when the VHDL code is analyzed. It implies that the expression does not depend on any external signal and that no physical circuit should be inferred from the expression.

In a parameterized code, static expressions are commonly used to express the size of arrays and the range of for generate and for loop statements. They are also used to represent more involved indexing structures, as in the parallel-prefix reduced-xor-vector code of Listings 15.15. In a complex circuit structure, we sometimes use auxiliary static expressions to assist development of the parameterized VHDL codes. For example, the VHDL code of the binary encoder in Listing 15.11 first utilizes an auxiliary gen_or_mask function to generate the static mask and then applies the mask to the input signal (via the and operation) to disable the unneeded elements of the input signal. The function and the and operation are both static. Good synthesis software should be able to calculate the mask, propagate the constants through the and expression, and keep only the needed elements of the input signal for the final or expression.

15.6 SYNTHESIS GUIDELINES

- Portability of two-dimensional data type can be an issue since it is not defined in the RTL synthesis standard.

- User-defined genuine unconstrained two-dimensional data types are the most general type.

- User-defined array-of-arrays data types cannot have unconstrained elements and are not general enough to be used in a port declaration.

- Be aware of the difference between static and dynamic expressions. The former should not infer any physical logic during synthesis and can be of assistance in developing parameterized code.

- A one-dimensional cascading-chain structure should be avoided and replaced by more efficient two-dimensional alternatives.

15.7   BIBLIOGRAPHIC NOTES

While developing parameterized VHDL codes relies on an understanding of basic language constructs and some programming skills, developing *efficient* parameterized codes requires the insight and in-depth knowledge of the problem domain, as demonstrated by the parallel-prefix reduced-xor-vector circuit and carry-save multiplier. The parallel-prefix scheme is a class of algorithms that can be applied to a variety of operations. The dissertation, *Binary Adder Architectures for Cell-Based VLSI and Their Synthesis* by R. Zimmermann of Swiss Federal Institute of Technology, provides a detailed analysis on applying the algorithms to construct addition circuits. Implementing and synthesizing complex arithmetic circuits is an active research topic. The text, *Computer Arithmetic Algorithms* by I. Koren, gives a comprehensive coverage of the algorithm and construction of various arithmetic functions.

Problems

15.1   Consider the parameterized binary decoder in Section 15.3.2. Derive the VHDL code for a 1-to-2^1 decoder with an enable signal and rewrite the code using a generate statement and component instantiation.

15.2   The parameterized binary decoder can also be constructed using 2-to-2^2 decoders.
  (a) Derive the VHDL code for a 2-to-2^2 decoder with an enable signal.
  (b) Derive the VHDL code of the parameterized binary decoder using only the 2-to-2^2 decoders of part (a).

15.3   Repeat part (b) of Problem 15.2. Instead of being limited to 2-to-2^2 decoders, use a 1-to-2^1 decoder in the leftmost stage if the input of the parameterized decode has an odd number of bits.

15.4   Consider the parameterized multiplexer in Section 15.3.3. Redesign the multiplexer using 4-to-1 multiplexers and derive the VHDL code accordingly.

15.5   Extend the parameterized multiplexer code in Listing 15.3.3 to accommodate two-dimensional data. We need to define a three-dimensional data type for the internal signals.
  (a) Follow the definition of std_logic_2d and define a genuine three-dimensional data type. Derive the VHDL code using this data type.
  (b) Follow the discussion of the emulated two-dimensional array and define an index function to emulate a three-dimensional array. Derive the VHDL code using this method.

15.6   Consider the parameterized binary encoder in Section 15.3.4. Instead of using for loop statements, rewrite the VHDL code with for generate statements.

15.7   We want to extend the parameterized barrel shifter in Section 15.3.5 by adding one additional mode of shift operation, arithmetic shift right. In this mode, the MSB, instead of '0', will be shifted into the left portion of the output. Modify the VHDL code to include this mode.

15.8   The VHDL code in Listing 15.15, the number of input bits of the parallel-prefix reduced-xor-vector circuit is limited a power of 2. Revise the code so that the number of input bits can be any arbitrary number.

15.9   Discuss the circuit complexity (in terms of the number of two-input xor gates) of the two reduced-xor-vector circuits discussed in Section 15.4.1.

15.10   The code of the adder-based multiplier of Listing 15.17 has a feature parameter to insert pipeline registers to the circuit. The number of stages of the pipeline is the same as the width of the input operand. Modify the code to incorporate an additional parameter that specifies the number of desired pipeline stages.

15.11   In the discussion of the multiplier circuit, the widths of the two input operands (i.e., multiplier and multiplicand) are assumed to be identical. In some application the widths can be different. Let the number of bits of multiplier and multiplicand be MR and MD respectively. Modify the sequential multiplier code of Listing 15.16 for the new requirement.

15.12   Repeat Problem 15.11, but modify the adder-based multiplier of Listing 15.17.

15.13   Repeat Problem 15.11, but modify the cell-base carry-ripple multiplier of Listing 15.19.

15.14   Repeat Problem 15.11, but modify the cell-base carry-save multiplier of Listing 15.20.

15.15   Both the adder-based multiplier of Section 15.4.2 and the carry-save multiplier of Section 15.4.2 can be configured as a pipelined circuit. Assume that the ripple adders are used in the adder-based multiplier. Let both the input width and the number of pipelined stages be N. Compare the delay and bandwidth of the two circuits.

15.16   The parameterized LFSR of Section 15.4.3 can only circulate through $2^N - 1$ or 2^N patterns. Modify the design so that the LFSR can circulate through M patterns, where M is a separate parameter and $M < 2^N$. You can create a function that determines the Mth pattern in the LFSR sequence and load the initial value to the register when the LFSR reaches this pattern.

15.17   The register file of Section 15.4.5 has one read port. We want to revise the design so that the number of read ports can be specified by a parameter. To achieve this, the read ports need to be grouped as a single output with a two-dimensional data type. Use the std_logic_2d data type and derive the VHDL code.

15.18   The operation of a stack was discussed in Problem 9.11. Follow the design procedure in Section 15.4.5 to derive VHDL code for a parameterized stack.

15.19   The operation and design of a CAM was discussed in Section 9.3.3. Follow the design procedure in Section 15.4.5 to derive VHDL code for a parameterized CAM.

CHAPTER 16

CLOCK AND SYNCHRONIZATION: PRINCIPLE AND PRACTICE

The single most important design principle used in this book is the synchronous methodology, in which all registers are controlled by a common clock signal. Design and analysis so far are based on an ideal clocking scenario. We assume that the entire system can be driven by a single clock signal and that the sampling edge of this clock signal can reach all registers at the same time. In reality, this is hardly possible. We need to take into consideration a non-ideal clock signal and sometimes even have to divide a large system into subsystems with independent clock signals. This chapter discusses the modeling and effect of a non-ideal clock signal, the synchronization of an asynchronous signal, and the interface between two independent clock domains.

16.1 OVERVIEW OF A CLOCK DISTRIBUTION NETWORK

16.1.1 Physical implementation of a clock distribution network

The clock distribution network is the circuit that distributes the clock signal to all FFs in the system. Since the circuit does not perform any logic function, its design and analysis are mainly at the transistor level. In Section 6.5.1, we discussed the low-level model of gates and wires for propagation delay calculation. As shown in Figure 6.15, each input port of a gate and each wire introduce small values of resistance and capacitance. The output port of a cell has to charge or discharge (i.e., "drive") all capacitors when a signal switches state. The number of input ports driven by a cell is known as *fan-out*. The driving capability of

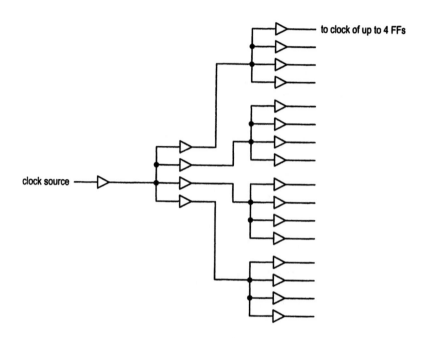

to clock of up to 4 FFs

clock source

Figure 16.1   Conceptual clock distribution network.

a cell depends on the electrical characteristics of the internal transistors. A typical cell can normally drive up to half a dozen cells.

While the basic transistor-level model of the clock distribution network is similar to that in Figure 6.15, the fan-out is much larger. Since all registers are connected to the same clock signal in a synchronous circuit, the fan-out of a clock signal is the number of FFs in the system. It may reach thousands or even tens of thousands in a large system. Thus, the physical implementation of a clock distribution network is very different from that of regular connection wires. Its construction is separated from the routing of regular logic and processed independently.

In addition to the clock signal, the reset signal is connected to all FFs of the system. Thus, construction of the reset network is somewhat similar to that of the clock distribution network. Because the reset signal does not impose many strict timing constraints, its implementation is simpler and less critical.

Clock synthesis of ASIC devices   In ASIC technology, the clock distribution network is constructed by a process known as *clock synthesis*, which is a step in the physical design. The clock synthesis uses multiple levels of buffers to increase the driving capability and applies a special routing algorithm to balance the distribution network and minimize the difference in propagation delays. A conceptual three-level clock distribution network is shown in Figure 16.1. We assume that each buffer can drive four input ports. The buffers are used to increase the driving capability and do not perform any logic function.

An example of idealized physical routing of the previous distribution network is shown in Figure 16.2. It is done by a two-level recursive H-shaped network so that the wire length from the clock source to each FF is about the same. While the propagation delay from the clock source to an FF is unavoidable, this routing helps to ensure that the clock signal reaches each FF at about the same time.

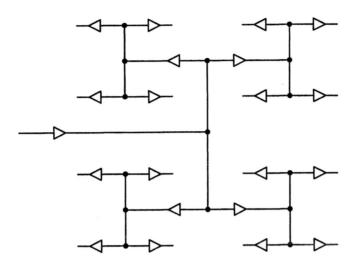

Figure 16.2   Idealized routing of a clock distribution network.

Clock distribution networks of FPGA devices   In FPGA technology, a chip usually has one or more prerouted and prefabricated clock distribution networks. If we develop the VHDL code in a disciplined way, the synthesis software can recognize the existence of the clock signal and automatically map it to a prefabricated clock distribution network.

16.1.2   Clock skew and its impact on synchronous design

The construction and analysis of a clock distribution network is essentially a task at the transistor level. At the gate and RT levels, the effect of the clock distribution network is modeled by propagation delays from the clock source to various registers. Because of the variation in buffering and routing, the propagation delays may be different, as shown in the simple example in Figure 16.3. The key characteristic is the difference between the arrival times of the sampling edges, which is known as the *clock skew*. For multiple registers, we consider the worst-case scenario and define the clock skew as the difference between the arrival times of the earliest and latest sampling edges.

As the size of a circuit and the number of FFs increase, the clock distribution network becomes larger and more complex. Controlling the arrival time of the clock's sampling edge to each FF becomes more difficult. This introduces larger variations over the arrival times, which, in turn, increase the clock skew. Thus, we can expect that the clock skew increases with the size of the circuit.

To accommodate the existence of clock skew, we have to modify the synchronous design methodology. The modification depends on the size and clock rate of the system. For a small circuit, propagation delays from the clock source to various FFs are small and almost identical, which implies that the rising edge of the clock signals arrives at the register at almost the same time. We can treat this as the ideal clocking scenario and ignore the clock skew.

For a moderately-sized system, the clock skew is normally a small fraction (a few percent) of the clock period. We can treat it as an ideal synchronous system and design it accordingly. However, during the analysis of setup time and hold time constraints, the

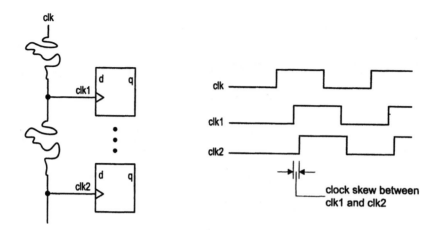

Figure 16.3 Clock skew.

clock skew must be taken into consideration. The skew usually introduces tighter timing requirements and reduces system performance. Current technology can support a clock distribution network with up to several tens of thousands of FFs with an acceptable clock skew. Recall that the proper partition size for synthesis is between 5000 and 50,000 gates. There are no problems applying synchronous design methodology inside each partition block. Section 16.2 discusses the effect of small clock skew in timing analysis.

For a fast, large-scale system, the skew may become comparable to the clock period and can no longer be treated as a small deviation of the arrival time. Because of the lack of a common clock, the synchronous design methodology can no longer be applied. One way to deal with this problem is to divide the system into several smaller subsystems and let each subsystem be controlled by an independent clock signal. Whereas the internal operation of a subsystem is synchronous, its interface to other subsystems is asynchronous. Because of the asynchronous interface, timing violations may occur. We need to use special synchronization schemes and protocols to ensure that the control signals and data can be transferred between subsystems reliably. These schemes and protocols are discussed in Sections 16.3 to 16.9.

16.2 TIMING ANALYSIS WITH CLOCK SKEW

Clock skew is the difference between the arrival times of the sampling edges of a clock signal. It has a significant impact on the timing of sequential circuits. In general, clock skew degrades system performance and imposes a tighter hold time margin. In Section 8.6, we analyzed the timing of sequential circuits with an ideal clock and showed conditions to meet setup time and hold time constraints. The following subsections repeat the analysis with the existence of clock skew.

16.2.1 Effect on setup time and maximal clock rate

For a digital system with an ideal clock, there is no clock skew. We bundle all registers together into a single element, as shown in the basic sequential circuit block diagram of Figure 8.5. To express the different arrival times, individual registers must be considered.

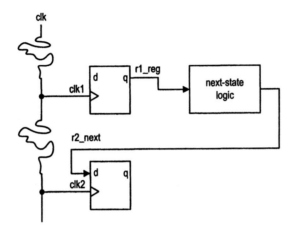

Figure 16.4   Next-state logic feedback path with positive clock skew.

A conceptual diagram of two registers and a single feedback path is shown in Figure 16.4. We assume that the lengthy routing wire introduces a delay of T_{skew} and thus the arrival times of the rising edges are different for the clk1 and clk2 signals. In this particular case, the arrival time of the clk2 signal is late by the amount T_{skew}. The late arrival is also known as a *positive* clock skew.

The timing diagram is shown in Figure 16.5. We follow the procedure used in Section 8.6.2 to analyze the new circuit. To satisfy the setup time constraint, the r2_next signal must be stabilized before t_4. This requirement translates into the condition

$$t_3 < t_4$$

From the timing diagram, we see that

$$t_3 = t_0 + T_{cq} + T_{next(max)}$$

and

$$t_4 = t_5 - T_{setup} = (t_0 + T_c + T_{skew}) - T_{setup}$$

After we substitute t_3 and t_4, the inequality equation is simplified to

$$T_{cq} + T_{next(max)} + T_{setup} - T_{skew} < T_c$$

and the minimal clock period is

$$T_{c(min)} = T_{cq} + T_{next(max)} + T_{setup} - T_{skew}$$

Note that in this particular case, the existence of clock skew reduces the minimal clock period and actually helps the performance.

The clock skew does not always mean late arrival of the sampling edge. For example, if we switch the locations of the two registers in the previous example, the arrival time of the clk2 signal is ahead of the arrival time of the clk1 signal by the amount T_{skew}. The early arrival is also known as a *negative* clock skew. In this case, t_4 becomes

$$t_4 = t_5 - T_{setup} = (t_0 + T_c - T_{skew}) - T_{setup}$$

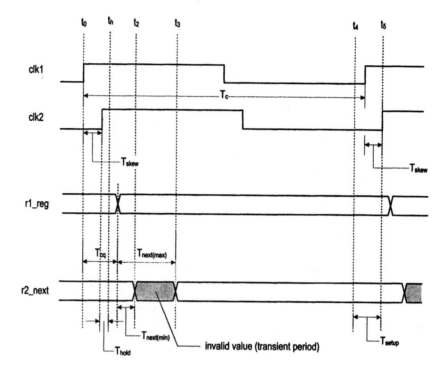

Figure 16.5 Timing analysis of positive clock skew.

and the minimal clock period becomes

$$T_{c(min)} = T_{cq} + T_{next(max)} + T_{setup} + T_{skew}$$

This implies that the clock period must be increased by the amount T_{skew}.

If there are multiple feedback paths, the effect of the clock skew can be positive for some paths and negative for the others. Consider that we add a feedback path from the r2 register to the r1 register, as shown in Figure 16.6. For simplicity, we assume that the propagation delays of the next-state logic are identical. The minimal clock periods of the two paths are

$$T_{cq} + T_{next(max)} + T_{setup} - T_{skew}$$

and

$$T_{cq} + T_{next(max)} + T_{setup} + T_{skew}$$

respectively. Since the design has to satisfy the worst-case scenario, the clock period of the system must be at least $T_{cq} + T_{next(max)} + T_{setup} + T_{skew}$.

While the positive clock skew can be used to reduce the minimal clock period in theory, it is very difficult to do this in real design because of the existence of multiple feedback paths and constraints on the placement of registers and routing of the clock distribution network. The clock skew usually has a negative effect on the clock period and thus imposes a penalty on performance.

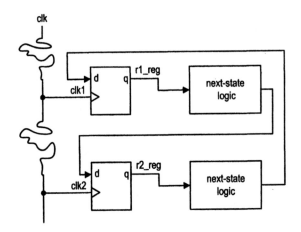

Figure 16.6   Circuit with multiple feedback paths.

16.2.2   Effect on hold time constraint

The hold time analysis is similar to that in Section 8.6.3. To satisfy the hold time constraint, we must ensure that

$$t_h < t_2$$

From the timing diagram, we see that

$$t_2 = t_0 + T_{cq} + T_{next(min)}$$

and

$$t_h = t_0 + T_{hold} + T_{skew}$$

After substitution, the inequality equation can be simplified to

$$T_{hold} < T_{cq} + T_{next(min)} - T_{skew}$$

Compared to the original inequality equation, the positive clock skew imposes a tighter margin for the hold time constraint.

In the worst-case scenario, $T_{next(min)}$ can be close to 0 (when the output of one FF is connected to the input of another FF, as in a shift register). The inequality equation becomes

$$T_{hold} < T_{cq} - T_{skew}$$

Recall that the device manufacture usually guarantees that $T_{hold} < T_{cq}$, and thus the inequality equation will always be satisfied if there is no clock skew. With a positive clock skew, the margin on the inequality equation will be reduced. It may even lead to a timing violation if T_{skew} is greater than the safety margin of $T_{cq} - T_{hold}$.

Unfortunately, there is no fix from the RT-level design. T_{hold}, T_{cq} and T_{skew} are the three parameters in the inequality equation. The first two are inherent characteristics of the device, and the third can only be obtained after synthesis of the clock distribution network. Although in theory we can add some redundant combinational logic (such as a pair of cascading invertors) to introduce artificial propagation delay, this approach is delay sensitive and is not recommended. This problem is better left for the physical design. After construction of the clock distribution network and completion of the placement and routing, accurate clock skew information can be obtained. The physical design software will check for hold time violations. It should correct the problem by rearranging the clock routing or inserting a delay element in the violated path.

The analysis of negative clock skew is similar. The inequality equation becomes

$$T_{hold} < T_{cq} + T_{next(min)} + T_{skew}$$

The clock skew provides extra slack. Since the device manufacturer usually guarantees that $T_{hold} < T_{cq}$, the extra margin provides no additional benefit.

16.3   OVERVIEW OF A MULTIPLE-CLOCK SYSTEM

When we apply the synchronous design methodology, all FFs of the system are controlled by a single global clock. However, as digital systems grow more complex, it becomes very difficult or even impossible to follow the pure synchronous design principle. Multiple clocks may exist or become necessary for several reasons:

- *Inherent multiple-clock sources.* A digital system frequently needs to interact with external systems, such as peripheral devices, or to exchange information through communication links. These external systems or links may not use the same clock signal.

- *Circuit size.* As discussed in Section 16.1, the clock skew increases with the size of the circuit and the number of FFs. When a circuit is large, it is not possible to maintain a single clock. We must divide the system into smaller subsystems and use separate clock signals in the subsystems.

- *Design complexity.* A large digital system is frequently composed of several small subsystems of different performance and power criteria. Applying pure synchronous design methodology may introduce unnecessary constraints. For example, consider a system with a 16-bit 20-MHz processor, a fast 100-MHz 1-bit serial network interface and several 1-MHz peripheral I/O controllers. If the pure synchronous design methodology is used, the system must be operated at 100 MHz to accommodate the highest clock rate, even though this clock rate is only used in the serial-to-parallel conversion of the serial network interface. It is clear that this system is "overdesigned" for the processor and I/O controllers. The artificial, unnecessarily high clock rate introduces a tighter timing constraint, complicates the design and synthesis process, and increases the hardware complexity. Utilizing separate clock signals can reduce the circuit complexity and simplify the design process.

- *Power consideration.* The dynamic power of a CMOS device is proportional to the switching frequency of transistors, which is correlated to the system clock frequency. An inflated system clock rate will unnecessarily increase the system's power consumption. If we consider the previous system, synchronous design methodology requires the entire system to be operated at 100 MHz. It will consume much more power than three subsystems with clock rates of 100, 20 and 1 MHz.

As discussed in Section 8.2, the synchronous methodology is the fundamental principle in today's digital system development, and most design and analysis schemes are based on

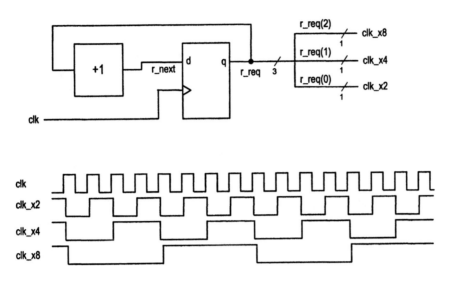

Figure 16.7   Poorly conceived clock divider.

this methodology. Thus, even with multiple clocks, we still want to apply this methodology as much as possible. The basic approach is to divide a system into multiple synchronous subsystems and design a special interface between the subsystems. This allows us to continuously apply the synchronous methodology to design a much larger system.

In a multiple-clock system, the subsystems can use either a *derived* clock signal or an *independent* clock signal. We briefly review the two schemes in the following subsections.

16.3.1   System with derived clock signals

A derived clock signal is a clock signal obtained from a known clock signal. A special clock generation circuit takes the original clock signal, generates new clock signals with different frequencies or phases, and routes them to different subsystems. Each subsystem then utilizes its own clock distribution network to distribute the clock signals to the registers within the subsystem.

In theory, we can apply general RT-level design technique to modify the frequency of a clock signal. For example, we can obtain three lower-frequency clock signals by tapping the output of 3-bit binary counter, as shown in Figure 16.7. There are two problems with this approach. First, because of the clock-to-q delay, there is a skew between the rising edges of original clock signal and the derived clock signals. Second, due to the variation of the clock-to-q delays of the FFs and the unknown wiring delays, it is difficult to determine the exact values of the skews among the three derived clock signals.

To control the skew between the clock signals, the clock generation circuit should be separated from regular logic, and manually analyzed and implemented. Special analog components, such as delay lines and buffers, or even a phase-locked loop (PLL), can be used to minimize the skew.

A system with derived clock signals is subjected to the same setup and hold time constraints. Once the clock skew between the two clock signals is known, we can apply the technique in Section 16.2 to analyze timing. The derivation procedure is similar except that we must take into consideration the difference in the clock periods.

In a multiple-clock system, we use the term *clock domain* to describe use of the clock signal. A clock domain is a block of circuitry in which the FFs are controlled by the same clock signal. Although a derived clock signal has a different clock frequency or phase, its relationship to the original clock signal is known. The design and analysis techniques of synchronous sequential circuit can be modified and applied in such a system. Because of this, we consider that subsystems with derived clock signals are in the same clock domain. Note that these derive clock signals need their own individual clock distribution networks even though they are in the same clock domain.

16.3.2 GALS system

Due to the clock skew of large circuit size or inherent I/O characteristics, it is sometimes difficult or impossible to maintain or find the relationship between the clock signals of subsystems. The clock signals in these subsystems are considered to be independent, and each subsystem constitutes its own clock domain.

Within a clock domain, the circuit operation is completely synchronous and its design follows the synchronous design methodology. Interface between the two clock domains involves two independent clocks and thus is asynchronous. This configuration is sometimes known as a *globally asynchronous locally synchronous (GALS)* system. After we develop a proper asynchronous interface, this scheme allows us to continuously apply the synchronous methodology to design a much larger system.

The major difficulty in designing a GALS system is the interface of clock domains; i.e., how to exchange information and transfer data between two clock domains (known as *domain crossing*). Since the circuit in one domain has no clock information about another domain, a signal may switch at the clock's sampling edge of another domain, which leads to a setup or hold time violation. Recall that one main advantage of the synchronous design methodology is that it provides a systematic way to *avoid a timing violation*. Since a timing violation in the domain crossing is not avoidable, the design must focus on what to do *after a timing violation occurs*.

The interface between clock domains is very different from a regular synchronous system or a system with derived clock signals. Its design cannot be automated and usually needs detailed manual analysis. It is more difficult and error-prone. Furthermore, the existence of multiple clock domains affects other processes in the development flow and complicates the static timing analysis, formal verification and test circuit synthesis. Thus, before adding an additional clock domain, we must carefully consider the trade-off between the benefits and potential complexities. In general, it is warranted only for a substantially sized subsystem or a critical high-performance subsystem. The subsequent sections discuss the nature of synchronization failure, the design of a synchronization circuit, and the design and implementation of data transfer protocols.

16.4 METASTABILITY AND SYNCHRONIZATION FAILURE

One fundamental timing constraint of a sequential circuit is the setup and hold times of an FF. It specifies that the input data to an FF must be stable in a decision window around the sampling edge of the clock signal. Consider the basic sequential circuit block diagram shown in Figure 8.5. The input of the register is the next-state logic's output, which is obtained from the register's output and an external input.

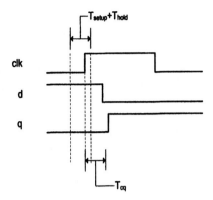

(a) Input stable during setup and hold time

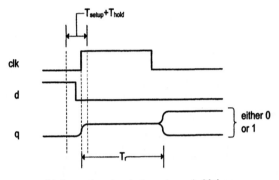

(b) Input changing during setup or hold time

Figure 16.8   Timing diagrams of a D FF.

Since the register's output is based on the sampling edge of the clock, we have timing information about this signal. Our timing analysis examines the closed loop formed by the register and next-state logic and ensures that no timing violation will occur. Similar analysis can be performed if the external input signal is generated in the same clock domain. On the other hand, if the external input signal comes from another clock domain, as in a GALS system, the subsystem has no information about the timing relation to its clock, and thus the signal is treated as an asynchronous signal. An asynchronous input signal can change any time, including inside the decision window, and cause a timing violation. The following subsections discuss the characteristics of a timing violation.

16.4.1 Nature of metastability

When an input data signal satisfies the timing constraint, the sampled value will be propagated to the FF's output after the clock-to-q (T_{cq}) delay, as shown in Figure 16.8(a). On the other hand, if the input signal changes during setup or hold time, it violates the timing constraint and the output response is very different. Assume that the input changes from '0' to '1' during the setup and hold time window. One of three scenarios happens:

- The output of the FF becomes '1'.
- The output of the FF becomes '0'.

- The FF enters a *metastable state*, and the output exhibits an in-between voltage value.

The first scenario is the desired result and causes no problem. The second scenario implies that the FF just sampled the previous value. If the input remains unchanged, the correct value will be sampled at the next rising edge. Since we make no assumption about the arrival time of the input signal, there will be no ill effect.

The third scenario is the troublesome one. In normal operation, an FF stays in one of the two stable states, and its output voltage is either high or low. They are interpreted as logic 1 or logic 0 if the positive logic is used. When an FF enters a metastable state, its output voltage is somewhere between the low and the high, and cannot be interpreted as either logic 0 or logic 1. If the output of the FF is used to drive other logic cells, the in-between value may propagate to downstream logic cells and lead the entire digital system into an unknown state.

As its name indicates, a metastable state is not really a stable state. A small noise or disturbance will offset its "balance" and force the FF to enter one of the stable states. In other words, the FF will eventually resolve to a stable state. The time required to reach a stable state is known as the *resolution time*, T_r. The timing diagram is shown in Figure 16.8(b). Theoretical study shows that a bistable device always has a metastable state, and this phenomenon is unavoidable. The only solution is to provide enough time to let the device resolve the situation and reach a stable state.

The resolution time, unfortunately, is not deterministic. It is characterized by a probability distribution function

$$P(T_r) = e^{-\frac{T_r}{\tau}}$$

In this equation, τ is the *decay time constant* and is determined by the electrical characteristics of the FF. A typical value of today's device technology is around a fraction of a nanosecond. The equation indicates the probability that the metastability condition persists beyond T_r after the clock edge. It can be interpreted as the probability that the metastability cannot be resolved within T_r seconds.

16.4.2 Analysis of MTBF(T_r)

Since the timing violation can occur in any asynchronous input, the goal of the design is to confine the metastable condition in an FF and to prevent the in-between value being propagated to the downstream logic. When an FF cannot resolve the metastability condition within the given time, it is known as a *synchronization failure*.

Because of the stochastic nature of the occurrence of a timing violation and resolution time, analysis of the metastable condition is characterized by a statistical average. We use the *average* time interval between two synchronization failures to express the reliability of the design. It is known as *mean time between synchronization failures (MTBF)* and is the main quantity used in metastability timing analysis. MTBF depends on many factors. However, in a realistic design scenario, most factors cannot be easily changed, and the only freedom we have is to provide proper resolution time. Thus, MTBF is frequently expressed as a function of T_r, as in MTBF(T_r).

We can derive the MTBF by calculating the average rate of synchronization failures, AF, which is the reciprocal of MTBF. AF is defined as the average number of synchronization failures occurring in a 1-second interval. It is determined by two factors:

- R_{meta}: The average rate at which an FF enters the metastable state.
- $P(T_r)$: The probability that an FF cannot resolve the metastability condition within T_r.

R_{meta} is determined by the formula

$$R_{meta} = w * f_{clk} * f_d$$

In this formula, w is the *susceptible time window*, which is a constant determined by the electrical characteristics of the FF. It can be interpreted as a metastability susceptible time interval associated with the triggering edge of the clock signal. For current device technology, the typical value of w is from few picoseconds to a fraction of a nanosecond. The f_{clk} parameter is the frequency of the clock signal, which is defined as the number of clock cycles per second. During the 1-second interval, there are f_{clk} triggering edges, and thus the $w * f_{clk}$ portion of 1 second is susceptible for the metastability. The f_d parameter is the rate of change in input data, which is defined as the number of input changes per second. We assume that the input data is independent of the clock and that the change can occur at any time. The probability of a single change occurring within a metastability susceptible interval is $w * f_{clk}$. Since there are f_d changes in 1 second, the FF will enter the metastability state $w * f_{clk} * f_d$ times per second, as shown in the equation above.

Once an FF enters the metastable state, it takes a certain amount of time to resolve to a stable state. The discussion in Section 16.4.1 shows that the probability that the FF cannot resolve the metastable condition within the given resolution time of T_r is

$$P(T_r) = e^{-\frac{T_r}{\tau}}$$

In other words, when an FF enters the metastable state, it may resolve the condition within the given resolution time. Only $P(T_r)$ of the events persists over T_r and leads to synchronization failure. Since the FF enters the metastability state R_{meta} times per second on average and only $P(T_r)$ of the entries leads to synchronization failure for the given T_r, the average number of synchronization failures per second is $R_{meta} * P(T_r)$; that is,

$$AF(T_r) = R_{meta} * P(T_r) = w * f_{clk} * f_d * e^{-\frac{T_r}{\tau}}$$

For a given T_r, MTBF(T_r) becomes

$$\text{MTBF}(T_r) = \frac{1}{AF(T_r)} = \frac{e^{\frac{T_r}{\tau}}}{w * f_{clk} * f_d}$$

Note that the f_{clk}, f_d, w and τ parameters are associated with original system specifications and device technology, and revising them can lead to significant design modification. The only freedom we have is to adjust the resolution time (T_r) in the synchronization circuit. That is why we normally express MTBF as a function of T_r, as in MTBF(T_r).

In the MTBF calculation, τ and w depend on the electrical characteristics of the device, and their values can be found in the manufacturer's data sheet. Note that some FPGA manufacturers define the resolution time as the additional time needed after the regular clock-to-q delay (T_{cq}). If we call this time T_{r2}, the relationship between T_r and T_{r2} is

$$T_r = T_{cq} + T_{r2}$$

After simple mathematical manipulation, we can easily convert MTBF(T_{r2}) to MTBF(T_r), and vice versa.

Table 16.1   Sample MTBF(T_r) computation

T_r	MTBF
0.0 ns	$4.00 * 10^{-05}$ sec (0.04 msec)
2.5 ns	$5.94 * 10^{-03}$ sec (5.94 msec)
5.0 ns	$8.81 * 10^{-01}$ sec (0.88 sec)
7.5 ns	$1.31 * 10^{+02}$ sec (131 sec)
10.0 ns	$1.94 * 10^{+04}$ sec (5.39 hours)
12.5 ns	$2.88 * 10^{+06}$ sec (3.33 days)
15.0 ns	$4.27 * 10^{+08}$ sec (1.36 years)
17.5 ns	$6.34 * 10^{+10}$ sec ($2.01 * 10^3$ years)
20.0 ns	$9.42 * 10^{+12}$ sec ($2.99 * 10^5$ years)
22.5 ns	$1.40 * 10^{+15}$ sec ($4.43 * 10^7$ years)
25.0 ns	$2.07 * 10^{+17}$ sec ($6.58 * 10^9$ years)
27.5 ns	$3.08 * 10^{+19}$ sec ($9.76 * 10^{11}$ years)
30.0 ns	$4.57 * 10^{+21}$ sec ($1.45 * 10^{14}$ years)
32.5 ns	$6.78 * 10^{+23}$ sec ($2.15 * 10^{16}$ years)
35.0 ns	$1.01 * 10^{+26}$ sec ($3.19 * 10^{18}$ years)

16.4.3   Unique characteristics of MTBF(T_r)

We have examined various timing parameters, such as propagation delay, setup time and hold time. The metastability resolution time is very different. It is not deterministic and not even bounded, and thus must be characterized by a probability distribution function. Note that the resolution time is random in nature, and MTBF, as its name shows, is an average value. When a system has an MTBF value of 1 year, it does not mean that the synchronization failure always happens once a year. It means that the synchronization failure happens once a year *on average*. The actual interval can be 1 month, 6 months, 1 year, 2 years, 5 years and so on. A system may fail in a year regardless of whether its MTBF value is 1 year, 10 years or 1000 years. However, the probability of failure for the system with a 1000-year MTBF is much smaller.

Another observation about the resolution time relates to its highly non-linear characteristics. Note that T_r is in the exponent position of the MTBF(T_r) formula. A small variation over T_r leads to drastic change in the value of MTBF. For example, consider an FF with a w of 0.1 ns and a τ of 0.5 ns and assume that the system clock frequency (f_{clk}) is 50 MHz and the data rate (f_d) is $0.1f_{clk}$. The resolution time of a synchronizer is normally ranged between a fraction of a clock period to one or two clock periods (discussed in the next section). Table 16.1 lists the MTBF values of T_r from 0 to 35 ns at increments of 2.5 ns. Note that the period of a 50-MHz clock signal is 20 ns. When no resolution time is provided (i.e., $T_r = 0$), the MTBF is an unacceptable 0.04 ms. If we can use a T_r value of half a clock period (i.e., 10 ns), the MTBF becomes about 5 hours. Because of the exponential rate, each extra 2.5 ns can increase the MTBF more than 100 times. When T_r reaches 17.5 ns, the MTBF reaches about 2000 years. If we provide 1.5 times the clock period (i.e., 30 ns), the MTBF becomes about 10^{14} years (for comparison, the age the universe is on the order of 10^{11} years, and the appearance of the human being is on the order of 10^5 years).

This phenomenon is a mixed blessing. On the positive side, while the synchronizing failure cannot be eliminated, we can make the probability of occurrence extremely small.

On the negative side, because of the sensitivity of the resolution time, a small decrease in the resolution time can significantly degrade the value of MTBF. Thus, minor revisions in the system, such as the slight increment of the system clock rate or use of an FF with a slightly larger setup time, may lead to a drastic consequence.

16.5 BASIC SYNCHRONIZER

When an asynchronous input causes a setup or hold time violation, the FF may enter the metastable state and its output exhibits an in-between value. If not blocked, the in-between value will be passed to the next stage and gradually propagated through the entire system.

As its name shows, a *synchronization circuit* (or a *synchronizer*) is to synchronize an asynchronous input with the system clock. As we learned from the previous sections, no circuit can prevent the occurrence of the metastability of a bistable device. The purpose of a synchronization circuit is to stop the propagation of the in-between value and confine the metastability condition within the synchronizer. Since the metastability condition will eventually resolve itself, the task of a synchronizer is just to provide enough time for the FF to reach a stable state.

The following subsections analyze various configurations of a synchronizer and their MTBFs. In our examples, we assume that the circuit utilizes the FF of Section 16.4.3, and has same clock frequency and data rate; i.e., $w = 0.1$ ns, $\tau = 0.5$ ns, $f_{clk} = 50$ MHz and $f_d = 0.1 f_{clk}$.

16.5.1 The danger of no synchronizer

We first consider a sequential circuit that has no synchronizer for its asynchronous input, as shown in Figure 16.9(a). If the asynchronous signal causes a timing violation, the system register may enter the metastable state, and the in-between value will be propagated to the next-state logic circuit. We can analyze how frequently the system enters the metastable state using the previous MTBF formula. Since there is no synchronizer, no resolution time is provided (i.e., $T_r = 0$). Substituting this value into the formula, we have MTBF(0) = 0.04 ms. This failure rate is clearly unacceptable.

16.5.2 One-FF synchronizer and its deficiency

The first design of a synchronizer is to use a single D FF, as shown in Figure 16.9(b). Let T_c, T_{setup} and T_{comb} be the clock period of the system, the setup time of the FF and the propagation delay of the combinational circuit respectively. Consider the path from the synchronizer D FF to the system D FF. The synchronizer provides one clock period for the out_sync signal to resolve, propagate through the combinational logic and satisfy the setup time constraint of the system D FF. The required time for the latter two is $T_{comb} + T_{setup}$, and the remaining balance can be used to resolve the metastability condition, which is

$$T_r = T_c - (T_{comb} + T_{setup})$$

Assume that T_{setup} of the system register is 2.5 ns. The resolution time of this circuit becomes

$$T_r = 20 - (T_{comb} + 2.5) = 17.5 - T_{comb}$$

T_r and MTBF depend on T_{comb}, the propagation delay of the combinational circuit. For a simple combination circuit, the T_r will be relatively large. For example, if T_{comb} is 1 ns,

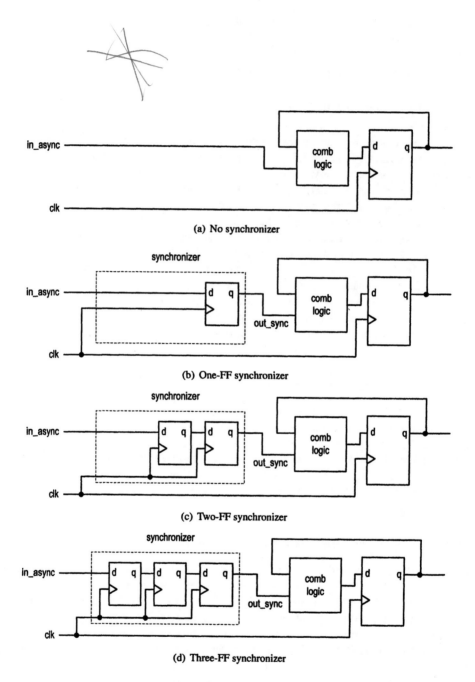

Figure 16.9   Synchronizers.

T_r becomes 16.5 ns and MTBF(16.5 ns) is about 272 years. On the other hand, a complex combination circuit can drastically reduce the MTBF value. If T_{comb} is 12.5 ns, T_r becomes 5 ns and MTBF(5 ns) is dropped to about 0.88 second.

As discussed earlier, MTBF is extremely sensitive to T_r, and a small variation leads to a huge swing in MTBF value. The value of T_{comb} depends on the logic function of the combinational circuit, device technology as well as synthesis and placement and routing process, and thus cannot be determined in advance. A minor modification in the combinational logic, the synthesis process or the placement and routing process can lead to a significant reduction in MTBF and make the system susceptible to synchronization failure. Therefore, this is not a reliable design. A better alternative is to use two D FFs for the synchronizer.

16.5.3 Two-FF synchronizer

The previous analysis shows that a maximal resolution time can be obtained if T_{comb} is 0. Since the function of the combinational logic is defined by the original system, we cannot modify it arbitrarily. Instead, we can insert another D FF to form a two-FF synchronizer, as shown in Figure 16.9(c). The resolution time provided by the two FFs inside the synchronizer is

$$T_r = T_c - T_{setup}$$

If T_{setup} is 2.5 ns, the resolution time becomes

$$T_r = 20 - 2.5 = 17.5 \text{ ns}$$

The MTBF(17.5 ns) is about 3000 years. In addition to providing more resolution time, this design is also more robust since no logic function or synthesis is involved. The only uncertain factor in this design is the wiring delay, which can be substantial if the two D FFs are located far apart. To minimize this delay, the two D FFs must be placed as close as possible. In physical design, we may need to manually perform the placement and routing for the synchronizer.

The VHDL code for the synchronizer is straightforward, following the block diagram of Figure 16.9(c). The code is shown in Listing 16.1.

Listing 16.1 Two-FF synchronizer

```
library ieee;
use ieee.std_logic_1164.all;
entity synchronizer is
   port(
5       clk, reset: in std_logic;
        in_async: in std_logic;
        out_sync: out std_logic
   );
end synchronizer;
10
architecture two_ff_arch of synchronizer is
   signal meta_reg, sync_reg: std_logic;
   signal meta_next, sync_next: std_logic;
begin
15  -- two D FFs
   process(clk,reset)
```

```
       begin
          if (reset='1') then
             meta_reg <= '0';
20           sync_reg <= '0';
          elsif (clk'event and clk='1') then
             meta_reg <= meta_next;
             sync_reg <= sync_next;
          end if;
25     end process;
       -- next-state logic
       meta_next <= in_async;
       sync_next <= meta_reg;
       -- output
30     out_sync <= sync_reg;
    end two_ff_arch;
```

Because of its simplicity and robustness, the two-FF configuration is the most widely used synchronizer. It is satisfactory in most applications. However, the regular D FF occasionally may not be able to provide sufficient T_r. For example, if we increase the system clock by one-third to 66.7 MHz, the clock period is reduced to 15 ns and T_r becomes 12.5 ns. The MTBF is reduced to 3.33 days. To overcome this problem, many ASIC technologies have a special *metastability-hardened* D FF cell in their libraries. The functionality of this D FF is identical to that of a regular D FF, but its w, τ and T_{setup} are made smaller to increase MTBF. We can use component instantiation in VHDL code to instantiate this type of D FF cell in a synchronizer. Due to its internal implementation, the circuit size of a metastability-hardened D FF cell is several times larger than that of a regular D FF cell and thus should not be used in regular sequential circuits.

16.5.4 Three-FF synchronizer

If the device technology does not provide a metastability-hardened D FF cell, we can increase the resolution time by cascading more D FF cells or artificially enlarging the clock period of the synchronizer. The three-FF synchronizer is shown in Figure 16.9(d). An extra D FF is cascaded with a two-FF synchronizer. The idea behind this design is to use the extra D FF to provide an additional opportunity to resolve the metastability condition.

We can follow the procedure in Section 16.4.2 to calculate the MTBF of this circuit. Recall that R_{meta}, the average rate at which the first D FF enters a metastable state, is

$$R_{meta} = w * f_{clk} * f_d$$

Once the FF enters the metastable state, it has a time interval of $T_c - T_{setup}$ to resolve the situation. The probability that the metastability condition persists beyond the current clock cycle is

$$P1 = e^{-\frac{T_c - T_{setup}}{\tau}}$$

If this situation happens, the metastability condition is sampled and passed to the second FF. The second FF, again, has a time interval of $T_c - T_{setup}$ to resolve the situation, and the probability that the metastability condition persists beyond this clock cycle is

$$P2 = e^{-\frac{T_c - T_{setup}}{\tau}}$$

The MTBF of this circuit becomes

$$\text{MTBF} = \frac{1}{R_{meta} * P1 * P2} = \frac{e^{\frac{2(T_c - T_{setup})}{\tau}}}{w * f_{clk} * f_d}$$

If we compare this equation to the two-FF synchronizer, which is

$$\text{MTBF} = \frac{1}{R_{meta} * P1} = \frac{e^{\frac{(T_c - T_{setup})}{\tau}}}{w * f_{clk} * f_d}$$

We can interpret that the three-FF synchronizer increases the resolution time from $T_c - T_{setup}$ to $2(T_c - T_{setup})$.

Since this term is in the exponent of the equation, its impact is very significant. If T_c is 20 ns and T_{setup} is 2.5 ns, the resolution time increases from 17.5 ns to 35 ns, and the MTBF increases from 2000 thousand years to 10^{18} years. If T_c is 15 ns, the resolution time increases from 12.5 ns to 25 ns and the MTBF increases from 3 days to about 6 billion years, which is a pretty safe number.

The disadvantage of the three-buffer synchronizer is the delay for the input signal. The extra D FF increases the delay from two clock cycles to three clock cycles. When possible, we should use a metastability-hardened D FF cell rather than using an additional D FF.

We can cascade more D FFs to increase the MTBF. However, because of the effect of the exponential decay, this is seldom needed in reality.

16.5.5 Proper use of a synchronizer

The function of a synchronizer is to provide a non-metastable output value. We must use it properly to obtain a reliable synchronized result. Good design practices can help us to achieve this goal and avoid subtle errors:

- Use a glitch-free signal for synchronization.
- Synchronize a signal in a single place.
- Avoid synchronizing multiple "related" signals.
- Reanalyze the synchronizer after each design change.

These practices are discussed in the following paragraphs.

Use a glitch-free signal for synchronization   The asynchronous input signal normally comes from another clock domain. Since the synchronizer has no knowledge about the clock signal in another domain, it can sample the asynchronous input any time. If a glitch exists in the input signal, it may be sampled and synchronized incorrectly as a legitimate value. It is important to pass a glitch-free signal for synchronization. This can be achieved by adding an output buffer when the signal is generated.

Synchronize a signal in a single place   The function of a synchronizer is to generate a stable output value. The synchronizer, however, cannot guarantee which value will be reached. For example, if a timing violation occurs when the input changes from '0' to '1', the synchronized input value can be '0' or '1' at the current sampling clock edge. Assume that the input signal does not change. It will be sampled again at the next rising edge of the clock and a stable '1' will be obtained. This implies that the arrival time of a synchronized asynchronous input signal may exhibit a random one-clock delay. We must take the random delay into consideration when using a synchronizer.

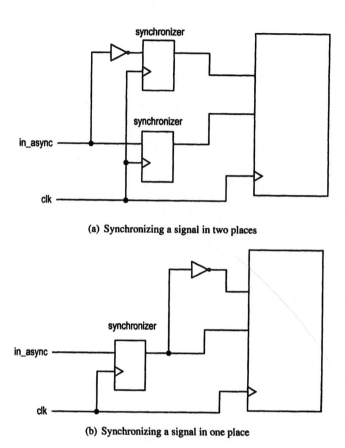

(a) Synchronizing a signal in two places

(b) Synchronizing a signal in one place

Figure 16.10   Synchronization at multiple places.

An asynchronous input signal may be used in multiple places in a clock domain. It should be synchronized in a single entry point. An example of a poor design is shown in Figure 16.10(a), in which the in_async signal and its derivative are synchronized by two individual synchronizers. The potentially random one-clock delay may introduce inconsistent values to the system and lead to incorrect operation. A better alternative is shown in Figure 16.10(b). The signal is synchronized by a single synchronizer, and the system is always fed with the same value.

Avoid synchronizing related signals   A similar issue is to synchronize related signals. *Related signals* means that a group of signals are combined to represent a command, state and so on. For example, we may use two signals to represents four possible actions. Because of the random one-clock delay, synchronizing related signals may to lead to uncertain results. For example, consider that two related signals changes from "00" to "11". If the two signals switch at about the same time and both transitions cause timing violations, the resolved results can be "00", "01", "10" or "11" for one clock cycle. Although the signal will eventually be settled to "11" in the next clock cycle, the "01" and "10" conditions may exist for one clock cycle. This may cause a serious problem for some applications.

There are two ways to correct the problem. The first is to apply special coding patterns, such as Gray code, to ensure that only one bit changes during the transition. One example is given in Section 16.9.1. A better, more systematic alternative is to bundle all signals and pass them as a single data item. The data transfer between two clock domains is discussed in Section 16.8.

Reanalyze the synchronizer after each design change   MTBF is extremely sensitive to the available resolution time, and a small variation can lead to drastic change. For example, consider the two-FF synchronizer discussed in Section 16.5.3. If the original system is running at 50 MHz, the MTBF is about 3000 years. Assume that we upgrade the design using faster functional units and the new system can run at 66.7 MHz, about 33% faster. Since the same device technology is used for the D FFs, w and τ remain unchanged. The MTBF is reduced to a mere 3 days, which is only 0.0002% of the original value. The example demonstrates the sensitivity of the synchronizing circuit. It is good practice to examine the synchronizer after each design modification.

16.6   SINGLE ENABLE SIGNAL CROSSING CLOCK DOMAINS

In a GALS system, clock domains are driven by independent clock signals. The clock frequencies and data processing rates of these domains may not be identical. A subsystem can communicate with another subsystem whose clock frequency is 10 times faster or 10 times slower. The function of a synchronizer is to prevent the subsystems from entering the metastable state. Additional control schemes are needed to coordinate the information exchange between the two clock domains. We show how to propagate an enable pulse signal from one clock domain to another clock domain in this section and Section 16.7 and discuss the data transfer in Sections 16.8 and 16.9.

16.6.1   Edge detection scheme

A digital system frequently includes a control signal in the form of an enable pulse, which activates the desired action for a single time. The enable signal of a counter and the start signal of a sequential multiplier are signals of this type. An enable pulse should be sampled by exactly one clock edge. A longer duration may cause errors. For example, a counter may count twice for a single event or a multiplier may load the incorrect operands.

While using an enable pulse between two synchronous subsystems is straightforward, it is much harder to pass the pulse crossing the clock domains. We must consider the synchronization and the difference in clock rates. The following subsections discuss several ad hoc edge detection schemes to regenerate an enable pulse from a slow or a fast clock domain. A more robust scheme that involves feedback signal is discussed in the next section.

Wide enable signal   If an enable pulse is generated from a slow clock domain, its duration may last for several clock cycles in the current clock domain, and the signal appears as a very wide pulse. A rising-edge detection circuit is needed to regenerate a shorter, synchronized enable pulse in the current clock domain. The block diagram is shown in Figure 16.11(a), which includes a synchronizer and an edge detection circuit.

The rising-edge detection circuit can be designed by using an FSM or direct implementation, as discussed in Section 10.4.1. We use the implementation shown in Figure 10.19 of Section 10.8.1, and its VHDL code is repeated in Listing 16.2.

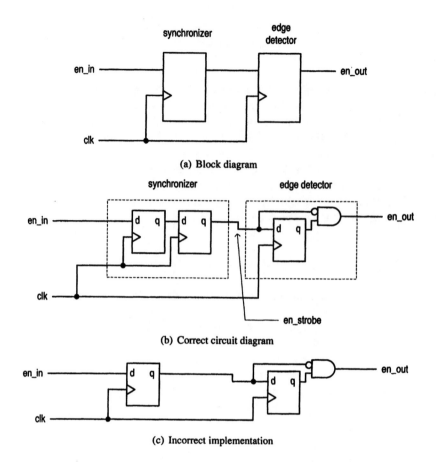

(a) Block diagram

(b) Correct circuit diagram

(c) Incorrect implementation

Figure 16.11   Regeneration of a wide enable signal.

Listing 16.2   Rising-edge detection circuit

```
library ieee;
use ieee.std_logic_1164.all;
entity rising_edge_detector is
    port(
        clk, reset: in std_logic;
        strobe: in std_logic;
        pulse: out std_logic
    );
end rising_edge_detector;

architecture direct_arch of rising_edge_detector is
    signal delay_reg: std_logic;
begin
    -- delay register
    process(clk,reset)
    begin
        if (reset='1') then
            delay_reg <= '0';
```

```
            elsif (clk'event and clk='1') then
20              delay_reg <= strobe;
            end if;
        end process;
        -- decoding logic
        pulse <= (not delay_reg) and strobe;
25 end direct_arch;
```

After substituting the gate-level implementation in the blocks, we can obtain a more detailed circuit diagram, as shown in Figure 16.11(b).

The VHDL code for the complete enable pulse regeneration circuit is shown in Listing 16.3. We intentionally use the component instantiation and create two component instances in the top-level description to highlight the use of a synchronizer and to differentiate it from a regular sequential circuit. After each design change, the synchronizer instance must be reexamined and, if needed, replaced, to ensure the proper MTBF.

Listing 16.3   Enable pulse regenerator for a wide enable signal

```
library ieee;
use ieee.std_logic_1164.all;
entity sync_en_pulse is
    port(
5       clk, reset: in std_logic;
        en_in: in std_logic;
        en_out: out std_logic
    );
end sync_en_pulse;
10
architecture slow_en_arch of sync_en_pulse is
    component synchronizer
        port(
            clk, reset: in std_logic;
15          in_async: in std_logic;
            out_sync: out std_logic
        );
    end component;
    component rising_edge_detector
20      port(
            clk, reset: in std_logic;
            strobe: in std_logic;
            pulse: out std_logic
        );
25  end component;
    signal en_strobe: std_logic;
begin
    sync: synchronizer
        port map (clk=>clk, reset=>reset, in_async=>en_in,
30                 out_sync=>en_strobe);
    edge_detect: rising_edge_detector
        port map (clk=>clk, reset=>reset, strobe=>en_strobe,
                   pulse=>en_out);
end slow_en_arch;
```

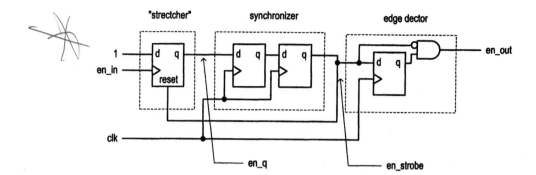

Figure 16.12 Regeneration of a narrow enable signal.

We may be tempted to use the second D FF of the synchronizer to function as the edge detection circuit to save a D FF and to reduce the propagation delay, as shown in Figure 16.11(c). This is a poor design since the unresolved signal may leak through the and cell and propagate to the downstream logic.

Narrow enable signal Handling an enable pulse from a fast clock domain is more difficult. For example, if the pulse is generated from a domain whose clock frequency is eight times faster than the frequency of the current clock domain, the duration of the enable pulse is only one-eighth of the period of the current clock signal. The sampling edge of the D FF of the synchronizer is likely to miss the narrow pulse.

Since the signal cannot be sampled by the clock edge, no synchronous design method can solve this problem. We must turn to ad hoc techniques to "stretch" the pulse until it is sampled by the current clock. One possible design is shown in Figure 16.12. In this design, the enable pulse is used as the clock for the stretcher D FF. When a pulse arrives, the stretcher D FF is loaded with '1'. The output of the D FF is then passed to the synchronizer. After the pulse is synchronized, the asserted synchronizer output clears the first D FF via the asynchronous reset. Due to the random one-clock delay of the synchronizer, the duration of the synchronized output can be one or two clock periods, and thus an edge detection circuit is needed to ensure correct operation. Because the first D FF is driven by a different clock signal, it should be excluded for the regular timing analysis and testing circuit. The VHDL code of the revised architecture body is shown in Listing 16.4.

Listing 16.4 Enable pulse regenerator for a narrow enable signal

```
architecture fast_en_arch of sync_en_pulse is
    component synchronizer
      port(
        clk, reset: in std_logic;
        in_async: in std_logic;
        out_sync: out std_logic
      );
    end component;
    component rising_edge_detector
      port(
        clk, reset: in std_logic;
        strobe: in std_logic;
        pulse: out std_logic
      );
```

```
15  end component;
    signal en_strobe: std_logic;
    signal en_q: std_logic;
  begin
    -- ad hoc stretcher
20  process(en_in,en_strobe)
    begin
       if (en_strobe='1') then
          en_q <= '0';
       elsif (en_in'event and en_in='1') then
25        en_q <= '1';
       end if;
    end process;
    -- slow enable pulse regenerator
    sync: synchronizer
30     port map (clk=>clk, reset=>reset, in_async=>en_q,
                 out_sync=>en_strobe);
    edge_detect: rising_edge_detector
       port map (clk=>clk, reset=>reset, strobe=>en_strobe,
                 pulse=>en_out);
35 end fast_en_arch;
```

Since the function of the first D FF depends only on the rising edge, not on the duration, of the incoming pulse, this scheme can be applied to a wide enable pulse as well. Note that the incoming enable pulse must be glitch-free to prevent false triggering.

16.6.2 Level-alternation scheme

An alternative to the ad hoc pulse-stretching circuit is to slightly modify the interface between the two clock domains and use the alternation of the output level to carry the information. In this scheme, the sending subsystem toggles the output value when an enable pulse is generated and thus embeds the pulse information into the signal transition edges. The block diagram is shown in Figure 16.13(a). The circuit is a T FF, which toggles its output after each time the en signal is asserted. When an enable pulse arrives, the en_level signal switches state, as shown in the top and middle parts of the timing diagram in Figure 16.13(c). The corresponding VHDL segment is

```
    . . .
    -- D FF
    process(clk, reset)
    begin
       if (reset='1') then
          t_next <= '0';
       elsif (clk'event and clk='1') then
          t_reg <= t_next;
       end if;
    end process;
    -- next-state logic
    t_next <= not (t_reg) when en='1' else
              t_reg;
    -- output logic
    en_level <= t_reg
```

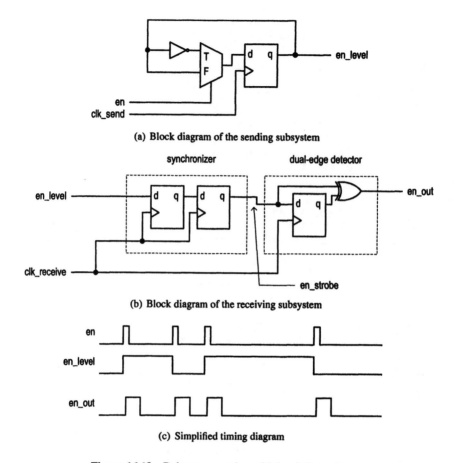

(a) Block diagram of the sending subsystem

(b) Block diagram of the receiving subsystem

(c) Simplified timing diagram

Figure 16.13 Pulse regeneration with level alternation.

In the domain that receives the enable pulse, it needs a synchronizer and a dual-edge detection circuit that can detect both the rising and falling edges of an input signal. The edge detection circuit senses the change in signal level and converts it back to a single one-clock-period pulse. We can derive the dual-edge detection circuit by using an FSM or direct implementation. One possible direct implementation is to perform an xor operation over the current input value and the previous input value stored in a D FF, as shown in Figure 16.13(b). The output waveform of the regenerator is demonstrated in the bottom part of the timing diagram in Figure 16.13(c). For clarity, the synchronizer delay is not included in the diagram.

The VHDL codes for the dual-edge detection circuit and the architecture body of the revised enable pulse regeneration circuit are shown in Listings 16.5 and 16.6 respectively. Note that the new dual_edge_dector component is used in the architecture body.

Listing 16.5 Dual-edge detection circuit

```
library ieee;
use ieee.std_logic_1164.all;
entity dual_edge_detector is
    port(
```

```
s        clk, reset: in std_logic;
         strobe: in std_logic;
         pulse: out std_logic
     );
  end dual_edge_detector;
10
  architecture direct_arch of dual_edge_detector is
    signal delay_reg: std_logic;
  begin
     -- delay register
15    process(clk,reset)
      begin
         if (reset='1') then
            delay_reg <= '0';
         elsif (clk'event and clk='1') then
20          delay_reg <= strobe;
         end if;
      end process;
      -- decoding logic
      pulse <= delay_reg xor strobe;
25 end direct_arch;
```

Listing 16.6   Enable pulse regenerator using the level-alternation scheme

```
  architecture level_arch of sync_en_pulse is
    component synchronizer
       port(
          clk, reset: in std_logic;
s         in_async: in std_logic;
          out_sync: out std_logic
       );
    end component;
    component dual_edge_detector
10     port(
          clk, reset: in std_logic;
          strobe: in std_logic;
          pulse: out std_logic
       );
15    end component;
      signal en_strobe: std_logic;
  begin
    sync: synchronizer
       port map (clk=>clk, reset=>reset, in_async=>en_in,
20               out_sync=>en_strobe);
    edge_detect: dual_edge_detector
       port map (clk=>clk, reset=>reset, strobe=>en_strobe,
                 pulse=>en_out);
  end level_arch;
```

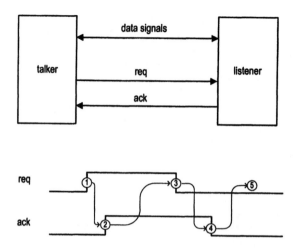

Figure 16.14   Basic conceptual and timing diagrams of the four-phase handshaking protocol.

16.7   HANDSHAKING PROTOCOL

While the pulse regeneration schemes of Section 16.6 can handle an enable signal with different widths, they cannot control the *rate* at which the enable pulses are generated. For example, consider a sending subsystem with a clock frequency that is eight times faster than that of a receiving subsystem. The previous schemes can regenerate the enable pulse in the receiving subsystem even when the input's width is only one-eighth that of the clock period. However, if the enable pulse is generated every four clock cycles, the rate is too fast for the receiving subsystem to process, and some pulses will be lost when crossing the domains. In order to function properly, the sending subsystem needs some knowledge of the receiving subsystem and issues the enable pulse accordingly.

To develop a more robust scheme, we must utilize a feedback signal from the receiving subsystem to communicate its status and establish a rule, which is known as a *protocol*, between the two subsystems. The following subsections discuss a four-phase and a two-phase handshaking protocols. While these protocols can be used to regulate the rate of the arriving enable pulses, their major applications are associated with the data transfer between two clock domains. This subject is discussed in the next section.

16.7.1   Four-phase handshaking protocol

The most commonly used scheme to coordinate operations between two clock domains is the four-phase handshaking protocol. This protocol makes no assumptions about the relative data processing rates between the clock domains and thus can accommodate a wide range of applications. In this protocol, the two subsystems are designated as the *talker* and the *listener* respectively. The talker and the listener exchange information via the req signal, which is the request signal from the talker to the listener, and the ack signal, which is the acknowledge signal from the listener to the talker. The simplified block diagram is shown in Figure 16.14(a).

The basic operation sequence (i.e., the handshaking procedure) of the four-phase handshaking protocol is illustrated in Figure 16.14(b). It consists of the following steps:

1. The talker activates the req signal.

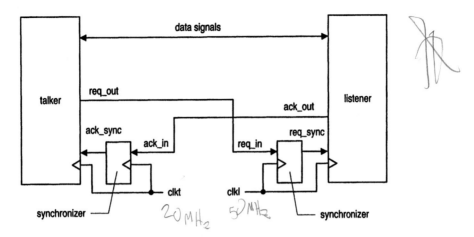

Figure 16.15   Handshaking system with synchronizers.

2. When the listener detects activation of the req signal, it activates the ack signal to inform the talker.
3. When the talker senses activation of the ack signal, it deactivates the req signal.
4. After the listener detects deactivation of the req signal, it deactivates the ack signal accordingly.
5. Once the talker senses deactivation of the ack signal, it returns to the initial state. The talker can issue a new request if needed.

In this protocol, the listener provides feedback information via the ack signal to let the talker know that a change is detected in the req signal, and the talker can respond accordingly. Note that there is no assumption about the operation speed of the listener and the talker. The talker must keep the req signal asserted until the ack signal is activated. The talker does not need to make any assumptions about the operation speed or the clock rate and can send a signal to a subsystem with unknown characteristics.

Note that we can combine the talker and the listener and treat them as a single system. The values of the req and ack signals define the "system state." When the req and ack signals are "00", the system is in the idle or initial state. During the handshaking process, they change to "10", "11" and "01" and eventually return to "00", the original state. We call the protocol *four-phase handshaking* because the sequence progresses through four distinctive states.

Since the req and ack signals cross the clock domains, two synchronizers are needed in the actual implementation. The more detailed block diagram of the handshaking scheme is shown in Figure 16.15. In the actual implementation, we use the _in, _out and _sync suffixes to indicate that the corresponding signal is an asynchronous input signal, output signal and synchronized input signal respectively.

The protocol can be implemented by two separate FSMs, one for the talker and one for the listener. Their ASM charts are shown in Figure 16.16. We assume that the talker FSM also has an input command, start, and an output status, ready. The FSM initializes the handshaking operation when the start signal is activated and asserts the ready signal when it is in the idle state. When the sending subsystem wants to issue an enable pulse across the clock domain, it checks the ready signal to ensure that the talker FSM is idle and then activates the start signal for one clock cycle. After the talker FSM senses the

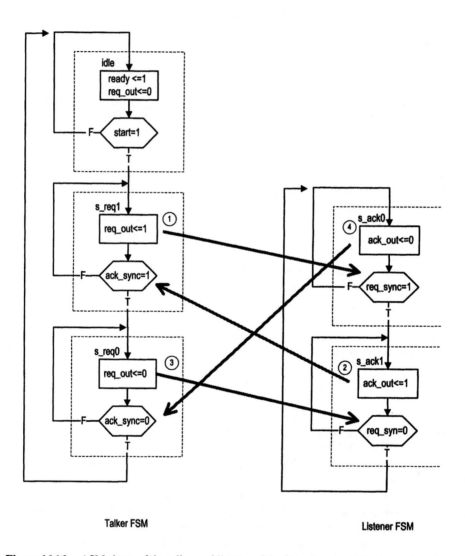

Talker FSM

Listener FSM

Figure 16.16   ASM charts of the talker and listener of the four-phase handshaking protocol.

start signal, it moves to the s_req1 state, in which the req_out signal is activated. The FSM then stays in the s_req1 state until activation of the acknowledge signal, ack_sync. It then moves to the s_req0 state and deactivates the req_out signal. The FSM returns to the idle state after it senses deactivation of the ack_sync signal.

The listener FSM is similar to the talker FSM except that it contains no start signal and thus can only respond to the talker FSM.

Because the ack_out and req_out signals are to be synchronized by a different clock domain, they must be glitch-free. This can be achieved by adding proper output buffers. Since they are designed as Moore outputs in the FSMs, we use the look-ahead output buffer scheme discussed in Section 10.7.2. The VHDL codes of the two FSMs are shown in Listings 16.7 and 16.8 respectively.

Listing 16.7   Talker FSM of the four-phase handshaking protocol

```vhdl
library ieee;
use ieee.std_logic_1164.all;
entity talker_fsm is
    port(
        clk, reset: in std_logic;
        start, ack_sync: in std_logic;
        ready: out std_logic;
        req_out: out std_logic
    );
end talker_fsm;

architecture arch of talker_fsm is
    type t_state_type is (idle, s_req1, s_req0);
    signal state_reg, state_next: t_state_type;
    signal req_buf_reg, req_buf_next: std_logic;
begin
    -- state register and output buffer
    process(clk,reset)
    begin
        if (reset='1') then
            state_reg <= idle;
            req_buf_reg <='0';
        elsif (clk'event and clk='1') then
            state_reg <= state_next;
            req_buf_reg <=req_buf_next;
        end if;
    end process;
    -- next-state logic
    process(state_reg,start,ack_sync)
    begin
        ready <='0';
        state_next <= state_reg;
        case state_reg is
            when idle =>
                if start='1' then
                    state_next <= s_req1;
                end if;
                ready <= '1';
            when s_req1 =>
```

```
40              if ack_sync='1' then
                    state_next <= s_req0;
                end if;
            when s_req0 =>
                if ack_sync='0' then
45                      state_next <= idle;
                end if;
        end case;
    end process;
    -- look-ahead output logic
50  process(state_next)
    begin
        case state_next is
            when idle =>
                req_buf_next <= '0';
55          when s_req1 =>
                req_buf_next <= '1';
            when s_req0 =>
                req_buf_next <= '0';
        end case;
60  end process;
    req_out <= req_buf_reg;
end arch;
```

Listing 16.8 Listener FSM of the four-phase handshaking protocol

```
library ieee;
use ieee.std_logic_1164.all;
entity listener_fsm is
    port(
5       clk, reset: in std_logic;
        req_sync: in std_logic;
        ack_out: out std_logic
    );
end listener_fsm;
10
architecture arch of listener_fsm is
    type l_state_type is (s_ack0, s_ack1);
    signal state_reg, state_next: l_state_type;
    signal ack_buf_reg, ack_buf_next: std_logic;
15 begin
    -- state register and output buffer
    process(clk,reset)
    begin
        if (reset='1') then
20          state_reg <= s_ack0;
            ack_buf_reg <='0';
        elsif (clk'event and clk='1') then
            state_reg <= state_next;
            ack_buf_reg <= ack_buf_next;
25      end if;
    end process;
    -- next-state logic
```

```
     process(state_reg,req_sync)
     begin
30       state_next <= state_reg;
         case state_reg is
             when s_ack0 =>
                 if req_sync='1' then
                     state_next <= s_ack1;
35               end if;
             when s_ack1 =>
                 if req_sync='0' then
                     state_next <= s_ack0;
                 end if;
40           end case;
     end process;
     -- look-ahead output logic
     process(state_next)
     begin
45       case state_next is
             when s_ack0 =>
                 ack_buf_next <= '0';
             when s_ack1 =>
                 ack_buf_next <= '1';
50           end case;
     end process;
     ack_out <= ack_buf_reg;
  end arch;
```

To complete the design, synchronizers are needed for the input signals. Again, to emphasize the unique characteristics of the synchronization circuits, we separate them from the regular sequential circuits and instantiate the synchronizers in the top level. The VHDL codes of the complete design follow the block diagram in Figure 16.15 and are shown in Listings 16.9 and 16.10 respectively.

Listing 16.9   Talker of the four-phase handshaking protocol

```
  library ieee;
  use ieee.std_logic_1164.all;
  entity talker_str is
     port(
5        clkt: in std_logic;
         resett: in std_logic;
         ack_in: in std_logic;
         start: in std_logic;
         ready: out std_logic;
10       req_out: out std_logic
     );
  end talker_str;

  architecture str_arch of talker_str is
15   signal ack_sync: std_logic;
     component synchronizer
         port(
             clk: in std_logic;
             in_async: in std_logic;
```

```
20          reset: in std_logic;
            out_sync: out std_logic
        );
    end component;
    component talker_fsm
25      port(
            ack_sync: in std_logic;
            clk: in std_logic;
            reset: in std_logic;
            start: in std_logic;
30          ready: out std_logic;
            req_out: out std_logic
        );
    end component;
  begin
35  sync_unit: synchronizer
        port map (clk=>clkt, reset=>resett, in_async=>ack_in,
                  out_sync=>ack_sync);
    fsm_unit: talker_fsm
        port map (clk=>clkt, reset=>resett, start=>start,
40                ack_sync=>ack_sync, ready=>ready,
                  req_out=>req_out);
  end str_arch;
```

Listing 16.10   Listener of the four-phase handshaking protocol

```
library ieee;
use ieee.std_logic_1164.all;
entity listener_str is
    port(
5       clkl: in std_logic;
        resetl: in std_logic;
        req_in: in std_logic;
        ack_out: out std_logic
    );
10 end listener_str;

  architecture str_arch of listener_str is
    signal req_sync: std_logic;
    component listener_fsm
15      port (
            clk: in std_logic;
            req_sync: in std_logic;
            reset: in std_logic;
            ack_out: out std_logic
20      );
    end component;
    component synchronizer
        port (
            clk: in std_logic;
25          in_async: in std_logic;
            reset: in std_logic;
            out_sync: out std_logic
```

```
        );
    end component;
30  begin
    sync_unit: synchronizer
        port map (clk=>clkl, reset=>resetl, in_async=>req_in,
                  out_sync=> req_sync);
    fsm_unit: listener_fsm
35      port map (clk=>clkl, reset=>resetl, req_sync=>req_sync,
                  ack_out => ack_out);
    end str_arch;
```

We can use this protocol to pass an enable pulse across the clock domain by connecting the en signal to the start signal of the talker FSM. When an enable pulse arrives, the talker initiates the handshaking operation. When the listener detects the activation edge of the req_in signal, it can also generate an output pulse, which corresponds to the regenerated enable pulse in the new clock domain. Since the sending subsystem cannot generate another enable pulse until the handshaking operation is completed, the sending subsystem will not overrun the the receiving subsystem.

At first glance, the four phases may appear to be somewhat redundant. We may be tempted to discard the second half of the handshaking to simplify the FSMs and let the talker and listener return to the initial state automatically. Let us consider what happens if this is done. Assume that the talker and the listener deactivate the req and ack signals automatically after the system reaches the "11" phase. There will be no problem if the deactivations are done simultaneously, as shown in the timing diagram of Figure 16.17(a). However, since the two subsystems are driven by different clocks, this is hardly possible. If the talker is much slower, the listener may be fooled into thinking that the asserted req signal is the initiation of a new request, as shown in the timing diagram of Figure 16.17(b). At time t_2, the listener deactivates the ack signal. It then senses the activation of the req signal and mistakenly treats the condition as a new round of handshaking and responds accordingly. Thus, the same incoming request pulse will be incorrectly regenerated again. On the other hand, if the listener is too slow, the talker may start to send a new request when the ack signal is still asserted, as shown in the timing diagram of Figure 16.17(c). At time t_3, the talker mistakenly thinks that the handshaking is completed and starts a new round shortly after. Since the listener still processes the first request, the new request will be lost. These examples show that all steps are needed in the original four-phase handshaking protocol.

16.7.2 Two-phase handshaking protocol

In the four-phase handshaking protocol, the talker and the listener exchange information on two separate occasions. One is during the first half of the handshaking, activation and acknowledgment of the req signal, and the other is during the second half, deactivation and acknowledgment of the req signal. Some applications, such as sending an enable pulse across the domain, require only a single exchange of information. In these applications, the req signal (e.g., the enable signal) has already been successfully detected and regenerated in the first half. The purpose of the second half is to ensure that the system can return safely to the initial state.

We can make the handshaking scheme more efficient by including only a single information exchange in the protocol. In this scheme, we do not require the system to return to the original state and define that the system is idle when the req and ack signals are

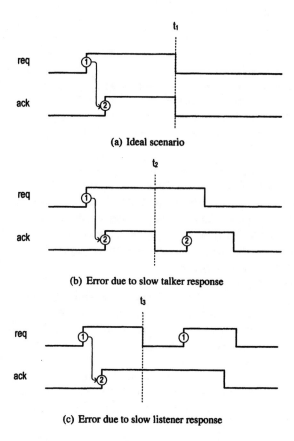

(a) Ideal scenario

(b) Error due to slow talker response

(c) Error due to slow listener response

Figure 16.17 Timing diagrams of an erroneous protocol.

both '0' or both '1'. The system will alternate between the two representations of the idle state.

The operation sequence of the new protocol includes the following steps:

1. The talker activates the req signal.
2. When the listener detects activation of the req signal, it activates the ack signal to inform the talker.
3. After the talker senses activation of the ack signal, it knows that the handshaking is completed and the system reaches the idle state.

Note that the values of the req and ack signals are "11". When a new round of handshaking is initiated, the system starts from the "11" state and the steps are:

1. The talker deactivates the req signal.
2. When the listener detects deactivation of the req signal, it deactivates the ack signal to inform the talker.
3. After the the talker senses deactivation of the ack signal, it knows that the handshaking is completed and the system reaches the idle state.

Note that the values of the req and ack signals are "00" now, and thus the system returns to its initial state. The timing diagram is shown in Figure 16.18. Although the appearance of the four-phase and two-phase timing diagrams are similar, interpretation of the req and ack signals (i.e., system state) is very different.

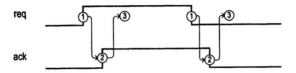

Figure 16.18 Timing diagram of the two-phase handshaking protocol.

We can follow the previous procedure to derive the talker and listener FSMs for the two-phase handshaking protocol. The revised the talker and listener ASM charts are shown in Figure 16.19. Note that the talker FSM stays in the s_req1 and s_req0 states until a new round of handshaking is initiated (i.e., when the start signal is '1'). Closer observation shows that the idle and s_req0 states of the talker FSM are equivalent, and we can merge the two states and remove the idle state.

As in the four-phase handshaking system, two synchronizers are needed for the acknowledge and request signals in the final implementation.

16.8 DATA TRANSFER CROSSING CLOCK DOMAINS

Data transfer between synchronous subsystems is just passing data from one register to another register, and the operation takes one clock cycle. Data transfer between two clock domains is more complicated. As in passing a single enable signal, it involves two issues, which are synchronization of the data signals and regulation of the data transfer rate.

In most applications, the interface between clock domains includes command signals, data lines and address lines. As we discussed in Section 16.5.5, synchronizing related signals is difficult and error-prone. A better alternative is to bundle all signals and use an enable signal to coordinate the access of the bundled signals. Basically, the sending subsystem activates the bundled signals, waits until they are stabilized, and then activates an enable signal to inform the receiving subsystem to access the bundled signals. Since the bundled signals are stabilized when accessed, no timing violation will occur. Only the enable signal is subjected to the metastability condition and needs to be synchronized. Instead of worrying about the synchronization of all signals, we need only focus on the enable signal.

Since clock frequencies and data processing rates are likely to be different in two clock domains, resolving the synchronization problem alone cannot guarantee reliable data transfer. We also need a mechanism to control the rate of data transfer to ensure that no information is lost or duplicated during the transaction. We can incorporate the data transfer into the earlier handshaking protocols and divide the transfer into three categories:

- Four-phase handshaking transfer
- Two-phase handshaking transfer
- One-phase transfer

The four-phase handshaking transfer has the highest overhead but is most robust. It assumes that the two subsystems have minimal information about each other. One-phase transfer uses a single enable signal with no feedback. It involves minimal overhead, but its operation is based on the assumption that the two subsystems have prior knowledge of the other's timing characteristics.

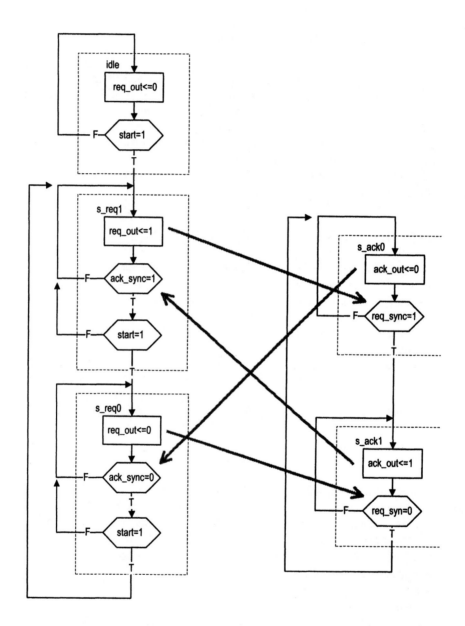

Figure 16.19   ASM charts of the talker and listener of the two-phase handshaking protocol.

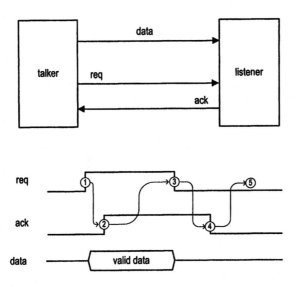

Figure 16.20   Push operation using the four-phase handshaking protocol.

For an asynchronous subsystem, storing data into another subsystem is known as a *push* operation and retrieving data from another subsystem is known as a *pull* operation. Many applications process data stage after stage, and thus the push operation is more common.

16.8.1   Four-phase handshaking protocol data transfer

The req and ack signals of the handshaking protocol form a special signaling mechanism and can be associated with various operations in the talker and listener. They can be used to perform push, pull or combined operations.

Basic one-direction data transfer   Let us first consider the basic push operation, in which the talker transfers one data word to the listener. The conceptual block diagram and a representative timing diagram are shown in Figure 16.20. The basic handshaking sequence remains the same, and the talker places data on the data bus according to activation and deactivation of the req signal. The operation follows the basic handshaking sequence:

1. The talker activates the req signal and also places the data on the data bus.
2. The listener detects activation of the req signal and understands that data is available. After retrieving and processing the data, it activates the ack signal.
3. When the talker senses activation of the ack signal, it deactivates the req signal and removes the data from the data bus.
4. The listener deactivates the ack signal accordingly.
5. Once the talker senses deactivation of the ack signal, it knows the data transfer is completed and a new one can be initiated.

A possible implementation of the talker and listener is shown in Figure 16.21. We assume that the data line is a tri-state bus. The talker can place the data word on the bus by asserting the tri_oe signal, the enable signal of the tri-state buffer. As discussed above, the data is placed on the bus when the req signal is asserted. This can be achieved by asserting the tri_oe signal in the s_req1 state of the talker FSM. Note that when the data is on the bus,

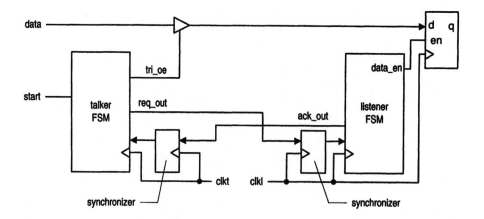

Figure 16.21   Block diagram of the push operation.

the req_out signal is also asserted. We can actually use the req_out signal to control the tri-state buffer.

The listener has a register for the input data and retrieves the data word by asserting the data_en signal, the enable signal of the register. The data_en signal can be asserted when the listener detects activation of the req_syn signal. Since the req_syn signal is delayed by two D FFs of the synchronizer, its activation is at least one clock cycle later than activation of the req and data signals. Thus, the data signal should be stabilized when the req_sync signal is activated and thus no timing violation will occur. We can modify the code of the listener FSM in Listing 16.8 to include the data_en signal as an output signal:

```
    . . .
state_next <= state_reg;
data_en <='0';
case state_reg is
    when s_ack0 =>
        if req_sync='1' then
            state_next <= s_ack1;
            data_en <='1'; -- activate enable signal
        end if;
    when s_ack1 =>
        . . .
end case;
```

Note that this design is only for demonstration purposes. Since the data transfer is not bidirectional, the tri-state buffer is not actually needed. The push operation should function properly as long as the desired data is placed on the data bus when the req_out signal is asserted.

The basic pull operation is similar to the push operation except that the listener provides the data and the talker retrieves the data. The simplified block diagram and timing diagram are shown in Figure 16.22. After sensing activation of the req signal, the listener places the data on the data bus and activates the ack signal. Once detecting activation of the ack signal, the talker retrieves the data and deactivates the req signal. The listener then removes the data and deactivates the ack signal accordingly. Again, because the ack_syn signal is

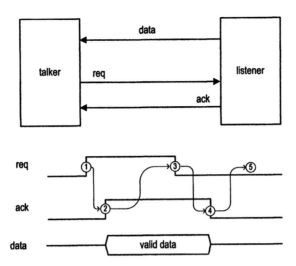

Figure 16.22   Pull operation using the four-phase handshaking protocol.

delayed by the synchronizer, the data signal should be stabilized when the ack_sync signal is activated.

Bidirectional data transfer   The four-phase handshaking protocol can also incorporate more sophisticated operation. The talker can bundle additional information, such as the commands and address lines, push them to the listener and later pull the result back. The listener retrieves the bundled signals, processes the data according to the command and activates the ack signal when the operation is done. The following example illustrates the use of handshaking to access an eight-word register file in a different clock domain. We assume that a system consists of a processor and an I/O controller, which reside in different clock domains. The processor can read data from or write data to the eight-word register file of the I/O controller through an asynchronous interface based on the four-phase handshaking protocol. The talker and listener are in the processor's clock domain and the I/O controller's clock domain respectively. The basic block diagram is shown in Figure 16.23. To reduce the clutter, only the main components and connections of the data paths are shown.

In this system, the processor first checks the ready signal to ensure that the talker is not busy and then initiates the access by activating the start signal of the talker accordingly. When asserting the start signal, the processor also uses the rw signal to indicate the type of operation ('1' for read and '0' for write), places the address of the register file on the addr line and, in the case of a write operation, places data on the data line. After detecting the start signal, the talker of the asynchronous interface loads the address, the rw control signal and data (if needed) into its internal registers and starts the handshaking and data transfer operation.

The bundled signals include a 3-bit address line, a control signal, pull, and an 8-bit data line. Since the pull and push operations are mutually exclusive, the data line can be shared and thus is bidirectional.

The data path of the talker includes a register for the address, a register for the rw signal and two data registers to store the transmitted and received data. The data path of the listener is an eight-word register file. In a realistic scenario, the I/O controller should also be able

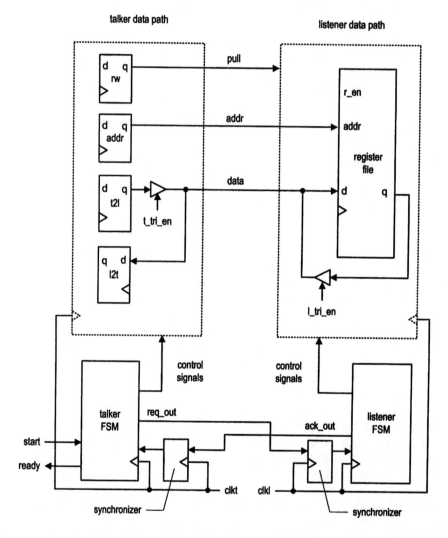

Figure 16.23   Block diagram of a push-and-pull system using the four-phase handshaking protocol.

to access the register file, and thus all signals should be multiplexed. For simplicity, the signals from the I/O controller to the register file are not shown.

The basic handshaking sequence of this circuit remains the same, and thus the state transition is similar to the FSMs of Section 16.7.1. The talker and listener FSMs also function as the control paths that control operation of the two data paths. In the idle state, the talker FSM checks the start signal. If it is asserted, the FSM moves to the s_req1 state and stores the relevant information to the registers. The remaining operation of the data path depends on the type of access. Let us first consider the push operation. In the s_req1 state, the talker activates the req_out signal and enables the tri-state buffer. The data is placed in the data line accordingly. Note that since the address line and the pull signal are not shared, they are connected to the listener data path during the entire operation.

When the listener FSM detects activation of the req_sync signal, it also checks the pull signal, whose '0' value indicates a push operation. At the next rising edge of the clock, the FSM moves to the s_ack1 state, and the data will be stored into the location specified by the addr line. Note that the req_sync signal is delayed by the D FFs of the synchronizer. All other signals are already stabilized when its activation is detected. The FSM also activates the ack_out signal when entering the s_ack1 state.

After the talker FSM senses activation of the ack_sync signal, it moves to the s_req0 state, deactivates the req_out signal and disables the tri-state buffer. The talker and listener then proceed as in regular four-phase handshaking protocol to return to the initial state.

For the pull operation, the tri-state buffer of the talker is always disabled. When the listener FSM detects activation of the req_sync signal and assertion of the pull signal, it knows that the transaction is a pull operation. At the next rising edge of the clock, the listener FSM moves to the s_ack1 state, activates the ack_out signal, and enables the tri-state buffer to place the register's output on the data line. When the talker FSM senses activation of the ack_sync signal, it knows that the data is also available. At the next rising edge of the clock, the talker FSM moves to the s_req0 state, deactivates the req_out signal and stores the data into the 12t register. The talker and listener then proceed to return to the initial state.

The VHDL codes for the talker and listener interfaces are shown in Listings 16.11 and 16.12 respectively.

Listing 16.11 Talker interface of a push-and-pull system

```
library ieee;
use ieee.std_logic_1164.all;
entity talker_interface is
    port(
5       clkt, resett: in std_logic;
        start, rw: in std_logic;   --read or write to i/o
        ack_sync: in std_logic;
        ready: out std_logic;
        req_out: out std_logic;
10      d_t2l: in std_logic_vector(7 downto 0);
        addr_in: in std_logic_vector(1 downto 0);
        d_l2t: out std_logic_vector(7 downto 0);
        pull: out std_logic;
        addr: out std_logic_vector(1 downto 0);
15      d: inout std_logic_vector(7 downto 0)
    );
  end talker_interface;
```

```
    architecture arch of talker_interface is
20     type t_state_type is (idle, s_req1, s_req0);
       signal state_reg, state_next: t_state_type;
       signal req_buf_reg, req_buf_next: std_logic;
       signal t_tri_en: std_logic;
       signal l2t_next, l2t_reg: std_logic_vector(7 downto 0);
25     signal t2l_next, t2l_reg: std_logic_vector(7 downto 0);
       signal rw_next, rw_reg: std_logic;
       signal addr_next, addr_reg: std_logic_vector(1 downto 0);
    begin
       --=================
30     -- talker FSM
       --=================
       -- state register and output buffer
       process(clkt,resett)
       begin
35        if (resett='1') then
              state_reg <= idle;
              req_buf_reg <='0';
           elsif (clkt'event and clkt='1') then
              state_reg <= state_next;
40            req_buf_reg <=req_buf_next;
           end if;
       end process;
       -- next-state logic
       process(state_reg,start,ack_sync)
45     begin
          ready <='0';
          state_next <= state_reg;
          case state_reg is
             when idle =>
50              if start='1' then
                   state_next <= s_req1;
                end if;
                ready <= '1';
             when s_req1 =>
55              if ack_sync='1' then
                   state_next <= s_req0;
                end if;
             when s_req0 =>
                if ack_sync='0' then
60                 state_next <= idle;
                end if;
          end case;
       end process;
       -- look-ahead output logic
65     process(state_next)
       begin
          case state_next is
             when idle =>
                req_buf_next <= '0';
70           when s_req1 =>
```

```
                        req_buf_next <= '1';
                when s_req0 =>
                        req_buf_next <= '0';
            end case;
75   end process;
     req_out <= req_buf_reg;
     --==================
     -- talker data path
     --==================
80   -- data register
     process(clkt,resett)
     begin
        if (resett='1') then
            t2l_reg <= (others=>'0');
85          l2t_reg <= (others=>'0');
            addr_reg <= (others=>'0');
            rw_reg <= '0';
        elsif (clkt'event and clkt='1') then
            t2l_reg <= t2l_next;
90          l2t_reg <= l2t_next;
            addr_reg <= addr_next;
            rw_reg <= rw_next;
        end if;
     end process;
95   -- data path next-state logic
     process(state_reg,t2l_reg,l2t_reg,addr_reg,rw_reg,d_t2l,
             addr_in,d,rw,start,ack_sync)
     begin
        t2l_next <= t2l_reg;
100      l2t_next <= l2t_reg;
        addr_next <= addr_reg;
        rw_next <= rw_reg;
        t_tri_en <= '0';
        case state_reg is
105         when idle =>
                rw_next <= rw;
                addr_next <= addr_in;
                if (start='1' and rw='0') then  -- write to i/o
                    t2l_next <= d_t2l;
110             end if;
            when s_req1 =>
                if (rw_reg='0') then  -- write to i/o
                    t_tri_en <= '1';
                end if;
115             if (ack_sync='1') and (rw_reg='1') then
                    l2t_next <= d;
                end if;
            when s_req0 =>
                t_tri_en <= '0';
120     end case;
     end process;
     -- output
     d <= t2l_reg when t_tri_en='1' else (others=>'Z');
```

```
      pull <= rw_reg;
125   d_l2t <= l2t_reg;
      addr <= addr_reg;
    end arch;
```

Listing 16.12   Listener interface of a push-and-pull system

```
  library ieee;
  use ieee.std_logic_1164.all;
  use ieee.numeric_std.all;
  entity listener_interface is
5   port(
        clkl, resetl: in std_logic;
        req_sync: in std_logic;
        ack_out: out std_logic;
        pull: in std_logic;
10      addr: in std_logic_vector(1 downto 0);
        d: inout std_logic_vector(7 downto 0)
    );
  end listener_interface;

15 architecture arch of listener_interface is
      type l_state_type is (s_ack0, s_ack1);
      signal state_reg, state_next: l_state_type;
      signal ack_buf_reg, ack_buf_next: std_logic;
      signal l_tri_en, r_en: std_logic;
20    type r_file_type is array (3 downto 0) of
            std_logic_vector(7 downto 0);
      signal r_file_reg: r_file_type;
    begin
      --=================
25    -- listener FSM
      --=================
      -- state register and output buffer
      process(clkl,resetl)
      begin
30      if (resetl='1') then
            state_reg <= s_ack0;
            ack_buf_reg <='0';
        elsif (clkl'event and clkl='1') then
            state_reg <= state_next;
35          ack_buf_reg <= ack_buf_next;
        end if;
      end process;
      -- next-state logic
      process(state_reg,req_sync)
40    begin
        state_next <= state_reg;
        case state_reg is
          when s_ack0 =>
            if req_sync='1' then
45            state_next <= s_ack1;
            end if;
```

```
               when s_ack1 =>
                  if req_sync='0' then
                     state_next <= s_ack0;
50                 end if;
               end case;
        end process;
        -- look-ahead output logic
        process(state_next)
55      begin
           case state_next is
               when s_ack0 =>
                  ack_buf_next <= '0';
               when s_ack1 =>
60                ack_buf_next <= '1';
           end case;
        end process;
        ack_out <= ack_buf_reg;
        --====================
65      -- listener data path
        --====================
        -- register file
        process(clkl,resetl)
        begin
70         if (resetl='1') then
               for i in 0 to 3 loop
                  r_file_reg(i) <= (others=>'0');
               end loop;
           elsif (clkl'event and clkl='1') then
75             if r_en='1' then
                  r_file_reg(to_integer(unsigned(addr))) <= d;
               end if;
           end if;
        end process;
80      -- enable logic
        process(state_reg,req_sync,pull)
        begin
           l_tri_en <= '0';
           r_en <= '0';
85         case state_reg is
               when s_ack0 =>
                  if (req_sync='1')  then
                     if (pull='0') then -- push
                        r_en <= '1';
90                   end if;
                  end if;
               when s_ack1 =>
                  if (pull='1')then
                     l_tri_en <='1';
95                end if;
           end case;
        end process;
         -- output
        d <= r_file_reg(to_integer(unsigned(addr)))
```

```
100                                     when l_tri_en='1' else
          (others=>'Z');
  end arch;
```

As in the handshaking code of Section 16.7.1, we must add two synchronizers for the request and acknowledge signals to complete the implementation.

Performance of four-phase handshaking data transfer   The strength of four-phase handshaking is that it makes a minimal assumption about the two subsystems. It will function properly even if a subsystem has no knowledge of the clock frequency and the data processing rate of other subsystems. However, there is a high overhead associated with this protocol. Assume that the clock period of the talker and listener are T_{c_t} and T_{c_l} respectively. We can estimate the required time to complete one data transfer. During a data transfer, each FSM traverses all its states and then returns to the initial state. Since the talker and listener FSMs have three and two states respectively, it takes $3T_{c\_t} + 2T_{c\_l}$. Because both the ack and req signals cross the clock domain, synchronizers are needed. If we assume that two-FF synchronizers are used, the synchronization requires up to two clock cycles whenever a signal is synchronized. The ack signal is used twice in the talker FSM, and the synchronization introduces an overhead of $4T_{c\_t}$. Similarly, synchronization of the req signal introduces an overhead of $4T_{c\_l}$. Thus, it takes $7T_{c\_t} + 6T_{c\_l}$ to complete one data transfer, which is very slow compared with the one-clock synchronous data transfer.

16.8.2   Two-phase handshaking data transfer

The two-phase handshaking protocol can reduce the overhead by half. However, since only a single handshaking occurs in the protocol, this scheme is less flexible and imposes certain constraints on the data transfer.

Let us first consider the push operation. The data transfer can be embedded in the two-phase handshaking protocol as follows:

1. The talker activates the req signal and places data on the data bus.
2. The listener detects activation of the req signal. It retrieves the data and activates the ack signal.
3. Once the talker senses activation of the ack signal, it removes the data from the data bus.

The first push operation is done at this point. Note that both the req and ack signals are '1'. When the talker wants to push the next data, the handshaking continues from this state:

1. The talker deactivates the req signal and places data on the data bus.
2. The listener detects deactivation of the req signal. It retrieves the data and deactivates the ack signal.
3. Once the talker senses deactivation of the ack signal, it removes the data from the data bus.

Note that after two push operations, the req and ack signals will be '0' and the system returns to the original state.

The block diagram for the two-phase push operation is identical to the four-phase push operation, as shown in Figure 16.20(a). A representative timing diagram is shown in Figure 16.24(a). Unlike the four-phase push operation, the req signal remains unchanged when the talker removes the data from the data bus.

Using the two-phase handshaking protocol to perform a pull operation is more difficult. The two-phase operation only allows the listener to signal the talker that it has placed the

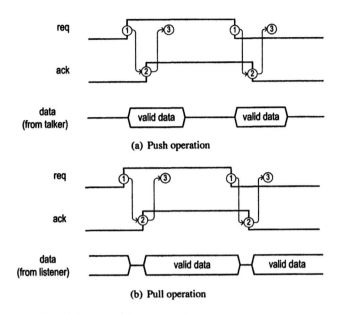

req

ack

data
(from talker)   valid data   valid data

(a) Push operation

req

ack

data
(from listener)   valid data   valid data

(b) Pull operation

Figure 16.24   Timing diagrams of push and pull operations using two-phase handshaking protocol.

data on the data bus. There is no explicit signaling mechanism to let the listener know when the data is retrieved and when the data can be removed from the bus. One way to overcome the problem is to embed this information in the next operation. When the talker initiates a new data transfer, it implicitly indicates that the data from the previous pull operation has been retrieved. Thus, when the listener detects the transition of the req signal of the next operation, it can safely remove the data from the data bus. The timing diagram is shown in Figure 16.24(b). Note that data must stay on the data bus for a long time if the two pull operations are far apart.

In the four-phase handshaking protocol, the two handshaking operations are used to indicate the initiation and completion of the data transfer. The values of the req and ack signals represent the system state and can be used to indicate the status of the data line as well. The talker and the listener, or any other subsystems that have access to the two signals, can determine the status of the data line via the two signals. This feature is important if the data line is shared. For example, in the push-and-pull design of Section 16.8.1, the push and pull are done in the same data line. The talker and listener need to know the status of the line to avoid bus fighting. On the other hand, in the two-phase handshaking protocol, handshaking operation is used only for initiation of the data transfer. The status of the data line cannot be determined by the req and ack signals, and thus the line cannot be shared. If we want to use the two-phase handshaking protocol for the previous push–pull design, separate data lines are needed for the push and pull operations.

16.8.3   One-phase data transfer

If the characteristics of the listener are known in advance, we can customize the data transfer timing and eliminate the acknowledge signal. Since there is no feedback, the request signal behaves like the enable pulse discussed in Section 16.6.

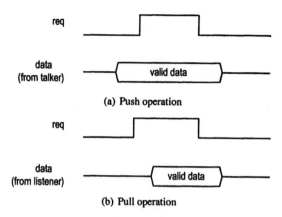

Figure 16.25 Timing diagrams of push and pull operations using one-phase protocol.

Let us first consider the push operation. The req signal now functions as an enable signal to inform the listener of the availability of the data. Since there is no feedback from the listener, the talker relies on prior knowledge about the listener to calculate the minimal assertion time for the data signal. The talker asserts the req signal and places the data on the data bus for a predetermined interval. The listener will detect activation of the req signal and retrieve the data within this interval. We can use the schemes discussed in Section 16.6 to regenerate an enable pulse from the req signal. A representative timing diagram is shown in Figure 16.25(a). If we assume that the listener is always available, the listener needs about three clock cycles (i.e., $3T_{c\_l}$) to store the data into a register. The interval includes two clock cycles to synchronize the req signal and one clock cycle to store the data.

The basic pull operation can be done in a similar fashion. The listener knows in advance how long the data should be put on the data line, and the talker knows when the data should be available. After activating the req signal, the talker waits for a predetermined amount of time and then retrieves data from the data line. A representative timing diagram is shown in Figure 16.25(b).

16.9 DATA TRANSFER VIA A MEMORY BUFFER

Although the handshaking protocol provides a reliable mechanism to transfer data across clock domains, it is not an efficient scheme. Each transaction involves a large overhead, and thus this method is good only for small, random exchanges of information between two subsystems. It is not an effective way to move a large amount of data between the two clock domains. A better alternative is to use a memory buffer as temporary storage. Instead of direct interactions, the two subsystems store and retrieve data via the memory buffer. Two common configurations are the *asynchronous FIFO buffer* and *shared memory*. These configurations cannot eliminate the metastable condition but can significantly reduce the overhead associated with data transfer.

16.9.1 FIFO buffer

A FIFO buffer is like a one-directional pipe. The sending subsystem puts the data in one end of the pipe, and the receiving subsystem retrieves the data from the other end of the pipe. In Section 9.3.2, we discussed the operation and design of a synchronous FIFO buffer, in which the two subsystems are controlled by the same clock signal. The operation of an asynchronous FIFO is similar, but the sending and receiving subsystems are controlled by clocks from different clock domains.

In an asynchronous FIFO, the read pointer (counter) is controlled by the clock signal from the receiving subsystem and the write pointer (counter) is controlled by the clock signal from the sending subsystem. Since the operation of these counters only involves the clock signal from its own domain, the counters impose no synchronization problem. The difficulty comes from the `full` and `empty` status signals. As discussed in Section 9.3.2, there are several different methods to obtain the status. These methods need information from both the sending and receiving subsystems and thus involve the signals from two clock domains. The main task of implementing an asynchronous FIFO is to design a circuit that generates reliable, properly synchronized status signals.

One possible implementation is to follow the synchronous FIFO organization discussed in Figure 9.14. For a synchronous FIFO with an n-bit address space (i.e., 2^n words), it is constructed as follows:

- Use two $(n + 1)$-bit binary counters as the pointers, one for the read pointer and one for the write pointer.
- Use two n-bit binary counters (which are the n LSBs of the $(n + 1)$-bit counters) as the read and write addresses to access the designated element of the memory array.
- Compare the two $(n + 1)$-bit counters to obtain full and empty status.

To use this scheme in an asynchronous environment, we must ensure that the comparison circuit can generate the full and empty status signals that are synchronized with their respective clock domains. To accomplish this, we must revise the design as follows:

- Add a synchronizer in the comparison circuit to synchronize the pointer from the other clock domain.
- Replace $(n + 1)$-bit and n-bit binary counters with the $(n + 1)$-bit and n-bit Gray counters.

In synchronous FIFO, the read and write pointers are implemented by binary counters or LFSRs. In these counters, there may be multiple bit changes in a transition. For example, consider a 4-bit binary counter. When the counter wraps around from "1111" to "0000", all four bits change. As discussed in Section 16.5.5, synchronizing multiple changing bits may lead to the capture of erroneous, intermediate transition values, and thus these counters can cause problems if the values are passed to a different clock domain. To prevent this, we must use a Gray counter for the pointer, in which only one bit is changed in a transition. The circulation pattern of a 4-bit counter is shown in the first column of Table 16.2.

In Section 9.3.2, we add an extra bit in the binary counter and use this bit (the MSB of the counter) to distinguish whether the FIFO is full or empty. In this approach, we use two $(n + 1)$-bit binary counters as the read and write pointers, and use two n-bit binary counters for the write and read addresses of the memory array. Note that the MSB of the Gray counter is the same as the MSB of the binary counter, and thus it can be used to distinguish whether the FIFO is empty or full. As in the binary counter–based implementation, we need two $(n + 1)$-bit Gray counters as the pointers and two n-bit Gray counters as the addresses.

It is straightforward to obtain the n-bit binary counting patterns since they are the same as the n LSBs of the $(n + 1)$-bit binary counter. It is more difficult for the Gray counter. For example, the counting patterns of three LSBs of the 4-bit Gray counter and the counting patterns of a 3-bit Gray counter are shown in the second and third columns of Table 16.2. Their patterns are different in the bottom half. Although the patterns are not identical, there

Table 16.2   Circulation pattern of 4-bit and 3-bit Gray counters

4-bit Gray counter	3 LSBs of 4-bit Gray counter	3-bit Gray counter
0000	000	000
0001	001	001
0011	011	011
0010	010	010
0110	110	110
0111	111	111
0101	101	101
0100	100	100
1100	100	000
1101	101	001
1111	111	011
1110	110	010
1010	010	110
1011	011	111
1001	001	101
1000	000	100

is no need to construct a separate n-bit Gray counter from scratch. Closer observation shows that the $(n - 1)$ LSBs of the $(n + 1)$-bit Gray counter and n-bit Gray counter are identical, and the MSB of the n-bit Gray counter can be obtained by performing an xor operation on the two MSBs of the $(n + 1)$-bit Gray counter. In other words, let a_n, a_{n-1}, $\ldots$, a_0 be the bits of an $(n + 1)$-bit Gray counter, and b_{n-1}, b_{n-2}, $\ldots$, b_0 be the bits of an n-bit Gray counter. We can derive the n-bit counting pattern by using

$$b_i = \begin{cases} a_{i+1} \oplus a_i & \text{if } i = n - 1 \\ a_i & \text{otherwise} \end{cases}$$

The block diagram of an n-bit asynchronous FIFO control circuit is shown in Figure 16.26. In the write control part, an $(n + 1)$-bit Gray counter is used as the write pointer and the derived n-bit Gray counter is used for the write address. The read pointer is obtained from the read control part. It is first synchronized by an $(n + 1)$-bit synchronizer. The comparing circuit derives the n-bit read address and compares it to the write address. If the read and write addresses are the same and the MSBs of the read and write pointers are different, the FIFO is full and the full signal is asserted accordingly. Since all inputs of the comparing circuits are synchronized with the write controller's clock, the full signal will not cause a timing violation when used. The read control part essentially mirrors the write control except for the minor difference in the comparing circuit. The empty signal will be asserted when the read and write addresses are the same and the MSBs of the read and write pointers are the same.

The VHDL codes of the write port control and the read port control are shown in Listings 16.13 and 16.14 respectively. The code of the Gray counter is similar to the code discussed in Section 7.5.1. We use a generic, N, to express the number of bits of the FIFO control circuit. The code of a generic n-bit two-FF synchronizer is shown in Listing 16.15.

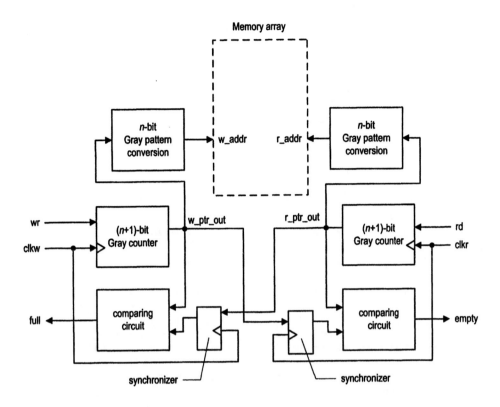

Figure 16.26   Block diagram of an asynchronous FIFO controller.

The complete asynchronous FIFO control circuit follows the basic block diagram, and its VHDL code is shown in Listing 16.16.

Listing 16.13   Write port control of an asynchronous FIFO

```
library ieee;
use ieee.std_logic_1164.all;
use ieee.numeric_std.all;
entity fifo_write_ctrl is
  generic(N: natural);
  port(
    clkw, resetw: in std_logic;
    wr: in std_logic;
    r_ptr_in: in std_logic_vector(N downto 0);
    full: out std_logic;
    w_ptr_out: out std_logic_vector(N downto 0);
    w_addr: out std_logic_vector(N-1 downto 0)
  );
end fifo_write_ctrl;

architecture gray_arch of fifo_write_ctrl is
  signal w_ptr_reg, w_ptr_next:
    std_logic_vector(N downto 0);
  signal gray1, bin, bin1: std_logic_vector(N downto 0);
```

```
20      signal waddr_all: std_logic_vector(N-1 downto 0);
        signal waddr_msb, raddr_msb: std_logic;
        signal full_flag: std_logic;
     begin
        -- register
25      process(clkw,resetw)
        begin
           if (resetw='1') then
               w_ptr_reg <= (others=>'0');
           elsif (clkw'event and clkw='1') then
30             w_ptr_reg <= w_ptr_next;
           end if;
        end process;
        -- (N+1)-bit Gray counter
        bin <= w_ptr_reg xor ('0' & bin(N downto 1));
35      bin1 <= std_logic_vector(unsigned(bin) + 1);
        gray1 <= bin1 xor ('0' & bin1(N downto 1));
        -- update write pointer
        w_ptr_next <= gray1 when wr='1' and full_flag='0' else
                      w_ptr_reg;
40      -- N-bit Gray counter
        waddr_msb <= w_ptr_reg(N) xor w_ptr_reg(N-1);
        waddr_all <= waddr_msb & w_ptr_reg(N-2 downto 0);
        -- check for FIFO full
        raddr_msb <= r_ptr_in(N) xor r_ptr_in(N-1);
45      full_flag <=
           '1' when r_ptr_in(N) /=w_ptr_reg(N) and
               r_ptr_in(N-2 downto 0)=w_ptr_reg(N-2 downto 0) and
               raddr_msb = waddr_msb else
           '0';
50      -- output
        w_addr <= waddr_all;
        w_ptr_out <= w_ptr_reg;
        full <= full_flag;
     end gray_arch;
```

Listing 16.14   Read port control of an asynchronous FIFO

```
library ieee;
use ieee.std_logic_1164.all;
use ieee.numeric_std.all;
entity fifo_read_ctrl is
5    generic(N: natural);
     port(
        clkr, resetr: in std_logic;
        w_ptr_in: in std_logic_vector(N downto 0);
        rd: in std_logic;
10      empty: out std_logic;
        r_ptr_out: out std_logic_vector(N downto 0);
        r_addr: out std_logic_vector(N-1 downto 0)
     );
  end fifo_read_ctrl;
15
```

```
architecture gray_arch of fifo_read_ctrl is
   signal r_ptr_reg, r_ptr_next: std_logic_vector(N downto 0);
   signal gray1, bin, bin1: std_logic_vector(N downto 0);
   signal raddr_all: std_logic_vector(N-1 downto 0);
   signal raddr_msb,waddr_msb: std_logic;
   signal empty_flag: std_logic;
begin
   -- register
   process(clkr,resetr)
   begin
      if (resetr='1') then
         r_ptr_reg <= (others=>'0');
      elsif (clkr'event and clkr='1') then
         r_ptr_reg <= r_ptr_next;
      end if;
   end process;
   -- (N+1)-bit Gray counter
   bin <= r_ptr_reg xor ('0' & bin(N downto 1));
   bin1 <= std_logic_vector(unsigned(bin) + 1);
   gray1 <= bin1 xor ('0' & bin1(N downto 1));
   -- update read pointer
   r_ptr_next <= gray1 when rd='1' and empty_flag='0' else
                 r_ptr_reg;
   -- N-bit Gray counter
   raddr_msb <= r_ptr_reg(N) xor r_ptr_reg(N-1);
   raddr_all <= raddr_msb & r_ptr_reg(N-2 downto 0);
   waddr_msb <= w_ptr_in(N) xor w_ptr_in(N-1);
   -- check for FIFO empty
   empty_flag <=
      '1' when w_ptr_in(N)=r_ptr_reg(N) and
           w_ptr_in(N-2 downto 0)=r_ptr_reg(N-2 downto 0) and
           raddr_msb = waddr_msb else
      '0';
   -- output
   r_addr <= raddr_all;
   r_ptr_out <= r_ptr_reg;
   empty <= empty_flag;
end gray_arch;
```

Listing 16.15 *n*-bit synchronizer

```
library ieee;
use ieee.std_logic_1164.all;
entity synchronizer_g is
   generic(N: natural);
   port(
      clk, reset: in std_logic;
      in_async: in std_logic_vector(N-1 downto 0);
      out_sync: out std_logic_vector(N-1 downto 0)
   );
end synchronizer_g;

architecture two_ff_arch of synchronizer_g is
```

```
     signal meta_reg, sync_reg: std_logic_vector(N-1 downto 0);
     signal meta_next, sync_next:
15      std_logic_vector(N-1 downto 0);
   begin
     -- two registers
     process(clk,reset)
     begin
20      if (reset='1') then
          meta_reg <= (others=>'0');
          sync_reg <= (others=>'0');
        elsif (clk'event and clk='1') then
          meta_reg <= meta_next;
25        sync_reg <= sync_next;
        end if;
     end process;
     -- next-state logic
     meta_next <= in_async;
30   sync_next <= meta_reg;
     -- output
     out_sync <= sync_reg;
   end two_ff_arch;
```

Listing 16.16 Top-level structural description of an asynchronous FIFO control circuit

```
library ieee;
use ieee.std_logic_1164.all;
use ieee.numeric_std.all;
entity fifo_async_ctrl is
5    generic(DEPTH: natural);
     port(
        clkw: in std_logic;
        resetw: in std_logic;
        wr: in std_logic;
10      full: out std_logic;
        w_addr: out std_logic_vector (DEPTH-1 downto 0);
        clkr: in std_logic;
        resetr: in std_logic;
        rd: in std_logic;
15      empty: out std_logic;
        r_addr: out std_logic_vector (DEPTH-1 downto 0)
     );
   end fifo_async_ctrl ;

20 architecture str_arch of fifo_async_ctrl is
     signal r_ptr_in: std_logic_vector(DEPTH downto 0);
     signal r_ptr_out: std_logic_vector(DEPTH downto 0);
     signal w_ptr_in: std_logic_vector(DEPTH downto 0);
     signal w_ptr_out: std_logic_vector(DEPTH downto 0);
25   -- component declarations
     component fifo_read_ctrl
        generic(N: natural);
        port(
           clkr: in std_logic;
```

```
30          rd: in std_logic;
            resetr: in std_logic;
            w_ptr_in: in std_logic_vector (N downto 0);
            empty: out std_logic;
            r_addr: out std_logic_vector (N-1 downto 0);
35       r_ptr_out: out std_logic_vector (N downto 0)
         );
      end component;
      component fifo_write_ctrl
         generic(N: natural);
40       port(
            clkw: in std_logic;
            r_ptr_in: in std_logic_vector (N downto 0);
            resetw: in std_logic;
            wr: in std_logic;
45          full: out std_logic;
            w_addr: out std_logic_vector (N-1 downto 0);
            w_ptr_out: out std_logic_vector (N downto 0)
         );
      end component;
50    component synchronizer_g
         generic(N: natural);
         port(
            clk: in std_logic;
            in_async: in std_logic_vector (N-1 downto 0);
55          reset: in std_logic;
            out_sync: out std_logic_vector (N-1 downto 0)
         );
      end component;
   begin
60    read_ctrl: fifo_read_ctrl
         generic map(N=>DEPTH)
         port map (clkr=>clkr, resetr=>resetr, rd=>rd,
                   w_ptr_in=>w_ptr_in, empty=>empty,
                   r_ptr_out=>r_ptr_out, r_addr=>r_addr);
65    write_ctrl: fifo_write_ctrl
         generic map(N =>DEPTH)
         port map(clkw=>clkw, resetw=>resetw, wr=>wr,
                  r_ptr_in=>r_ptr_in, full=>full,
                  w_ptr_out=>w_ptr_out, w_addr=>w_addr);
70    sync_w_ptr: synchronizer_g
         generic map(N=>DEPTH+1)
         port map(clk=>clkw, reset=>resetw,
                  in_async=>w_ptr_out, out_sync=>w_ptr_in);
      sync_r_ptr: synchronizer_g
75       generic map(N=>DEPTH+1)
         port map(clk=>clkr, reset=>resetr,
                  in_async=>r_ptr_out, out_sync =>r_ptr_in);
   end str_arch;
```

Because of the synchronizer, the onset of the empty and full signals may be delayed.
For example, assume that the FIFO is originally empty. After a write operation is performed,
the write pointer changes and the FIFO contains one data item. It takes two clock cycles

to propagate the change through the two cascading FFs of the synchronizer, and thus the onset of the `empty` signal is delayed by two clock cycles. During this interval, the reading subsystem will falsely assume that there is no data to retrieve and will stay idle. While the delay causes late data retrieval, the functionality of the FIFO remains intact and no invalid data item is retrieved though the buffer. The same situation happens to the `full` signal. The deassertion of the `full` signal is delayed by two clock cycles. During this interval, the sending subsystem falsely assumes that the FIFO is full and suspends the write operation.

The delayed `empty` and `full` signals force the subsystems to be idle unnecessarily and thus penalize the performance. The penalty is essentially due to the overhead associated with the synchronization of two clock domains and cannot be avoided. However, the idle situation occurs only when the FIFO is almost empty or almost full. There is no overhead or extra delay when the FIFO is partially full. In comparison, the handshaking scheme involves the overhead in every data transaction, and thus the FIFO buffer is more efficient.

16.9.2 Shared memory

Another frequently used buffering scheme is shared memory. The basic idea is to allow multiple subsystems to access a common memory. The sending subsystem can first write the data into the memory, and the receiving subsystem then obtains the data by reading the same memory location. This scheme is best suited when a large chunk of data, such as a high-resolution image, has to be transferred.

The shared memory scheme can be implemented by using a regular single-port memory or a special dual-port memory. For a single-port memory configuration, we can treat the memory as the shared resource and use an arbiter to resolve the conflicting requests and coordinate the memory usage. The basic design of the arbiter is similar to that in Section 10.8.2. Because the interactions are between different clock domains, synchronization circuits are needed for all request and grant signals.

The request-grant process is somewhat like the handshaking procedure and has a similar overhead. However, the overhead is associated with each resource arbitration, not each data transaction. Since one round of arbitration allows any amount of data to be transferred (up to the size of the shared memory), the average overhead of a single data transaction becomes very small.

A better alternative is to use dual-port memory. A dual-port memory has two independent access ports, each containing its own address line, data line and control signals. The multiplexing and decoding circuits are duplicated inside the memory module. Two memory accesses can be performed simultaneously as long as the memory addresses are different. A conflict occurs when two memory operations access the same memory location (i.e., two operations have the same memory addresses). An arbiter is used to resolve the condition. When there is no clock in the regular dual-port memory, the internal arbitrator is an asynchronous sequential circuit. It may enter a metastable state if the two request signals are asserted too closely. As in the synchronizer, the arbitration circuit needs to provide some time to resolve the metastable condition and thus introduces similar overhead.

As with other memory modules, dual-port memory cannot be synthesized from scratch in RT-level code. We must instantiate the predesigned module from the target device technology.

16.10 SYNTHESIS OF A MULTIPLE-CLOCK SYSTEM

The synchronous design is the most important methodology and the cornerstone of the entire design and fabrication process. The synthesis, timing analysis, verification and testing of synchronous systems are well understood, and many EDA software tools have been developed to automate the tasks.

One major motivation behind the synchronous methodology is to provide a systematic way to satisfy the timing constraints. The objective of synthesis and verification is to *identify* and *prevent* timing violations. The EDA software tools are developed to assist designers in achieving this objective. On the other hand, the metastability analysis and synchronization circuit are to deal with the scenario that a timing violation *has already occurred.* This is essentially a transistor-level phenomenon, and its behavior cannot easily be modeled at the gate or RT level. The EDA tools are not able to handle metastability, and the analysis and synthesis process cannot be automated. Most software can only detect and warn the onset of a time violation but cannot model or analyze what happens afterward. For example, in the std_logic data type, the 'X' value is used to model the output after a timing violation. After a timing violation occurs, the output 'X' value will be permanent and propagated to the downstream circuit. This is very different from the actual timing violation, in which the output signal may become '0' or '1', or be resolved to a stable value after a random period of time.

While the design considerations for a multiple-clock system are different from those of a synchronous system, it is not wise to abandon the synchronous design methodology and start from scratch. Instead, we want to incorporate the methodology into the new design flow and utilize the previous techniques and EDA tools as much as possible.

To achieve this goal, a multiple-clock system should be divided into synchronous subsystems and crossing-domain interfaces. A synchronous subsystem is within the same clock domain, and thus we can design it just as a regular synchronous system. On the other hand, the crossing domain interface involves the synchronization and data transfer protocol. Its analysis and design are very different from those of the synchronous system, and very few EDA tools are available for these tasks. We usually have to manually analyze, design and verify the interface circuit and protocols. Since the synchronization circuit depends on the device characteristics and the data transfer protocol sometimes depends on the clock rates of the domains, the interface is usually device dependent and is not portable.

The general design approach for a multiple-clock system can be summarized as follows:

1. Partition the system into locally synchronous subsystems.
2. Design and verify these subsystems following the synchronous methodology.
3. Develop protocol to pass data and exchange information between clock domains.
4. Manually analyze and design the necessary synchronization circuits between clock domains.
5. Verify operation of the entire system.

A representative top-level partition of a system with two clock domains is shown in Figure 16.27. It is a good idea to treat the synchronization circuit and data transfer interface as separate modules and instantiate them individually in VHDL code. These modules are normally device dependent and may need to be reanalyzed and redesigned when the system is ported to a different device technology or operation environment (e.g., a different clock rate).

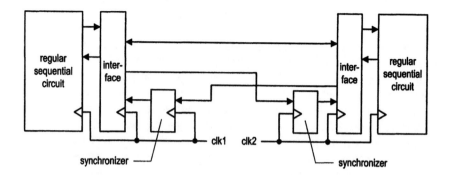

Figure 16.27   Partition of a system with two clock domains.

16.11   SYNTHESIS GUIDELINES

16.11.1   Guidelines for general use of a clock

- Do not manipulate the clock signal in regular RT-level design and synthesis.

- Minimize the number of clock signals in a system.

- Minimize the number of clock domains (i.e., the number of independent clock signals). Use a derived clock signal when possible.

- If a derived clock signal is needed, manually derive and instantiate the circuit and separate it from the regular synthesis.

16.11.2   Guidelines for a synchronizer

- Synchronize a signal in a single place.

- Avoid synchronizing related signals.

- Use a glitch-free signal for synchronization.

- Reanalyze and examine the synchronizer and MTBF when the device is changed or the clock rate is revised.

16.11.3   Guidelines for an interface between clock domains

- Clearly identify the boundary of the clock domain and the signals that cross the domain.

- Separate the synchronization circuits and asynchronous interface from the synchronous subsystems and instantiate them as individual modules.

- Use a reliable synchronizer design to provide sufficient metastability resolution time.

- Analyze the data transfer protocol over a wide range of scenarios, including faster and slower clock frequencies and different data rates.

16.12 BIBLIOGRAPHIC NOTES

The construction of the clock network involves the transmission and distribution of electronic signals. It is normally covered under the subjects of signal integration and high-speed design. Two texts by Howard Johnson, *High-Speed Digital Design: A Handbook of Black Magic* and *High-Speed Signal Propagation: Advanced Black Magic*, provide comprehensive coverage of this topic.

The study of metastability and synchronization failure relies on transistor-level model and analysis. The text, *Digital Systems Engineering* by William J. Dally and John W. Poulton, covers the theoretical foundation of this subject. The book also includes a discussion of the design of asynchronous circuit.

The more practical design materials on the synchronizer and asynchronous interface can be found in manufacturers' application notes or articles from technical conferences. Articles by Clifford E. Cummings, *Synthesis and Scripting Techniques for Designing Multi-Asynchronous Clock Designs*, *Simulation and Synthesis Techniques for Asynchronous FIFO Design* and *Simulation and Synthesis Techniques for Asynchronous FIFO Design with Asynchronous Pointer Comparisons*, provide many practical design examples and good advice.

Problems

16.1 Assume that a sequential system with an ideal clock signal can operate at a maximal clock rate of 100 MHz. If the physical clock distribution network introduces a 1.5-ns clock skew, what is the new maximal clock rate?

16.2 Consider a D FF with w and τ.
  (a) If we improve the D FF by reducing w by 10%, discuss the effect on MTBF.
  (b) If we improve the D FF by reducing τ by 10%, discuss the effect on MTBF.

16.3 At the end of Section 16.4.2, we discuss the difference between T_r and T_{r2}. Assume that w and τ are identical for the calculation of MTBF(T_r) and MTBF(T_{r2}). Derive MTBF(T_r) in terms of MTBF(T_{r2}), w and τ.

16.4 For the two-FF synchronizer discussed in Section 16.5.3, determine the new MTBF for the following:
  (a) The placement and routing process adds a 2.5-ns wiring delay.
  (b) The system clock rate is decreased by 10%.
  (c) The setup time of the D FF is reduced by 10%.
  (d) The setup time of the D FF is reduced by 25%.
  (e) The τ of the D FF is reduced by 10%.
  (f) The τ of the D FF is reduced by 25%.

16.5 We want to regenerate the enable pulse in the listener's clock domain using the four-phase handshaking protocol. In this scheme, the listener has an output signal that is asserted once during the handshaking process.
  (a) Revise the listener ASM chart of Figure 16.16 to add a Mealy output signal.
  (b) Modify the VHDL code to reflect the revised ASM chart.

16.6 Repeat Problem 16.5, but add a Moore output signal.

16.7 Repeat Problem 16.5 for the two-phase handshaking protocol of Figure 16.19.

16.8   Repeat Problem 16.5, but add a Moore output signal for the two-phase handshaking protocol of Figure 16.19.

16.9   Revise the talker ASM chart of the two-phase handshaking protocol of Figure 16.19 to eliminate the `idle` state.

16.10   In a handshaking protocol, we like to include a `ready` signal in talker to indicate that the system is idle and ready to accept another operation. Revise the talker ASM chart of the two-phase handshaking protocol of Figure 16.19 to include the `ready` output signal.

16.11   We want to design a four-phase handshaking asynchronous interface for the sequential multiplier in Section 11.6. The operand width is 8 bits and the data is passed by a 16-bit bidirectional bus. After sensing the `start` signal, the talker of the sending subsystem places the data on the data bus and activates the handshaking operation. Once the receiving subsystem detects the request, it retrieves the data and performs the multiplication operation. When the operation is completed, the listener of the receiving subsystem places the result on the data bus and asserts the acknowledge signal, and the talker retrieves the result accordingly. Draw the block diagram and derive VHDL code for this system.

16.12   Repeat Problem 16.11, but use an 8-bit data bus. Since the operation involves two data transfers, we need a master control FSM to coordinate the operation. Derive VHDL code for this system.

16.13   Modify the push-and-pull system of Section 16.8.1 using the two-phase handshaking protocol (additional data lines are needed). Derive the revised block diagram and VHDL code.

16.14   Repeat Problem 16.11, but use the two-phase handshaking protocol and two 16-bit unidirectional data buses. Derive the VHDL code.

16.15   Consider the one-phase push operation in Section 16.8.3. Derive VHDL code for the sending and receiving subsystems with the following clock rates.
  (a) $f_{send} = 10$ MHz and $f_{receive} = 10$ MHz
  (b) $f_{send} = 10$ MHz and $f_{receive} = 40$ MHz
  (c) $f_{send} = 10$ MHz and $f_{receive} = 2.5$ MHz

16.16   Repeat Problem 16.15 for the one-phase pull operation.

16.17   Consider the FIFO buffer of Section 16.9.1, and let the clock periods of the writing and reading subsystems be T_w and T_r respectively. Assume that the sending subsystem has 10 words to pass through the FIFO buffer. Determine the total time to complete the operation with the following buffer sizes:
  (a) One word.
  (b) Two words.
  (c) Four words.
  (d) Eight words.
  (e) 16 words.

REFERENCES

1. P. Alfke, "Efficient Shift Registers, LFSR Counters, and Long Pseudo-Random Sequence Generators," Xilinx Application Note XAPP-052, 1996.

2. M. G. Arnold, *Verilog Digital Computer Design*, Prentice Hall, 1998.

3. P. J. Ashenden, *The Designer's Guide to VHDL, 2nd ed.*, Morgan Kaufmann, 2001.

4. P. H. Bardell, *Built In Test for VLSI: Pseudorandom Techniques*, Wiley-Interscience, 1987.

5. L. Bening and H. D. Foster, *Principles of Verifiable RTL Design, 2nd ed.*, Springer-Verlag, 2001.

6. J. Bergeron, *Writing Testbenches: Functional Verification of HDL Models*, Springer-Verlag, 2003.

7. M. D. Ciletti, *Advanced Digital Design with the Verilog HDL*, Prentice Hall, 2003.

8. M. D. Ciletti, *Starter's Guide to Verilog 2001*, Prentice Hall, 2003.

9. C. E. Cummings, "Coding and Scripting Techniques for FSM Designs with Synthesis-Optimized, Glitch-Free Outputs," *SNUG (Synopsys Users Group conference)*, Boston, 2000.

10. C. E. Cummings, "Synthesis and Scripting Techniques for Designing Multi-Asynchronous Clock Designs," *SNUG (Synopsys Users Group conference)*, San Jose, 2001.

11. C. E. Cummings, "Simulation and Synthesis Techniques for Asynchronous FIFO Design," *SNUG (Synopsys Users Group conference)*, San Jose, 2002.

12. C. E. Cummings and P. Alfke, "Simulation and Synthesis Techniques for Asynchronous FIFO Design with Asynchronous Pointer Comparisons," *SNUG (Synopsys Users Group conference)*, San Jose, 2002.

13. W. J. Dally and J. W. Poulton, *Digital Systems Engineering*, Cambridge University Press, 1998.

14. G. De Micheli, *Synthesis and Optimization of Digital Circuits*, McGraw-Hill, 1994.

15. S. Devadas et al., *Logic Synthesis*, McGraw-Hill Professional, 1994.

16. M. D. Ercegovac and T. Lang, *Digital Arithmetic*, Morgan Kaufmann, 2003.

17. D. D. Gajski, *Principles of Digital Design*, Prentice Hall, 1997.

18. D. D. Gajski, *High-Level Synthesis: Introduction to Chip and System Design*, Springer-Verlag, 1992.

19. S. Ghosh, *Hardware Description Languages: Concepts and Principles*, Wiley-IEEE Press, 1999.

20. IEEE, *IEEE Standard for Verilog Hardware Description Language, (IEEE Std 1364-2001)*, Institute of Electrical and Electronics Engineers, 2001.

21. IEEE, *IEEE Standard VHDL Language Reference Manual (IEEE Std 1076-2001)*, Institute of Electrical and Electronics Engineers, 2001.

22. IEEE, *IEEE Standard for VHDL Register Transfer Level (RTL) Synthesis, (IEEE Std 1076.6-1999)*, Institute of Electrical and Electronics Engineers, 2000.

23. IEEE, *IEEE Standard VHDL Synthesis Packages (IEEE Std 1076.3-1997)*, Institute of Electrical and Electronics Engineers, 1997.

24. IEEE, *IEEE Standard Multivalue Logic System for VHDL Model Interoperability (IEEE Std 1164-1993)*, Institute of Electrical and Electronics Engineers, 1993.

25. H. Johnson, *High-Speed Digital Design: A Handbook of Black Magic*, Prentice Hall, 1993.

26. T. Kam, *Synthesis of Finite State Machines: Functional Optimization*, Kluwer Academic, 1997.

27. R. H. Katz and G. Borriello, *Contemporary Logic Design, 2nd ed.*, Prentice Hall, 2004.

28. M. Keating and P. Bricaud, *Methodology Manual for System-on-a-Chip Designs, 3rd ed.*, Springer-Verlag, 2002.

29. I. Koren, *Computer Arithmetic Algorithms, 2nd ed.*, A. K. Peters, 2002.

30. H. A. Landman, "Visualizing the Behavior of Logic Synthesis Algorithms," *SNUG (Synopsys Users Group conference)*, 1998.

31. C. M. Maxfield, *The Design Warrior's Guide to FPGAs*, Newnes, 2004.

32. S. Palnitkar, *Verilog HDL, 2nd ed.*, Prentice Hall, 2003.

33. D. A. Patterson and J. L. Hennessy, *Computer Organization and Design: The Hardware/Software Interface, 3rd ed.*, Morgan Kaufmann, 2004.

34. Jan M. Rabaey, *Digital Integrated Circuits, 2nd ed.*, Prentice Hall, 2002.

35. A. Rushton, *VHDL for Logic Synthesis, 2nd ed.*, John Wiley & Sons, 1998.

36. P. Sinander, *VHDL Modelling Guidelines*, European Space Agency, 1994.

37. T. Villa et al., *Synthesis of Finite State Machines: Logic Optimization*, Kluwer Academic, 1997.

38. J. F. Wakerly, *Digital Design: Principles and Practices*, Prentice Hall, 2002.

39. W. Wolf, *FPGA-Based System Design*, Prentice Hall, 2004.

40. W. Wolf, *Modern VLSI Design: System-on-Chip Design, 3rd ed.*, Prentice Hall, 2002.

41. R. Zimmermann, *Binary Adder Architectures for Cell-Based VLSI and Their Synthesis*, PhD. thesis, Swiss Federal Institute of Technology (ETH), Zurich, 1998.

42. R. Zimmermann, "VHDL Library of Arithmetic Units," *First International Forum on Design Languages (FDL'98)*, 1998.

INDEX

abstraction, 9
  gate-level, 10
  processor-level, 12
  register-transfer (RT) level, 11
  transistor-level, 10
actual signal, 478
adder, 171, 201, 510
alias, 52
ALU, 75, 87, 104, 112
arbiter, 353
architecture body, 28, 46
array
  aggregate, 59
  array-of-arrays, 548
  constrained, 546
  emulated two-dimensional, 550
  two-dimensional, 546
  unconstrained, 503, 547
ASIC, 3
  full custom, 3
  gate array, 3
  standard-cell, 3, 143
ASM chart, 317
ASMD chart, 379
association
  named, 479
  positional, 480
asynchronous circuit, 216, 219
attribute
  'event, 222

'high, 502
'left, 502
'length, 502
'low, 502
'range, 502
'reverse_range, 502
'right, 502
array, 502
enum_encoding, 339
user, 339
barrel shifter, 178, 192, 566
bidirectional I/O, 134
big-O notation, 126
binary decoder, 73, 86, 104, 112, 513, 528, 558
binary encoder, 564
binding, 461
CAM (content addressable memory), 287
case statement, 112
clock
  derived, 262, 611
  distribution network, 603
  gated, 260
  skew, 605
combinational circuit, 69, 213
comment, 47
comparator, 173, 177
component, 30, 475
  declaration, 31, 475
  instantiation, 32, 477
computability, 126

667

computation complexity, 126
concurrent statement, 46
conditional signal assignment statement, 72, 105
configuration, 37, 46, 485, 526
    declaration, 486
    specification, 488
constant, 52
cost
    development, 7
    non-recurring engineering (NRE), 6
    part, 6
    time-to-market, 7
counter
    arbitrary-sequence, 232
    binary, 233, 247, 367, 482, 518, 521
    decade, 236, 476
    decimal, 272, 481
    Gray, 265
    mod-n, 483
    programmable, 237, 248, 252
    ring, 266, 511
critical path, 152
D FF, 214, 222, 226, 245
D latch, 214, 219, 221
data type, 53
    bit, 53
    bit_vector, 53
    boolean, 53
    realization, 133
    signed, 60
    std_logic, 56
    std_logic_vector, 56
    unsigned, 60
dataflow graph, 461
delay-sensitive, 159
delta delay (δ-delay), 30, 70
design reuse, 21, 218, 474
design unit, 44, 46
design-for-test, 16
development flow, 17, 38
difference circuit, 175
don't-care, 137
EDA (Electronic Design Automation), 16, 125, 148, 218
edge detection circuit, 326, 348, 623
entity declaration, 28, 44
equivalence check, 126
exit statement, 537
false path, 152
field programmable device, 4
    CPLD, 4
    FPGA, 4, 146
    simple, 4
FIFO buffer, 279, 591, 652
floor planning, 14
for loop statement, 118, 528
formal signal, 478
FSM, 219, 313, 379, 422
    output buffering, 342
    safe, 342
FSMD, 376, 385

function, 491
gate count, 11, 132
generate statement
    conditional, 517
    for, 512
generic, 481, 501
glitch, 156
globally asynchronous locally synchronous (GALS), 216, 612
Gray code, 196, 338
greatest common divisor (GCD), 445
Hamming distance circuit, 206
handshaking, 630
    four-phase, 630, 641
    two-phase, 637, 650
hardware emulation, 16, 218
HDL (hardware description language), 23
hold time, 216, 243, 609
identifier, 48
IEEE standard, 26
    1076 VHDL standard, 26
    1076.3 Synthesis Packages, 26
    1076.6 RTL Synthesis, 26
    1164 Multivalue Logic, 26
if statement, 103
intractable, 128
IP (Intellectual Property), 12
leading-zero counting circuit, 538
LFSR (linear feedback shift register), 269, 586
library, 46, 489
    ieee, 47
    work, 46
LUT (look-up table), 146
Manchester encoding circuit, 363
mask, 2
maximal clock rate, 240
Mealy output, 314, 327, 400
metastability, 613
micron, 2
mode, 45
    buffer, 45
    in, 45
    inout, 45, 135
    out, 45
Moore output, 314, 326
    look-ahead output buffer, 344
MTBF (mean time between synchronization failure), 614
multiplexer, 72, 78, 85, 103, 112, 532, 552, 560
    abstract, 77, 88, 90
multiplier
    cell-based carry-save, 581
    cell-based ripple-carry, 577
    combinational, 203
    pipelined, 297, 303, 305, 574
    repetitive-addition, 382
    sequential add-and-shift, 407, 572
netlist, 9
next statement, 540
number, 49
object, 51

one-shot pulse generator, 422
operator, 54
   overloaded, 57, 61, 65
   precedence, 55
   realization, 129
   sharing, 164, 396
   synthesis support, 130
package, 46, 492
parallel-prefix structure, 187, 570
parameter, 500
   feature, 501
   width, 500
partition, 495
physical design, 14
pipeline, 293
placement and routing, 14
population counter, 206, 534, 540
priority encoder, 74, 86, 104, 112, 187, 199, 530, 588
process, 33, 97
propagation delay, 132, 150
PWM (pulse width modulation), 275
RAM, 225
   controller, 430
reduced-and, 509, 537
reduced-xor, 181, 501, 505, 509, 514, 518, 528, 533, 555
reduced-xor-vector, 183, 570
register file, 276, 593
register transfer, 12
   RT methodology, 12, 219, 373, 375, 425
   RT-level abstraction, 12
register, 225
reserved words, 48
resolution time, 614
scheduling, 461
selected signal assignment statement, 85, 114
sensitivity list, 33, 98
sequential circuit, 69, 213
   combined, 219
   random, 219
   regular, 219, 424
sequential signal assignment statement, 100
sequential statement, 97
serial-to-parallel converter, 510, 515, 529
setup time, 216, 240, 606
shift register, 229
signal, 51
square-root approximation circuit, 460
state assignment, 338
state diagram, 315
string, 49
strobe generation circuit, 358
strongly typed language, 53

synchronizer, 617
synchronous circuit, 216–217
synthesis, 13
   behavioral, 469
   clock, 604
   flow, 139
   FSM, 337
   FSMD, 417
   gate-level, 13
   high-level, 13, 469
   iterative structure, 541
   logic, 142
   module generator, 141
   multilevel, 142
   multiple clock, 661
   pipelined circuit, 307
   retiming, 308
   RT-level, 13, 139
   sequential circuit, 253
   technology mapping, 14, 143
   two-level, 142
T FF, 228, 246
testbench, 35
testing, 16, 218
timing analysis, 15, 218
   clock skew, 606
   FSM, 324
   FSMD, 404
   synchronous sequential circuit, 239
timing hazards, 156
   dynamic hazards, 156
   static hazards, 156
tractable, 128
tri-state buffer, 133
tri-state bus, 135
type casting, 63
type conversion, 53, 57, 62
UART, 455
variable assignment statement, 101
variable, 51, 250
verification, 14
   formal, 15
   functional, 14
   timing, 15
Verilog, 25
VHDL, 25
   analysis, 47
   elaboration, 47
view
   behavioral, 8, 33
   physical, 9
   structural, 9, 30
wait statement, 99